D1378071

Robust Statistics

WILEY SERIES IN PROBABILITY AND STATISTICS

ESTABLISHED BY WALTER A. SHEWHART AND SAMUEL S. WILKS

A complete list of the titles in this series appears at the end of this volume.

Robust Statistics
Theory and Methods

Ricardo A. Maronna
Universidad Nacional de La Plata, Argentina

R. Douglas Martin
University of Washington, Seattle, USA

Víctor J. Yohai
University of Buenos Aires, Argentina

John Wiley & Sons, Ltd

Telephone (+44) 1243 779777

Email (for orders and customer service enquiries): cs-books@wiley.co.uk
Visit our Home Page on www.wiley.com

Reprinted with corrections November 2006

Other Wiley Editorial Offices

John Wiley & Sons Inc., 111 River Street, Hoboken, NJ 07030, USA

Jossey-Bass, 989 Market Street, San Francisco, CA 94103-1741, USA

Wiley-VCH Verlag GmbH, Boschstr. 12, D-69469 Weinheim, Germany

John Wiley & Sons Australia Ltd, 42 McDougall Street, Milton, Queensland 4064, Australia

John Wiley & Sons (Asia) Pte Ltd, 2 Clementi Loop #02-01, Jin Xing Distripark, Singapore 129809

John Wiley & Sons Canada Ltd, 22 Worcester Road, Etobicoke, Ontario, Canada M9W 1L1

Wiley also publishes its books in a variety of electronic formats. Some content that appears in print may not be available in electronic books.

British Library Cataloguing in Publication Data

A catalogue record for this book is available from the British Library

ISBN-13 978-0-470-01092-1 (HB)
ISBN-10 0-470-01092-4 (HB)

Typeset in 10/12pt Times by TechBooks, New Delhi, India
Printed and bound in Great Britain by TJ International, Padstow, Cornwall
This book is printed on acid-free paper responsibly manufactured from sustainable forestry in which at least two trees are planted for each one used for paper production.

To Susana, Jean, Julia, Livia and Paula

and

with recognition and appreciation of the foundations laid by the founding fathers of robust statistics: John Tukey, Peter Huber and Frank Hampel

Contents

Preface

Why robust statistics are needed

All statistical methods rely explicitly or implicitly on a number of assumptions. These assumptions generally aim at formalizing what the statistician knows or conjectures about the data analysis or statistical modeling problem he or she is faced with, and at the same time aim at making the resulting model manageable from the theoretical and computational points of view. However, it is generally understood that the resulting formal models are simplifications of reality and that their validity is at best approximate. The most widely used model formalization is the assumption that the observed data have a *normal* (Gaussian) distribution. This assumption has been present in statistics for two centuries, and has been the framework for all the classical methods in regression, analysis of variance and multivariate analysis. There have been attempts to justify the assumption of normality with theoretical arguments, such as the central limit theorem. These attempts, however, are easily proven wrong. The main justification for assuming a normal distribution is that it gives an approximate representation to many real data sets, and at the same time is theoretically quite convenient because it allows one to derive explicit formulas for optimal statistical methods such as maximum likelihood and likelihood ratio tests, as well as the sampling distribution of inference quantities such as t-statistics. We refer to such methods as *classical* statistical methods, and note that they rely on the assumption that normality holds *exactly*.The classical statistics are by modern computing standards quite easy to compute. Unfortunately theoretical and computational convenience does not always deliver an adequate tool for the practice of statistics and data analysis, as we shall see throughout this book.

It often happens in practice that an assumed normal distribution model (e.g., a location model or a linear regression model with normal errors) holds approximately in that it describes the majority of observations, but some observations follow a different pattern or no pattern at all. In the case when the randomness in the model is assigned to observational errors—as in astronomy, which was the first instance of the use of the least-squares method—the reality is that while the behavior of many sets of data appeared rather normal, this held only approximately, with the main discrepancy being that a small proportion of observations were quite atypical by virtue of being far from the bulk of the data. Behavior of this type is common across the entire spectrum of data analysis and statistical modeling applications. Such atypical data are called *outliers*, and even a single outlier can have a large distorting influence on a classical statistical method that is optimal under the assumption of normality or linearity. The kind of "approximately" normal distribution that gives rise to outliers is one that has a normal shape in the central region, but has tails that are heavier or "fatter" than those of a normal distribution.

One might naively expect that if such approximate normality holds, then the results of using a normal distribution theory would also hold approximately. This is unfortunately not the case. If the data are assumed to be normally distributed but their actual distribution has heavy tails, then estimates based on the maximum likelihood principle not only cease to be "best" but may have unacceptably low statistical efficiency (unnecessarily large variance) if the tails are symmetric and may have very large bias if the tails are asymmetric. Furthermore, for the classical tests their level may be quite unreliable and their power quite low, and for the classical confidence intervals their confidence level may be quite unreliable and their expected confidence interval length may be quite large.

The robust approach to statistical modeling and data analysis aims at deriving methods that produce reliable parameter estimates and associated tests and confidence intervals, not only when the data follow a given distribution exactly, but also when this happens only approximately in the sense just described. While the emphasis of this book is on approximately normal distributions, the approach works as well for other distributions that are close to a nominal model, e.g., approximate gamma distributions for asymmetric data. A more informal data-oriented characterization of robust methods is that they fit the bulk of the data well: if the data contain no outliers the robust method gives approximately the same results as the classical method, while if a small proportion of outliers are present the robust method gives approximately the same results as the classical method applied to the "typical" data. As a consequence of fitting the bulk of the data well, robust methods provide a very reliable method of detecting outliers, even in high-dimensional multivariate situations.

We note that one approach to dealing with outliers is the *diagnostic* approach. Diagnostics are statistics generally based on classical estimates that aim at giving numerical or graphical clues for the detection of data departures from the assumed model. There is a considerable literature on outlier diagnostics, and a good outlier diagnostic is clearly better than doing nothing. However, these methods present two drawbacks. One is that they are in general not as reliable for detecting outliers as examining departures from a robust fit to the data. The other is that, once suspicious observations have been flagged, the actions to be taken with them remain the analyst's personal decision, and thus there is no objective way to establish the properties of the result of the overall procedure.

Robust methods have a long history that can be traced back at least to the end of the nineteenth century with Simon Newcomb (see Stigler, 1973). But the first great steps forward occurred in the 1960s, and the early 1970s with the fundamental work of John Tukey (1960, 1962), Peter Huber (1964, 1967) and Frank Hampel (1971, 1974). The applicability of the new robust methods proposed by these researchers was made possible by the increased speed and accessibility of computers. In the last four decades the field of robust statistics has experienced substantial growth as a research area, as evidenced by a large number of published articles. Influential books have been written by Huber (1981), Hampel, Ronchetti, Rousseeuw and Stahel (1986), Rousseeuw and Leroy (1987) and Staudte and Sheather (1990). The research efforts of the current book's authors, many of which are reflected in the various chapters, were stimulated by the early foundation results, as well as work by many other contributors to the field, and the emerging computational opportunities for delivering robust methods to users.

The above body of work has begun to have some impact outside the domain of robustness specialists, and there appears to be a generally increased awareness of the dangers posed by atypical data values and of the unreliability of exact model assumptions. Outlier detection methods are nowadays discussed in many textbooks on classical statistical methods, and implemented in several software packages. Furthermore, several commercial statistical software packages currently offer some robust methods, with that of the robust library in S-PLUS being the currently most complete and user friendly. In spite of the increased awareness of the impact outliers can have on classical statistical methods and the availability of some commercial software, robust methods remain largely unused and even unknown by most communities of applied statisticians, data analysts, and scientists that might benefit from their use. It is our hope that this book will help to rectify this unfortunate situation.

Purpose of the book

This book was written to stimulate the routine use of robust methods as a powerful tool to increase the reliability and accuracy of statistical modeling and data analysis. To quote John Tukey (1975a), who used the terms *robust* and *resistant* somewhat interchangeably:

> It is perfectly proper to use both classical and robust/resistant methods routinely, and only worry when they differ enough to matter. But when they differ, you should think hard.

For each statistical model such as location, scale, linear regression, etc., there exist several if not many robust methods, and each method has several variants which an applied statistician, scientist or data analyst must choose from. To select the most appropriate method for each model it is important to understand how the robust methods work, and their pros and cons. The book aims at enabling the reader to select and use the most adequate robust method for each model, and at the same time to understand the theory behind the method: that is, not only the "how" but also the "why". Thus for each of the models treated in this book we provide:

- Conceptual and statistical theory explanations of the main issues
- The leading methods proposed to date and their motivations
- A comparison of the properties of the methods
- Computational algorithms, and S-PLUS implementations of the different approaches
- Recommendations of preferred robust methods, based on what we take to be reasonable trade-offs between estimator theoretical justification and performance, transparency to users and computational costs.

Intended audience

The intended audience of this book consists of the following groups of individuals among the broad spectrum of data analysts, applied statisticians and scientists: (1) those who will be quite willing to apply robust methods to their problems once they are aware of the methods, supporting theory and software implementations; (2) instructors who want to teach a graduate-level course on robust statistics; (3) graduate students wishing to learn about robust statistics; (4) graduate students and faculty who wish to pursue research on robust statistics and will use the book as background study.

General prerequisites are basic courses in probability, calculus and linear algebra, statistics and familiarity with linear regression at the level of Weisberg (1985), Montgomery, Peck and Vining (2001) and Seber and Lee (2003). Previous knowledge of multivariate analysis, generalized linear models and time series is required for Chapters 6, 7 and 8, respectively.

Organization of the book

There are many different approaches for each model in robustness, resulting in a huge volume of research and applications publications (though perhaps fewer of the latter than we might like). Doing justice to all of them would require an encyclopedic work that would not necessarily be very effective for our goal. Instead we concentrate on the methods we consider most sound according to our knowledge and experience.

Chapter 1 is a data-oriented motivation chapter. Chapter 2 introduces the main methods in the context of location and scale estimation; in particular we concentrate on the so-called M-estimates that will play a major role throughout the book. Chapter 3 discusses methods for the evaluation of the robustness of model parameter estimates, and derives "optimal" estimates based on robustness criteria. Chapter 4 deals with linear regression for the case where the predictors contain no outliers, typically because they are fixed nonrandom values, including for example fixed balanced designs. Chapter 5 treats linear regression with general random predictors which mainly contain outliers in the form of so-called "leverage" points. Chapter 6 treats robust estimation of multivariate location and dispersion, and robust principal components. Chapter 7 deals with logistic regression and generalized linear models. Chapter 8 deals with robust estimation of time series models, with a main focus on AR and ARIMA. Chapter 9 contains a more detailed treatment of the iterative algorithms for the numerical computation of M-estimates. Chapter 10 develops the asymptotic theory of some robust estimates, and contains proofs of several results stated in the text. Chapter 11 contains detailed instructions on the use of robust procedures written in S-PLUS. Chapter 12 is an appendix containing descriptions of most data sets used in the book.

All methods are introduced with the help of examples with real data. The problems at the end of each chapter consist of both theoretical derivations and analysis of other real data sets.

How to read this book

Each chapter can be read at two levels. The main part of the chapter explains the models to be tackled and the robust methods to be used, comparing their advantages and shortcomings through examples and avoiding technicalities as much as possible. Readers whose main interest is in applications should read enough of each chapter to understand what is the currently preferred method, and the reasons it is preferred. The theoretically oriented reader can find proofs and other mathematical details in appendices and in Chapter 9 and Chapter 10. Sections marked with an asterisk may be skipped at first reading.

Computing

A great advantage of classical methods is that they require only computational procedures based on well-established numerical linear algebra methods which are generally quite fast algorithms. On the other hand, computing robust estimates requires solving highly nonlinear optimization problems that typically involve a dramatic increase in computational complexity and running time. Most current robust methods would be unthinkable without the power of today's standard personal computers. Fortunately computers continue getting faster, have larger memory and are cheaper, which is good for the future of robust statistics.

Since the behavior of a robust procedure may depend crucially on the algorithm used, the book devotes considerable attention to algorithmic details for all the methods proposed. At the same time, in order that robust statistics be widely accepted by a wide range of users, the methods need to be readily available in commercial software. Robust methods have been implemented in several available commercial statistical packages, including S-PLUS and SAS. In addition many robust procedures have been implemented in the public-domain language R, which is similar to S. References for free software for robust methods are given at the end of Chapter 11. We have focused on S-PLUS because it offers the widest range of methods, and because the methods are accessible from a user-friendly menu and dialog user interface as well as from the command line.

For each method in the book, instructions are given in Chapter 11 on how to compute it using S-PLUS. For each example, the book gives the reference to the respective data set and the S-PLUS code that allow the reader to reproduce the example. Datasets and codes are to be found on the book's Web site

http://www.wiley.com/go/robust_statistics.

This site will also contain corrections to any errata we subsequently discover, and clarifying comments and suggestions as needed. We will appreciate any feedback from readers that will result in posting additional helpful material on the web site.

S-PLUS software download

A time-limited version of S-PLUS for Windows software, which expires after 150 days, is being provided by Insightful for this book. To download and install the S-PLUS software, follow the instructions at

http://www.insightful.com/support/splusbooks/robstats.

To access the web page, the reader must provide a password. The password is the web registration key provided with this book as a sticker on the inside back cover. In order to activate S-PLUS for Windows the reader must use the web registration key.

Acknowledgements

The authors thank Elena Martínez, Ana Bianco, Mahmut Kutlukaya, Débora Chan, Isabella Locatelli and Chris Green for their helpful comments. Special thanks are due to Ana Julia Villar, who detected a host of errors and also contributed part of the computer code.

This book could not have been written without the incredible patience of our wives and children for the many hours devoted to it and our associated research over the years. Untold thanks to Susana, Livia, Jean, Julia and Paula.

One of us (RDM) wishes to acknowledge his fond memory of and deep indebtedness to John Tukey for introducing him to robustness and arranging a consulting appointment with Bell Labs, Murray Hill, that lasted for ten years, and without which he would not be writing this book and without which S-PLUS would not exist.

1

Introduction

1.1 Classical and robust approaches to statistics

This introductory chapter is an informal overview of the main issues to be treated in detail in the rest of the book. Its main aim is to present a collection of examples that illustrate the following facts:

- Data collected in a broad range of applications frequently contain one or more *atypical observations* called *outliers*; that is, observations that are well separated from the majority or "bulk" of the data, or in some way deviate from the general pattern of the data.
- Classical estimates such as the sample mean, the sample variance, sample covariances and correlations, or the least-squares fit of a regression model, can be very adversely influenced by outliers, even by a single one, and often fail to provide good fits to the bulk of the data.
- There exist *robust* parameter estimates that provide a good fit to the bulk of the data when the data contain outliers, as well as when the data are free of them. A direct benefit of a good fit to the bulk of data is the reliable detection of outliers, particularly in the case of multivariate data.

In Chapter 3 we shall provide some formal probability-based concepts and definitions of robust statistics. Meanwhile it is important to be aware of the following performance distinctions between classical and robust statistics at the outset. Classical statistical inference quantities such as confidence intervals, t-statistics and p-values, R^2 values and model selection criteria in regression can be very adversely influenced by the presence of even one outlier in the data. On the other hand, appropriately constructed robust versions of those inference quantities are not much influenced by outliers. Point estimate predictions and their confidence intervals based on classical

Robust Statistics – Theory and Methods Ricardo A. Maronna, R. Douglas Martin and Víctor J. Yohai

statistics can be spoiled by outliers, while predictive models fitted using robust statistics do not suffer from this disadvantage.

It would, however, be misleading to always think of outliers as "bad" data. They may well contain unexpected relevant information. According to Kandel (1991, p. 110):

> The discovery of the ozone hole was announced in 1985 by a British team working on the ground with "conventional" instruments and examining its observations in detail. Only later, after reexamining the data transmitted by the TOMS instrument on NASA's Nimbus 7 satellite, was it found that the hole had been forming for several years. Why had nobody noticed it? The reason was simple: the systems processing the TOMS data, designed in accordance with predictions derived from models, which in turn were established on the basis of what was thought to be "reasonable", had rejected the very ("excessively") low values observed above the Antarctic during the Southern spring. As far as the program was concerned, there must have been an operating defect in the instrument.

In the next sections we present examples of classical and robust estimates to data containing outliers for the estimation of mean and standard deviation, linear regression and correlation, Except in Section 1.2, we do not describe the robust estimates in any detail, and return to their definitions in later chapters.

1.2 Mean and standard deviation

Let $\mathbf{x} = (x_1, x_2, \ldots, x_n)$ be a set of observed values. The sample mean $\overline{x}$ and sample standard deviation (SD) s are defined by

$$\overline{x} = \frac{1}{n}\sum_{i=1}^{n} x_i, \; s^2 = \frac{1}{n-1}\sum_{i=1}^{n}(x_i - \overline{x})^2. \tag{1.1}$$

The sample mean is just the arithmetic average of the data, and as such one might expect that it provides a good estimate of the *center* or *location* of the data. Likewise, one might expect that the sample SD would provide a good estimate of the *dispersion* of the data. Now we shall see how much influence a single outlier can have on these classical estimates.

Example 1.1 *Consider the following 24 determinations of the copper content in wholemeal flour (in parts per million), sorted in ascending order (Analytical Methods Committee, 1989):*

2.20	2.20	2.40	2.40	2.50	2.70	2.80	2.90
3.03	3.03	3.10	3.37	3.40	3.40	3.40	3.50
3.60	3.70	3.70	3.70	3.70	3.77	5.28	28.95

The value 28.95 immediately stands out from the rest of the values and would be considered an outlier by almost anyone. One might conjecture that this inordinately large value was caused by a misplaced decimal point with respect to a "true" value of 2.895. In any event, it is a highly influential outlier as we now demonstrate.

The values of the sample mean and SD for the above data set are $\overline{x} = 4.28$ and $s = 5.30$, respectively. Since $\overline{x} = 4.28$ is larger than all but two of the data values,

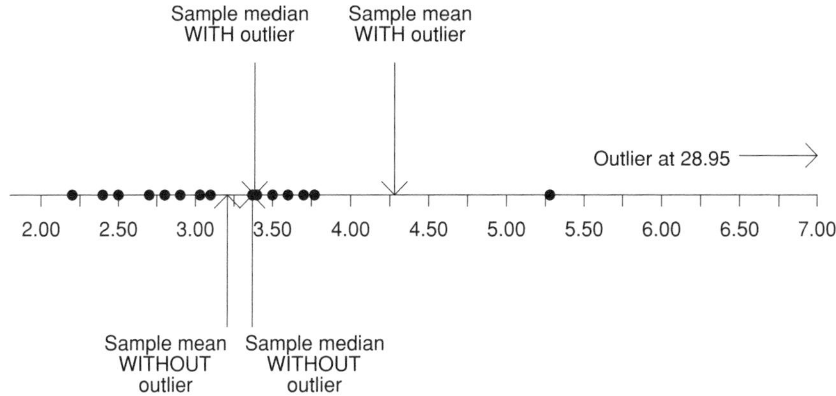

Figure 1.1 Copper content of flour data with sample mean and sample median estimates

it is not among the bulk of the observations and as such does not represent a good estimate of the center of the data. If one deletes the suspicious value 28.95, then the values of the sample mean and sample SD are changed to $\overline{x} = 3.21$ and $s = 0.69$. Now the sample mean does provide a good estimate of the center of the data, as is clearly revealed in Figure 1.1, and the SD is over seven times smaller than it was with the outlier present. See the leftmost upward pointing arrow and the rightmost downward-pointing arrow in Figure 1.1.

Let us consider how much influence a single outlier can have on the sample mean and sample SD. For example, suppose that the value 28.95 is replaced by an arbitrary value x for the 24-th observation x_{24}. It is clear from the definition of the sample mean that by varying x from $-\infty$ to $+\infty$ the value of the sample mean changes from $-\infty$ to $+\infty$. It is an easy exercise to verify that as x ranges from $-\infty$ to $+\infty$ sample SD ranges from some positive value smaller than that based on the first 23 observations to $+\infty$. Thus we can say that a single outlier has an *unbounded influence* on these two classical statistics.

An outlier may have a serious adverse influence on confidence intervals. For the flour data the classical interval based on the t-distribution with confidence level 0.95 is (2.05, 6.51), while after removing the outlier the interval is (2.91, 3.51). The impact of the single outlier has been to considerably lengthen the interval in an asymmetric way.

The above example suggests that a simple way to handle outliers is to detect them and remove them from the data set. There are many methods for detecting outliers (see for example Barnett and Lewis, 1998). Deleting an outlier, although better than doing nothing, still poses a number of problems:

- When is deletion justified? Deletion requires a subjective decision. When is an observation "outlying enough" to be deleted?

- The user or the author of the data may think that "an observation is an observation" (i.e., observations should speak for themselves) and hence feel uneasy about deleting them.
- Since there is generally some uncertainty as to whether an observation is really atypical, there is a risk of deleting "good" observations, which results in underestimating data variability.
- Since the results depend on the user's subjective decisions, it is difficult to determine the statistical behavior of the complete procedure.

We are thus led to another approach: why use the sample mean and SD? Maybe there are other better possibilities?

One very old method for estimating the "middle" of the data is to use the sample *median*. Any number t such that the numbers of observations on both sides of it are equal is called a *median* of the data set: t is a median of the data set $\mathbf{x} = (x_1, \ldots, x_n)$, and will be denoted by

$$t = \mathrm{Med}(\mathbf{x}), \text{ if } \#\{x_i > t\} = \#\{x_i < t\},$$

where $\#\{A\}$ denotes the number of elements of the set A. It is convenient to define the sample median in terms of the *order statistics* $(x_{(1)}, x_{(2)}, \ldots, x_{(n)})$, obtained by sorting the observations $\mathbf{x} = (x_1, \ldots, x_n)$ in increasing order so that

$$x_{(1)} \leq \ldots \leq x_{(n)}. \tag{1.2}$$

If n is odd, then $n = 2m - 1$ for some integer m, and in that case $\mathrm{Med}(\mathbf{x}) = x_{(m)}$. If n is even, then $n = 2m$ for some integer m, and then any value between $x_{(m)}$ and $x_{(m+1)}$ satisfies the definition of a sample median, and it is customary to take

$$\mathrm{Med}(\mathbf{x}) = \frac{x_{(m)} + x_{(m+1)}}{2}.$$

However, in some cases (e.g., in Section 4.5.1) it may be more convenient to choose $x_{(m)}$ or $x_{(m+1)}$ ("low" and "high" medians, respectively).

The mean and the median are approximately equal if the sample is symmetrically distributed about its center, but not necessarily otherwise.

In our example the median of the whole sample is 3.38, while the median without the largest value is 3.37, showing that the median is not much affected by the presence of this value. See the locations of the sample median with and without the outlier present in Figure 1.1 above. Notice that for this sample, the value of the sample median with the outlier present is relatively close to the sample mean value of 3.21 with the outlier deleted.

Suppose again that the value 28.95 is replaced by an arbitrary value x for the 24-th observation $x_{(24)}$. It is clear from the definition of the sample median that when x ranges from $-\infty$ to $+\infty$ the value of the sample median does not change from $-\infty$ to $+\infty$ as was the case for the sample mean. Instead, when x goes to $-\infty$ the sample median undergoes the small change from 3.38 to 3.23 (the latter being the average of $x_{(11)} = 3.10$ and $x_{(12)} = 3.37$ in the original data set), and when x goes to

$+\infty$ the sample median goes to the value 3.38 given above for the original data Since the sample median fits the bulk of the data well with or without the outlier and is not much influenced by the outlier, it is a good robust alternative to the sample mean.

Likewise, one robust alternative to the SD is the *median absolute deviation about the median (MAD)*, defined as

$$\mathrm{MAD}(\mathbf{x}) = \mathrm{MAD}(x_1, x_2, \ldots, x_n) = \mathrm{Med}\{|\mathbf{x} - \mathrm{Med}\,(\mathbf{x})|\}\,.$$

This estimator uses the sample median twice, first to get an estimate of the center of the data in order to form the set of absolute residuals about the sample median, $\{|\mathbf{x} - \mathrm{Med}\,(\mathbf{x})|\}$, and then to compute the sample median of these absolute residuals. To make the MAD comparable to the SD, we define the *normalized MAD (“MADN”)* as

$$\mathrm{MADN}(\mathbf{x}) = \frac{\mathrm{MAD}(\mathbf{x})}{0.6745}.$$

The reason for this definition is that 0.6745 is the MAD of a standard normal random variable, and hence a $\mathrm{N}(\mu, \sigma^2)$ variable has $\mathrm{MADN} = \sigma$.

For the above data set one gets $\mathrm{MADN} = 0.53$, as compared with $s = 5.30$. Deleting the large outlier yields $\mathrm{MADN} = 0.50$, as compared to the somewhat higher sample SD value of $s = 0.69$. The MAD is clearly not influenced very much by the presence of a large outlier, and as such provides a good robust alternative to the sample SD.

So why not always use the median and MAD? An informal explanation is that if the data contain no outliers, these estimates have statistical performance which is poorer than that of the classical estimates $\overline{x}$ and s. The ideal solution would be to have “the best of both worlds”: estimates that behave like the classical ones when the data contain no outliers, but are insensitive to outliers otherwise. This is the data-oriented idea of robust estimation. A more formal notion of robust estimation based on statistical models, which will be discussed in the following chapters, is that the statistician always has a statistical model in mind (explicitly or implicitly) when analyzing data, e.g., a model based on a normal distribution or some other idealized parametric model such as an exponential distribution. The classical estimates are in some sense “optimal” when the data are exactly distributed according to the assumed model, but can be very suboptimal when the distribution of the data differs from the assumed model by a “small” amount. Robust estimates on the other hand maintain approximately optimal performance, not just under the assumed model, but under “small” perturbations of it too.

1.3 The “three-sigma edit” rule

A traditional measure of the “outlyingness” of an observation x_i with respect to a sample is the ratio between its distance to the sample mean and the sample SD:

$$t_i = \frac{x_i - \overline{x}}{s}. \tag{1.3}$$

Observations with $|t_i| > 3$ are traditionally deemed as suspicious (the "three-sigma rule"), based on the fact that they would be "very unlikely" under normality, since $\mathrm{P}(|x| \geq 3) = 0.003$ for a random variable x with a standard normal distribution. The largest observation in the flour data has $t_i = 4.65$, and so is suspicious. Traditional "three-sigma edit" rules result in either discarding observations for which $|t_i| > 3$, or adjusting them to one of the values $\overline{x} \pm 3s$, whichever is nearer.

Despite its long tradition, this rule has some drawbacks that deserve to be taken into account:

- In a very large sample of "good" data, some observations will be declared suspicious and altered. More precisely, in a large normal sample about three observations out of 1000 will have $|t_i| > 3$. For this reason, normal Q–Q plots are more reliable for detecting outliers (see example below).
- In very small samples the rule is ineffective: it can be shown that

$$|t_i| < \frac{n-1}{\sqrt{n}}$$

for all possible data sample values, and hence if $n \leq 10$ then $|t_i| < 3$ always. The proof is left to the reader (Problem 1.3).
- When there are several outliers, their effects may interact in such a way that some or all of them remain unnoticed (an effect called *masking*), as the following example shows.

Example 1.2 *The following data (Stigler, 1977) are 20 determinations of the time (in microseconds) needed for light to travel a distance of 7442 m. The actual times are the table values* $\times$ 0.001 + 24.8.

28	26	33	24	34	-44	27	16	40	-2
29	22	24	21	25	30	23	29	31	19

The normal Q–Q plot in Figure 1.2 reveals the two lowest observations (-44 and -2) as suspicious. Their respective t_i's are -3.73 and -1.35 and so the value of $|t_i|$ for the observation -2 does not indicate that it is an outlier. The reason that -2 has such a small $|t_i|$ value is that both observations pull $\overline{x}$ to the left and inflate s; it is said that the value -44 "masks" the value -2.

To avoid this drawback it is better to replace $\overline{x}$ and s in (1.3) by robust location and dispersion measures. A robust version of (1.3) can be defined by replacing the sample mean and SD by the median and MADN, respectively:

$$t_i' = \frac{x_i - \mathrm{Med}(\mathbf{x})}{\mathrm{MADN}(\mathbf{x})}. \tag{1.4}$$

The t_i's for the two leftmost observations are now -11.73 and -4.64 and hence the "robust three-sigma edit rule", with t' instead of t, pinpoints both as suspicious. This suggests that even if we only want to detect outliers—rather than to estimate parameters—detection procedures based on robust estimates are more reliable.

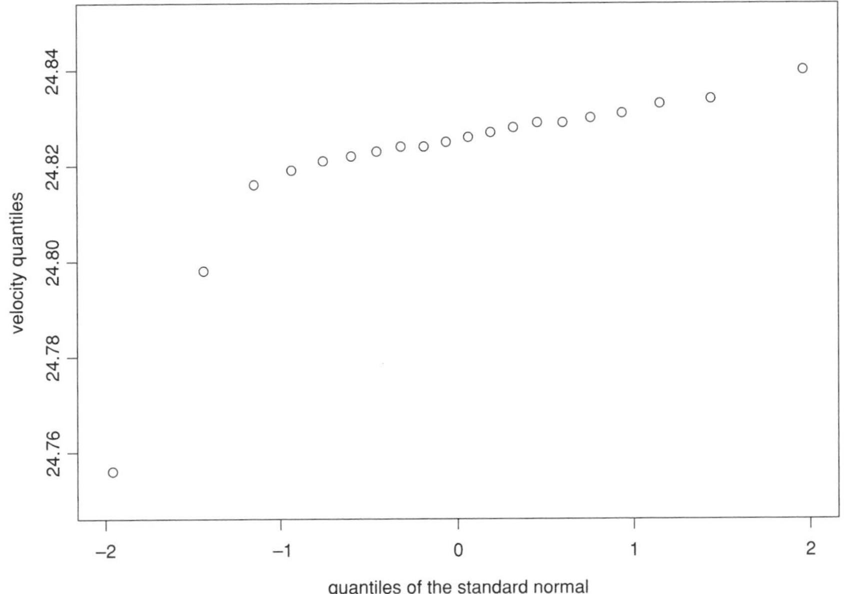

Figure 1.2 Velocity of light: Q–Q plot of observed times

A simple robust location estimate could be defined by deleting all observations with $|t_i'|$ larger than a given value, and taking the average of the rest. While this procedure is better than the three-sigma edit rule based on t, it will be seen in Chapter 3 that the estimates proposed in this book handle the data more smoothly, and can be tuned to possess certain desirable robustness properties that this procedure lacks.

1.4 Linear regression

1.4.1 Straight-line regression

First consider fitting a straight-line regression model to the data set $\{(x_i, y_i) : i = 1, \ldots, n\}$

$$y_i = \alpha + x_i\beta + u_i, \ i = 1, \ldots, n$$

where x_i and y_i are the predictor and response variable values, respectively, and u_i are random errors. The time-honored classical way of fitting this model is to estimate the parameters α and β with the least-squares (LS) estimates

$$\widehat{\beta} = \frac{\sum_{i=1}^{n}(x_i - \overline{x})(y_i - \overline{y})}{\sum_{i=1}^{n}(x_i - \overline{x})^2}$$

$$\widehat{\alpha} = \overline{y} - \overline{x}\widehat{\beta}.$$

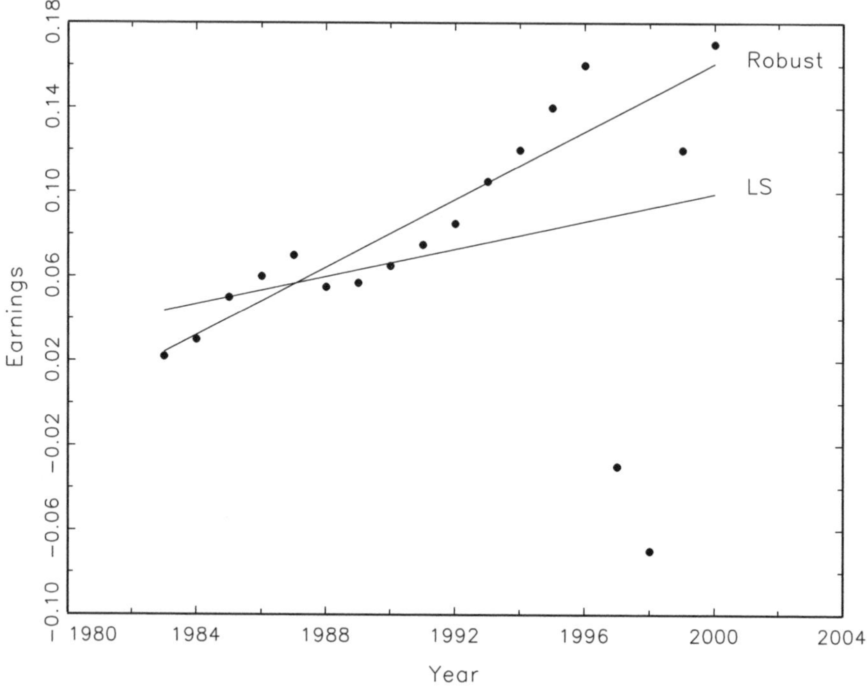

Figure 1.3 EPS data with robust and LS fits

As an example of how influential two outliers can be on these estimates, Figure 1.3 plots the earnings per share (EPS) versus time in years for the company with stock exchange ticker symbol IVENSYS, along with the straight-line fits of the LS estimate and of a robust regression estimate (called an MM-estimate) that has desirable theoretical properties to be described in detail in Chapter 5.

The two unusually low EPS values in 1997 and 1998 have caused the LS line to fit the data very poorly, and one would not expect the line to provide a good prediction of EPS in 2001. By way of contrast, the robust line fits the bulk of the data well, and is expected to provide a reasonable prediction of EPS in 2001.

The above EPS example was brought to one of the author's attention by an analyst in the corporate finance organization of a large well-known company. The analyst was required to produce a prediction of next year's EPS for several hundred companies, and at first he used the LS line fit for this purpose. But then he noticed a number of firms for which the data contained outliers that distorted the LS parameter estimates, resulting in a very poor fit and a poor prediction of next year's EPS. Once he discovered the robust estimate, and found that it gave him essentially the same results as the LS estimate when the data contained no outliers, while at the same time providing a better fit and prediction than LS when outliers were present, he began routinely using the robust estimate for his task.

It is important to note that automatically flagging large differences between a classical estimate (in this case LS) and a robust estimate provides a useful diagnostic alert that outliers may be influencing the LS result.

1.4.2 Multiple linear regression

Now consider fitting a multiple linear regression model

$$y_i = \sum_{j=1}^{p} x_{ij}\beta_j + u_i, \ i = 1, \ldots, n$$

where the response variable values are y_i, and there are p predictor variables $x_{ij},\ j = 1, \ldots, p$, and p regression coefficients β_j. Not surprisingly, outliers can also have an adverse influence on the LS estimate $\widehat{\beta}$ for this general linear model, a fact which is illustrated by the following example that appears in Hubert and Rousseeuw (1997).

Example 1.3 *The response variable values y_i are the rates of unemployment in various geographical regions around Hanover, Germany, and the predictor variables $x_{ij},\ j = 1, \ldots, p$, are as follows:*

PA:	*percentage engaged in production activities*
GPA:	*growth in PA*
HS:	*percentage engaged in higher services*
GHS:	*growth in HS*
Region:	*geographical region around Hanover (21 regions)*
Period:	*time period (three periods: 1979–1982, 1983–1988, 1989–1992)*

Note that the categorical variables Region and Period require 20 and 2 parameters respectively, so that, including an intercept, the model has 27 parameters, and the number of response observations is 63, one for each region and period. The following set of displays shows the results of LS and robust fitting in a manner that facilitates easy comparison of the results. The robust fitting is done by a special type of "M-estimate" that has desirable theoretical properties, and is described in detail in Section 5.15.

For a set of estimated parameters $\left(\widehat{\beta}_1, \ldots, \widehat{\beta}_p\right)$, with fitted values $\widehat{y}_i = \sum_{j=1}^{p} x_{ij}\widehat{\beta}_j$, residuals $\widehat{u}_i = y_i - \widehat{y}_i$ and residuals dispersion estimate $\widehat{\sigma}$, Figure 1.4 shows the standardized residuals $\widetilde{u}_i = \widehat{u}_i/\widehat{\sigma}$ plotted versus the observations' index values i. Standardized residuals that fall outside the horizontal dashed lines at ± 2.33, which occurs with probability 0.02, are declared suspicious. The display for the LS fit does not reveal any outliers while that for the robust fit clearly reveals 10 to 12 outliers among 63 observations. This is because the robust regression has found a linear relationship that fits the majority of the data points well, and consequently is able to reliably identify the outliers. The LS estimate instead attempts to fit all data points and so is heavily influenced by the outliers. The fact that all of the LS standardized residuals lie inside the horizontal dashed lines is because the outliers have inflated

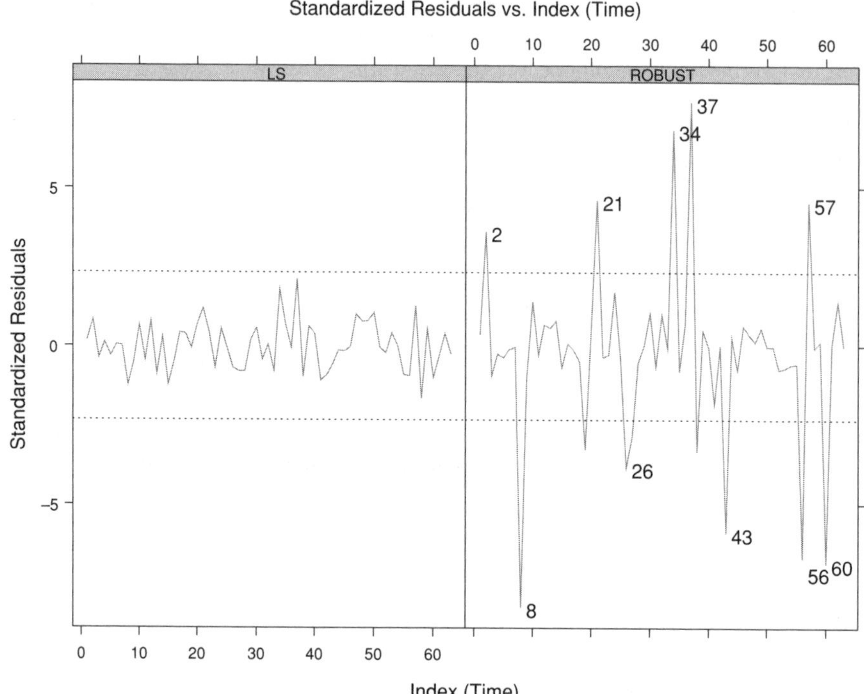

Figure 1.4 Standardized residuals for LS and robust fits

the value of $\widetilde{\sigma}$ computed in the classical way based on the sum of squared residuals, while a robust estimate $\widetilde{\sigma}$ used for the robust regression is not much influenced by the outliers.

Figure 1.5 shows normal Q–Q plots of the residuals for the LS and robust fits, with light dotted lines showing the 95% simulated pointwise confidence regions to help one judge whether or not there are significant outliers and potential nonnormality. These plots may be interpreted as follows. If the data fall along the straight line (which itself is fitted by a robust method) with no points outside the 95% confidence region then one is moderately sure that the data are normally distributed.

Making only the LS fit, and therefore looking only at the normal Q–Q plot in the left-hand plot above, would lead to the conclusion that the residuals are indeed quite normally distributed with no outliers. The normal Q–Q plot of residuals for the robust fit in the right-hand panel of Figure 1.5 clearly shows that such a conclusion is wrong. This plot shows that the bulk of the residuals is indeed quite normally distributed, as is evidenced by the compact linear behavior in the middle of the plot, and at the same time clearly reveals the outliers that were evident in the plot of standardized residuals (Figure 1.4).

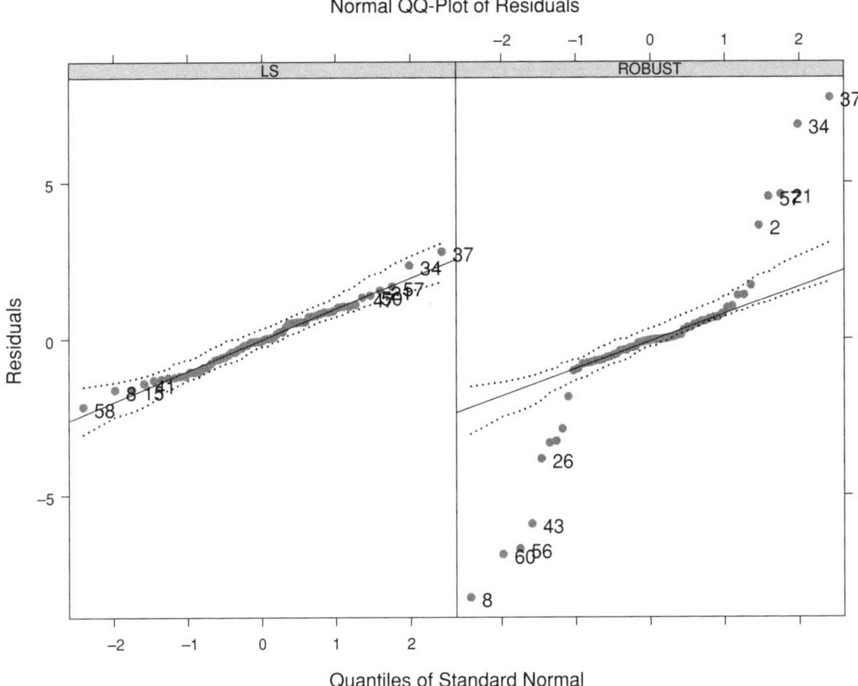

Figure 1.5 Normal Q–Q plots for LS and robust fits

1.5 Correlation coefficients

Let $\{(x_i, y_i)\}$, $i = 1, \ldots, n$, be a bivariate sample. The most popular measure of association between the x's and the y's is the sample correlation coefficient defined as

$$\widehat{\rho} = \frac{\sum_{i=1}^{n}(x_i - \overline{x})(y_i - \overline{y})}{\left(\sum_{i=1}^{n}(x_i - \overline{x})^2\right)^{1/2}\left(\sum_{i=1}^{n}(y_i - \overline{y})^2\right)^{1/2}}$$

where $\overline{x}$ and $\overline{y}$ are the sample means of the x_i's and y_i's.

The sample correlation coefficient is highly sensitive to the presence of outliers. Figure 1.6 shows a scatterplot of the gain (increase) in telephones versus the annual difference in new housing starts for a period of 15 years in a geographical region within New York City in the 1960s and 1970s, in coded units.

There are two outliers in this bivariate (two-dimensional) data set that are clearly separated from the rest of the data. It is important to notice that these two outliers are not one-dimensional outliers; they are not even the largest or smallest values in any of the two coordinates. This observation illustrates an extremely important point:

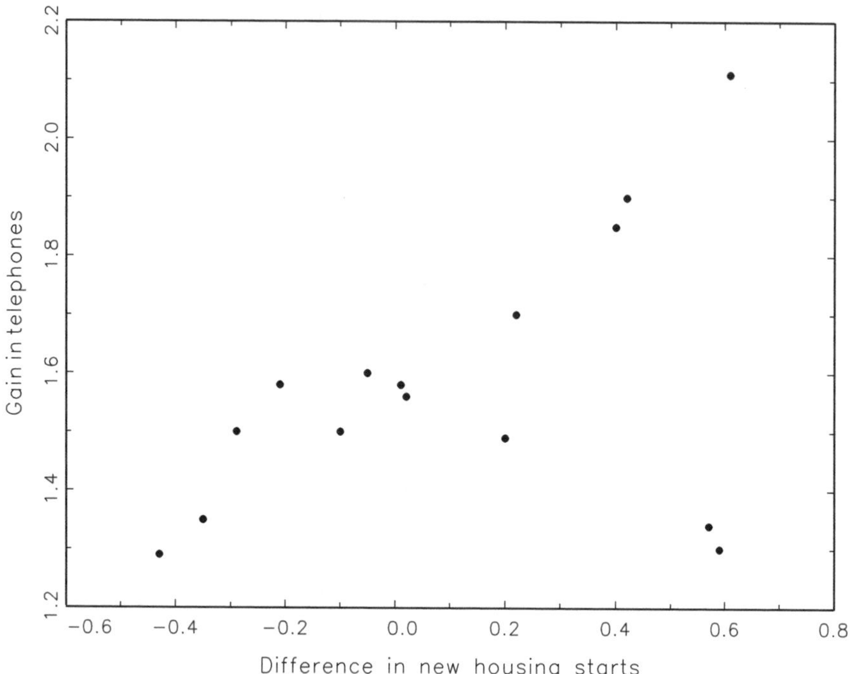

Figure 1.6 Gain in telephones versus difference in new housing starts

two-dimensional outliers cannot be reliably detected by examining the values of bivariate data one-dimensionally, i.e., one variable at a time!

The value of the sample correlation coefficient for the main-gain data is $\widehat{\rho} = 0.44$, and deleting the two outliers yields $\widehat{\rho} = 0.91$, which is quite a large difference and in the range of what an experienced user might expect for the data set with the two outliers removed. The data set with the two outliers deleted can be seen as roughly elliptical with a major axis sloping up and to the right and the minor axis direction sloping up and to the left With this picture in mind one can see that the two outliers lie in the minor axis direction, though offset somewhat from the minor axis. The impact of the outliers is to decrease the value of the sample correlation coefficient by the considerable amount of 0.44 from its value of 0.91 with the two outliers deleted. This illustrates a general biasing effect of outliers on the sample correlation coefficient: outliers that lie along a minor axis direction of data that is otherwise positively correlated negatively influence the sample correlation coefficient. Similarly, outliers that lie along a minor axis direction of data that is otherwise negatively correlated will increase the sample correlation coefficient. Outliers that lie along a major axis direction of the rest of the data will increase the absolute value of the sample correlation coefficient, making it more positive in the case where the bulk of the data is positively correlated.

If one uses a robust correlation coefficient estimate it will not make much difference whether the outliers in the main-gain data are present or deleted. Using a good robust method $\widehat{\rho}_{Rob}$ for estimating covariances and correlations on the main-gain data yields $\widehat{\rho}_{Rob} = 0.85$ for the entire data set and $\widehat{\rho}_{Rob} = 0.90$ with the two outliers deleted. For the robust correlation coefficient the change due to deleting the outlier is only 0.05, compared to 0.47 for the classical estimate. A detailed description of robust correlation and covariance estimates is provided in Chapter 6.

When there are more than two variables, examining all pairwise scatterplots for outliers is hopeless unless the number of variables is relatively small. But even looking at all scatterplots or applying a robust correlation estimate to all pairs does not suffice, for in the same way that there are bivariate outliers which do not stand out in any univariate representation, there may be multivariate outliers that heavily influence the correlations and do not stand out in any bivariate scatterplot. Robust methods deal with this problem by estimating all the correlations simultaneously, in such a manner that points far away from the bulk of the data are automatically downweighted. Chapter 6 treats these methods in detail.

1.6 Other parametric models

We do not want to leave the reader with the impression that robust estimation is only concerned with outliers in the context of an assumed normal distribution model. Outliers can cause problems in fitting other simple parametric distributions such as an exponential, Weibull or gamma distribution, where the classical approach is to use a nonrobust maximum likelihood estimate (MLE) for the assumed model. In these cases one needs robust alternatives to the MLE in order to obtain a good fit to the bulk of the data.

For example, the exponential distribution with density

$$f(x;\lambda) = \frac{1}{\lambda}e^{-x/\lambda}, \quad x \geq 0$$

is widely used to model random inter-arrival times and failure times, and it also arises in the context of times series spectral analysis (see Section 8.14). It is easily shown that the parameter λ is the expected value of the random variable x, i.e., $\lambda = \mathrm{E}(x)$, and that the sample mean is the MLE. We already know from the previous discussion that the sample mean lacks robustness and can be greatly influenced by outliers. In this case the data are nonnegative so one is only concerned about large positive outliers that cause the value of the sample mean to be inflated in a positive direction. So we need a robust alternative to the sample mean, and one naturally considers use of the sample median $\mathrm{Med}\,(\mathbf{x})$. It turns out that the sample median is an *inconsistent* estimate of λ, i.e., it does not approach λ when the sample size increases, and hence a correction is needed. It is an easy calculation to check that the median of the exponential distribution has value $\lambda \log 2$, where log stands for natural logarithm, and so one can use $\mathrm{Med}\,(\mathbf{x}) / \log 2$ as a simple robust estimate of λ that is consistent with the assumed model. This estimate turns out to have desirable robustness properties that are described in Problem 3.15.

The methods of robustly fitting Weibull and gamma distributions are much more complicated than the above use of the adjusted median for the exponential distribution.

We present one important application of robust fitting a gamma distribution due to Marazzi, Paccaud, Ruffieux and Beguin (1998). The gamma distribution has density

$$f(x;\alpha,\sigma) = \frac{1}{\Gamma(\alpha)\sigma^\alpha} x^{\alpha-1} e^{-x/\sigma}, \; x \geq 0$$

and the mean of this distribution is known to be $\mathrm{E}(x) = \alpha\sigma$. The problem has to do with estimating the length of stay (LOS) of 315 patients in a hospital. The mean LOS is a quantity of considerable economic importance, and some patients whose hospital stays are much longer than those of the majority of the patients adversely influence the MLE fit of the gamma distribution. The MLE values turn out to be $\widehat{\alpha}_{MLE} = 0.93$ and $\widehat{\sigma}_{MLE} = 8.50$, while the robust estimates are $\widehat{\alpha}_{Rob} = 1.39$ and $\widehat{\sigma}_{Rob} = 3.64$, and the resulting mean LOS estimates are $\widehat{\mu}_{MLE} = 7.87$ and $\widehat{\mu}_{Rob} = 4.97$. Some patients with unusually long LOS values contribute to an inflated estimate of the mean LOS for the majority of the patients. A more complete picture is obtained with the following graphical displays.

Figure 1.7 shows a histogram of the data along with the MLE and robust gamma density fit to the LOS data. The MLE underestimates the density for small values of LOS and overestimates the density for large values of LOS thereby resulting in a larger MLE estimate of the mean LOS, while the robust estimate provides a

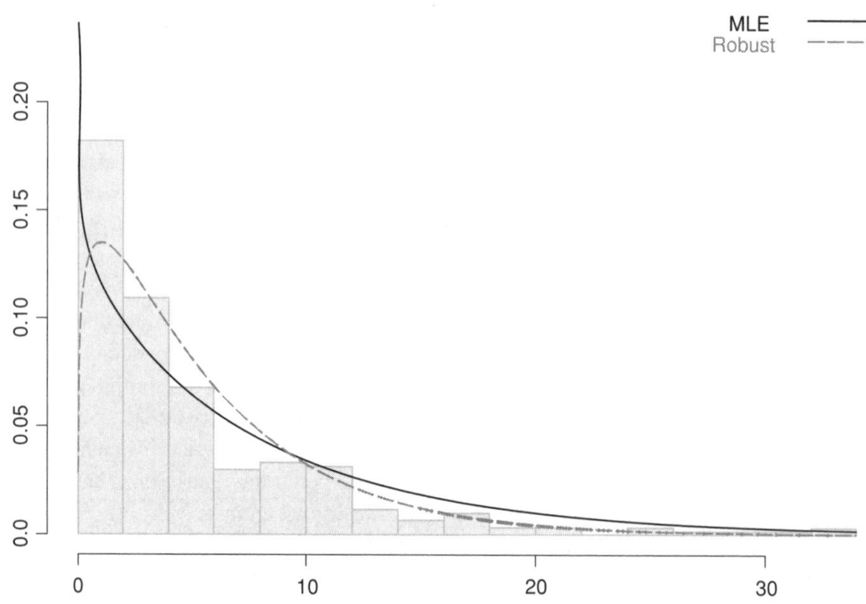

Figure 1.7 MLE and robust fits of a gamma distribution to LOS data

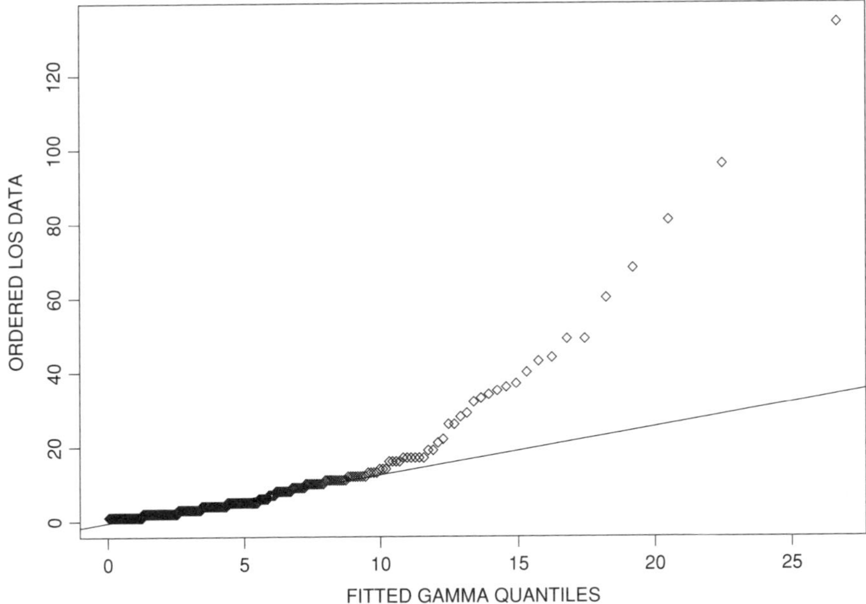

Figure 1.8 Fitted gamma Q–Q plot of LOS data

better overall fit and a mean LOS that better describes the majority of the patients. Figure 1.8 shows a gamma Q–Q plot based on the robustly fitted gamma distribution. This plot reveals that the bulk of the data is well fitted by the robust method, while approximately 30 of the largest values of LOS appear to come from a sub-population of the patients characterized by longer LOS values that is properly modeled separately by another distribution, possibly another gamma distribution with different values of the parameters α and σ.

1.7 Problems

1.1 Show that if a value x_0 is added to a sample $\mathbf{x} = \{x_1, \ldots, x_n\}$, when x_0 ranges from $-\infty$ to $+\infty$ the standard deviation of the enlarged sample ranges between a value smaller than SD$(\mathbf{x})$ and infinity.

1.2 Consider the situation of the former problem.

(a) Show that if n is even, the maximum change in the sample median when x_0 ranges from $-\infty$ to $+\infty$ is the distance from Med$(\mathbf{x})$ to the next order statistic, the farthest from Med$(\mathbf{x})$.

(b) What is the maximum change in the case when n is odd?

1.3 Show for t_i defined in (1.3) that $|t_i| < (n-1)/\sqrt{n}$ for all possible data sets of size n, and hence for all data sets $|t_i| < 3$ if $n \leq 10$.

1.4 The interquartile range (IQR) is defined as the difference between the third and the first quartiles.

(a) Calculate the IQR of the $\mathrm{N}(\mu, \sigma^2)$ distribution.

(b) Consider the sample interquartile range

$$\mathrm{IQR}(\mathbf{x}) = \mathrm{IQR}(x_1, x_2, \ldots, x_n) = x_{(\lfloor 3n/4 \rfloor)} - x_{(\lfloor n/4 \rfloor)}$$

as a measure of dispersion. It is known that sample quantiles tend to the respective distribution quantiles if these are unique. Based on this fact determine the constant c such that the normalized interquartile range $\mathrm{IQRN}(\mathbf{x}) = \mathrm{IQR}(\mathbf{x})/c$ is a consistent estimate of σ when the data have a $\mathrm{N}(\mu, \sigma^2)$ distribution.

(c) Can you think of a reason why you would prefer $\mathrm{MADN}(\mathbf{x})$ to $\mathrm{IQRN}(\mathbf{x})$ as a robust estimate of dispersion?

1.5 Show that the median of the exponential distribution is $\lambda \log 2$, and hence $\mathrm{Med}\,(\mathbf{x})\,/\,\log 2$ is a consistent estimate of λ.

2

Location and Scale

2.1 The location model

For a systematic treatment of the situations considered in the Introduction, we need to represent them by probability-based statistical models. We assume that the outcome x_i of each observation depends on the "true value" μ of the unknown parameter (in Example 1.1, the copper content of the whole flour lot) and also on some random error process. The simplest assumption is that the error acts additively, i.e.,

$$x_i = \mu + u_i \ \ (i = 1, \ldots, n) \tag{2.1}$$

where the errors $u_1, \ldots, u_n$ are random variables. This is called the *location model.*

If the observations are independent replications of the same experiment under equal conditions, it may be assumed that

- $u_1, \ldots, u_n$ have the same distribution function F_0.
- $u_1, \ldots, u_n$ are independent.

It follows that $x_1, \ldots, x_n$ are independent with common distribution function

$$F(x) = F_0(x - \mu) \tag{2.2}$$

and we say that the x_i's are *i.i.d.*—independent and identically distributed—random variables .

The assumption that there are no systematic errors can be formalized as

- u_i and $-u_i$ have the same distribution, and consequently $F_0(x) = 1 - F_0(-x)$.

An *estimate* $\widehat{\mu}$ is a function of the observations: $\widehat{\mu} = \widehat{\mu}(x_1, \ldots, x_n) = \widehat{\mu}(\mathbf{x})$. We are looking for estimates such that in some sense $\widehat{\mu} \approx \mu$ with high probability. One

Robust Statistics – Theory and Methods Ricardo A. Maronna, R. Douglas Martin and Víctor J. Yohai

way to measure the approximation is with *mean squared error* (MSE):

$$\text{MSE}(\widehat{\mu}) = \text{E}(\widehat{\mu} - \mu)^2$$

(other measures will be developed later). The MSE can be decomposed as

$$\text{MSE}(\widehat{\mu}) = \text{Var}(\widehat{\mu}) + \text{Bias}(\widehat{\mu})^2,$$

with

$$\text{Bias}(\widehat{\mu}) = \text{E}\widehat{\mu} - \mu$$

where "E" stands for the expectation.

Note that if $\widehat{\mu}$ is the sample mean and c is any constant, then

$$\widehat{\mu}(x_1 + c, \ldots, x_n + c) = \widehat{\mu}(x_1, \ldots, x_n) + c \tag{2.3}$$

and

$$\widehat{\mu}(cx_1, \ldots, cx_n) = c\widehat{\mu}(x_1, \ldots, x_n). \tag{2.4}$$

The same holds for the median. These properties are called respectively *shift* (or location) and *scale equivariance* of $\widehat{\mu}$. They imply that, for instance, if we express our data in degrees Celsius instead of Fahrenheit, the estimate will automatically adapt to the change of units.

A traditional way to represent "well-behaved" data, i.e. data without outliers, is to assume F_0 is normal with mean 0 and unknown variance σ^2, which implies

$$F = \mathcal{D}(x_i) = \text{N}(\mu, \sigma^2),$$

where $\mathcal{D}(x)$ denotes the distribution of the random variable x, and $\text{N}(\mu, v)$ is the normal distribution with mean μ and variance v. Classical methods assume that F belongs to an *exactly* known parametric family of distributions. If the data were *exactly* normal, the mean would be an "optimal" estimate: it is the maximum likelihood estimate (MLE) (see next section), and minimizes the MSE among unbiased estimates, and also among equivariant ones (Bickel and Doksum, 2001; Lehmann and Casella, 1998). But data are seldom so well behaved.

Figure 2.1 shows the normal Q–Q plots of the observations in Example 1.1. We see that the *bulk* of the data may be described by a normal distribution, but not the whole of it. The same feature can be observed in the Q–Q plot of Figure 1.2. In this sense, we may speak of F as being only *approximately normal*, with normality failing at the tails. We may thus state our initial goal as: looking for estimates that are almost as good as the mean when F is exactly normal, but that are also "good" in some sense when F is only approximately normal.

At this point it may seem natural to think that an adequate procedure could be to test the hypothesis that the data are normal; if it is not rejected, we use the mean, otherwise, the median; or, better still, fit a distribution to the data, and then use the MLE for the fitted one. But this has the drawback that very large sample sizes are needed to distinguish the true distribution, especially since here the *tails*—precisely the regions with fewer data—are most influential.

To formalize the idea of approximate normality, we may imagine that a proportion $1 - \epsilon$ of the observations is generated by the normal model, while a proportion ϵ

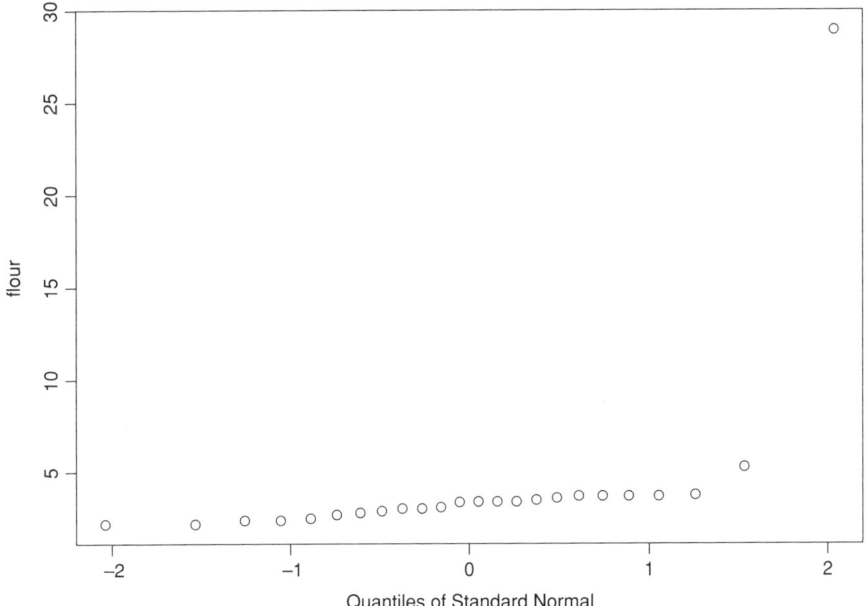

Figure 2.1 Q–Q plot of flour data

is generated by an unknown mechanism. For instance, repeated measurements are made of some magnitude, which are 95% of the time correct, but 5% of the time the apparatus fails or the experimenter makes a wrong transcription. This may be described by supposing

$$F = (1 - \epsilon)G + \epsilon H \tag{2.5}$$

where $G = \mathrm{N}(\mu, \sigma^2)$ and H may be any distribution; for instance, another normal with a larger variance and a possibly different mean. This is called a *contaminated normal distribution*. An early example of the use of these distributions to show the dramatic lack of robustness of the SD was given by Tukey (1960). In general, F is called a *mixture* of G and H, and is called a *normal mixture* when both G and H are normal.

To justify (2.5), let A be the event "the apparatus fails", which has $\mathrm{P}(A) = \varepsilon$, and A' its complement. We are assuming that our observation x has distribution G conditional on A' and H conditional on A. Then by the total probability rule

$$\begin{aligned} F(t) &= \mathrm{P}(x \le t) = \mathrm{P}(x \le t | A')P(A') + \mathrm{P}(x \le t | A)P(A) \\ &= G(t)(1 - \varepsilon) + H(t)\varepsilon. \end{aligned}$$

If G and H have densities g and h, respectively, then F has density

$$f = (1 - \varepsilon)g + \varepsilon h. \tag{2.6}$$

It must be emphasized that—as in the ozone layer example of Section 1.1—atypical values are not necessarily due to erroneous measurements: they simply reflect an unknown change in the measurement conditions in the case of physical measurements, or more generally the behavior of a sub-population of the data. An important example of the latter is that normal mixture distributions have been found to often provide quite useful models for the stock returns, i.e., the relative change in price from one time period to the next, with the mixture components corresponding to different volatility regimes of the returns.

Another model for outliers is the so-called *heavy-tailed* or *fat-tailed* distributions, i.e., distributions whose density tails tend to zero more slowly than the normal density tails. An example is the so-called *Cauchy* distribution , with density

$$f(x) = \frac{1}{\pi(1+x^2)}. \tag{2.7}$$

It is bell shaped like the normal, but its mean does not exist. It is a particular case of the family of *Student* (or t) densities with $\nu > 0$ degrees of freedom, given by

$$f_\nu(x) = c_\nu \left(1 + \frac{x^2}{\nu}\right)^{-(\nu+1)/2} \tag{2.8}$$

where c_υ is a constant:

$$c_\upsilon = \frac{\Gamma((\upsilon+1)/2)}{\sqrt{\upsilon\pi}\,\Gamma(\upsilon/2)},$$

where Γ is the gamma function. This family contains all degrees of heavy-tailedness. When $\nu \to \infty$, f_ν tends to the standard normal density; for $\nu = 1$ we have the Cauchy distribution.

Figure 2.2 shows the densities of N(0, 1), the Student distribution with 4 degrees of freedom, and the contaminated distribution (2.6) with $g = \mathrm{N}(0, 1)$, $h = \mathrm{N}(0, 100)$ and $\varepsilon = 0.10$, denoted by N, T4 and CN respectively. To make comparisons more clear, the three distributions are normalized to have the same interquartile range.

If $F_0 = \mathrm{N}(0, \sigma^2)$ in (2.2), then $\overline{x}$ is $\mathrm{N}(\mu, \sigma^2/n)$. As we shall see later, the sample median is approximately $\mathrm{N}(\mu, 1.57\sigma^2/n)$, so the sample median has a 57% increase in variance relative to the sample mean. We say that the median has a *low efficiency* at the normal distribution.

On the other hand, assume that 95% of our observations are well behaved, represented by $G = \mathrm{N}(\mu, 1)$, but that 5% of the times the measuring system gives an erratic result, represented by a normal distribution with the same mean but a 10-fold increase in the standard deviation. We thus have the model (2.5) with $\epsilon = 0.05$ and $H = \mathrm{N}(\mu, 100)$. In general, under the model

$$F = (1-\varepsilon)\mathrm{N}(\mu, 1) + \varepsilon \mathrm{N}(\mu, \tau^2) \tag{2.9}$$

we have (see (2.85), (2.26) and Problem 2.3)

$$\mathrm{Var}(\overline{x}) = \frac{(1-\varepsilon)+\varepsilon\tau^2}{n}, \quad \mathrm{Var}(\mathrm{Med}(\mathbf{x})) \approx \frac{\pi}{2n\,(1-\varepsilon+\varepsilon/\tau)^2}. \tag{2.10}$$

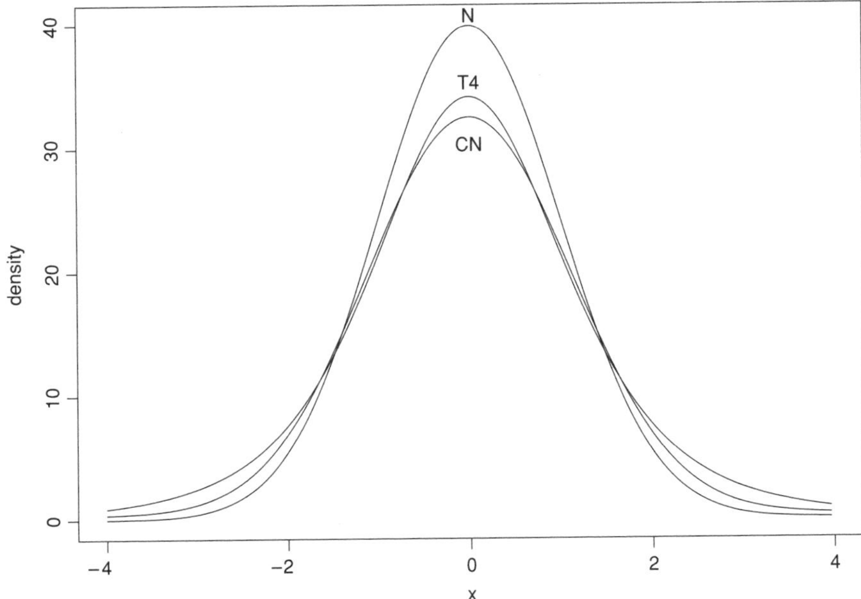

Figure 2.2 Standard normal (N), Student (T4) and contaminated normal (CN) densities, scaled to equal interquartile range

Note that Var(Med(**x**)) above means "the *theoretical* variance of the *sample* median of **x**". It follows that for $\epsilon = 0.05$ and $H = \mathrm{N}(\mu, 100)$, the variance of $\overline{x}$ increases to 5.95, while that of the median is only 1.72. The gain in robustness due to using the median is paid for by an increase in variance ("a loss in efficiency") at the normal distribution.

Table 2.1 shows the approximations for large n of n times the variances of the mean and median for different values of τ. It is seen that the former increases rapidly with τ, while the latter stabilizes.

In the sequel we shall develop estimates which combine the low variance of the mean at the normal with the robustness of the median under contamination. For

Table 2.1 Variances ($\times n$) of mean and median for large n

ε	0.05		0.10	
τ	$n\mathrm{Var}(\overline{x})$	$n\mathrm{Var}(\mathrm{Med})$	$n\mathrm{Var}(\overline{x})$	$n\mathrm{Var}(\mathrm{Med})$
3	1.40	1.68	1.80	1.80
4	1.75	1.70	2.50	1.84
5	2.20	1.70	3.40	1.86
6	2.75	1.71	4.50	1.87
10	5.95	1.72	10.90	1.90
20	20.9	1.73	40.90	1.92

introductory purposes we deal only with *symmetric* distributions. The distribution of the variable x is symmetric about μ if $x - \mu$ and $\mu - x$ have the same distribution. If x has a density f, symmetry about μ is equivalent to $f(\mu + x) = f(\mu - x)$. Symmetry implies that $\text{Med}(x) = \mu$, and if the expectation exists, also that $\text{E}x = \mu$. Hence if the data have a symmetric distribution, there is no bias and only the variability is at issue. In Chapter 3 general contamination will be addressed.

Two primitive ideas to obtain robust estimates are based on *deleting* and *truncating* atypical data. Assume that we define an interval $[a, b]$ (depending on the data) containing supposedly "typical" observations, such as $a = \overline{x} - 2s, b = \overline{x} + 2s$. Then deletion means using a modified sample, obtained by omitting all points outside $[a, b]$. Truncation means replacing all $x_i < a$ by a and all $x_i > b$ by b, and not altering the other points, i.e., atypical values are shifted to the nearest typical ones. Naive uses of these ideas are not necessarily good, but some of the methods we shall study are elaborate versions of them.

2.2 M-estimates of location

We shall now develop a general family of estimates that contains the mean and the median as special cases.

2.2.1 Generalizing maximum likelihood

Consider again the location model (2.1). Assume that F_0, the distribution function of u_i, has a density $f_0 = F_0'$. Then the joint density of the observations (the *likelihood function*) is

$$L(x_1, \ldots, x_n; \mu) = \prod_{i=1}^{n} f_0(x_i - \mu)$$

The *maximum likelihood estimate* (MLE) of μ is the value $\widehat{\mu}$—depending on $x_1, \ldots, x_n$— that maximizes $L(x_1, \ldots, x_n; \mu)$:

$$\widehat{\mu} = \widehat{\mu}(x_1, \ldots, x_n) = \arg\max_{\mu} L(x_1, \ldots, x_n; \mu) \tag{2.11}$$

where "arg max" stands for "the value maximizing".

If we knew F_0 exactly, the MLE would be "optimal" in the sense of attaining the lowest possible asymptotic variance among a "reasonable" class of estimates (see Section 10.8). But since we know F_0 only approximately, our goal will be to find estimates which are

(A) "nearly optimal" when F_0 is exactly normal,
and also
(B) "nearly optimal" when F_0 is approximately normal (e.g. contaminated normal).

If f_0 is everywhere positive, since the logarithm is an increasing function, (2.11) can be written as

$$\widehat{\mu} = \arg\min_{\mu} \sum_{i=1}^{n} \rho\,(x_i - \mu) \tag{2.12}$$

where

$$\rho = -\log f_0. \tag{2.13}$$

If $F_0 = \mathrm{N}(0, 1)$, then

$$f_0(x) = \frac{1}{\sqrt{2\pi}} e^{-x^2/2} \tag{2.14}$$

and apart from a constant, $\rho(x) = x^2/2$. Hence (2.12) is equivalent to

$$\widehat{\mu} = \arg\min_{\mu} \sum_{i=1}^{n} (x_i - \mu)^2. \tag{2.15}$$

If F_0 is the double exponential distribution

$$f_0(x) = \frac{1}{2} e^{-|x|} \tag{2.16}$$

then $\rho(x) = |x|$, and (2.12) is equivalent to

$$\widehat{\mu} = \arg\min_{\mu} \sum_{i=1}^{n} |x_i - \mu|. \tag{2.17}$$

We shall see below that the solutions to (2.15) and (2.17) are the sample mean and median, respectively.

If ρ is differentiable, differentiating (2.12) with respect to μ yields

$$\sum_{i=1}^{n} \psi(x_i - \widehat{\mu}) = 0 \tag{2.18}$$

with $\psi = \rho'$. If ψ is discontinuous, solutions to (2.18) might not exist, and in this case we shall interpret (2.18) to mean that the left-hand side changes sign at μ. Note that if f_0 is symmetric, then ρ is even and hence ψ is odd.

If $\rho(x) = x^2/2$, then $\psi(x) = x$, and (2.18) becomes

$$\sum_{i=1}^{n} (x_i - \widehat{\mu}) = 0$$

which has $\widehat{\mu} = \overline{x}$ as solution.

For $\rho(x) = |x|$, it will be shown that any median of x is a solution of (2.17). In fact, the derivative of $\rho(x)$ exists for $x \neq 0$, and is given by the sign function:

$\psi(x) = \text{sgn}(x)$, where

$$\text{sgn}(x) = \begin{cases} -1 & \text{if} \quad x < 0 \\ 0 & \text{if} \quad x = 0 \\ 1 & \text{if} \quad x > 0. \end{cases} \tag{2.19}$$

Since the function to be minimized in (2.17) is continuous, it suffices to find the values of μ where its derivative changes sign. Note that

$$\text{sgn}(x) = \text{I}(x > 0) - \text{I}(x < 0) \tag{2.20}$$

where I(.) stands for the *indicator function*, i.e.,

$$\text{I}(x > 0) = \begin{cases} 1 & \text{if} \quad x > 0 \\ 0 & \text{if} \quad x \leq 0. \end{cases}$$

Applying (2.20) to (2.18) yields

$$\begin{aligned} \sum_{i=1}^{n} \text{sgn}(x_i - \mu) &= \sum_{i=1}^{n} (\text{I}(x_i - \mu > 0) - \text{I}(x_i - \mu < 0)) \\ &= \#(x_i > \mu) - \#(x_i < \mu) = 0 \end{aligned}$$

and hence $\#(x_i > \mu) = \#(x_i < \mu)$, which implies that μ is any sample median.

From now on, the average of a data set $\mathbf{z} = \{z_1, \ldots, z_n\}$ will be denoted by ave($\mathbf{z}$), or by $\text{ave}_i(z_i)$ when necessary, i.e.,

$$\text{ave}(\mathbf{z}) = \text{ave}_i(z_i) = \frac{1}{n} \sum_{i=1}^{n} z_i,$$

and its median by Med($\mathbf{z}$) or $\text{Med}_i(z_i)$. If c is a constant, $\mathbf{z} + c$ and $c\mathbf{z}$ will denote the data sets $(z_1 + c, \ldots, z_n + c)$ and $(cz_1, \ldots, cz_n)$. If x is a random variable with distribution F, the mean and median of a function $g(x)$ will be denoted by $\text{E}_F g(x)$ and $\text{Med}_F g(x)$, dropping the subscript F when there is no ambiguity.

Given a function ρ, an *M-estimate of location* is a solution of (2.12). We shall henceforth study estimates of this form, which need not be MLEs for any distribution. The function ρ will be chosen in order to ensure the goals (A) and (B) above.

Assume ψ is monotone nondecreasing, with $\psi(-\infty) < 0 < \psi(\infty)$. Then it is proved in Theorem 10.1 that (2.18)—and hence (2.12)—always has a solution. If ψ is continuous and increasing, the solution is unique, otherwise the set of solutions is either a point or an interval (throughout this book, we shall call any function g *increasing* (*nondecreasing*) if $a < b$ implies $g(a) < g(b)$ $(g(a) \leq g(b))$). More details on uniqueness are given in Section 10.1.

It is easy to show that M-estimates are shift equivariant as defined in (2.3) (Problem 2.5). The mean and median are scale equivariant, but this does not hold in general for M-estimates in their present form. This drawback will be overcome in Section 2.6.

2.2.2 The distribution of M-estimates

In order to evaluate the performance of M-estimates, it is necessary to calculate their distributions. Except for the mean and the median (see (10.47)), there are no explicit expressions for the distribution of M-estimates in finite sample sizes, but approximations can be found and a heuristic derivation is given in Section 2.9.2 (a rigorous treatment is given in Section 10.3).

Assume ψ is increasing. For a given distribution F, define $\mu_0 = \mu_0(F)$ as the solution of

$$\mathrm{E}_F \psi(x - \mu_0) = 0. \tag{2.21}$$

For the sample mean, $\psi(x) = x$, and (2.21) implies $\mu_0 = \mathrm{E}x$, i.e., the population mean. For the sample median, (2.20) and (2.21) yield

$$\mathrm{P}(x > \mu_0) - \mathrm{P}(x < \mu_0) = 2F(\mu_0) - 1 = 0$$

which implies $F(\mu_0) = 1/2$, which corresponds to $\mu_0 = \mathrm{Med}(x)$, i.e., a population median. In general if F is symmetric then μ_0 coincides with the center of symmetry (Problem 2.6).

It can be shown (see Section 2.9.2) that when $n \to \infty$,

$$\widehat{\mu} \to_p \mu_0 \tag{2.22}$$

where "$\to_p$" stands for "tends in probability" and μ_0 is defined in (2.21) (we say that $\widehat{\mu}$ is "*consistent* for μ_0"), and the distribution of $\widehat{\mu}$ is approximately

$$\mathrm{N}\left(\mu_0, \frac{v}{n}\right) \text{ with } v = \frac{\mathrm{E}_F\left(\psi(x - \mu_0)^2\right)}{(\mathrm{E}_F \psi'(x - \mu_0))^2}. \tag{2.23}$$

Note that under model (2.2) v does not depend on μ_0, i.e.,

$$v = \frac{\mathrm{E}_{F_0}\left(\psi(x)^2\right)}{\left(\mathrm{E}_{F_0} \psi'(x)\right)^2}. \tag{2.24}$$

If the distribution of an estimate $\widehat{\mu}$ is approximately $\mathrm{N}(\mu_0, v/n)$ for large n, we say that $\widehat{\mu}$ is *asymptotically normal*, with asymptotic value μ_0 and asymptotic variance v. The *asymptotic efficiency* of $\widehat{\mu}$ is the ratio

$$\mathrm{Eff}(\widehat{\mu}) = \frac{v_0}{v}, \tag{2.25}$$

where v_0 is the asymptotic variance of the MLE, and measures how near $\widehat{\mu}$ is to the optimum. The expression for v in (2.23) is called the *asymptotic variance* of $\widehat{\mu}$.

To understand the meaning of efficiency, consider two estimates with asymptotic variances v_1 and v_2. Since their distributions are approximately normal with variances v_1/n and v_2/n, if for example $v_1 = 3v_2$ then the first estimate requires three times as many observations to attain the same variance as the second.

For the sample mean, $\psi' \equiv 1$ and hence $v = \mathrm{Var}(x)$. For the sample median, the numerator of v is one. Here ψ' does not exist, but if x has a density f, it is shown in Section 10.3 that the denominator is $2f(\mu_0)$, and hence

$$v = \frac{1}{4f(\mu_0)^2}. \tag{2.26}$$

Thus for $F = \mathrm{N}(0, 1)$ we have

$$v = \frac{2\pi}{4} = 1.571.$$

It will be seen that a type of ρ- and ψ-functions with important properties is the family of *Huber functions*, plotted in Figure 2.3:

$$\rho_k(x) = \begin{cases} x^2 & \text{if} \quad |x| \leq k \\ 2k\,|x| - k^2 & \text{if} \quad |x| > k \end{cases} \tag{2.27}$$

with derivative $2\psi_k(x)$, where

$$\psi_k(x) = \begin{cases} x & \text{if} \quad |x| \leq k \\ \mathrm{sgn}(x)k & \text{if} \quad |x| > k. \end{cases} \tag{2.28}$$

It is seen that ρ_k is quadratic in a central region, but increases only linearly to infinity. The M-estimates corresponding to the limit cases $k \to \infty$ and $k \to 0$ are the mean and the median, and we define $\psi_0(x)$ as $\mathrm{sgn}(x)$.

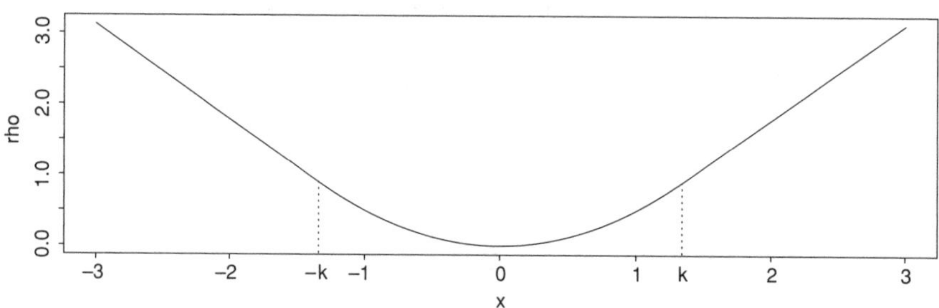

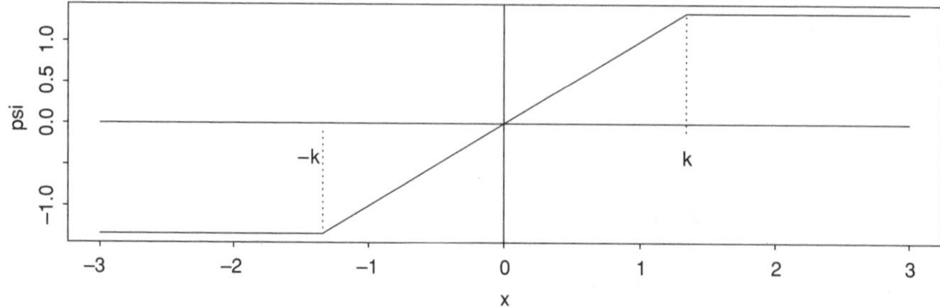

Figure 2.3 Huber ρ- and ψ-functions

Table 2.2 Asymptotic variances of Huber M-estimate

k	ε=0	ε=0.05	ε=0.10
0	1.571	1.722	1.897
0.7	1.187	1.332	1.501
1.0	1.107	1.263	1.443
1.4	1.047	1.227	1.439
1.7	1.023	1.233	1.479
2.0	1.010	1.259	1.550
∞	1.000	5.950	10.900

The value of k is chosen in order to ensure a given asymptotic variance—hence a given asymptotic efficiency—at the normal distribution. Table 2.2 gives the asymptotic variances of the estimate at model (2.5) with $G = \mathrm{N}(0, 1)$ and $H = \mathrm{N}(0, 10)$, for different values of k.

Here we see the trade-off between robustness and efficiency: when $k = 1.4$, the variance of the M-estimator at the normal is only 4.7% larger than that of $\overline{x}$ (which corresponds to $k = \infty$) and much smaller than that of the median (which corresponds to $k = 0$), while for contaminated normals it is clearly smaller than both.

Huber's ψ is one of the few cases where the asymptotic variance at the normal distribution can be calculated analytically. Since $\psi_k'(x) = \mathrm{I}(|x| \leq k)$, the denominator of (2.23) is $(\Phi(k) - \Phi(-k))^2$. The reader can verify that the numerator is

$$\mathrm{E}_\Phi \psi_k(x)^2 = 2\left[k^2\left(1 - \Phi\left(k\right)\right) + \Phi(k) - 0.5 - k\varphi(k)\right] \tag{2.29}$$

where φ and Φ are the standard normal density and distribution function, respectively (Problem 2.7). In Table 2.3 we give the values of k yielding prescribed asymptotic variances v. The last row gives values of the quantity $\alpha = 1 - \Phi(k)$ that will play a role in Section 2.3.

2.2.3 An intuitive view of M-estimates

A location M-estimate can be seen as a weighted mean. In most cases of interest, $\psi(0) = 0$ and $\psi'(0)$ exists, so that ψ is approximately linear at the origin. Let

$$W(x) = \begin{cases} \psi(x)/x & \text{if} \quad x \neq 0 \\ \psi'(0) & \text{if} \quad x = 0. \end{cases} \tag{2.30}$$

Table 2.3 Asymptotic variances for Huber's ψ-function

k	0.66	1.03	1.37
v	1.20	1.10	1.05
α	0.25	0.15	0.085

Then (2.18) can be written as

$$\sum_{i=1}^{n} W(x_i - \widehat{\mu})(x_i - \widehat{\mu}) = 0,$$

or equivalently

$$\widehat{\mu} = \frac{\sum_{i=1}^{n} w_i x_i}{\sum_{i=1}^{n} w_i}, \text{ with } w_i = W(x_i - \widehat{\mu}), \tag{2.31}$$

which expresses the estimate as a weighted mean. Since in general $W(x)$ is a nonincreasing function of $|x|$, outlying observations will receive smaller weights. Note that although (2.31) looks like an explicit expression for $\widehat{\mu}$, actually the weights on the right-hand side depend also on $\widehat{\mu}$. Besides its intuitive value, this representation of the estimate will be useful for its numeric computation in Section 2.7. The weight function corresponding to Huber's ψ is

$$W_k(x) = \min\left\{1, \frac{k}{|x|}\right\} \tag{2.32}$$

which is plotted in the upper panel of Figure 2.4.

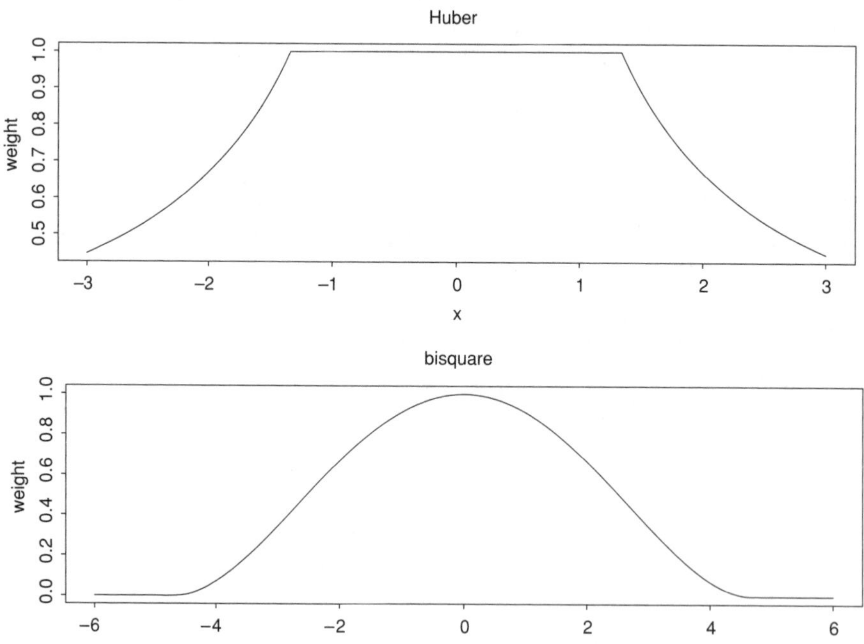

Figure 2.4 Huber and bisquare weight functions

Another intuitive way to interpret an M-estimate is to rewrite (2.18) as

$$\widehat{\mu} = \widehat{\mu} + \frac{1}{n}\sum_{i=1}^{n}\psi(x_i - \widehat{\mu}) = \frac{1}{n}\sum_{i=1}^{n}\zeta(x_i, \widehat{\mu}), \tag{2.33}$$

where

$$\zeta(x, \mu) = \mu + \psi(x - \mu) \tag{2.34}$$

which for the Huber function takes the form

$$\zeta(x, \mu) = \begin{cases} \mu - k & \text{if} \quad x < \mu - k \\ x & \text{if} \quad \mu - k \le x \le \mu + k \\ \mu + k & \text{if} \quad x > \mu + k. \end{cases} \tag{2.35}$$

That is, $\widehat{\mu}$ may be viewed as an average of the modified observations $\zeta(x_i, \widehat{\mu})$ (called "pseudo-observations"):observations in the bulk of the data remain unchanged, while those too large or too small are truncated as described at the end of Section 2.1 (note that here the truncation interval depends on the data).

2.2.4 Redescending M-estimates

It is easy to show (Problem 2.15) that the MLE for the Student family of densities (2.8) has the ψ-function

$$\psi(x) = \frac{x}{x^2 + \nu}, \tag{2.36}$$

which tends to zero when $x \to \infty$. This suggests that for symmetric heavy-tailed distributions, it is better to use "redescending" ψ's that tend to zero at infinity. This implies that for large x, the respective ρ-function increases more slowly than Huber's ρ (2.27), which is linear for $x > k$.

Actually, we shall later discuss the advantages of using a *bounded* ρ. A popular choice of ρ- and ψ-functions is the *bisquare* (also called *biweight*) family of functions:

$$\rho(x) = \begin{cases} 1 - \left[1 - (x/k)^2\right]^3 & \text{if} \quad |x| \le k \\ 1 & \text{if} \quad |x| > k \end{cases} \tag{2.37}$$

with derivative $\rho'(x) = 6\psi(x)/k^2$ where

$$\psi(x) = x\left[1 - \left(\frac{x}{k}\right)^2\right]^2 \mathrm{I}(|x| \le k). \tag{2.38}$$

These functions are displayed in Figure 2.5. Note that ψ is everywhere differentiable and vanishes outside $[-k, k]$. M-estimates with ψ vanishing outside an interval are not MLEs for any distribution (Problem 2.12).

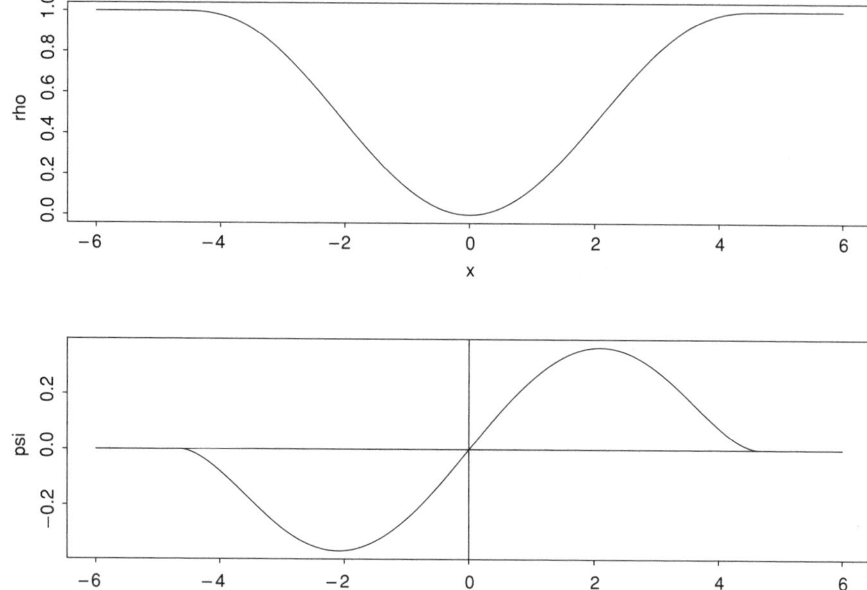

Figure 2.5 ρ- and ψ-functions for bisquare estimate

The weight function (2.30) for this family is

$$W(x) = \left[1 - \left(\frac{x}{k}\right)^2\right]^2 \mathrm{I}(|x| \leq k)$$

and is plotted in Figure 2.4.

If ρ is everywhere differentiable and ψ is monotonic, then the forms (2.12) and (2.18) are equivalent. If ψ is redescending, some solutions of (2.18)—usually called "bad solutions"—may not correspond to the absolute minimum of the criterion, which defines the M-estimate.

Estimates defined as solutions of (2.18) with monotone ψ will be called "monotone M-estimates" for short, while those defined by (2.12) when ψ is not monotone will be called "redescending M-estimates". Numerical computing of redescending location estimates is essentially no more difficult than that of monotone estimates (Section 2.7.1). It will be seen later that redescending estimates offer an increase in robustness toward large outliers.

The values of k for prescribed efficiencies (2.25) of the bisquare estimate are given in the table below:

eff.	0.80	0.85	0.90	0.95
k	3.14	3.44	3.88	4.68

(2.39)

If ρ has a nondecreasing derivative, it can be shown (Feller, 1971) that for all x, y

$$\rho(\alpha x + (1-\alpha)y) \leq \alpha\rho(x) + (1-\alpha)\rho(y) \ \forall \alpha \in [0,1]. \tag{2.40}$$

Functions verifying (2.40) are called *convex*.

We state the following definitions for later reference.

Definition 2.1 *Unless stated otherwise, a ρ-function will denote a function ρ such that*

R1 *$\rho(x)$ is a nondecreasing function of $|x|$*
R2 *$\rho(0) = 0$*
R3 *$\rho(x)$ is increasing for $x > 0$ such that $\rho(x) < \rho(\infty)$*
R4 *If ρ is bounded, it is also assumed that $\rho(\infty) = 1$.*

Definition 2.2 *A ψ-function will denote a function ψ which is the derivative of a ρ-function, which implies in particular that*

Ψ1 *ψ is odd and $\psi(x) \geq 0$ for $x \geq 0$.*

2.3 Trimmed means

Another approach to robust estimation of location would be to discard a proportion of the largest and smallest values. More precisely, let $\alpha \in [0, 1/2)$ and $m = [(n-1)\alpha]$ where [.] stands for the integer part, and define the *α-trimmed mean* as

$$\overline{x}_\alpha = \frac{1}{n-2m}\sum_{i=m+1}^{n-m} x_{(i)},$$

where $x_{(i)}$ denotes the order statistics (1.2).

The reader may think that we are again suppressing observations. Note, however, that no subjective choice has been made: the result is actually a function of *all* observations (even of those that have not been included in the sum).

The limit cases $\alpha = 0$ and $\alpha \to 0.5$ correspond to the sample mean and median, respectively. For the data at the beginning of Section 2.1, the α-trimmed means with $\alpha = 0.10$ and 0.25 are respectively 3.20 and 3.27. Deleting the largest observation changes them to 3.17 and 3.22, respectively.

The exact distribution of trimmed means is intractable. Its large-sample approximation is more complicated than that of M-estimates, and will be described in Section 10.7. It can be proved that for large n the distribution under model (2.1) is approximately normal, and for symmetrically distributed u_i $\mathcal{D}(\widehat{\mu}) \approx \mathrm{N}(\mu, v/n)$ where the asymptotic variance v is that of an M-estimate with Huber's function ψ_k, where k is the $(1-\alpha)$-quantile of u:

$$v = \frac{\mathrm{E}\,[\psi_k(u)]^2}{(1-2\alpha)^2}. \tag{2.41}$$

The values of α yielding prescribed asymptotic variances are given at the bottom of Table 2.3. Note that the asymptotic efficiency of $\overline{x}_{0.25}$ is 0.83, even though we seem to be "throwing away" 50% of the observations!

Note also that the asymptotic variance of a trimmed mean is not a trimmed variance! This would be so if the numerator of (2.41) were

$$\mathrm{E}\,[(x-\mu)\mathrm{I}(|x-\mu|)\leq k]^2 .$$

But it is instead a *truncated* variance in the sense explained at the end of Section 2.1.

A more general class of estimates, called *L-estimates*, is defined as linear combinations of order statistics:

$$\widehat{\mu}=\sum_{i=1}^{n} a_i x_{(i)}, \tag{2.42}$$

where the a_i's are given constants. For α-trimmed means,

$$a_i=\frac{1}{n-2m}\,\mathrm{I}(m+1\leq i\leq n-m). \tag{2.43}$$

It is easy to show (Problem 2.10) that if the coefficients of an L-estimate satisfy the conditions

$$a_i\geq 0,\ \sum_{i=1}^{n} a_i=1,\quad a_i=a_{n-i+1}, \tag{2.44}$$

then the estimate is shift and scale equivariant, and also fulfills the natural conditions

C1 If $x_i\geq 0$ for all i, then $\widehat{\mu}\geq 0$
C2 If $x_i=c$ for all i, then $\widehat{\mu}=c$
C3 $\widehat{\mu}(-x)=-\widehat{\mu}(x)$.

2.4 Dispersion estimates

The traditional way to measure the variability of a data set $\mathbf{x}$ is with the standard deviation (SD)

$$\mathrm{SD}(\mathbf{x})=\left[\frac{1}{n-1}\sum_{i=1}^{n}(x_i-\overline{x})^2\right]^{1/2}.$$

For any constant c the SD satisfies the *shift invariance* and *scale equivariance* conditions

$$\mathrm{SD}(\mathbf{x}+c)=\mathrm{SD}(\mathbf{x}),\qquad \mathrm{SD}(c\mathbf{x})=|c|\ \mathrm{SD}(\mathbf{x}). \tag{2.45}$$

Any statistic satisfying (2.45) will be called a *dispersion estimate*.

In Example 1.1 we observed the lack of robustness of the standard deviation, and we now look for robust alternatives to it. One alternative estimate proposed in the past is the *mean absolute deviation* (MD):

$$\mathrm{MD}(\mathbf{x}) = \frac{1}{n}\sum_{i=1}^{n}|x_i - \overline{x}| \tag{2.46}$$

which is also sensitive to outliers, although less so than the SD (Tukey, 1960). In the flour example, the MD with and without the largest observation is respectively 2.14 and 0.52, still a large difference.

Both the SD and MD are defined by first centering the data by subtracting $\overline{x}$ (which ensures shift invariance) and then taking a measure of "largeness" of the absolute values. A robust alternative is to subtract the median instead of the mean, and then take the median of the absolute values, which yields the MAD estimate introduced in the previous chapter:

$$\mathrm{MAD}(\mathbf{x}) = \mathrm{Med}(|\mathbf{x} - \mathrm{Med}(\mathbf{x})|) \tag{2.47}$$

which clearly satisfies (2.45). For the flour data with and without the largest observation, the MAD is 0.35 and 0.34, respectively.

In the same way as (2.46) and (2.47), we define the mean and the median absolute deviations of a random variable x as

$$\mathrm{MD}(x) = \mathrm{E}|x - \mathrm{E}x| \tag{2.48}$$

and

$$\mathrm{MAD}(x) = \mathrm{Med}(|x - \mathrm{Med}(x)|), \tag{2.49}$$

respectively.

Two other well-known dispersion estimates are the *range* defined as $\max(x) - \min(x) = x_{(n)} - x_{(1)}$ and the sample *interquartile range*

$$\mathrm{IQR}(x) = x_{(n-m+1)} - x_{(m)}$$

where $m = [n/4]$. Both are based on order statistics; the former is clearly very sensitive to outliers, while the latter is not.

Note that if $x \sim \mathrm{N}(\mu, \sigma^2)$ (where "$\sim$" stands for "is distributed as") then $\mathrm{SD}(x) = \sigma$ by definition, while $\mathrm{MD}(x)$, $\mathrm{MAD}(x)$ and $\mathrm{IQR}(x)$ are constant multiples of σ:

$$\mathrm{MD}(x) = c_1\sigma, \;\; \mathrm{MAD}(x) = c_2\sigma, \;\; \mathrm{IQR}(x) = 2c_2\sigma,$$

where

$$c_1 = 2\varphi(0) \text{ and } c_2 = \Phi^{-1}(0.75)$$

(Problem 2.11). Hence if we want a dispersion estimate that "measures the same thing" as the SD at the normal, we should normalize the MAD by dividing it by $c_2 \approx 0.675$. The "normalized MAD" (MADN) is thus

$$\mathrm{MADN}(x) = \frac{\mathrm{MAD}(x)}{0.675}. \tag{2.50}$$

Likewise, we should normalize the MD and the IQR by dividing them by c_1 and by $2c_2$ respectively.

Observe that for the flour data (which was found to be approximately normal) MADN $= 0.53$, which is not far from the standard deviation of the data without the outlier: 0.69.

Note that the first step in computing the SD, MAD and MD is "centering" the data: that is, subtracting a location estimate from the data values. The IQR does not use centering. A dispersion estimate that does not require centering and is more robust than the IQR (in a sense to be defined in the next chapter) was proposed by Croux and Rousseeuw (1992) and Rousseeuw and Croux (1993). The estimate, which they call Q_n, is based only on the differences between data values. Let $m = \binom{n}{2}$. Call $d_{(1)} \leq \ldots \leq d_{(m)}$ the ordered values of the m differences $d_{ij} = x_{(i)} - x_{(j)}$ with $i > j$. Then the estimate is defined as

$$Q_n = z_{(k)}, \quad k = \binom{[n/2]+1}{2}, \tag{2.51}$$

where [.] denotes the integer part. Since $k \approx m/4$, Q_n is approximately the first quartile of the d_{ij}'s. It is easy to verify that, for any k, Q_n is shift invariant and scale equivariant. It can be shown that, at the normal, Q_n has an efficiency of 0.82, and the estimate $2.222Q_n$ is consistent for the SD.

Martin and Zamar (1993b) studied another dispersion estimate that does not require centering and has interesting robustness properties (Problem 2.16b).

2.5 M-estimates of scale

In this section we discuss a situation that, while not especially important in itself, will play an important auxiliary role in the development of estimates for location, regression and multivariate analysis. Consider observations x_i satisfying the *multiplicative* model

$$x_i = \sigma u_i \tag{2.52}$$

where the u_i's are i.i.d with density f_0 and $\sigma > 0$ is the unknown parameter. The distributions of the x_i's constitute a *scale family*, with density

$$\frac{1}{\sigma} f_0\left(\frac{x}{\sigma}\right).$$

Examples are the exponential family with $f_0(x) = \exp(-x)\mathrm{I}(x > 0)$ and the normal scale family N$(0, \sigma^2)$ with f_0 given by (2.14).

The MLE of σ in (2.52) is

$$\widehat{\sigma} = \arg\max_{\sigma} \frac{1}{\sigma^n} \prod_{i=1}^{n} f_0\left(\frac{x_i}{\sigma}\right).$$

Taking logs and differentiating with respect to σ yields

$$\frac{1}{n}\sum_{i=1}^{n}\rho\left(\frac{x_i}{\widehat{\sigma}}\right)=1 \tag{2.53}$$

where $\rho(t)=t\psi(t)$, with $\psi=-f_0'/f_0$. If f_0 is N(0, 1) then $\rho(t)=t^2$, which yields $\widehat{\sigma}=\sqrt{\text{ave}(\mathbf{x}^2)}$ (the *root mean square*, RMS); if f is double-exponential defined in (2.16), then $\rho(t)=|t|$ which yields $\widehat{\sigma}=\text{ave}(|\mathbf{x}|)$. Note that if f_0 is even, so is ρ, and this implies that $\widehat{\sigma}$ depends only on the absolute values of the x's.

In general, any estimate satisfying an equation of the form

$$\frac{1}{n}\sum_{i=1}^{n}\rho\left(\frac{x_i}{\widehat{\sigma}}\right)=\delta, \tag{2.54}$$

where ρ is a ρ-function and δ is a positive constant, will be called an *M-estimate of scale.* Note that in order for (2.54) to have a solution we must have $0<\delta<\rho(\infty)$. Hence if ρ is bounded it will be assumed without loss of generality that

$$\rho(\infty)=1,\quad \delta\in(0,1).$$

In the rarely occurring event that $\#(x_i=0)>n(1-\delta)$ should happen, then (2.54) has no solution and in this case it is natural to define $\widehat{\sigma}(\mathbf{x})=0$. It is easy to verify that scale M-estimates are equivariant in the sense that $\widehat{\sigma}(c\mathbf{x})=c\widehat{\sigma}(\mathbf{x})$ for any $c>0$, and if ρ is even then

$$\widehat{\sigma}(c\mathbf{x})=|c|\widehat{\sigma}(\mathbf{x})$$

for any c. For large n, the sequence of estimates (2.54) converges to the solution of

$$\mathrm{E}\rho\left(\frac{x}{\sigma}\right)=\delta \tag{2.55}$$

if it is unique (Section 10.2); see Problem 10.6

The reader can verify that the scale MLE for the Student distribution is equivalent to

$$\rho(t)=\frac{t^2}{t^2+\upsilon}\ \text{and}\ \delta=\frac{1}{\nu+1}. \tag{2.56}$$

A frequently used scale estimate is the *bisquare scale*, where ρ is given by (2.37) with $k=1$, i.e.,

$$\rho(x)=\min\left\{1-\left(1-x^2\right)^3,1\right\} \tag{2.57}$$

and $\delta=0.5$. It is easy to verify that (2.56) and (2.57) satisfy the conditions for a ρ-function in Definition 2.1.

When ρ is the step function

$$\rho(t)=\mathrm{I}(|t|>c), \tag{2.58}$$

where c is a positive constant, and $\delta=0.5$, we have $\widehat{\sigma}=\text{Med}(|\mathbf{x}|)/c$. The argument in Problem 2.12 shows that it is not the scale MLE for any distribution.

Most often we shall use a ρ that is quadratic near the origin, i.e., $\rho'(0) = 0$ and $\rho''(0) > 0$, and in such cases an M-scale estimate can be represented as a weighted RMS estimate. We define the weight function as

$$W(x) = \begin{cases} \rho(x)/x^2 & \text{if} \quad x \neq 0 \\ \rho''(0) & \text{if} \quad x = 0 \end{cases} \tag{2.59}$$

and then (2.54) is equivalent to

$$\widehat{\sigma}^2 = \frac{1}{n\delta} \sum_{i=1}^{n} W\left(\frac{x_i}{\widehat{\sigma}}\right) x_i^2. \tag{2.60}$$

It follows that $\widehat{\sigma}$ can be seen as a weighted RMS estimate. For the Student MLE

$$W(x) = \frac{1}{\nu + x^2}, \tag{2.61}$$

and for the bisquare scale

$$W(x) = \min\left\{3 - 3x^2 + x^4, 1/x^2\right\}. \tag{2.62}$$

It is seen that larger x's receive smaller weights.

Note that using $\rho(x/c)$ instead of $\rho(x)$ in (2.54) yields $\widehat{\sigma}/c$ instead of $\widehat{\sigma}$. This can be used to normalize $\widehat{\sigma}$ to have a given asymptotic value, as was done at the end of Section 2.4. If we want $\widehat{\sigma}$ to coincide asymptotically with $\mathrm{SD}(x)$ when x is normal, then (recalling (2.55)) we have to take c as the solution of $\mathrm{E}\rho(x/c) = \delta$ with $x \sim \mathrm{N}(0, 1)$, which can be obtained numerically. For the bisquare scale, the solution is $c = 1.56$.

Although scale M-estimates play an auxiliary role here, their importance will be seen in Chapters 5 and 6.

2.6 M-estimates of location with unknown dispersion

Estimates defined by (2.12) are not scale equivariant, which implies that our results may depend heavily on our measurement units. To fix ideas, assume we want to estimate μ in model (2.1) where F is given by the mixture (2.5) with $G = \mathrm{N}(\mu, \sigma^2)$. *If* σ were known, it would be natural to divide (2.1) by σ to reduce the problem to the case $\sigma = 1$, which implies estimating μ by

$$\widehat{\mu} = \arg\min_{\mu} \sum_{i=1}^{n} \rho\left(\frac{x_i - \mu}{\sigma}\right).$$

It is easy to verify that as in (2.23) for large n the approximate distribution of $\widehat{\mu}$ is $\mathrm{N}(\mu, v/n)$ where

$$v = \sigma^2 \frac{\mathrm{E}\psi((x-\mu)/\sigma)^2}{(\mathrm{E}\psi'((x-\mu)/\sigma))^2}. \tag{2.63}$$

2.6.1 Previous estimation of dispersion

To obtain scale equivariant M-estimates of location, an intuitive approach is to use

$$\widehat{\mu} = \arg\min_{\mu} \sum_{i=1}^{n} \rho\left(\frac{x_i - \mu}{\widehat{\sigma}}\right), \tag{2.64}$$

where $\widehat{\sigma}$ is a previously computed dispersion estimate. It is easy to verify that $\widehat{\mu}$ is indeed scale equivariant. Since $\widehat{\sigma}$ does not depend on μ, (2.64) implies that $\widehat{\mu}$ is a solution of

$$\sum_{i=1}^{n} \psi\left(\frac{x_i - \widehat{\mu}}{\widehat{\sigma}}\right) = 0. \tag{2.65}$$

It is intuitive that $\widehat{\sigma}$ must itself be robust. In Example 1.2, using (2.64) with bisquare ψ with $k = 4.68$, and $\widehat{\sigma} = \text{MADN}(\mathbf{x})$, yields $\widehat{\mu} = 25.56$; using $\widehat{\sigma} = \text{SD}(\mathbf{x})$ instead gives $\widehat{\mu} = 25.12$. Now add to the data set three copies of the lowest value -44. The results change to 26.42 and 17.19. The reason for this change is that the outliers "inflate" the SD, and hence the location estimate attributes to them too much weight.

Note that since k is chosen in order to ensure a given efficiency at the unit normal, if we want $\widehat{\mu}$ to attain the same efficiency at any normal, $\widehat{\sigma}$ must "estimate the SD at the normal", in the sense that if the data are $\text{N}(\mu, \sigma^2)$, then when $n \to \infty$, $\widehat{\sigma}$ tends in probability to σ. This is why we use the normalized median absolute deviation MADN described previously, rather than the un-normalized version MAD.

If a number $m > n/2$ of data values are concentrated at a single value x_0, we have $\text{MAD}(\mathbf{x}) = 0$, and hence the estimate is not defined. In this case we define $\widehat{\mu} = x_0 = \text{Med}(\mathbf{x})$. Besides being intuitively plausible, this definition can be justified by a limit argument. Let the n data values be different, and let m of them tend to x_0. Then it is not difficult to show that in the limit the solution of (2.64) is x_0.

It can be proved that if F is symmetric, then for n large $\widehat{\mu}$ behaves as if $\widehat{\sigma}$ were constant, in the following sense: if $\widehat{\sigma}$ tends in probability to σ, then the distribution of $\widehat{\mu}$ is approximately normal with variance (2.63) (for asymmetric F the asymptotic variance is more complicated; see Section 10.6). Hence the efficiency of $\widehat{\mu}$ does not depend on that of $\widehat{\sigma}$. In Chapter 3 it will be seen, however, that its robustness does depend on that of $\widehat{\sigma}$.

2.6.2 Simultaneous M-estimates of location and dispersion

An alternative approach is to consider a *location–dispersion* model with two unknown parameters

$$x_i = \mu + \sigma u_i \tag{2.66}$$

where u_i has density f_0, and hence x_i has density

$$f(x) = \frac{1}{\sigma} f_0\left(\frac{x-\mu}{\sigma}\right). \tag{2.67}$$

In this case σ is the scale parameter of the random variables σu_i, but it is a dispersion parameter for the x_i.

We now derive the simultaneous MLE of μ and σ in model (2.67), i.e.,

$$(\widehat{\mu}, \widehat{\sigma}) = \arg\max_{\mu,\sigma} \frac{1}{\sigma^n} \prod_{i=1}^{n} f_0\left(\frac{x_i-\mu}{\sigma}\right)$$

which can be written as

$$(\widehat{\mu}, \widehat{\sigma}) = \arg\min_{\mu,\sigma} \left\{ \frac{1}{n} \sum_{i=1}^{n} \rho_0\left(\frac{x_i-\mu}{\sigma}\right) + \log\sigma \right\} \tag{2.68}$$

with $\rho_0 = -\log f_0$. The main interest here is on μ, while σ is a "nuisance parameter".

Proceeding as in the derivations of (2.18) and (2.54) it follows that the MLEs satisfy the system of equations

$$\sum_{i=1}^{n} \psi\left(\frac{x_i-\widehat{\mu}}{\widehat{\sigma}}\right) = 0 \tag{2.69}$$

$$\frac{1}{n}\sum_{i=1}^{n} \rho_{\text{scale}}\left(\frac{x_i-\widehat{\mu}}{\widehat{\sigma}}\right) = \delta, \tag{2.70}$$

where

$$\psi(x) = -\rho_0', \ \ \rho_{\text{scale}}(x) = x\psi(x), \ \ \delta = 1. \tag{2.71}$$

The reason for notation "ρ_{scale}" is that in all instances considered in this book, ρ_{scale} is a ρ-function in the sense of Definition 2.1, and this characteristic is exploited later in Section 5.6.1. The notation will be used whenever it is necessary to distinguish this ρ_{scale} used for scale from the ρ in (2.13) for location; otherwise, we shall write just ρ.

We shall deal in general with simultaneous estimates $(\widehat{\mu}, \widehat{\sigma})$ defined as solutions of systems of equations of the form (2.69)–(2.70) which need not correspond to the MLE for any distribution. It can be proved (see Section 10.5) that for large n the distributions of $\widehat{\mu}$ and $\widehat{\sigma}$ are approximately normal. If F is symmetric then $\mathcal{D}(\widehat{\mu}) \approx \mathrm{N}(\mu, v/n)$ with v given by (2.63), where μ and σ are the solutions of the system

$$\mathrm{E}\,\psi\left(\frac{x-\widehat{\mu}}{\sigma}\right) = 0 \tag{2.72}$$

$$\mathrm{E}\,\rho_{\text{scale}}\left(\frac{x-\mu}{\widehat{\sigma}}\right) = \delta. \tag{2.73}$$

We may choose Huber's or the bisquare function for ψ. A very robust choice for ρ_{scale} is (2.58) with $c = 0.675$ to make it consistent for the SD at the normal, which yields

$$\widehat{\sigma} = \frac{1}{0.675}\text{Med}(|\mathbf{x}-\widehat{\mu}|). \tag{2.74}$$

Although this looks similar to using the previously computed MADN, it will be seen in Chapter 6 that the latter yields more robust results.

In general, estimation with a previously computed dispersion is more robust than simultaneous estimation. However, simultaneous estimation will be useful in more general situations, as will be seen in Chapter 6.

2.7 Numerical computation of M-estimates

There are several methods available for computing M-estimates of location and/or scale. In principle one could use any of the general methods for equation solving such as the Newton–Raphson algorithm, but methods based on derivatives may be unsafe with the types of ρ- and ψ-functions that yield good robustness properties (see Chapter 9). Here we shall describe a computational method called *iterative reweighting* that takes special advantage of the characteristics of the problem.

2.7.1 Location with previously computed dispersion estimation

For the solution of the robust location estimation optimization problem (2.64), the weighted average expression (2.31) suggests an iterative procedure. Start with a robust dispersion estimate $\widehat{\sigma}_0$ (for instance, the MADN) and some initial estimate $\widehat{\mu}_0$ (for instance, the sample median). Given $\widehat{\mu}_k$ compute

$$w_{k,i} = W\left(\frac{x_i - \widehat{\mu}_k}{\widehat{\sigma}}\right) \quad (i = 1, \ldots, n) \tag{2.75}$$

where W is the function in (2.30) and let

$$\widehat{\mu}_{k+1} = \frac{\sum_{i=1}^{n} w_{k,i}x_i}{\sum_{i=1}^{n} w_{k,i}}. \tag{2.76}$$

Results to be proved in Section 9.1 imply that if $W(x)$ is bounded and nonincreasing for $x > 0$, then the sequence $\widehat{\mu}_k$ converges to a solution of (2.64).

The algorithm, which requires a stopping rule based on a tolerance parameter ε, is thus

1. Compute $\widehat{\sigma} = \text{MADN}(x)$ and $\mu_0 = \text{Med}(\mathbf{x})$.
2. For $k = 0, 1, 2, \ldots$, compute the weights (2.75) and then $\widehat{\mu}_{k+1}$ in (2.76).
3. Stop when $\left|\widehat{\mu}_{k+1} - \widehat{\mu}_k\right| < \varepsilon\widehat{\sigma}$.

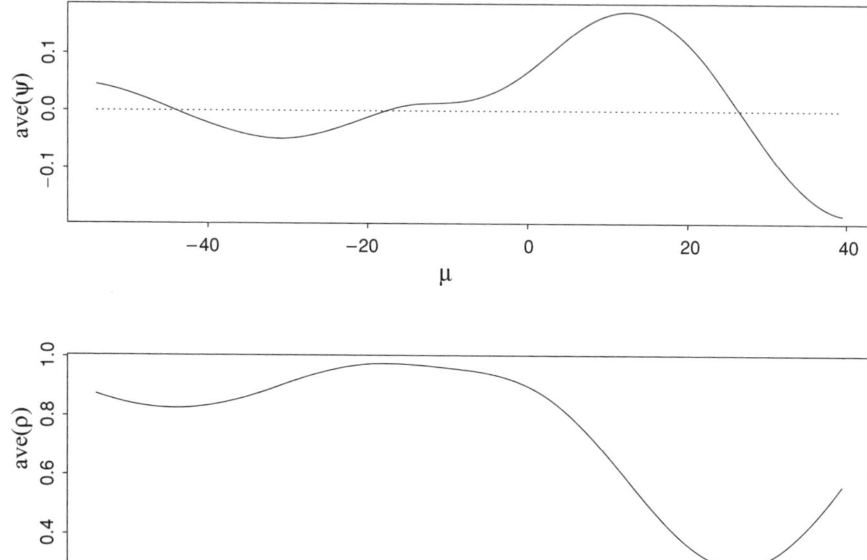

Figure 2.6 Averages of $\psi(x-\mu)$ and $\rho(x-\mu)$ as a function of μ

If ψ is increasing the solution is unique, and the starting point $\widehat{\mu}_0$ influences only the number of iterations. If ψ is redescending then $\widehat{\mu}_0$ must be robust in order to insure convergence to a "good" solution. Choosing $\widehat{\mu}_0 = \text{Med}(\mathbf{x})$ suffices for this purpose.

Figure 2.6 shows the averages of $\psi\left((\mathbf{x}-\mu)/\widehat{\sigma}\right)$ and of $\rho\left((\mathbf{x}-\mu)/\widehat{\sigma}\right)$ as a function of μ, where ψ and ρ correspond to the bisquare estimate with efficiency 0.95, and $\widehat{\sigma} = \text{MADN}$, for the data of Example1.2, to which three extra values of the outlier -44 were added. Three roots of the estimating equation (2.65) are apparent, one of which corresponds to the absolute minimum of (2.64) while the other two correspond to a relative minimum and a relative maximum. This effect occurs also with the original data, but is less visible.

2.7.2 Scale estimates

For solving (2.54), the expression (2.60) suggests an iterative procedure. Start with some $\widehat{\sigma}_0$, for instance, the normalized MAD (MADN). Given $\widehat{\sigma}_k$ compute

$$w_{k,i} = W\left(\frac{x_i}{\widehat{\sigma}_k}\right) \quad (i = 1, \ldots, n) \tag{2.77}$$

where W is the weight function in (2.59) and let

$$\widehat{\sigma}_{k+1} = \sqrt{\frac{1}{n\delta}\sum_{i=1}^{n} w_{k,i}\, x_i^2}. \tag{2.78}$$

Then if $W(x)$ is bounded, even, continuous and nonincreasing for $x > 0$, the sequence σ_N converges to a solution of (2.60) and hence of (2.54) (for a proof see Section 9.4). The algorithm is thus

1. For $k = 0, 1, 2, \ldots,$ compute the weights (2.77) and then $\widehat{\sigma}_{k+1}$ in (2.78).
2. Stop when $|\widehat{\sigma}_{k+1}/\widehat{\sigma}_k - 1| < \varepsilon$.

2.7.3 Simultaneous estimation of location and dispersion

The procedure for (2.69)–(2.70) is a combination of the two former ones. Compute starting values $\widehat{\mu}_0, \widehat{\sigma}_0$, and given $\widehat{\mu}_k, \widehat{\sigma}_k$ compute for $i = 1, \ldots, n$

$$r_{k,i} = \frac{x_i - \widehat{\mu}_k}{\widehat{\sigma}_k}$$

and

$$w_{1k,i} = W_1\left(r_{k,i}\right), \quad w_{2k,i} = W_2(r_{k,i})$$

where W_1 is the weight functions W in (2.30) and W_2 is the W in (2.59) corresponding to ρ_{scale}. Then at the k-th iteration

$$\widehat{\mu}_{k+1} = \frac{\sum_{i=1}^{n} w_{1k,i}x_i}{\sum_{i=1}^{n} w_{1k,i}}, \quad \widehat{\sigma}^2_{k+1} = \frac{\widehat{\sigma}^2_k}{n\delta}\sum_{i=1}^{n} w_{2k,i}\, r_i^2.$$

2.8 Robust confidence intervals and tests

2.8.1 Confidence intervals

Since outliers affect both the sample mean $\overline{x}$ and the sample standard deviation s, confidence intervals for $\mu = \mathrm{E}(x)$ based on normal theory may be unreliable. Outliers may displace $\overline{x}$ and/or "inflate" s, resulting in one or both of the following degradations in performance: (1) the true coverage probability may be much lower than the nominal one; (2) the coverage probability may be either close to or higher than the nominal one, but at the cost of a loss of precision in the form of an inflated expected confidence interval length. We briefly elaborate on these points.

Recall that the usual Student confidence interval, justified by the assumption of a normal distribution for i.i.d. observations, is based on the "t-statistic"

$$T = \frac{\overline{x} - \mu}{s/\sqrt{n}}. \tag{2.79}$$

From this one gets the usual two-sided confidence intervals for μ with level $1-\alpha$

$$\overline{x} \pm t_{n-1,1-\alpha/2}\frac{s}{\sqrt{n}},$$

where $t_{m,\beta}$ is the β-quantile of the t-distribution with m degrees of freedom.

The simplest situation is when the distribution of the data is symmetric about $\mu = \mathrm{E}x$. Then $\mathrm{E}\overline{x} = \mu$ and the confidence interval is centered. However, heavy tails in the distribution will cause the value of s to be inflated, and hence the interval length will be inflated, possibly by a large amount. Thus in the case of symmetric heavy-tailed distributions the price paid for maintaining the target confidence interval error rate α will often be unacceptably long confidence interval lengths. If the data have a mixture distribution $(1-\varepsilon)\mathrm{N}(\mu,\sigma^2)+\varepsilon H$, where H is not symmetric about μ, then the distribution of the data is not symmetric about μ and $\mathrm{E}\overline{x} \neq \mu$. Then the t confidence interval with purported confidence level $1-\alpha$ will not be centered and will not have the error rate α, and will lack robustness of both level and length. If the data distribution is both heavy tailed and asymmetric, then the t confidence interval can fail to have the target error rate and at the same time have unacceptably large interval lengths. Thus the classic t confidence interval lacks robustness of both error rate (confidence level) and length, and we need confidence intervals with both types of robustness.

Approximate confidence intervals for a parameter of interest can be obtained from the asymptotic distribution of a parameter estimate. Robust confidence intervals that are not much influenced by outliers can be obtained by imitating the form of the classical Student t confidence interval, but replacing the average and SD by robust location and dispersion estimates. Consider the M-estimates $\widehat{\mu}$ in Section 2.6, and recall that if $\mathcal{D}(x)$ is symmetric then for large n the distribution of $\widehat{\mu}$ is approximately $\mathrm{N}(\mu, v/n)$ with v given by (2.63). Since v is unknown, an estimate $\hat{v}$ may be obtained by replacing the expectations in (2.63) by sample averages, and the parameters by their estimates:

$$\hat{v} = \widehat{\sigma}^2 \frac{\mathrm{ave}[\psi((\mathbf{x}-\widehat{\mu})/\widehat{\sigma})]^2}{(\mathrm{ave}[\psi'((\mathbf{x}-\widehat{\mu})/\widehat{\sigma})])^2}. \tag{2.80}$$

A robust approximate t-statistic ("Studentized M-estimate") is then defined as

$$T = \frac{\widehat{\mu}-\mu}{\sqrt{\widehat{v}/n}} \tag{2.81}$$

and its distribution is approximately normal N(0, 1) for large n . Thus a robust approximate interval can then be computed as

$$\widehat{\mu} \pm z_{1-\alpha/2}\sqrt{\frac{\widehat{v}}{n}}$$

where z_β denotes the β-quantile of N(0, 1).

Table 2.4 Confidence intervals for flour data

Estimate	$\widehat{\mu}$	$\sqrt{\widehat{v}(\widehat{\mu})}$	Interval	
Mean	4.280	1.081	2.161	6.400
Bisquare M	3.144	0.130	2.885	3.404
$\overline{x}_{0.25}$	3.269	0.085	3.103	3.435

A similar procedure can be used for the trimmed mean. Recall that the asymptotic variance of the α-trimmed mean for symmetric F is (2.41). We can estimate v with

$$\widehat{v} = \frac{1}{n-2m}\left(\sum_{i=m+1}^{n-m}(x_{(i)}-\widehat{\mu})^2 + m(x_{(m)}-\widehat{\mu})^2 + m(x_{(n-m+1)}-\widehat{\mu})^2\right). \tag{2.82}$$

An approximate t-statistic is then defined as (2.81). Note again that the variance of the trimmed mean is not a trimmed variance: all values larger than $x_{(n-m)}$ or smaller than $x_{(m+1)}$ are not omitted but replaced by $x_{(n-m+1)}$ or $x_{(m)}$, respectively.

Table 2.4 gives for the data of Example 1.1 the location estimates, their estimated asymptotic SDs and the respective confidence intervals with level 0.95.

2.8.2 Tests

It appears that many applied statisticians have the impression that t-tests are sufficiently "robust" that there is nothing to worry about when using such tests. Again, this impression no doubt comes from the fact—which is a consequence of the central limit theorem—that it suffices for the data to have finite variance for the classical t-statistic (2.79) to be approximately N(0, 1) in large samples. See for example the discussion to this effect in the introductory text by Box, Hunter and Hunter (1978). This means that in large samples the Type 1 error rate of a level α is in fact α for testing a null hypothesis about the value of μ. However, this fact is misleading, as we now demonstrate.

Recall that the t-test with level α for the null hypothesis $H_0 = \{\mu = \mu_0\}$ rejects H_0 when the t-interval with confidence level $1 - \alpha$ does not contain μ_0. According to the former discussion on the behavior of the t-intervals under contamination, we conclude that if the data are symmetric but heavy tailed, the intervals will be longer than necessary, with the consequence that the actual Type 1 error rate may be much smaller than α, but the Type 2 error rate may be too large, i.e., the test will have low power. If the contaminated distribution is asymmetric and heavy tailed, both errors may become unacceptably high.

Robust tests can be derived from a "robust t-statistic" (2.81) in the same way as was done with confidence intervals. The tests of level α for the null hypothesis $\mu = \mu_0$ against the two-sided alternative $\mu \neq \mu_0$ and the one-sided alternative $\mu > \mu_0$ have the rejection regions

$$|\widehat{\mu} - \mu_0| > \sqrt{\widehat{v}} z_{1-\alpha/2} \text{ and } \widehat{\mu} > \mu_0 + \sqrt{\widehat{v}} z_{1-\alpha}, \tag{2.83}$$

respectively.

The robust t-like confidence intervals and test are easy to apply. They have, however, some drawbacks when the contamination is asymmetric due to the bias of the estimate. Procedures that ensure a given probability of coverage or Type 1 error probability for a contaminated parametric model were given by Huber (1965, 1968), Huber-Carol (1970), Rieder (1978, 1981) and Fraiman, Yohai and Zamar (2001). Yohai and Zamar (2004) developed tests and confidence intervals for the median which are "nonparametric" in the sense that their level is valid for arbitrary distributions. Further references on robust tests will be given in Section 4.7.

2.9 Appendix: proofs and complements

2.9.1 Mixtures

Let the density f be given by

$$f = (1 - \varepsilon)g + \varepsilon h. \tag{2.84}$$

This is called a *mixture* of g and h. If the variable x has density f and q is any function, then

$$\mathrm{E}q(x) = \int_{-\infty}^{\infty} q(x)f(x)dx = (1 - \varepsilon)\int_{-\infty}^{\infty} q(x)g(x)dx + \varepsilon \int_{-\infty}^{\infty} q(x)h(x)dx.$$

With this expression we can calculate $\mathrm{E}x$; the variance is obtained from

$$\mathrm{Var}(x) = \mathrm{E}(x^2) - (\mathrm{E}x)^2 .$$

If $g = \mathrm{N}(0, 1)$ and $h = \mathrm{N}(a, b^2)$ then

$$\mathrm{E}x = \varepsilon a \text{ and } \mathrm{E}x^2 = (1 - \varepsilon) + \varepsilon(a^2 + b^2),$$

and hence

$$\mathrm{Var}(x) = (1 - \varepsilon)(1 + \varepsilon a^2) + \varepsilon b^2. \tag{2.85}$$

Evaluating the performance of robust estimates requires simulating distributions of the form (2.84). This is easily accomplished: generate u with uniform distribution in $(0, 1)$; if $u \geq \varepsilon$, generate x with distribution g, else generate x with distribution h.

2.9.2 Asymptotic normality of M-estimates

In this section we give a heuristic proof of (2.23). To this end we begin with an intuitive proof of (2.22). Define the functions

$$\lambda(s) = \mathrm{E}\psi(x-s),\ \widehat{\lambda}_n(s) = \frac{1}{n}\sum_{i=1}^{n}\psi(x_i-s),$$

so that $\widehat{\mu}$ and μ_0 verify respectively

$$\widehat{\lambda}_n(\widehat{\mu}) = 0,\ \lambda(\mu_0) = 0.$$

For each s, the random variables $\psi(x_i - s)$ are i.i.d. with mean $\lambda(s)$, and hence the law of large numbers implies that when $n \to \infty$

$$\widehat{\lambda}_n(s) \to_p \lambda(s)\ \forall s.$$

It is intuitive that also the solution of $\widehat{\lambda}_n(s) = 0$ should tend to that of $\lambda(s) = 0$. This can in fact be proved rigorously (see Theorem 10.5).

Now we prove (2.23). Taking the Taylor expansion of order 1 of (2.18) as a function of $\widehat{\mu}$ about μ_0 yields

$$0 = \sum_{i=1}^{n}\psi(x_i-\mu_0) - (\widehat{\mu}-\mu_0)\sum_{i=1}^{n}\psi'(x_i-\mu_0) + o(\widehat{\mu}-\mu_0) \tag{2.86}$$

where the last ("second-order") term is such that

$$\lim_{t\to 0}\frac{o(t)}{t} = 0.$$

Dropping the last term in (2.86) yields

$$\sqrt{n}\,(\widehat{\mu}-\mu_0) \approx \frac{A_n}{B_n}, \tag{2.87}$$

with

$$A_n = \sqrt{n}\,\mathrm{ave}(\psi(\mathbf{x}-\mu_0)),\ B_n = \mathrm{ave}(\psi'(\mathbf{x}-\mu_0)).$$

The random variables $\psi(x_i - \mu_0)$ are i.i.d. with mean 0 because of (2.21). The central limit theorem implies that the distribution of A_n tends to $\mathrm{N}(0, a)$ with $a = \mathrm{E}\psi(x-\mu_0)^2$, and the law of large numbers implies that B_n tends in probability to $b = \mathrm{E}\psi'(x-\mu_0)$. Hence by Slutsky's lemma (see Section 2.9.3) A_n/B_n can be replaced for large n by A_n/b, which tends in distribution to $\mathrm{N}(0, a/b^2)$, as stated. A rigorous proof will be given in Theorem 10.7.

Note that we have shown that $\sqrt{n}(\widehat{\mu}-\mu_0)$ converges in distribution; this is expressed by saying that "$\widehat{\mu}$ has order $n^{-1/2}$ consistency".

2.9.3 Slutsky's lemma

Let u_n and v_n be two sequences of random variables such that u_n tends in probability to a constant u, and the distribution of v_n tends to the distribution of a variable v (abbreviated "$v_n \to_d v$"). Then

$$u_n + v_n \to_d u + v \text{ and } u_n v_n \to_d uv.$$

The proof can be found in Bickel and Doksum (2001, p. 467) or Shao (2003, p. 60).

2.9.4 Quantiles

For $\alpha \in (0, 1)$ and F a continuous and increasing distribution function, the α-quantile of F is the unique value $q(\alpha)$ such that $F(q(\alpha)) = \alpha$. If F is discontinuous, such a value might not exist. For this reason we define $q(\alpha)$ in general as a value where $F(t) - \alpha$ changes sign, i.e.,

$$\operatorname{sgn}\left\{\lim_{t \uparrow q(\alpha)} (F(t) - \alpha)\right\} \neq \operatorname{sgn}\left\{\lim_{t \downarrow q(\alpha)} (F(t) - \alpha)\right\},$$

where "$\uparrow$" and "$\downarrow$" denote the limits from the left and from the right, respectively. It is easy to show that such a value always exists. It is unique if F is increasing. Otherwise, it is not necessarily unique, and hence we may speak of *an* α-quantile.

If x is a random variable with distribution function $F(t) = \mathrm{P}(x \leq t)$, $q(\alpha)$ will also be considered as an α-quantile of the variable x, and in this case is denoted by x_α.

If g is a monotonic function, and $y = g(x)$, then

$$g(x_\alpha) = \begin{cases} y_\alpha & \text{if} \quad g \text{ is increasing} \\ y_{1-\alpha} & \text{if} \quad g \text{ is decreasing,} \end{cases} \tag{2.88}$$

in the sense that, for example, if z is *an* α-quantile of x, then z^3 is *an* α-quantile of x^3.

When the α-quantile is not unique, there exists an interval $[a, b)$ such that $F(t) = \alpha$ for $t \in [a, b)$. We may obtain uniqueness by defining $q(\alpha)$ as a—the smallest α-quantile—and then (2.88) remains valid. It seems more symmetric to define it as the midpoint $(a + b)/2$, but then (2.88) ceases to hold.

2.9.5 Alternative algorithms for M-estimates

The Newton–Raphson procedure

The Newton–Raphson procedure is a widely used iterative method for the solution of nonlinear equations. To solve the equation $h(t) = 0$, at each iteration h is "linearized", i.e., replaced by its Taylor expansion of order 1 about the current approximation. Thus, if at iteration m we have the approximation t_m, then the next value t_{m+1} is the solution of

$$h(t_m) + h'(t_m)(t_{m+1} - t_m) = 0,$$

i.e.,

$$t_{m+1} = t_m - \frac{h(t_m)}{h'(t_m)}. \tag{2.89}$$

If the procedure converges, the convergence is very fast; but it is not guaranteed to converge. If h' is not bounded away from zero, the denominator in (2.89) may become very small, making the sequence t_m unstable unless the initial value t_0 is very near to the solution.

This happens in the case of a location M-estimate, where we must solve the equation $h(\mu) = 0$ with $h(\mu) = \text{ave}\{\psi(\mathbf{x} - \mu)\}$. Here the iterations are

$$\mu_{m+1} = \mu_m + \frac{\sum_{i=1}^n \psi(x_i - \mu_m)}{\sum_{i=1}^n \psi'(x_i - \mu_m)}. \tag{2.90}$$

If ψ is bounded, its derivative ψ' tends to zero at infinity, and hence the denominator is not bounded away from zero, which makes the procedure unreliable.

For this reason we prefer the algorithms based on iterative reweighting, which are guaranteed to converge.

However, although the result of iterating the Newton–Raphson process indefinitely may be unreliable, the result of a *single* iteration may be a robust and efficient estimate, if the initial value μ_0 is robust but not necessarily efficient, like the median. See Problem 3.16.

Iterative pseudo-observations

The expression (2.33) of an M-estimate as a function of the pseudo-observations (2.34) can be used as the basis for an iterative procedure to compute a location estimate with previous dispersion $\widehat{\sigma}$. Starting with an initial $\widehat{\mu}_0$, define

$$\widehat{\mu}_{m+1} = \frac{1}{n}\sum_{i=1}^{n} \zeta(x_i, \widehat{\mu}_m, \widehat{\sigma}), \tag{2.91}$$

where

$$\zeta(x, \mu, \sigma) = \mu + \sigma\psi\left(\frac{x-\mu}{\sigma}\right). \tag{2.92}$$

It can be shown that μ_m converges under very general conditions to the solution of (2.65) (Huber, 1981). However, the convergence is much slower than that corresponding to the reweighting procedure.

2.10 Problems

2.1. Show that in a sample of size n from a contaminated distribution (2.5), the number of observations from H is random, with binomial distribution Bi (n, ε).

2.2. For the data of Example 1.2, compute the mean and median, the 25% trimmed mean and the M-estimate with previous dispersion and Huber's ψ with $k = 1.37$. Use the last to derive a 90% confidence interval for the true value.

2.3. Verify (2.10) using (2.26).

2.4. For which values of ν has the Student distribution moments of order k?

2.5. Show that if μ is a solution of (2.18), then $\mu + c$ is a solution of (2.18) with $x_i + c$ instead of x_i.

2.6. Show that if $x = \mu_0 + u$ where the distribution of u is symmetric about zero, then μ_0 is a solution of (2.21).

2.7. Verify (2.29) [hint: use $\varphi'(x) = -x\varphi(x)$ and integration by parts]. From this, find the values of k which yield variances $1/\alpha$ with $\alpha = 0.90$, 0.95 and 0.99 (by using an equation solver, or just trial and error).

2.8. Compute the α-trimmed means with $\alpha = 0.10$ and 0.25 for the data of Example 1.2.

2.9. Show that if ψ is odd, then the M-estimate $\widehat{\mu}$ with fixed σ satisfies conditions C1–C2–C3 at the end of Section 2.2.

2.10. Show using (2.44) that L-estimates are shift and scale equivariant [recall that the order statistics of $y_i = -x_i$ are $y_{(i)} = -x_{(n-i+1)}$!] and fulfill also C1–C2–C3 of Section 2.3.

2.11. If $x \sim \mathrm{N}(\mu, \sigma^2)$, calculate MD($x$), MAD($x$) and IQR($x$).

2.12. Show that if $\psi = \rho'$ vanishes identically outside an interval, there is no density verifying (2.13).

2.13. Define the sample α-quantile of $x_1, \ldots, x_n$—with $\alpha \in (1/n, 1 - 1/n)$—as $x_{(k)}$, where k is the smallest integer $\geq n\alpha$ and $x_{(i)}$ are the order statistics (1.2). Let

$$\psi(x) = \alpha \mathrm{I}(x > 0) - (1 - \alpha)\mathrm{I}(x < 0).$$

Show that $\widehat{\mu} = x_{(k)}$ is a solution (not necessarily unique) of (2.18). Use this fact to derive the asymptotic distribution of sample quantiles, assuming that $\mathcal{D}(x_i)$ has a unique α-quantile. Note that this ψ is not odd!.

2.14. Show that the M-scale (2.54) with $\rho(t) = \mathrm{I}(|t| > 1)$ is the h-th order statistic of the $|x_i|$ with $h = n - [n\delta]$.

2.15. Verify (2.36), (2.56) and (2.61).

2.16. Let $[a, b]$, where a and b depend on the data, be the shortest interval containing at least half of the data.

(a) The *Shorth* ("shortest half") location estimate is defined as the midpoint $\widehat{\mu} = (a + b)/2$. Show that $\widehat{\mu} = \arg\min_\mu \mathrm{Med}(|\mathbf{x} - \mu|)$.

(b) Show that the difference $b - a$ is a dispersion estimate.

(c) For a distribution F, let $[a, b]$ be the shortest interval with probability 0.5. Find this interval for $F = \mathrm{N}\left(\mu, \sigma^2\right)$.

2.17. Let $\widehat{\mu}$ be a location M-estimator. Show that if the distribution of the x_i's is symmetric about μ, so is the distribution of $\widehat{\mu}$, and that the same happens with trimmed means.

2.18. Verify numerically that the constant c at the end of Section 2.5 that makes the bisquare scale consistent at the normal is indeed equal to 1.56.

2.19. Show that

(a) if the sequence μ_m in (2.90) converges, then the limit is a solution of (2.18)

(b) if the sequence in (2.91) converges, then the limit is a solution of (2.65).

3

Measuring Robustness

We have seen in Chapter 2 that while in the classical approach to statistics one aims at estimates which have desirable properties at an exactly specified model, the aim of robust methods is loosely speaking to develop estimates which have a "good" behavior in a "neighborhood" of a model. This notion will now be made precise.

To gain some insight before giving more formal definitions, we use an artificial data set x by generating $n = 20$ random N(0,1) numbers. To measure the effect of different locations of an outlier, we add an extra data point x_0 which is allowed to range on the whole line. The *sensitivity curve* of the estimate $\widehat{\mu}$ for the sample $x_1, \ldots, x_n$ is the difference

$$\widehat{\mu}(x_1, \ldots, x_n, x_0) - \widehat{\mu}(x_1, \ldots, x_n)$$

as a function of the location x_0 of the outlier.

Figure 3.1 plots the sensitivity curves of the median, the 25% trimmed mean $\overline{x}_{0.25}$, the Huber M-estimate with $k = 1.37$ using both the SD and the MADN as previously computed dispersion estimates, and the bisquare M-estimate with $k = 4.68$ using the MADN as dispersion.

We see that all curves are bounded, except the one corresponding to the Huber estimate with the SD as dispersion, which grows without bound with x_0. The same unbounded behavior (not shown in the figure) occurs with the bisquare estimate with the SD as dispersion. This shows the importance of a robust previous dispersion. All curves are nondecreasing for positive x_0, except the one for the bisquare estimate. Loosely speaking, we say that the bisquare M-estimate *rejects* extreme values, while the others do not. The curve for the trimmed mean shows that it does not reject large observations, but just limits their influence. The curve for the median is very steep at the origin.

Robust Statistics – Theory and Methods Ricardo A. Maronna, R. Douglas Martin and Víctor J. Yohai

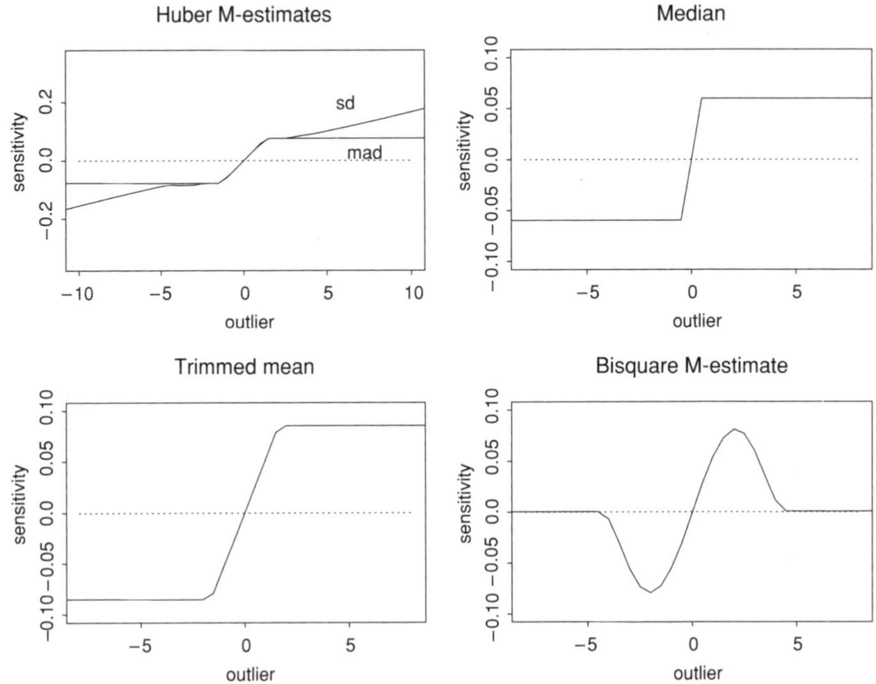

Figure 3.1 Sensitivity curves of location estimates

Figure 3.2 shows the sensitivity curves of the SD along with the normalized MD, MAD and IQR. The SD and MD have unbounded sensitivity curves while those of the normalized MAD and IQR are bounded.

Imagine now that instead of adding a single point at a variable location, we replace m points by a fixed value $x_0 = 1000$. Table 3.1 shows the resulting "biases"

$$\widehat{\mu}(x_0, x_0, \ldots, x_0, x_{m+1}, .., x_n) - \widehat{\mu}(x_1, .., x_n)$$

as a function of m for the following location estimates: the median; the Huber estimate with $k = 1.37$ and three different dispersions, namely, previously estimated MAD (denoted by MADp), simultaneous MAD ("MADs") and previous SD; the trimmed mean with $\alpha = 0.085$; and the bisquare estimate. Also we provide the biases for the normalized MAD and IQR dispersion estimates. The choice of k and α was made in order that both the Huber estimates and the trimmed mean have the same asymptotic variance at the normal distribution.

The mean deteriorates immediately when $m = 1$ as expected, and since $[\alpha n] = [0.085 \times 20] = 1$ the trimmed mean $\overline{x}_\alpha$ deteriorates when $m = 2$, as could be expected. The H(MADs) deteriorates rapidly starting at $m = 8$, while H(SD) is already quite bad at $m = 1$. By contrast the median, H(MADp) and M-Bisq do so only when

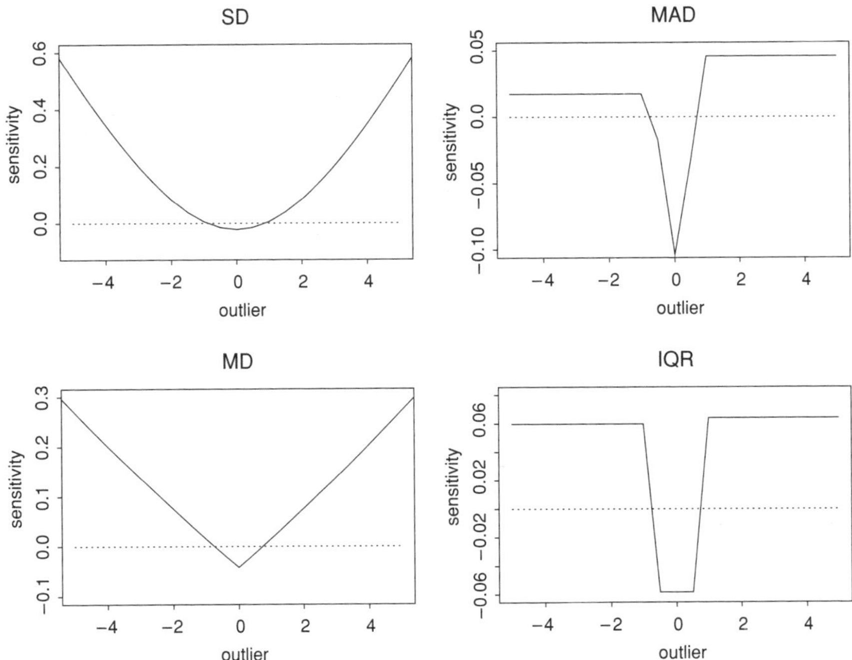

Figure 3.2 Sensitivity curves of dispersion estimates

$m = n/2$, with M-Bisq having smaller bias than H(MADp) and the median (Med) having small biases comparable to those of the M-Bisq (only slightly higher bias than M-Bisq at $m = 4, 5, 7, 9$).

To formalize these notions, it will be easier to study the behavior of estimates when the sample size tends to infinity ("asymptotic behavior"). Consider an estimate $\widehat{\theta}_n = \widehat{\theta}_n(x)$ depending on a sample $\mathbf{x} = \{x_1, \ldots, x_n\}$ of size n of i.i.d. variables with distribution F. In all cases of practical interest, there is a value depending on F,

Table 3.1 The effect of increasing contamination on a sample of size 20

m	Mean	Med	H(MADp)	H(MADs)	H(SD)	$\overline{x}_\alpha$	M-Bisq	MAD	IQR
1	50	0.00	0.03	0.04	16.06	0.04	−0.02	0.12	0.08
2	100	0.01	0.10	0.11	46.78	55.59	0.04	0.22	0.14
4	200	0.21	0.36	0.37	140.5	166.7	0.10	0.46	0.41
5	250	0.34	0.62	0.95	202.9	222.3	0.15	0.56	370.3
7	350	0.48	1.43	42.66	350.0	333.4	0.21	1.29	740.3
9	450	0.76	3.23	450.0	450.0	444.5	0.40	2.16	740.2
10	500	500.5	500.0	500.0	500.0	500.0	500.0	739.3	740.2

$\widehat{\theta}_\infty = \widehat{\theta}_\infty(F)$, such that

$$\widehat{\theta}_n \to_p \widehat{\theta}_\infty(F).$$

$\widehat{\theta}_\infty(F)$ is the *asymptotic value* of the estimate at F.

If $\widehat{\theta}_n = \overline{x}$ (the sample mean) then $\widehat{\theta}_\infty(F) = \mathrm{E}_F x$ (the distribution mean), and if $\widehat{\theta}_n(\mathbf{x}) = \mathrm{Med}(\mathbf{x})$ (the sample median) then $\widehat{\theta}_\infty(F) = F^{-1}(0.5)$ (the distribution median). If $\widehat{\theta}_n$ is a location M-estimator given by (2.18) with ψ monotonic, it was stated in Section 2.9.2 that $\widehat{\theta}_\infty(F)$ is the solution of

$$\mathrm{E}_F \psi(x - \theta) = 0.$$

A proof is given in Theorem 10.5. The same reasoning shows that if $\widehat{\theta}_n$ is a scale M-estimator (2.54), then $\widehat{\theta}_\infty(F)$ is the solution of

$$\mathrm{E}_F \rho\left(\frac{x}{\theta}\right) = \delta.$$

It can also be shown that if $\widehat{\theta}_n$ is a location M-estimator given by (2.12), then $\widehat{\theta}_\infty(F)$ is the solution of

$$\mathrm{E}_F \rho(x - \theta) = \min.$$

See Section 6.2 of Huber (1981). Asymptotic values exist also for the trimmed mean (Section 10.7).

The typical distribution of data depends on one or more unknown parameters. Thus in the location model (2.2) the data have distribution function $F_\mu(x) = F_0(x - \mu)$, and in the location–dispersion model (2.66) the distribution is $F_\theta(x) = F_0((x - \mu)/\sigma)$ with $\theta = (\mu, \sigma)$. These are called *parametric models*. In the location model we have seen in (2.22) that if the data are symmetric about μ and $\widehat{\mu}$ is an M-estimate, then $\widehat{\mu} \to_p \mu$ and so $\widehat{\mu}_\infty(F_\mu) = \mu$. An estimator $\widehat{\theta}$ of the parameter(s) of a parametric family F_θ will be called *consistent* if

$$\widehat{\theta}_\infty(F_\theta) = \theta. \tag{3.1}$$

Since we assume F to be only approximately known, we are interested in the behavior of $\widehat{\theta}_\infty(F)$ when F ranges over a "neighborhood" of a distribution F_0. There are several ways to characterize neighborhoods. The easiest to deal with are *contamination neighborhoods*:

$$\mathcal{F}(F, \varepsilon) = \{(1 - \varepsilon)F + \varepsilon G : G \in \mathcal{G}\} \tag{3.2}$$

where $\mathcal{G}$ is a suitable set of distributions, often the set of all distributions but in some cases the set of point mass distributions, where the "point mass" δ_{x_0} is the distribution such that $\mathrm{P}(x = x_0) = 1$.

3.1 The influence function

The *influence function* (IF) of an estimator (Hampel, 1974) is an asymptotic version of its sensitivity curve. It is an approximation to the behavior of $\widehat{\theta}_\infty$ when the sample contains a small fraction ε of identical outliers. It is defined as

$$\mathrm{IF}_{\widehat{\theta}}(x_0, F) = \lim_{\varepsilon \downarrow 0} \frac{\widehat{\theta}_\infty\left((1-\varepsilon)F + \varepsilon\delta_{x_0}\right) - \widehat{\theta}_\infty(F)}{\varepsilon} \tag{3.3}$$

$$= \frac{\partial}{\partial \varepsilon} \widehat{\theta}_\infty\left((1-\varepsilon)F + \varepsilon\delta_0\right)\Big|_{\varepsilon\downarrow 0}, \tag{3.4}$$

where δ_{x_0} is the point-mass at x_0 and "$\downarrow$ " stands for "limit from the right". If there are p unknown parameters, then $\widehat{\theta}_\infty$ is a p-dimensional vector and so is its IF. Henceforth, the argument of $\widehat{\theta}_\infty(F)$ will be dropped if there is no ambiguity.

The quantity $\widehat{\theta}_\infty\left((1-\varepsilon)F + \varepsilon\delta_{x_0}\right)$ is the asymptotic value of the estimate when the underlying distribution is F and a fraction ε of outliers is equal to x_0. Thus if ε is small this value can be approximated by

$$\widehat{\theta}_\infty\left((1-\varepsilon)F + \varepsilon\delta_{x_0}\right) \approx \widehat{\theta}_\infty(F) + \varepsilon \mathrm{IF}_{\widehat{\theta}}(x_0, F)$$

and the *bias* $\widehat{\theta}_\infty\left((1-\varepsilon)F + \varepsilon\delta_{x_0}\right) - \widehat{\theta}_\infty(F)$ is approximated by $\varepsilon\mathrm{IF}_{\widehat{\theta}}(x_0, F)$.

The IF may be considered as a "limit version" of the sensitivity curve, in the following sense. When we add the new observation x_0 to the sample $x_1, \ldots, x_n$ the fraction of contamination is $1/(n+1)$, and so we define the *standardized sensitivity curve* (SC) as

$$\mathrm{SC}_n(x_0) = \frac{\widehat{\theta}_{n+1}(x_1, \ldots, x_n, x_0) - \widehat{\theta}_n(x_1, \ldots, x_n)}{1/(n+1)},$$

$$= (n+1)\left(\widehat{\theta}_{n+1}(x_1, \ldots, x_n, x_0) - \widehat{\theta}_n(x_1, \ldots, x_n)\right)$$

which is similar to (3.3) with $\varepsilon = 1/(n+1)$. One would expect that if the x_i's are i.i.d. with distribution F, then $\mathrm{SC}_n(x_0) \approx \mathrm{IF}(x_0, F)$ for large n. This notion can be made precise. Note that for each x_0, $\mathrm{SC}_n(x_0)$ is a random variable. Croux (1998) has shown that if $\widehat{\theta}$ is a location M-estimate with a bounded and continuous ψ-function, or is a trimmed mean, then for each x_0

$$\mathrm{SC}_n(x_0) \rightarrow_{a.s.} \mathrm{IF}_{\widehat{\theta}}(x_0, F), \tag{3.5}$$

where "a.s." denotes convergence with probability 1 ("almost sure" convergence). This result is extended to general M-estimates in Section 10.4. See, however, the remarks in Section 3.1.1.

It will be shown in Section 3.8.1 that for a location M-estimate $\widehat{\mu}$

$$\mathrm{IF}_{\widehat{\mu}}(x_0, F) = \frac{\psi(x_0 - \widehat{\mu}_\infty)}{\mathrm{E}\psi'(x - \widehat{\mu}_\infty)} \tag{3.6}$$

and for a scale M-estimate $\widehat{\sigma}$ (Section 2.5)

$$\mathrm{IF}_{\widehat{\sigma}}(x_0, F) = \widehat{\sigma}_\infty \frac{\rho\left(x_0/\widehat{\sigma}_\infty\right) - \delta}{\mathrm{E}\left(x/\widehat{\sigma}_\infty\right)\rho'\left(x/\widehat{\sigma}_\infty\right)}. \tag{3.7}$$

For the median estimate the denominator is to be interpreted as in (2.26). The similarity between the IF and the SC of a given estimator can be seen by comparing Figure 3.1 to Figures 2.3 and 2.5. The same thing happens with Figure 3.2.

We see above that the IF of an M-estimate is proportional to its ψ-function (or an offset ρ-function in the case of the scale estimate), and this behavior holds in general for M-estimates. Given a parametric model F_θ, a *general M-estimate* $\widehat{\theta}$ is defined as a solution of

$$\sum_{i=1}^{n} \Psi(x_i, \widehat{\theta}) = 0. \tag{3.8}$$

For location $\Psi(x,\theta) = \psi(x-\theta)$, and for scale $\Psi(x,\theta) = \rho(x/\theta) - \delta$. It is shown in Section 10.2 that the asymptotic value $\widehat{\theta}_\infty$ of the estimate at F satisfies

$$\mathrm{E}_F \Psi(x, \widehat{\theta}_\infty) = 0. \tag{3.9}$$

It is shown in Section 3.8.1 that the IF of a general M-estimate is

$$\mathrm{IF}_{\widehat{\theta}}(x_0, F) = -\frac{\Psi(x_0, \widehat{\theta}_\infty)}{B(\widehat{\theta}_\infty, \Psi)} \tag{3.10}$$

where

$$B(\theta, \Psi) = \frac{\partial}{\partial\theta} \mathrm{E}\Psi(x,\theta) \tag{3.11}$$

and thus the IF is proportional to the ψ-function $\Psi(x_0, \widehat{\theta}_\infty)$.

If Ψ is differentiable with respect to θ, and the conditions that allow the interchange of derivative and expectation hold, then

$$B(\theta, \Psi) = \mathrm{E}\dot{\Psi}(x,\theta) \tag{3.12}$$

where

$$\dot{\Psi}(x,\theta) = \frac{\partial \Psi(x,\theta)}{\partial\theta}. \tag{3.13}$$

The proof is given in Section 3.8.1. Then if $\widehat{\theta}$ is consistent for the parametric family F_θ, (3.10) becomes

$$\mathrm{IF}_{\widehat{\theta}}(x_0, F_\theta) = -\frac{\Psi(x_0, \theta)}{\mathrm{E}_F \dot{\Psi}(x,\theta)}.$$

Consider now an M-estimate $\widehat{\mu}$ of location with known dispersion σ, where the asymptotic value $\widehat{\mu}_\infty$ satisfies

$$\mathrm{E}_F \psi\left(\frac{x - \widehat{\mu}_\infty}{\sigma}\right) = 0.$$

It is easy to show, by applying (3.6) to the estimate defined by the function $\psi^*(x) = \psi(x/\sigma)$, that the IF of $\widehat{\mu}$ is

$$\mathrm{IF}_{\widehat{\mu}}(x_0, F) = \sigma \frac{\psi\left((x_0 - \widehat{\mu}_\infty)/\sigma\right)}{\mathrm{E}_F \psi'\left((x - \widehat{\mu}_\infty)/\sigma\right)}. \tag{3.14}$$

Now consider location estimation with a previously computed dispersion estimate $\widehat{\sigma}$ as in (2.64). In this case the IF is much more complicated than the one above, and depends on the IF of $\widehat{\sigma}$. But it can be proved that if F is symmetric, the IF simplifies to (3.14):

$$\mathrm{IF}_{\widehat{\mu}}(x_0, F) = \widehat{\sigma}_\infty \frac{\psi\left(\left(x_0 - \widehat{\mu}_\infty\right)/\widehat{\sigma}_\infty\right)}{\mathrm{E}_F\psi'\left(\left(x - \widehat{\mu}_\infty\right)/\widehat{\sigma}_\infty\right)}. \tag{3.15}$$

The IF for simultaneous estimation of μ and σ is more complicated, and can be derived from (3.47) in Section 3.6.

It can be shown that the IF of an α-trimmed mean $\widehat{\mu}$ at a symmetric F is proportional to Huber's ψ-function:

$$\mathrm{IF}_{\widehat{\mu}}(x_0, F) = \frac{\psi_k(x - \widehat{\mu}_\infty)}{1 - 2\alpha} \tag{3.16}$$

with $k = F^{-1}(1 - \alpha)$. Hence the trimmed mean and the Huber estimate in the example at the beginning of the chapter have not only the same asymptotic variances, but also the same IF. However, Table 3.1 shows that they have very different degrees of robustness.

Comparing (3.6) to (2.23) and (3.16) to (2.41), one sees that the asymptotic variance v of these M-estimates satisfies

$$v = \mathrm{E}_F \mathrm{IF}(x, F)^2. \tag{3.17}$$

It is shown in Section 3.7 that (3.17) holds for a general class of estimates called Fréchet-differentiable estimates, which includes M-estimates with bounded Ψ. However, the relationship (3.17) does not hold in general. For instance, the Shorth location estimate the midpoint of the shortest half of the data, see Problem 2.16(a) has a null IF (Problem 12). At the same time, its rate of consistency is $n^{-1/3}$ rather than the usual rate $n^{-1/2}$. Hence the left-hand side of (3.17) is infinite and the right-hand one is zero.

3.1.1 *The convergence of the SC to the IF

The plot in the upper left panel of Figure 3.1 for the Huber estimate using the SD as the previously computed dispersion estimate seems to contradict the convergence of $\mathrm{SC}_n(x_0)$ to $\mathrm{IF}(x_0)$. Note, however, that (3.5) asserts only the convergence for *each* x_0. This means that $\mathrm{SC}_n(x_0)$ will be near $\mathrm{IF}(x_0)$ for a given x_0 when n is sufficiently large, but the value of n will in general depend on x_0, i.e., the convergence will not be *uniform.* Rather than the convergence at an isolated point, what matters is being able to compare the influence of outliers at different locations; that is, the behavior of the *whole curve* corresponding to the SC. Both curves will be similar along their whole range only if the convergence is uniform. This does not happen with H(SD).

On the other hand Croux (1998) has shown that when $\widehat{\theta}$ is the median, the distribution of $\mathrm{SC}_n(x_0)$ does not converge in probability to any value, and hence (3.5) does not hold. This would seem to contradict the upper right panel of Figure 3.1. However, the *form* of the curve converges to the right limit in the sense that for each x_0

$$\frac{\mathrm{SC}_n(x_0)}{\max_x |\mathrm{SC}_n(x)|} \rightarrow_{a.s.} \frac{\mathrm{IF}(x_0)}{\max_x |\mathrm{IF}(x)|} = \mathrm{sgn}\,(x - \mathrm{Med}\,(x)). \tag{3.18}$$

The proof is left to the reader (Problem 3.2).

3.2 The breakdown point

Table 3.1 has shown the effects of replacing several data values by outliers. Roughly speaking, the breakdown point (BP) of an estimate $\widehat{\theta}$ of the parameter θ is the largest amount of contamination (proportion of atypical points) that the data may contain such that $\widehat{\theta}$ still gives some information about θ, i.e., about the distribution of the "typical" points.

Let θ range over a set Θ. In order for the estimate $\widehat{\theta}$ to give some information about θ the contamination should not be able to drive $\widehat{\theta}$ to infinity or to the boundary of Θ when it is not empty. For example, for a scale or dispersion parameter we have $\Theta = [0, \infty]$, and the estimate should remain bounded, and also bounded away from 0, in the sense that the distance between $\widehat{\theta}$ and 0 should be larger than some positive value.

Definition 3.1 *The asymptotic contamination BP of the estimate $\widehat{\theta}$ at F, denoted by $\varepsilon^*(\widehat{\theta}, F)$, is the largest $\varepsilon^* \in (0, 1)$ such that for $\varepsilon < \varepsilon^*$, $\widehat{\theta}_\infty\,((1-\varepsilon)\,F + \varepsilon G)$ as a function of G remains bounded, and also bounded away from the boundary of Θ.*

The definition means that there exists a bounded and closed set $K \subset \Theta$ such that $K \cap \partial\Theta = \emptyset$ (where $\partial\Theta$ denotes the boundary of Θ) such that

$$\widehat{\theta}_\infty\,((1-\varepsilon)\,F + \varepsilon G) \in K \forall \varepsilon < \varepsilon^* \text{and} \forall G. \tag{3.19}$$

It is helpful to extend the definition to the case when the estimate is not uniquely defined, e.g., when it is the solution of an equation that may have multiple roots. In this case, the boundedness of the estimate means that *all* solutions remain in a bounded set.

The BP for each type of estimate has to be treated separately. Note that it is easy to find estimates with high BP. For instance, the "estimate" identically equal to zero has $\varepsilon^* = 1$! However, for "reasonable" estimates it is intuitively clear that there must be more "typical" than "atypical" points and so $\varepsilon^* \leq 1/2$. Actually, it can be proved (Section 3.8.2) that all shift equivariant location estimates as defined in (2.3) have $\varepsilon^* \leq 1/2$.

3.2.1 Location M-estimates

It will be convenient first to treat the case of a monotonic but not necessarily odd ψ. Assume that

$$k_1 = -\psi(-\infty),\ k_2 = \psi(\infty)$$

are finite. Then it is shown in Section 3.8.3 that

$$\varepsilon^* = \frac{\min(k_1, k_2)}{k_1 + k_2}. \tag{3.20}$$

It follows that if ψ is odd, then $k_1 = k_2$ and the bound $\varepsilon^* = 0.5$ is attained. Define

$$\varepsilon_j^* = \frac{k_j}{k_1 + k_2} \, (j = 1, 2). \tag{3.21}$$

Then (3.20) is equivalent to

$$\varepsilon^* = \min\left(\varepsilon_1^*, \varepsilon_2^*\right).$$

The proof of (3.20) shows that ε_1^* and ε_2^* are respectively the BPs to $+\infty$ and to $-\infty$. It can be shown that redescending estimates also attain the bound $\varepsilon^* = 0.5$, but the proof is more involved since one has to deal not with equation (2.18) but with the minimization (2.12).

3.2.2 Scale and dispersion estimates

We deal first with scale estimates. Note that while a high proportion of atypical points with large values (outliers) may cause the estimate $\widehat{\sigma}$ to overestimate the true scale, a high proportion of data near zero ("inliers") may result in underestimation of the true scale. Thus it is desirable that the estimate remains bounded away from zero ("implosion") as well as away from infinity ("explosion"). This is equivalent to keeping the *logarithm* of $\widehat{\sigma}$ bounded.

Note that a scale M-estimate with ρ-function ρ may be written as a location M-estimate "in the log scale". Put

$$y = \log|x|, \ \mu = \log\sigma, \ \psi(t) = \rho(e^t) - \delta.$$

Since ρ is even and $\rho(0) = 0$, then

$$\rho\left(\frac{x}{\sigma}\right) - \delta = \rho\left(\frac{|x|}{\sigma}\right) - \delta = \psi(y - \mu),$$

and hence $\widehat{\sigma} = \exp(\widehat{\mu})$ where $\widehat{\mu}$ verifies ave $(\psi(\mathbf{y} - \widehat{\mu})) = 0$, and hence $\widehat{\mu}$ is a location M-estimate.

If ρ is bounded, we have $\rho(\infty) = 1$ by Definition 2.1. Then the BP ε^* of $\widehat{\sigma}$ is given by (3.20) with

$$k_1 = \delta, \ k_2 = 1 - \delta,$$

and so

$$\varepsilon^* = \min(\delta, 1 - \delta). \tag{3.22}$$

Since $\mu \to +\infty$ and $\mu \to -\infty$ are equivalent to $\sigma \to \infty$ and $\sigma \to 0$ respectively, it follows from (3.21) that δ and $1-\delta$ are respectively the BPs for explosion and for implosion.

As for dispersion estimates, it is easy to show that the BPs of the SD, the MAD and the IQR are 0, 1/2 and 1/4, respectively (Problem 3.3). In general, the BP of an equivariant dispersion estimate is ≤ 0.5 (Problem 3.5).

3.2.3 Location with previously computed dispersion estimate

In Table 3.1 we have seen the bad consequences of using an M-estimate $\widehat{\mu}$ with the SD as the previously computed dispersion estimate $\widehat{\sigma}$. The reason is that the outliers inflate this dispersion estimate, and hence outliers do not appear as such in the "standardized" residuals $(x_i - \widehat{\mu})/\widehat{\sigma}$. Hence the robustness of $\widehat{\sigma}$ is essential for that of $\widehat{\mu}$.

For monotone M-estimates with a bounded and odd ψ, it can be shown that $\varepsilon^*(\widehat{\mu}) = \varepsilon^*(\widehat{\sigma})$. Thus if $\widehat{\sigma}$ is the MAD then $\varepsilon^*(\widehat{\mu}) = 0.5$, but if $\widehat{\sigma}$ is the SD then $\varepsilon^*(\widehat{\mu}) = 0$.

Note that (3.15) implies that the location estimates using the SD and the MAD as previous dispersion have the same IF, while at the same time they have quite different BPs. By the way, this is an example of an estimate with a bounded IF but a zero BP.

For redescending M-estimates (2.64) with a bounded ρ the situation is more complex. Consider first the case of a fixed σ. Then it can be shown that $\varepsilon^*(\widehat{\mu})$ can be made arbitrarily small by taking σ small enough. This suggests that for the case of an estimated σ, it is not only the BP of $\widehat{\sigma}$ that matters but also the size of $\widehat{\sigma}$. Let $\widehat{\mu}_0$ be an initial estimate with BP $= 0.5$ (e.g., the median), and let $\widehat{\sigma}$ be an M-scale centered at $\widehat{\mu}_0$ as defined by

$$\frac{1}{n}\sum_{i=1}^{n} \rho_0\left(\frac{x_i - \widehat{\mu}_0}{\widehat{\sigma}}\right) = 0.5$$

where ρ_0 is another bounded ρ-function. If $\rho \leq \rho_0$, then $\varepsilon^*(\widehat{\mu}) = 0.5$ (a proof is given in Section 3.8.3).

Since the MAD has $\rho_0(x) = \mathrm{I}(x \geq 1)$, it does not fulfill $\rho \leq \rho_0$. In this case the situation is more complicated and the BP will in general depend on the distribution (or on the data in the case of the finite-sample BP introduced below). Huber (1984) calculated the BP for this situation, and it follows from his results that for the bisquare ρ with MAD scale, the BP is 1/2 for all practical purposes. Details are given in Section 3.8.3.

3.2.4 Simultaneous estimation

The BP for the estimates in Section 2.6.2 is much more complicated, requiring the solution of a nonlinear system of equations (Huber, 1981, p.141) . In general, the BP of $\widehat{\mu}$ is less than 0.5. In particular, using Huber's ψ_k with $\widehat{\sigma}$ given by (2.74) yields

$$\varepsilon^* = \min\left(0.5, \frac{0.675}{k + 0.675}\right),$$

so that with $k = 1.37$ we have $\varepsilon^* = 0.33$. This is clearly lower than the BP = 0.5 which corresponds to using a previously computed dispersion estimate treated above.

3.2.5 Finite-sample breakdown point

Although the asymptotic BP is an important theoretical concept, it may be more useful to define the notion of BP for a finite sample. Let $\widehat{\theta}_n = \widehat{\theta}_n(x)$ be an estimate defined for samples $\mathbf{x} = \{x_1, \ldots, x_n\}$. The *replacement finite-sample breakdown point* (FBP) of $\widehat{\theta}_n$ at x is the largest proportion $\varepsilon_n^*(\widehat{\theta}_n, \mathbf{x})$ of data points that can be arbitrarily replaced by outliers without $\widehat{\theta}_n$ leaving a set which is bounded and also bounded away from the boundary of Θ (Donoho and Huber, 1983). More formally, call $\mathcal{X}_m$ the set of all data sets $\mathbf{y}$ of size n having $n - m$ elements in common with $\mathbf{x}$:

$$\mathcal{X}_m = \{\mathbf{y} : \#(\mathbf{y}) = n,\ \#(\mathbf{x} \cap \mathbf{y}) = n - m\}.$$

Then

$$\varepsilon_n^*(\widehat{\theta}_n, \mathbf{x}) = \frac{m^*}{n}, \tag{3.23}$$

where

$$m^* = \max\left\{m \geq 0 : \widehat{\theta}_n(\mathbf{y}) \text{ bounded and also bounded away from } \partial\Theta\ \forall\, \mathbf{y} \in \mathcal{X}_m\right\}. \tag{3.24}$$

In most cases of interest, ε_n^* does not depend on $\mathbf{x}$, and tends to the asymptotic BP when $n \to \infty$. For equivariant location estimates, it is proved in Section 3.8.2 that

$$\varepsilon_n^* \leq \frac{1}{n}\left[\frac{n-1}{2}\right] \tag{3.25}$$

and that this bound is attained by M-estimates with an odd and bounded ψ. For the trimmed mean, it is easy to verify that $m^* = [n\alpha]$, so that $\varepsilon_n^* \approx \alpha$ for large n.

Another possibility is the *addition* FBP. Call $\mathcal{X}_m$ the set of all data sets of size $n + m$ containing $\mathbf{x}$:

$$\mathcal{X}_m = \{\mathbf{y} : \#(\mathbf{y}) = n + m,\ \mathbf{x} \subset \mathbf{y}\}.$$

Then

$$\varepsilon_n^{**}(\widehat{\theta}_n, \mathbf{x}) = \frac{m^*}{n + m},$$

where

$$m^* = \max\left\{m \geq 0 : \widehat{\theta}_{n+m}(\mathbf{y}) \text{ bounded and also bounded away from } \partial\,\Theta\ \forall\, y \in \mathcal{X}_m\right\}.$$

Both ε^* and ε^{**} give similar values for large n, but we prefer the former. The main reason for this is that the definition involves only the estimate for the given n, which makes it easier to generalize this concept to more complex cases, as will be seen in Section 4.6.

3.3 Maximum asymptotic bias

The IF and the BP consider extreme situations in the study of contamination. The first deals with "infinitesimal" values of ε, while the second deals with the largest ε an estimate can tolerate. Note that an estimate having a high BP means that $\widehat{\theta}_\infty(F)$ will remain in a bounded set when F ranges in an ε-neighborhood (3.2) with $\varepsilon \leq \varepsilon^*$, but this set may be very large. What we want to do now is, roughly speaking, to measure the worst behavior of the estimate for each given $\varepsilon < \varepsilon^*$.

We again consider F ranging in the ε-neighborhood

$$\mathcal{F}_{\varepsilon,\theta} = \{(1-\varepsilon)\, F_\theta + \varepsilon G : G \in \mathcal{G}\}$$

of an assumed parametric distribution F_θ, where $\mathcal{G}$ is a family of distribution functions. Unless otherwise specified, $\mathcal{G}$ will be the family of all distribution functions, but in some cases it will be more convenient to choose a more restricted family such as that of point mass distributions. The *asymptotic bias* of $\widehat{\theta}$ at any $F \in \mathcal{F}_{\varepsilon,\theta}$ is

$$\mathrm{b}_{\widehat{\theta}}(F, \theta) = \widehat{\theta}_\infty(F) - \theta$$

and the maximum bias (MB) is

$$\mathrm{MB}_{\widehat{\theta}}\,(\varepsilon, \theta) = \max\left\{\left|\mathrm{b}_{\widehat{\theta}}(F, \theta)\right| : F \in \mathcal{F}_{\varepsilon,\theta}\right\}.$$

In the case that the parameter space is the whole set of real numbers, the relationship between MB and BP is

$$\varepsilon^*(\widehat{\theta}, F_\theta) = \max\left\{\varepsilon \geq 0 : \mathrm{MB}_{\widehat{\theta}}\,(\varepsilon, \theta) < \infty\right\}.$$

Note that two estimates may have the same BP but different MBs (Problem 3.11).

The *contamination sensitivity* of $\widehat{\theta}$ at θ is defined as

$$\gamma_c(\widehat{\theta}, \theta) = \left[\frac{d}{d\varepsilon}\mathrm{MB}_{\widehat{\theta}}\,(\varepsilon, \theta)\right]_{\varepsilon=0}. \tag{3.26}$$

In the case that $\widehat{\theta}$ is consistent we have $\widehat{\theta}_\infty(F_\theta) = \theta$ and then $\mathrm{MB}_{\widehat{\theta}}(0, \theta) = b_{\widehat{\theta}}(F_\theta, 0) = 0$. Therefore γ_c gives an approximation to the MB for small ε:

$$\mathrm{MB}_{\widehat{\theta}}(\varepsilon, \theta) \approx \varepsilon\gamma_c(\widehat{\theta}, \theta). \tag{3.27}$$

Note, however, that since $\mathrm{MB}_{\widehat{\theta}}(\varepsilon^*, \theta) = \infty$ while the right-hand side of (3.27) always yields a finite result, this approximation will be quite unreliable for sufficiently large values of ε. Figure 3.3 shows $\mathrm{MB}_{\widehat{\theta}}(\varepsilon, \theta)$ at $F_\theta = \mathrm{N}(\theta, 1)$ and its approximation (3.27)

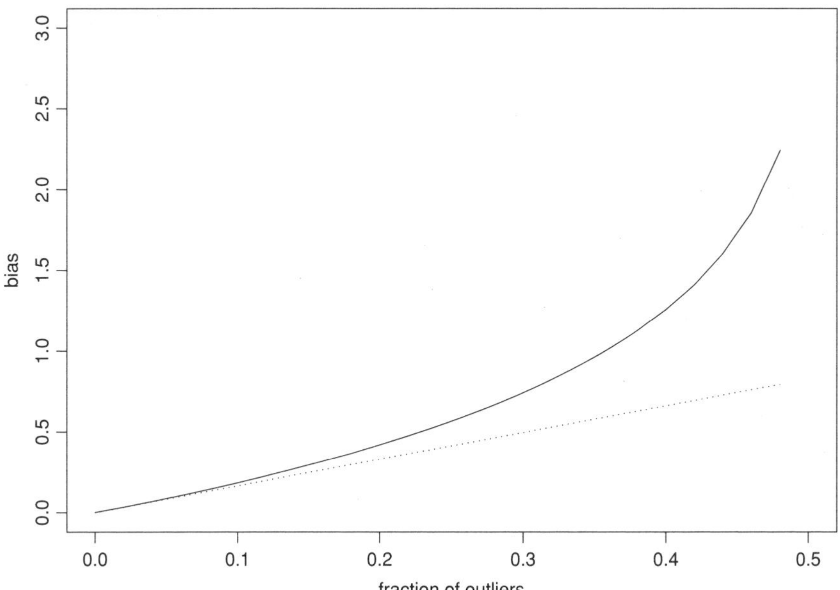

Figure 3.3 MB of Huber estimate (—) and its linear approximation (.....) as a function of ε

for the Huber location estimate with $k = 1.37$ (note that the bias does not depend on θ due to the estimate's shift equivariance).

The *gross-error sensitivity* (GES) of $\widehat{\theta}$ at θ is

$$\gamma^*(\widehat{\theta}, \theta) = \max_{x_0} \left|\mathrm{IF}_{\widehat{\theta}}(x_0, F_\theta)\right|. \tag{3.28}$$

Since $(1-\varepsilon)\,F_\theta + \varepsilon\delta_{x_0} \in \mathcal{F}_{\varepsilon,\theta}$, we have for all x_0

$$\left|\widehat{\theta}_\infty\left((1-\varepsilon)\,F_\theta + \varepsilon\delta_{x_0}\right) - \widehat{\theta}_\infty(F_\theta)\right| \le \mathrm{MB}_{\widehat{\theta}}(\varepsilon, \theta).$$

So dividing by ε and taking the limit we get

$$\gamma^* \le \gamma_c. \tag{3.29}$$

Equality above holds for M-estimates with bounded ψ-functions, but not in general. For instance, we have seen in Section 3.2.3 that the IF of the Huber estimate with the SD as previous dispersion is bounded, but since $\varepsilon^* = 0$ we have $\mathrm{MB}_{\widehat{\theta}}(\varepsilon, \theta) = \infty$ for all $\varepsilon > 0$ and so the right-hand side of (3.29) is infinite.

For location M-estimates $\widehat{\mu}$ with odd ψ and $k = \psi(\infty)$, and assuming a location model $F_\mu(x) = F_0(x - \mu)$, we have

$$\gamma^*(\widehat{\mu}, \mu) = \frac{k}{\mathrm{E}_{F_\mu}\psi'(x - \widehat{\mu}_\infty)} = \frac{k}{\mathrm{E}_{F_0}\psi'(x)} \tag{3.30}$$

so that $\gamma^*(\widehat{\mu}, \mu)$ does not depend on μ.

In general for equivariant estimates $\mathrm{MB}_{\widehat{\theta}}(\varepsilon, \theta)$ does not depend on θ. In particular, the MB for a bounded location M-estimate is given in Section 3.8.4, where it is shown that the median minimizes the MB for M-estimates at symmetric models.

3.4 Balancing robustness and efficiency

In this section we consider a parametric model F_θ and an estimate $\widehat{\theta}$ which is consistent for θ and such that the distribution of $\sqrt{n}(\widehat{\theta}_n - \theta)$ under F_θ tends to a normal distribution with mean 0 and variance $v = v(\widehat{\theta}, \theta)$. This is the most frequent case and contains most of the situations considered in this book.

Under the preceding assumptions $\widehat{\theta}$ has no asymptotic bias and we care only about its variability. Let $v_{\min} = v_{\min}(\theta)$ be the smallest possible asymptotic variance within a "reasonable" class of estimates (e.g., equivariant). Under reasonable regularity conditions $v_{\min}$ is the asymptotic variance of the MLE for the model (Section 10.8). Then the *asymptotic efficiency* of $\widehat{\theta}$ at θ is defined as $v_{\min}(\theta)/v(\widehat{\theta}, \theta)$.

If instead F does not belong to the family F_θ but is in a neighborhood of F_θ, the squared bias will dominate the variance component of MSE for all sufficiently large n. To see this let $b = \widehat{\theta}_\infty(F) - \theta$ and note that in general under F the distribution of $\sqrt{n}(\widehat{\theta}_n - \widehat{\theta}_\infty)$ tends to a normal with mean 0 and variance v . Then the distribution of $\widehat{\theta}_n - \theta$ is approximately $\mathrm{N}(b, w/n)$, so that the variance tends to zero while the bias does not. Thus we must balance the efficiency of $\widehat{\theta}$ at the model F_θ with the bias in a neighborhood of it.

We have seen that location M-estimates with a bounded ψ and previously computed dispersion estimate with BP = 1/2 attain the maximum BP of 1/2. To choose among them we must compare their biases for a given efficiency. We consider the Huber and bisquare estimates with previously computed MAD dispersion and efficiency 0.95. Their maximum biases for the model $F_{\varepsilon,\theta} = \{(1-\varepsilon)\, F_\theta + \varepsilon G : G \in \mathcal{G}\}$ with $F_\theta = \mathrm{N}(0,1)$ and a few values of ε are as follows:

ε	0.05	0.10	0.20
Huber	0.087	0.184	0.419
Bisq.	0.093	0.197	0.450

Figure 3.4 shows the respective biases for point contamination at K with $\varepsilon = 0.1$, as a function of the outlier location K. It is seen that although the maximum bias of the bisquare is higher, the difference is very small and its bias remains below that of the Huber estimate for the majority of the values. This shows that, although the maximum bias contains much more information than the BP, it is not informative enough to discriminate among estimates and that one should look at the whole bias behavior when possible

To study the behavior of the estimates under symmetric heavy-tailed distributions, we computed the asymptotic variances of the Huber and bisquare estimates, and of the

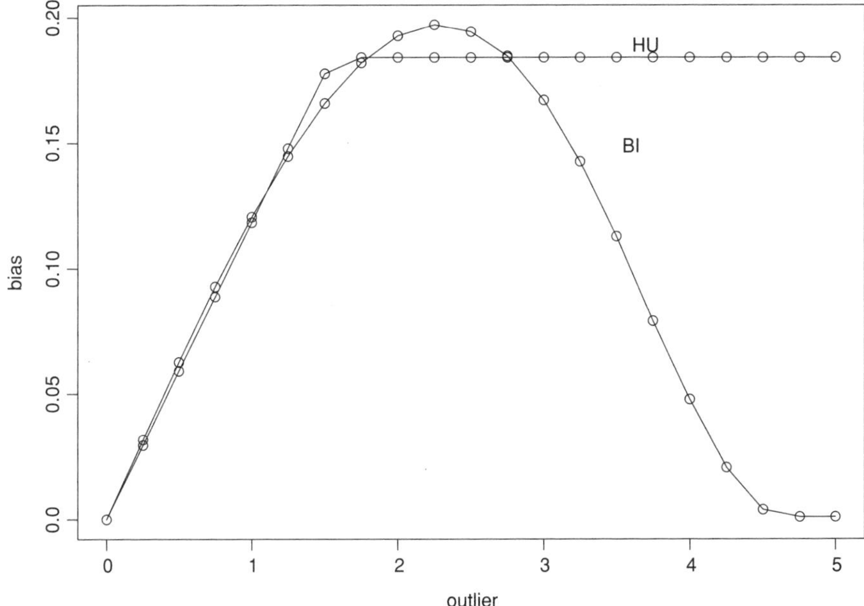

Figure 3.4 Asymptotic biases of Huber and bisquare estimates for 10% contamination as functions of the outlier location K

Cauchy MLE ("CMLE"), with simultaneous dispersion (Section 2.6.2) at the normal and Cauchy distributions, the latter of which can be considered an extreme case of heavy-tailed behavior. The efficiencies are given below:

	Huber	Bisq.	CMLE
Normal	0.95	0.95	0.60
Cauchy	0.57	0.72	1.00

It is seen that the bisquare estimate yields the best trade-off between the efficiencies at the two distributions.

For all the above reasons we recommend for estimating location the bisquare M-estimate with previously computed MAD.

3.5 *"Optimal" robustness

In this section we consider different way in which an "optimal" estimate may be defined.

3.5.1 Bias and variance optimality of location estimates

Minimax bias

If we pay attention only to bias, the quest for an "optimal" location estimate is simple: Huber (1964) has shown that the median has the smallest maximum bias ("minimax bias") among *all* shift equivariant estimates if the underlying distribution is symmetric and unimodal. See Section 3.8.5 for a proof.

Minimax variance

Huber (1964) studied location M-estimates in neighborhoods (3.2) of a symmetric F with symmetric contamination, so that there is no bias problem. The dispersion is assumed known. Call $v(\widehat{\theta}, H)$ the asymptotic variance of the estimate $\widehat{\theta}$ at the distribution H, and

$$v_\varepsilon(\widehat{\theta}) = \sup_{H \in \mathcal{F}(F,\varepsilon)} v(\widehat{\theta}, H),$$

where $\mathcal{F}(F, \varepsilon)$ is the neighborhood (3.2) with G ranging over all *symmetric* distributions. Assume that F has a density f and that $\psi_0 = -f'/f$ is nondecreasing. Then the M-estimate minimizing $v_\varepsilon(\widehat{\theta})$ has

$$\psi(x) = \begin{cases} \psi_0(x) & \text{if} & |\psi_0(x)| \leq k \\ k \text{ sgn(x)} & \text{otherwise} \end{cases}$$

where k depends on F and ε. For normal F, this is the Huber ψ_k. Since ψ_0 corresponds to the MLE for f, the result may be described as a truncated MLE.

The same problem with unknown dispersion was treated by Li and Zamar (1991).

3.5.2 Bias optimality of scale and dispersion estimates

The problem of minimax bias scale estimation for positive random variables was treated by Martin and Zamar (1989), who showed that for the case of a nominal exponential distribution the scaled median Med(**x**)/0.693, (as we will see in Problem 3.15, this estimate also minimizes the GES), was an excellent approximation to the minimax bias optimal estimate for a wide range of $\varepsilon < 0.5$. Minimax bias dispersion estimates were treated by Martin and Zamar (1993b) for the case of a nominal normal distribution and two separate families of estimates: (a) for simultaneous estimation of location and scale/dispersion with the monotone ψ-function, the minimax bias estimate is well approximated by the MAD for all $\varepsilon < 0.5$, thereby providing a theoretical rationale for an otherwise well-known high-BP estimate; (b) for M-estimates of scale with a general location estimate that includes location M-estimates with redescending ψ-functions, the minimax bias estimate is well approximated by the Shorth dispersion estimate (the shortest half of the data, see Problem 2.16b) for a wide range of $\varepsilon < 0.5$. This is an intuitively appealing estimate with BP $= 1/2$.

3.5.3 The infinitesimal approach

Several criteria have been proposed to define an optimal balance between bias and variance. The treatment can be simplified if ε is assumed to be "very small". Then the maximum bias can be approximated through the gross-error sensitivity (GES) (3.28). We first treat the simpler problem of minimizing the GES. Let F_θ be a parametric family with densities or frequency functions $f_\theta(x)$. Call E_θ the expectation with respect to F_θ: that is, if the random variable $z \sim F_\theta$ and h is any function,

$$\mathrm{E}_\theta h(z) = \begin{cases} \int h(x) f_\theta(x) dx & (z \text{ continuous}) \\ \sum_x h(x) f_\theta(x) & (z \text{ discrete}). \end{cases}$$

We shall deal with general M-estimates $\widehat{\theta}_n$ defined by (3.8), where Ψ is usually called the *score function*. An M-estimate is called *Fisher-consistent* for the family F_θ if

$$\mathrm{E}_\theta \Psi(x, \theta) = 0. \tag{3.31}$$

In view of (3.9), a Fisher-consistent M-estimate is consistent in the sense of (3.1).

It is shown in Section 10.3 that if $\widehat{\theta}_n$ is Fisher-consistent, then

$$n^{1/2}(\widehat{\theta}_n - \theta) \to_d \mathrm{N}(0, v(\Psi, \theta)),$$

with

$$v(\Psi, \theta) = \frac{A(\theta, \Psi)}{B(\theta, \Psi)^2},$$

where B is defined in (3.11) and

$$A(\theta, \Psi) = \mathrm{E}_\theta \left(\Psi(x, \theta)^2\right). \tag{3.32}$$

It follows from (3.10) that the GES of an M-estimate is

$$\gamma^*(\widehat{\theta}, \theta) = \frac{\max_x |\Psi(x, \theta)|}{|B(\theta, \Psi)|}.$$

The MLE is the M-estimate with score function

$$\Psi_0(x, \theta) = -\frac{\dot{f}_\theta(x)}{f_\theta(x)}, \text{ with } \dot{f}_\theta(x) = \frac{\partial f_\theta(x)}{\partial \theta}. \tag{3.33}$$

It is shown in Section 10.8 that this estimate is Fisher-consistent, i.e.,

$$\mathrm{E}_\theta \Psi_0(x, \theta) = 0, \tag{3.34}$$

and has the minimum asymptotic variance among Fisher-consistent M-estimates.

We now consider the problem of minimizing γ^* among M-estimates. To ensure that the estimates considered "estimate the right parameter", we consider only Fisher-consistent estimates.

Call Med_θ the median under F_θ, i.e., if $z \sim F_\theta$ and h is any function, then $\mathrm{Med}_\theta(h(z))$ is the value t where

$$\int \mathrm{I}\{h(x) \le t\}\, f_\theta(x)dx - 0.5$$

changes sign.

Define

$$M(\theta) = \mathrm{Med}_\theta \Psi_0(x, \theta).$$

It is shown in Section 3.8.6 that the M-estimate $\widetilde{\theta}$ with score function

$$\widetilde{\Psi}(x, \theta) = \mathrm{sgn}(\Psi_0(x, \theta) - M(\theta)). \tag{3.35}$$

is Fisher-consistent and is the M-estimate with smallest γ^* in that class.

This estimate has a clear intuitive interpretation. Recall that the median is a location M-estimate with ψ-function equal to the sign function. Likewise, $\widetilde{\theta}$ is the solution θ of

$$\mathrm{Med}\,\{\Psi_0(x_1, \theta), \ldots, \Psi_0(x_n, \theta)\} = \mathrm{Med}_\theta \Psi_0(x, \theta). \tag{3.36}$$

Note that, in view of (3.34), the MLE may be written as the solution of

$$\frac{1}{n}\sum_{i=1}^{n} \Psi_0(x_i, \theta) = \mathrm{E}_\theta \Psi_0(x, \theta). \tag{3.37}$$

Hence (3.36) can be seen as a version of (3.37), in which the average on the left-hand side is replaced by the sample median, and the expectation on the right is replaced by the distribution median.

3.5.4 The Hampel approach

Hampel (1974) stated the balance problem between bias and efficiency for general estimates as minimizing the asymptotic variance under a bound on the GES. For a symmetric location model, his result coincides with Huber's. It is remarkable that both approaches coincide at the location problem, and furthermore the result has a high BP.

To simplify notation, we shall in this section write $\gamma^*(\Psi, \theta)$ for the GES $\gamma^*(\widehat{\theta}, \theta)$ of an M-estimate $\widehat{\theta}$ with score function Ψ.

Hampel proposed to choose an M-estimate combining efficiency and robustness by finding Ψ such that subject to (3.31)

$$v(\Psi, \theta) = \min \text{with}\, \gamma^*(\Psi, \theta) \le G(\theta), \tag{3.38}$$

where $G(\theta)$ is a given bound expressing the desired degree of robustness. It is clear that a higher robustness means a lower $G(\theta)$, but that this implies a higher $v(\Psi, \theta)$. We call this optimization problem Hampel's *direct* problem.

We can also consider a *dual* Hampel problem where we look for a function Ψ such that

$$\gamma^*(\Psi, \theta) = \min \text{ with } v(\Psi, \theta) \leq V(\theta), \tag{3.39}$$

with given V. It is easy to see that both problems are equivalent in the following sense: if Ψ^* is optimal for the direct Hampel problem, then it is also optimal for the dual problem with $V(\theta) = v(\Psi^*, \theta)$. Similarly if Ψ^* is optimal for the dual problem, it is also optimal for the direct problem with $G(\theta) = \gamma^*(\Psi^*, \theta)$.

The solution to the direct and dual problems was given by Hampel (1974). The optimal score functions for both problems are of the following form:

$$\Psi^*(x, \theta) = \psi_{k(\theta)}(\Psi_0(x, \theta) - r(\theta)) \tag{3.40}$$

where Ψ_0 is given by (3.33), ψ_k is Huber's ψ-function (2.28), and $r(\theta)$ and $k(\theta)$ are chosen so that that Ψ^* satisfies (3.31).

A proof is given in Section 3.8.7. It is seen that the optimal score function is obtained from Ψ_0 by first centering through r and then bounding its absolute value by k.

Note that (3.35) is the limit case of (3.40) when $k \to 0$. Note also that for a solution to exist, $G(\theta)$ must be larger than the minimum GES $\gamma^*(\widetilde{\Psi}, \theta)$, and $V(\theta)$ must be larger than the asymptotic variance of the MLE: $v(\Psi_0, \theta)$.

It is not clear which one may be a practical rule for the choice of $G(\theta)$ for the direct Hampel problem. But for the second problem a reasonable criterion is to choose $V(\theta)$ as

$$V(\theta) = \frac{v(\Psi_0, \theta)}{1 - \alpha}, \tag{3.41}$$

where $1 - \alpha$ is the desired asymptotic efficiency of the estimate with respect to the MLE.

Finding k for a given V or G may be complicated. The problem simplifies considerably when F_θ is a location or a scale family, for in these cases the MLE is location (or scale) equivariant. We shall henceforth deal with bounds (3.41). We shall see that k may be chosen as a constant, which can then be found numerically.

For the location model we know from (2.18) that

$$\Psi_0(x, \xi) = \xi_0(x - \theta) \text{ with } \xi_0(x) = -\frac{f_0'(x)}{f_0(x)}, \tag{3.42}$$

Hence $v(\Psi_0, \theta)$ does not depend on θ, and

$$\Psi^*(x, \theta) = \psi_k(\xi_0(x - \theta) - r(\theta)). \tag{3.43}$$

If $k(\theta)$ is constant, then the $r(\theta)$ that fulfills (3.31) is constant too, which implies that $\Psi^*(x, \theta)$ depends only on $x - \theta$, and hence the estimate is location equivariant. This implies that $v(\Psi^*, \theta)$ does not depend on θ either, and depends only on k, which can be found numerically to attain equality in (3.39).

In particular, if f_0 is symmetric, it is easy to show that $r = 0$. When $f_0 = \mathrm{N}(0, 1)$ we obtain the Huber score function.

For a scale model it follows from (2.53) that

$$\Psi_0(x, \xi) = \frac{x}{\theta}\xi_0\left(\frac{x}{\theta}\right) - 1,$$

with ξ_0 as in (3.42). It follows that $v(\Psi_0, \theta)$ is proportional to θ, and that Ψ^* has the form

$$\Psi^*(x, \theta) = \psi_k\left(\frac{x}{\theta}\xi_0\left(\frac{x}{\theta}\right) - r(\theta)\right). \tag{3.44}$$

If k is constant, then the $r(\theta)$ that fulfills (3.31) is proportional to θ, which implies that $\Psi^*(x, \theta)$ depends only on x/θ, and hence the estimate is scale equivariant. This implies that $v(\Psi^*, \theta)$ is also proportional to θ^2, and hence k which can be found numerically to attain equality in (3.39).

The case of the exponential family is left for the reader in Problem 3.15.

Extensions of this approach when there is more than one parameter may be found in Hampel, Ronchetti, Rousseeuw and Stahel (1986).

3.5.5 Balancing bias and variance: the general problem

More realistic results are obtained by working with a positive (not "infinitesimal") ε. Martin and Zamar (1993a) found the location estimate minimizing the asymptotic variance under a given bound on the maximum asymptotic bias for a given $\varepsilon > 0$. Fraiman, Yohai and Zamar (2001) derived the location estimates minimizing the MSE of a given function of the parameters in an ε-contamination neighborhood. This allowed them to derive "optimal" confidence intervals which retain the asymptotic coverage probability in a neighborhood.

3.6 Multidimensional parameters

We now consider the estimation of p parameters $\theta_1, \ldots, \theta_p$ (e.g., location and dispersion), represented by the vector $\boldsymbol{\theta} = (\theta_1, \ldots, \theta_p)'$. Let $\widehat{\boldsymbol{\theta}}_n$ be an estimate with asymptotic value $\widehat{\boldsymbol{\theta}}_\infty$. Then the asymptotic bias is defined as

$$\mathrm{b}_{\widehat{\theta}}(F, \theta) = \mathrm{disc}(\widehat{\boldsymbol{\theta}}_\infty(F), \theta),$$

where $\mathrm{disc}(\mathbf{a}, \mathbf{b})$ is a measure of the discrepancy between the vectors $\mathbf{a}$ and $\mathbf{b}$, which depends on the particular situation. In many cases one may take the Euclidean distance $\|\mathbf{a} - \mathbf{b}\|$, but in other cases it may be more complex (as in Section 6.6).

We now consider the efficiency. Assume $\widehat{\boldsymbol{\theta}}_n$ is asymptotically normal with covariance matrix $\mathbf{V}$. Let $\widetilde{\boldsymbol{\theta}}_n$ be the MLE, with asymptotic covariance matrix $\mathbf{V}_0$. For $\mathbf{c} \in R^p$ the asymptotic variances of linear combinations $\mathbf{c}'\widehat{\boldsymbol{\theta}}_n$ and $\mathbf{c}'\widetilde{\boldsymbol{\theta}}_n$ are respectively $\mathbf{c}'\mathbf{V}\mathbf{c}$ and $\mathbf{c}'\mathbf{V}_0\mathbf{c}$, and their ratio would yield an efficiency measure for each $\mathbf{c}$. To express them though a single number, we take the worst situation, and define the asymptotic efficiency of $\widehat{\boldsymbol{\theta}}_n$ as

$$\mathrm{eff}(\widehat{\boldsymbol{\theta}}_n) = \min_{\mathbf{c}\neq\mathbf{0}} \frac{\mathbf{c}'\mathbf{V}_0\mathbf{c}}{\mathbf{c}'\mathbf{V}\mathbf{c}}.$$

It is easy to show that

$$\mathrm{eff}(\widehat{\boldsymbol{\theta}}_n) = \lambda_1(\mathbf{V}^{-1}\mathbf{V}_0), \tag{3.45}$$

where $\lambda_1(\mathbf{M})$ denotes the largest eigenvalue of the matrix $\mathbf{M}$.

In many situations (as in Section 4.4) $\mathbf{V} = a\mathbf{V}_0$ where a is a constant, and then the efficiency is simply $1/a$.

Consider now simultaneous M-estimators of location and dispersion (Section 2.6.2). Here we have two parameters, μ and σ, which satisfy a system of two equations. Put $\boldsymbol{\theta} = (\mu, \sigma)$, and

$$\Psi_1(x, \boldsymbol{\theta}) = \psi\left(\frac{x-\mu}{\sigma}\right) \text{ and } \Psi_2(x, \boldsymbol{\theta}) = \rho_{\text{scale}}\left(\frac{x-\mu}{\sigma}\right) - \delta.$$

Then the estimates satisfy

$$\sum_{i=1}^{n} \boldsymbol{\Psi}(x_i, \widehat{\boldsymbol{\theta}}) = \mathbf{0}, \tag{3.46}$$

with $\boldsymbol{\Psi} = (\Psi_1, \Psi_2)$. Given a parametric model $F_{\boldsymbol{\theta}}$ where $\boldsymbol{\theta}$ is a multidimensional parameter of dimension p, a general M-estimate is defined by (3.46) where $\boldsymbol{\Psi} = (\Psi_1, \ldots, \Psi_p)$.

Then (3.10) can be generalized by showing that the IF of $\widehat{\boldsymbol{\theta}}$ is

$$\mathrm{IF}_{\widehat{\theta}}(x_0, F) = -\mathbf{B}^{-1}\boldsymbol{\Psi}(x_0, \widehat{\boldsymbol{\theta}}_\infty), \tag{3.47}$$

were the matrix $\mathbf{B}$ has elements

$$B_{jk} = \mathrm{E}\left\{ \left. \frac{\partial \Psi_j(x, \boldsymbol{\theta})}{\partial \theta_k} \right|_{\theta = \widehat{\theta}_{\infty(F)}} \right\}.$$

M-estimates of multidimensional parameters are further treated in Section 10.5. It can be shown that they are asymptotically normal with asymptotic covariance matrix

$$\mathbf{V} = \mathbf{B}^{-1}\left(\mathrm{E}\boldsymbol{\Psi}(x, \boldsymbol{\theta})\boldsymbol{\Psi}(x, \boldsymbol{\theta})'\right)\mathbf{B}^{-1\prime}, \tag{3.48}$$

and hence they verify the analog of (3.17):

$$\mathbf{V} = \mathrm{E}\left\{\mathrm{IF}(x, F)\mathrm{IF}(x, F)'\right\}. \tag{3.49}$$

The results in this section hold also when the observations x are multidimensional.

3.7 *Estimates as functionals

The mean value may be considered as a "function" that attributes to each distribution F its expectation (when it exists); and the sample mean may be considered as a function attributing to each sample $\{x_1, \ldots, x_n\}$ its average $\overline{x}$. The same can be said of the median. This correspondence between distribution and sample values can be made systematic in the following way. Define the *empirical distribution function* of a sample $\mathbf{x} = \{x_1, \ldots, x_n\}$ as

$$\widehat{F}_{n,\mathbf{x}}(t) = \frac{1}{n}\sum_{i=1}^{n} \mathrm{I}(x_i \le t)$$

(the argument $\mathbf{x}$ will be dropped when there is no ambiguity). Then for any continuous function g

$$\mathrm{E}_{\widehat{F}_n} g(x) = \frac{1}{n}\sum_{i=1}^{n} g(x_i).$$

Define a "function" T whose argument is a distribution (a *functional*) as

$$T(F) = \mathrm{E}_F x = \int x dF(x).$$

It follows that $T(\widehat{F}_n) = \overline{x}$. If $\mathbf{x}$ is an i.i.d. sample from F, the law of large numbers implies that $T(\widehat{F}_n) \to_p T(F)$ when $n \to \infty$.

Likewise, define the functional $T(F)$ as the 0.5 quantile of F; if it is not unique, define $T(F)$ as the midpoint of 0.5 quantiles (see Section 2.9.4). Then $T(F) = \mathrm{Med}(x)$ for $x \sim F$, and $T(\widehat{F}_n) = \mathrm{Med}(x_1, \ldots, x_n)$. If $\mathbf{x}$ is a sample from F and $T(F)$ is unique, then $T(\widehat{F}_n) \to_p T(F)$.

More generally, M-estimates can be cast in this framework. For a given Ψ, define the functional $T(F)$ as the solution θ (assumed unique) of

$$\mathrm{E}_F \Psi(x, \theta) = 0. \tag{3.50}$$

Then $T(\widehat{F}_n)$ is a solution of

$$\mathrm{E}_{\widehat{F}_n} \Psi(x, \theta) = \frac{1}{n}\sum_{i=1}^{n} \Psi(x_i, \theta) = 0. \tag{3.51}$$

We see that $T(\widehat{F}_n)$ and $T(F)$ correspond to the M-estimate $\widehat{\theta}_n$ and to its asymptotic value $\widehat{\theta}_\infty(F)$, respectively.

A similar representation can be found for L-estimates. In particular, the α-trimmed mean corresponds to the functional

$$T(F) = \frac{1}{1-2\alpha}\mathrm{E}_F x \mathrm{I}(\alpha \le F(x) \le 1-\alpha).$$

Almost all of the estimates considered in this book can be represented as functionals, i.e.,

$$\widehat{\theta}_n = T(\widehat{F}_n) \tag{3.52}$$

for some functional T. The intuitive idea of robustness is that "modifying a small proportion of observations causes only a small change in the estimate". Thus robustness is related to some form of *continuity*. Hampel (1971) gave this intuitive concept a rigorous mathematical expression. The following is an informal exposition of these ideas; mathematical details and further references can be found in Chapter 3 of Huber (1981).

The concept of continuity requires the definition of a measure of distance $d(F, G)$ between distributions. Some particular distances (the *Lévy, bounded Lipschitz*, and *Prokhorov* metrics) are adequate to express the intuitive idea of robustness, in the sense that if the sample $\mathbf{y}$ is obtained from the sample $\mathbf{x}$ by

- arbitrarily modifying a small proportion of observations, and/or
- slightly modifying all observations,

then $d\left(\widehat{F}_{n,\mathbf{x}}, \widehat{F}_{n,\mathbf{y}}\right)$ is "small". Hampel (1971) defined the concept of *qualitative robustness*. A simplified version of his definition is that an estimate corresponding to a functional T is said to be qualitatively robust at F if T is continuous at F according to the metric d; that is, for all ε there exists δ such that $d(F, G) < \delta$ implies $|T(F) - T(G)| < \varepsilon$.

It follows that robust estimates are consistent, in the sense that $T(\widehat{F}_n)$ converges in probability to $T(F)$. To see this, recall that if $\mathbf{x}$ is an i.i.d. sample from F, then the law of large numbers implies that $\widehat{F}_n(t) \rightarrow_p F(t)$ for all t. A much stronger result called the Glivenko–Cantelli theorem (Durrett, 1996) states that $\widehat{F}_n \rightarrow F$ uniformly with probability 1; that is,

$$\mathrm{P}\left(\sup_t \left|\widehat{F}_n(t) - F(t)\right| \rightarrow 0\right) = 1.$$

It can be shown that this implies $d(\widehat{F}_n, F) \rightarrow_p 0$ for the Lévy metric; and if T is continuous then

$$\widehat{\theta}_\infty = T(F) = T\left(\mathrm{plim}_{n\rightarrow\infty}\widehat{F}_n\right) = \mathrm{plim}_{n\rightarrow\infty}T(\widehat{F}_n) = \mathrm{plim}_{n\rightarrow\infty}\widehat{\theta}_n,$$

where "plim" stands for "limit in probability".

A general definition of BP can be given in this framework. For a given metric, define an ε-neighborhood of F as

$$\mathcal{U}(\varepsilon, F) = \{G : d(F, G) < \varepsilon\},$$

and the maximum bias of T at F as

$$b_\varepsilon = \sup\{|T(G) - T(F)| : G \in \mathcal{U}(\varepsilon, F)\}.$$

For all the metrics considered, we have $d(F, G) < 1$ for all F, G; hence $\mathcal{U}(1, F)$ is the set of all distributions, and $b_1 = \sup\{|T(G) - T(F)| : \text{all } G\}$. Then the BP of T at F is defined as

$$\varepsilon^* = \sup\{\varepsilon : b_\varepsilon < b_1\}.$$

In this context, the IF may be viewed as a derivative. It will help to review some concepts from calculus. Let $h(\mathbf{z})$ be a function of m variables, with $\mathbf{z} = (z_1, \ldots, z_m)$

$\in R^m$. Then h is *differentiable* at $\mathbf{z}_0 = (z_{01}, \ldots, z_{0m})$ if there exists a vector $\mathbf{d} = (d_1, \ldots, d_m)$ such that for all $\mathbf{z}$

$$h(\mathbf{z}) - h(\mathbf{z}_0) = \sum_{j=1}^{m} d_j (z_j - z_{0j}) + o(\|\mathbf{z} - \mathbf{z}_0\|), \tag{3.53}$$

where "o" is a function such that $\lim_{t \to 0} o(t)/t = 0$. This means that in a neighborhood of $\mathbf{z}_0$, h can be approximated by a linear function. In fact, if $\mathbf{z}$ is near $\mathbf{z}_0$ we have

$$h(\mathbf{z}) \approx h(\mathbf{z}_0) + L(\mathbf{z} - \mathbf{z}_0),$$

where the linear function L is defined as $L(\mathbf{z}) = \mathbf{d}'\mathbf{z}$. The vector $\mathbf{d}$ is called the *derivative* of h at $\mathbf{z}_0$, which will be denoted by $\mathbf{d} = D(h, \mathbf{z}_0)$.

The *directional derivative* of h at $\mathbf{z}_0$ in the direction $\mathbf{a}$ is defined as

$$D(h, \mathbf{z}_0, \mathbf{a}) = \lim_{t \to 0} \frac{h(\mathbf{z}_0 + t\mathbf{a}) - h(\mathbf{z}_0)}{t}.$$

If h is differentiable, directional derivatives exist for all directions, and it can be shown that

$$D(h, \mathbf{z}_0, \mathbf{a}) = \mathbf{a}' D(h, \mathbf{z}_0).$$

The converse is not true: there are functions for which $D(h, \mathbf{z}_0, \mathbf{a})$ exists for all $\mathbf{a}$, but $D(h, \mathbf{z}_0)$ does not exist.

For an estimate $\widehat{\theta}$ represented as (3.52), the IF may also be viewed as a directional derivative of T as follows. Since

$$(1 - \varepsilon) F + \varepsilon \delta_{x_0} = F + \varepsilon \left(\delta_{x_0} - F \right),$$

we have

$$\text{IF}_{\widehat{\theta}}(x_0, F) = \lim_{\varepsilon \to 0} \frac{1}{\varepsilon} \left\{ T \left[F + \varepsilon \left(\delta_{x_0} - F \right) \right] - T(F) \right\},$$

which is the derivative of T in the direction $\delta_{x_0} - F$.

In some cases, the IF may be viewed as a derivative in the stronger sense of (3.53). This means that $T(H) - T(F)$ can be approximated by a linear function of H for *all* H in a neighborhood of F, and not just along each single direction. For a given $\widehat{\theta}$ represented by (3.52) and a given F, put for brevity

$$\xi(x) = \text{IF}_{\widehat{\theta}}(x, F).$$

Then T is *Fréchet-differentiable* if for any distribution H

$$T(H) - T(F) = \text{E}_H \xi(x) + o(d(F, H)). \tag{3.54}$$

The class of Frêchet differentiable estimates contains M-estimates with a bounded score function. Observe that the function

$$H \longrightarrow \text{E}_H \xi(x) = \int \xi(x) dH(x)$$

is linear in H.

Putting $H = F$ in (3.54) yields

$$\mathrm{E}_F \xi(x) = 0. \tag{3.55}$$

Some technical definitions are necessary at this point. A sequence z_n of random variables is said to be *bounded in probability* (abbreviated as $z_n = O_p(1)$) if for each ε there exists K such that $\mathrm{P}(|z_n| > K) < \varepsilon$ for all n; in particular, if $z_n \to_d z$ then $z_n = O_p(1)$. We say that $z_n = O_p(u_n)$ if $z_n/u_n = O_p(1)$, and that $z_n = o_p(u_n)$ if $z_n/u_n \to_p 0$.

It is known that the distribution of $\sup_t\{\sqrt{n}|\widehat{F}_n(t) - F(t)|\}$ (the so-called Kolmogorov–Smirnov statistic) tends to a distribution (see Feller, 1971), so that $\sup\left|\widehat{F}_n(t) - F(t)\right| = O_p(n^{-1/2})$. For the Lévy metric mentioned above, this fact implies that also $d(\widehat{F}_n, F) = O_p(n^{-1/2})$. Then taking $H = F_n$ in (3.54) yields

$$\begin{aligned}\widehat{\theta}_n - \widehat{\theta}_\infty(F) = T(F_n) - T(F) &= \mathrm{E}_{F_n}\xi(x) + o\left(d\left(\widehat{F}_n, F\right)\right) \\ &= \frac{1}{n}\sum_{i=1}^{n}\xi(x_i) + o_p\left(n^{-1/2}\right).\end{aligned} \tag{3.56}$$

Estimates satisfying (3.56) (called a *linear expansion* of $\widehat{\theta}_\infty$) are asymptotically normal and verify (3.17). In fact, the i.i.d. variables $\xi(x_i)$ have mean 0 (by (3.55)) and variance

$$v = \mathrm{E}_F\xi(x)^2.$$

Hence

$$\sqrt{n}\left(\widehat{\theta}_n - \widehat{\theta}_\infty\right) = \frac{1}{\sqrt{n}}\sum_{i=1}^{n}\xi(x_i) + o_p(1),$$

which by the central limit theorem tends to N(0, v).

For further work in this area, see Fernholz (1983) and Clarke (1983).

3.8 Appendix: proofs of results

3.8.1 IF of general M-estimates

Assume for simplicity that $\dot{\Psi}$ exists. For a given x_0, put for brevity

$$F_\varepsilon = (1-\varepsilon)F + \varepsilon\delta_{x_0} \text{ and } \theta_\varepsilon = \widehat{\theta}_\infty(F_\varepsilon).$$

Recall that by definition

$$\mathrm{E}_F\Psi(x, \theta_0) = 0. \tag{3.57}$$

Then θ_ε verifies

$$0 = \mathrm{E}_{F_\varepsilon}\Psi(x, \theta_\varepsilon) = (1-\varepsilon)\mathrm{E}_F\Psi(x, \theta_\varepsilon) + \varepsilon\Psi(x_0, \theta_\varepsilon).$$

Differentiating with respect to ε yields

$$-\mathrm{E}_F\Psi(x,\theta_\varepsilon)+(1-\varepsilon)\frac{\partial\theta_\varepsilon}{\partial\varepsilon}\mathrm{E}_F\dot{\Psi}(x,\theta_\varepsilon)+\Psi(x_0,\theta_\varepsilon)+\varepsilon\dot{\Psi}(x_0,\theta_\varepsilon)\frac{\partial\theta_\varepsilon}{\partial\varepsilon}=0. \quad (3.58)$$

The first term vanishes at $\varepsilon=0$ by (3.57). Taking $\varepsilon\downarrow 0$ above yields the desired result.

Note that this derivation is heuristic, since it is taken for granted that $\partial\theta_\varepsilon/\partial\varepsilon$ exists and that $\theta_\varepsilon\to 0$. A rigorous proof may be found in Huber (1981).

The same approach serves to prove (3.47) (Problem 3.9).

3.8.2 Maximum BP of location estimates

It suffices to show that $\varepsilon<\varepsilon^*$ implies $1-\varepsilon>\varepsilon^*$. Let $\varepsilon<\varepsilon^*$. For $t\in R$ define $F_t(x)=F(x-t)$, and let

$$H_t=(1-\varepsilon)F+\varepsilon F_t\in\mathcal{F}_\varepsilon,\ \ H_t^*=\varepsilon F+(1-\varepsilon)F_{-t}\in\mathcal{F}_{1-\varepsilon},$$

with

$$\mathcal{F}_\varepsilon=\{(1-\varepsilon)\,F+\varepsilon G: G\in\mathcal{G}\},$$

where $\mathcal{G}$ is the set of all distributions. Note that

$$H_t(x)=H_t^*(x-t). \quad (3.59)$$

The equivariance of $\widehat{\mu}$ and (3.59) imply

$$\widehat{\mu}_\infty(H_t)=\widehat{\mu}_\infty(H_t^*)+t\ \forall\, t.$$

Since $\varepsilon<\varepsilon^*$, $\widehat{\mu}_\infty(H_t)$ remains bounded when $t\to\infty$, and hence $\widehat{\mu}_\infty(H_t^*)$ is unbounded; since $H_t^*\in\mathcal{F}_{1-\varepsilon}$, this implies $1-\varepsilon>\varepsilon^*$.

A similar approach proves (3.25). The details are left to the reader.

3.8.3 BP of location M-estimates

Proof of (3.20)

Put for a given G

$$F_\varepsilon=(1-\varepsilon)F+\varepsilon G \text{ and } \mu_\varepsilon=\widehat{\mu}_\infty(F_\varepsilon).$$

Then

$$(1-\varepsilon)\,\mathrm{E}_F\psi(x-\mu_\varepsilon)+\varepsilon\mathrm{E}_G\psi(x-\mu_\varepsilon)=0. \quad (3.60)$$

We shall prove first that ε^* is not larger than the right-hand side of (3.20). Let $\varepsilon<\varepsilon^*$. Then for some C, $|\mu_\varepsilon|\le C$ for all G. Take $G=\delta_{x_0}$, so that

$$(1-\varepsilon)\,\mathrm{E}_F\psi(x-\mu_\varepsilon)+\varepsilon\psi(x_0-\mu_\varepsilon)=0. \quad (3.61)$$

Let $x_0 \to \infty$. Since μ_ε is bounded, we have $\psi(x_0 - \mu_\varepsilon) \to k_2$. Since $\psi \geq -k_1$, (3.61) yields

$$0 \geq -k_1(1-\varepsilon) + \varepsilon k_2, \tag{3.62}$$

which implies $\varepsilon \leq k_1/(k_1+k_2)$. Letting $x_0 \to -\infty$ yields likewise $\varepsilon \leq k_2/(k_1+k_2)$.

We shall now prove the opposite inequality. Let $\varepsilon > \varepsilon^*$. Then there exists a sequence G_n such that

$$\mu_{\varepsilon,n} = \widehat{\mu}_\infty\left((1-\varepsilon)F + \varepsilon G_n\right)$$

is unbounded. Suppose it contains a subsequence tending to $+\infty$. Then for this subsequence, $x - \mu_{\varepsilon,n} \to -\infty$ for each x, and since $\psi \leq k_2$, (3.60) implies

$$0 \leq (1-\varepsilon)\lim_{n\to\infty} \mathrm{E}_F\psi(x - \mu_{\varepsilon,n}) + \varepsilon k_2,$$

and since the bounded convergence theorem (Section 10.3) implies

$$\lim_{n\to\infty} \mathrm{E}_F\psi(x-\mu_{\varepsilon,n}) = \mathrm{E}_F\left(\lim_{n\to\infty}\psi(x-\mu_{\varepsilon,n})\right)$$

we have

$$0 \leq -k_1(1-\varepsilon) + \varepsilon k_2,$$

i.e., the opposite inequality to (3.62), from which it follows that $\varepsilon \geq \varepsilon_1^*$ in (3.21). If instead the subsequence tends to $-\infty$, we have $\varepsilon \geq \varepsilon_2^*$. This concludes the proof.

Location with previously estimated dispersion

Consider first the case of monotone ψ. Since $\varepsilon < \varepsilon^*(\widehat{\sigma})$ is equivalent to $\widehat{\sigma}$ being bounded away from zero and infinity when the contamination rate is less than ε, the proof is similar to that of the former section.

Now consider the case of a bounded ρ. Assume $\rho \leq \rho_0$. We shall show that $\varepsilon^* = 0.5$. Let $\varepsilon < 0.5$ and let $\mathbf{y}_N = (y_{N1}, \ldots, y_{Nn})$ be a sequence of data sets having m elements in common with $\mathbf{x}$, with $m \geq n(1-\varepsilon)$. Call $\widehat{\mu}_{0N}$ the initial location estimate, $\widehat{\sigma}_N$ the previous scale and $\widehat{\mu}_N$ the final location estimate for $\mathbf{y}_N$. Then it follows from the definitions of $\widehat{\mu}_{0N}$, $\widehat{\sigma}_N$ and $\widehat{\mu}_N$ that

$$\frac{1}{n}\sum_{i=1}^n \rho\left(\frac{y_{Ni}-\widehat{\mu}_N}{\widehat{\sigma}_N}\right) \leq \frac{1}{n}\sum_{i=1}^n \rho\left(\frac{y_{Ni}-\widehat{\mu}_{0N}}{\widehat{\sigma}_N}\right) \leq \frac{1}{n}\sum_{i=1}^n \rho_0\left(\frac{y_{Ni}-\widehat{\mu}_{0N}}{\widehat{\sigma}_N}\right) = 0.5. \tag{3.63}$$

Since $\widehat{\mu}_0$ and $\widehat{\sigma}_0$ have BP $= 0.5 > \varepsilon$, $\widehat{\mu}_{0N}$—and hence $\widehat{\sigma}_N$—remains bounded for any choice of $\mathbf{y}_N$.

Assume now that there is a sequence $\mathbf{y}_N$ such that $\widehat{\mu}_N \to \infty$. Let $D_N = \{i : y_{Ni} = x_i\}$, hence

$$\lim_{N\to\infty}\frac{1}{n}\sum_{i=1}^n \rho\left(\frac{y_{Ni}-\widehat{\mu}_N}{\widehat{\sigma}_N}\right) \geq \lim_{N\to\infty}\frac{1}{n}\sum_{i\in D_N}\rho\left(\frac{x_i-\widehat{\mu}_N}{\widehat{\sigma}_N}\right) \geq 1-\varepsilon > 0.5,$$

which contradicts (3.63), and therefore $\widehat{\mu}_N$ must be bounded, which implies $\varepsilon^* \geq 0.5$.

We now deal with the case of bounded ρ when $\rho \leq \rho_0$ does not hold. Consider first the case of fixed σ. Huber (1984) calculated the finite BP for this situation. For the sake of simplicity we treat the asymptotic case with point mass contamination.

Put

$$\gamma = \mathrm{E}_F \rho\left(\frac{x-\mu_0}{\sigma}\right),$$

where F is the underlying distribution and $\mu_0 = \widehat{\mu}_\infty(F)$:

$$\mu_0 = \arg\min_{\mu} \mathrm{E}_F \rho\left(\frac{x-\mu}{\sigma}\right).$$

It will be shown that

$$\varepsilon^* = \frac{1-\gamma}{2-\gamma}. \tag{3.64}$$

Consider a sequence x_N tending to infinity, and let $F_N = (1-\varepsilon)F + \varepsilon\delta_{x_N}$. Put for $\mu \in R$

$$A_N(\mu) = \mathrm{E}_{F_N} \rho\left(\frac{x-\mu}{\sigma}\right) = (1-\varepsilon)\mathrm{E}_F \rho\left(\frac{x-\mu}{\sigma}\right) + \varepsilon\rho\left(\frac{x_N-\mu}{\sigma}\right).$$

Let $\varepsilon < \mathrm{BP}(\widehat{\mu})$ first. Then $\mu_N = \widehat{\mu}_\infty(F_N)$ remains bounded when $x_N \to \infty$. By the definition of μ_0,

$$A_N(\mu_N) \geq (1-\varepsilon)\gamma + \varepsilon\rho\left(\frac{x_N-\mu_N}{\sigma}\right).$$

Since $\widehat{\mu}_\infty(F_N)$ minimizes A_N, we have $A_N(\mu_N) \leq A_N(x_N)$, and the latter tends to $1-\varepsilon$. The boundedness of μ_N implies that $x_N - \mu_N \to \infty$, and hence we have in the limit

$$(1-\varepsilon)\gamma + \varepsilon \leq 1 - \varepsilon,$$

which is equivalent to $\varepsilon < \varepsilon^*$. The reverse inequality follows likewise.

When ρ is the bisquare with efficiency 0.95, $F = \mathrm{N}(0, 1)$ and $\sigma = 1$, we have $\varepsilon^* = 0.47$.

Note that ε^* is an increasing function of σ.

In the more realistic case that σ is previously estimated, the situation is more complicated; but intuitively it can be seen that the situation is actually more favorable, since the contamination implies a larger σ. The procedure used above can be used to derive numerical bounds for ε^*. For the same ρ and MAD dispersion, it can be shown that $\varepsilon^* > 0.49$ at the normal distribution.

3.8.4 Maximum bias of location M-estimates

Let $F_\mu(x) = F_0(x-\mu)$ where F_0 is symmetric about zero. Let ψ be a nondecreasing and bounded ψ-function and call $k = \psi(\infty)$. The asymptotic value of the estimate is

$\widehat{\mu}_\infty(F_\mu) = \mu$, and the bias for an arbitrary distribution H is $\widehat{\mu}_\infty(H) - \mu$. Define for brevity the function

$$g(b) = \mathrm{E}_{F_0}\psi(x+b),$$

which is odd. It will be assumed that g is increasing. This holds either if ψ is increasing, or if F_0 has an everywhere positive density.

Let $\varepsilon < 0.5$. Then it will be shown that the maximum bias is the solution b_ε of the equation

$$g(b) = \frac{k\varepsilon}{1-\varepsilon}. \tag{3.65}$$

Since the estimate is shift equivariant, it may be assumed without loss of generality that $\mu = 0$. Put for brevity $\mu_H = \widehat{\mu}_\infty(H)$. For a distribution $H = (1-\varepsilon)\,F_0 + \varepsilon G$ (with G arbitrary), μ_H is the solution of

$$(1-\varepsilon)g(-\mu_H) + \varepsilon \mathrm{E}_G\psi(x-\mu_H) = 0. \tag{3.66}$$

Since $|g\,(b)| \le k$, we have for any G

$$(1-\varepsilon)g(-\mu_H) - \varepsilon k \le 0 \le (1-\varepsilon)g(-\mu_H) + \varepsilon k,$$

which implies

$$-\frac{k\varepsilon}{1-\varepsilon} \le g(-\mu_H) \le \frac{k\varepsilon}{1-\varepsilon},$$

and hence $|\mu_H| \le b_\varepsilon$. By letting $G = \delta_{x_0}$ in (3.66) with $x_0 \to \pm\infty$, we see that the bound is attained. This complete the proof.

For the median, $\psi(x) = \mathrm{sgn}(x)$ and $k = 1$, and a simple calculation shows (recalling the symmetry of F_0) that $g(b) = 2F_0(b) - 1$, and therefore

$$b_\varepsilon = F_0^{-1}\left(\frac{1}{2(1-\varepsilon)}\right). \tag{3.67}$$

To calculate the contamination sensitivity γ_c, put $\dot{b}_\varepsilon = db_\varepsilon/d\varepsilon$, so that $\dot{b}_0 = \gamma_c$. Then differentiating (3.65) yields

$$g'(b_\varepsilon)\dot{b}_\varepsilon = \frac{k}{(1-\varepsilon)^2},$$

and hence (recalling $b_0 = 0$) $\gamma_c = k/g'(0)$. Since $g'(0) = \mathrm{E}_{H_0}\psi'(x)$, we see that this coincides with (3.30) and hence $\gamma_c = \gamma^*$.

3.8.5 The minimax bias property of the median

Let F_0 have a density $f_0(x)$ which is a nonincreasing function of $|x|$ (a symmetric *unimodal* distribution) . Call b_ε the maximum asymptotic bias of the median given in (3.67). Let $\widehat{\theta}$ be any location equivariant estimate. It will be shown that the maximum bias of $\widehat{\theta}$ in a neighborhood $\mathcal{F}(F_0, \varepsilon)$, defined in (3.2), is not smaller than b_ε.

Call F_+ the distribution with density

$$f_+(x) = \begin{cases} (1-\varepsilon) f_0(x) & \text{if } x \le b_\varepsilon \\ (1-\varepsilon) f_0(x - 2b_\varepsilon) & \text{otherwise.} \end{cases}$$

Then f_+ belongs to $\mathcal{F}(F_0, \varepsilon)$. In fact, it can be written as

$$f_+ = (1-\varepsilon) f_0 + \varepsilon g,$$

with

$$g(x) = \frac{1-\varepsilon}{\varepsilon} \left(f_0(x - 2b_\varepsilon) - f_0(x) \right) \mathrm{I}(x > b_\varepsilon).$$

We must show that g is a density. It is nonnegative, since $x \in (b_\varepsilon, 2b_\varepsilon)$ implies $|x - 2b_\varepsilon| \le |x|$, and the unimodality of f_0 yields $f_0(x - 2b_\varepsilon) \ge f_0(x)$; the same thing happens if $x > 2b_\varepsilon$. And its integral equals one, since by (3.67),

$$\int_{b_\varepsilon}^{\infty} \left(f_0(x - 2b_\varepsilon) - f_0(x) \right) dx = 2F_0(b_\varepsilon) - 1 = \frac{\varepsilon}{1-\varepsilon}.$$

Define

$$F_-(x) = F_+(x + 2b_\varepsilon),$$

which also belongs to $\mathcal{F}(F_0, \varepsilon)$ by the same argument. The equivariance of $\widehat{\theta}$ implies that

$$\widehat{\theta}_\infty(F_+) - \widehat{\theta}_\infty(F_-) = 2b_\varepsilon,$$

and hence $\left|\widehat{\theta}_\infty(F_+)\right|$ and $\left|\widehat{\theta}_\infty(F_-)\right|$ cannot both be less than b_ε.

3.8.6 Minimizing the GES

To avoid cumbersome technical details, we assume henceforth that $\Psi_0(x, \theta)$ has a continuous distribution for all θ. We prove first that the M-estimate $\widetilde{\theta}$ is Fisher-consistent. In fact, by the definition of the function M,

$$\begin{aligned} \mathrm{E}_\theta \widetilde{\Psi}(x, \theta) &= -\mathrm{P}_\theta(\Psi_0(x, \theta) \le M(\theta)) + \mathrm{P}_\theta(\Psi_0(x, \theta) > M(\theta)) \\ &= -\frac{1}{2} + \frac{1}{2} = 0. \end{aligned}$$

Since $\max_x |\widetilde{\Psi}(x, \theta)| = 1$, the estimate has GES

$$\gamma^*(\widetilde{\Psi}, \theta) = \frac{1}{\left| B(\theta, \widetilde{\Psi}) \right|}.$$

It will be shown first that for any Fisher-consistent Ψ,

$$B(\theta, \Psi) = \mathrm{E}_\theta \Psi(x, \theta) \Psi_0(x, \theta). \tag{3.68}$$

We give the proof for the continuous case; the discrete case is similar. Condition (3.31) may be written as

$$\mathrm{E}_\theta \Psi(x, \theta) = \int_{-\infty}^{\infty} \Psi(x, \theta) f_\theta(x) dx = 0.$$

Differentiating the above expression with respect to θ yields

$$B(\theta, \Psi) + \int_{-\infty}^{\infty} \Psi(x, \theta) \dot{f}_\theta(x) dx = 0,$$

and (3.32)–(3.33) yield

$$\begin{aligned} B(\theta, \Psi) &= -\int_{-\infty}^{\infty} \Psi(x, \theta) \dot{f}_\theta(x) dx \\ &= \int_{-\infty}^{\infty} \Psi(x, \theta) \Psi_0(x, \theta) f_\theta(x)\, dx = \mathrm{E}_\theta \Psi(x, \theta) \Psi_0(x, \theta), \end{aligned}$$

as stated. Note that $\partial \widetilde{\Psi} / \partial \theta$ does not exist, and hence we must define $B(\theta, \widetilde{\Psi})$ through (3.11) and not (3.12).

Now let $C = \{x : \Psi_0(x, \theta) > M(\theta)\}$, with complement C'. It follows from $\mathrm{I}(C') = 1 - \mathrm{I}(C)$ that

$$\mathrm{P}_\theta \left(\widetilde{\Psi} = \mathrm{I}(C) - \mathrm{I}(C') = 2\mathrm{I}(C) - 1 \right) = 1.$$

Using (3.34) and (3.35) we have

$$\begin{aligned} B(\theta, \widetilde{\Psi}) &= \mathrm{E}_\theta \widetilde{\Psi}(x, \theta) \Psi_0(x, \theta) \\ &= 2\mathrm{E}_\theta \Psi_0(x, \theta) \mathrm{I}(C) - \mathrm{E}_\theta \Psi_0(x, \theta) = 2\mathrm{E}_\theta \Psi_0(x, \theta) \mathrm{I}(C). \end{aligned}$$

Hence

$$\gamma^*(\widetilde{\Psi}, \theta) = \frac{1}{2\, |\mathrm{E}_\theta \Psi_0(x, \theta) \mathrm{I}(C)|}. \tag{3.69}$$

Consider a Fisher-consistent Ψ. Then

$$\gamma^*(\Psi, \theta) = \frac{\max_x |\Psi(x, \theta)|}{|B(\theta, \Psi)|}. \tag{3.70}$$

Using (3.31) and (3.68) we have

$$\begin{aligned} B(\theta, \Psi) &= \mathrm{E}_\theta \Psi(x, \theta) \Psi_0(x, \theta) \\ &= \mathrm{E}_\theta \Psi(x, \theta) (\Psi_0(x, \theta) - M(\theta)) \\ &= \mathrm{E}_\theta \Psi(x, \theta) (\Psi_0(x, \theta) - M(\theta)) \mathrm{I}(C) \\ &\quad + \mathrm{E}_\theta \Psi(x, \theta) (\Psi_0(x, \theta) - M(\theta)) \mathrm{I}(C'). \end{aligned} \tag{3.71}$$

Besides

$$\begin{aligned} &|\mathrm{E}_\theta \Psi(x, \theta) (\Psi_0(x, \theta) - M(\theta)) \mathrm{I}(C)| \\ &\leq \max_x |\Psi(x, \theta)| \mathrm{E}_\theta (\Psi_0(x, \theta) - M(\theta)) \mathrm{I}(C) \\ &= \max_x |\Psi(x, \theta)| \left(\mathrm{E}_\theta \Psi_0(x, \theta) \mathrm{I}(C) - \frac{M(\theta)}{2} \right). \end{aligned} \tag{3.72}$$

Similarly

$$\begin{aligned}&\left|\mathrm{E}_\theta \Psi(x,\theta)(\Psi_0(x,\theta)-M(\theta))\mathrm{I}(C')\right|\\ &\leq -\max_x |\Psi(x,\theta)|\mathrm{E}_\theta(\Psi_0(x,\theta)-M(\theta))\mathrm{I}(C')\\ &=\max_x |\Psi(x,\theta)|\left(\mathrm{E}_\theta \Psi_0(x,\theta)\mathrm{I}(C)+\frac{M(\theta)}{2}\right).\end{aligned} \tag{3.73}$$

Therefore by (3.71), (3.72) and (3.73) we get

$$|B(\theta,\Psi)| \leq 2\max_x |\Psi(x,\theta)|\mathrm{E}_\theta \Psi_0(x,\theta)\mathrm{I}(C).$$

Therefore, using (3.70) we have

$$\gamma^*(\Psi,\theta) \geq \frac{1}{2\,|\mathrm{E}_\theta \Psi_0(x,\theta)\mathrm{I}(C)|}. \tag{3.74}$$

And finally (3.69) and (3.74) yield

$$\gamma^*(\widetilde{\Psi},\theta) \leq \gamma^*(\Psi,\theta).$$

The case of a discrete distribution is similar, but the details are much more involved.

3.8.7 Hampel optimality

It will be shown first that estimates with score function (3.40) are optimal for Hampel problems with certain bounds.

Theorem 3.2 *Given $k(\theta)$, the function Ψ^* given by (3.40) and satisfying (3.31) is optimal for the direct Hampel problem with bound*

$$G(\theta) = \gamma^*(\Psi^*,\theta) = \frac{k(\theta)}{B(\theta,\Psi^*)},$$

and for the dual Hampel problem with bound

$$V(\theta) = v(\Psi^*,\theta).$$

Proof of Theorem 3.2: We shall show that Ψ^* solves Hampel's direct problem. Observe that Ψ^* satisfies the side condition in (3.38), since by definition $\gamma^*(\Psi^*,\theta) = G(\theta)$. Let Ψ now satisfy (3.31) and

$$\gamma^*(\Psi,\theta) \leq G(\theta). \tag{3.75}$$

We must show that

$$v(\Psi,\theta) \geq v(\Psi^*,\theta). \tag{3.76}$$

We prove (3.76) for a fixed θ. Since for any real number $\lambda \neq 0$, $\lambda\Psi$ defines the same estimate as Ψ, we can assume without loss of generality that

$$B(\theta, \Psi) = B(\theta, \Psi^*), \tag{3.77}$$

and hence

$$\gamma^*(\Psi, \theta) = \frac{\max_x(|\Psi(x, \theta)|)}{\left|B(\theta, \Psi^*)\right|}.$$

Then, condition (3.75) becomes

$$\max_x(|\Psi(x, \theta)|) \leq k(\theta) \tag{3.78}$$

and (3.76) becomes $A(\theta, \Psi) \geq A(\theta, \Psi^*)$, so that we have to prove

$$\mathrm{E}_\theta \Psi^2(x, \theta) \geq \mathrm{E}_\theta \Psi^{*2}(x, \theta) \tag{3.79}$$

for any Ψ satisfying (3.78).

Call Ψ_0^c the ML score function centered by r:

$$\Psi_0^c(x, \theta) = \Psi_0(x, \theta) - r(\theta).$$

It follows from (3.68) and (3.31) that

$$\mathrm{E}_\theta \Psi(x, \theta)\Psi_0^c(x, \theta) = B(\theta, \Psi).$$

We now calculate $\mathrm{E}_\theta \Psi^2(x, \theta)$. Recalling (3.77) we have

$$\begin{aligned}
\mathrm{E}_\theta \Psi^2(x, \theta) &= \mathrm{E}_\theta\{[\Psi(x, \theta) - \Psi_0^c(x, \theta)] + \Psi_0^c(x, \theta)\}^2 \\
&= \mathrm{E}_\theta(\Psi(x, \theta) - \Psi_0^c(x, \theta))^2 + \mathrm{E}_\theta \Psi_0^c(x, \theta)^2 \\
&\quad + 2\mathrm{E}_\theta \Psi(x, \theta)\Psi_0^c - 2\mathrm{E}_\theta \Psi_0^{c2}(x, \theta) \\
&= \mathrm{E}_\theta(\Psi(x, \theta) - \Psi_0^c(x, \theta))^2 - \mathrm{E}_\theta \Psi_0^c(x, \theta)^2 + 2B(\theta, \Psi^*). \quad (3.80)
\end{aligned}$$

Since $\mathrm{E}\Psi_0(x, \theta)^2$ and $B(\theta, \Psi^*)$ do not depend on Ψ, it suffices to prove that putting $\Psi = \Psi^*$ minimizes

$$\mathrm{E}_\theta(\Psi(x, \theta) - \Psi_0^c(x, \theta))^2$$

subject to (3.78). Observe that for any function $\Psi(x, \theta)$satisfying (3.78) we have

$$|\Psi(x, \theta) - \Psi_0^c(x, \theta)| \geq ||\Psi_0^c(x, \theta)| - k(\theta)|\mathrm{I}\left\{|\Psi_0^c(x, \theta)| > k(\theta)\right\},$$

and since

$$|\Psi^*(x, \theta) - \Psi_0^c(x, \theta)| = ||\Psi_0^c(x, \theta)| - k(\theta)|\mathrm{I}\left\{|\Psi_0^c(x, \theta)| > k(\theta)\right\},$$

we get

$$|\Psi(x, \theta) - \Psi_0^c(x, \theta)| \geq |\Psi^*(x, \theta) - \Psi_0^c(x, \theta)|.$$

Then

$$\mathrm{E}_\theta(\Psi(x,\theta)-\Psi_0^c(x,\theta))^2 \geq \mathrm{E}_\theta(\Psi^*(x,\theta)-\Psi_0^c(x,\theta))^2,$$

which proves the statement for the direct problem. The dual problem is treated likewise. ■

The former theorem proves optimality for a certain class of bounds. The following theorem shows that actually any feasible bounds can be considered.

Theorem 3.3 *Let*

$$G(\theta) \geq \gamma^*(\widetilde{\Psi},\theta),\ V(\theta) \geq v(\Psi_0,\theta) \text{ for all } \theta, \tag{3.81}$$

where Ψ_0 *and* $\widetilde{\Psi}$ *are defined in (3.33) and (3.35) respectively. Then the solutions to both the direct and the dual Hampel problems have the form (3.40) for a suitable function* $k(\theta)$.

Proof of Theorem 3.3: We treat the dual problem; the direct one is treated likewise.

We show first that given any k there exists r so that $\Psi^*(x,\theta)$ is Fisher-consistent. Let

$$\lambda(r) = \mathrm{E}_\theta \psi_k\left(\Psi_0(x,\theta)-r\right).$$

Then λ is continuous, and $\lim_{r\to\pm\infty}\lambda(r) = \mp k$. Hence by the intermediate value theorem, there exists some r such that $\lambda(r)=0$. Besides, it can be shown that $B(\theta,\Psi^*) \neq 0$. The proof is involved and can be found in Hampel et al. (1986).

In view of (3.68):

$$v(\overset{*}{\Psi}_{(k(\theta))},\theta) = \frac{\mathrm{E}_\theta \overset{*}{\Psi}_{(k(\theta))}(x,\theta)^2}{\left[\mathrm{E}_\theta \overset{*}{\Psi}_{(k(\theta))}(x,\theta)\Psi_0(x,\theta)\right]^2}, \tag{3.82}$$

where Ψ^* in (3.40) is written as $\overset{*}{\Psi}_{(k(\theta))}$ to stress its dependence on k. Recall that the limit cases $k\to 0$ and $k\to\infty$ yield $v(\widetilde{\Psi},\theta)$ (which may be infinite) and $v(\Psi_0,\theta)$, respectively. Let $V(\theta)$ be given and such that $V(\theta)\geq v(\Psi_0,\theta)$. Consider a fixed θ. If $V(\theta)\leq v(\widetilde{\Psi},\theta)$, then there exists a value $k(\theta)$ such that $v(\overset{*}{\Psi}_{(k(\theta))},\theta) = V(\theta)$. If $V(\theta) > v(\widetilde{\Psi},\theta)$, then putting $k(\theta)=0$ (i.e., $\Psi^* = \widetilde{\Psi}$) minimizes $\gamma^*(\overset{*}{\Psi}_{(k(\theta))},\theta)$ and satisfies $v(\overset{*}{\Psi}_{(k(\theta))},\theta)\leq V(\theta)$. ■

3.9 Problems

3.1. Verify (3.14).
3.2. Prove (3.18).
3.3. Verify that the breakdown points of the SD, the MAD and the IQR are 0, 1/2 and 1/4, respectively.
3.4. Show that the asymptotic BP of the α-trimmed mean is α.

3.5. Show that the BP of equivariant dispersion estimates is ≤ 0.5.
3.6. Show that the asymptotic BP of sample β-quantiles is $\min(\beta, 1-\beta)$ [recall Problem 13].
3.7. Prove (3.25).
3.8. Verify (3.45).
3.9. Prove (3.47).
3.10. Prove (3.45).
3.11. Consider the location M-estimate with Huber function ψ_k and the MADN as previously computed dispersion. Recall that it has BP = 1/2 for all k. Show, however, that for each given $\varepsilon < 0.5$, its maximum bias $\mathrm{MB}(\varepsilon)$ at a given distribution is an unbounded function of k.
3.12. Let the density $f(x)$ be a decreasing function of $|x|$. Show that the shortest interval covering a given probability is symmetric about zero. Use this result to calculate the IF of the Shorth estimate (Problem 2.16a) for data with distribution f.
3.13. Show that the BP of the estimate Q_n in (2.51) is 0.5. Calculate the BP for the estimate defined as the median of the differences; that is, with $k = m/2$ in (2.51).
3.14. Show the equivalence of the direct and dual Hampel problems (3.38)–(3.39).
3.15. For the exponential family $f_\theta(x) = \mathrm{I}(x \geq 0)\exp(-x/\theta)/\theta$:
   (a) Show that the estimate with smallest GES is $\mathrm{Med}(\mathbf{x}) / \log 2$.
   (b) Find the asymptotic distribution of this estimate and its efficiency with respect to the MLE.
   (c) Find the form of the Hampel-optimal estimate for this family.
   (d) Write a program to compute the Hampel-optimal estimate with efficiency 0.95.
3.16. Consider the estimate $\widehat{\mu}_1$ defined by the one-step Newton–Raphson procedure defined in Section 2.9.5. Assume that the underlying distribution is symmetric about μ, that ψ is odd and differentiable, and that the initial estimate $\widehat{\mu}_0$ is consistent for μ.
   (a) Show that $\widehat{\mu}_1$ is consistent for μ.
   (b) If ψ is twice differentiable, show that $\widehat{\mu}_1$ has the same influence function as the M-estimate $\widehat{\mu}$ defined by ave $\{\psi(\mathbf{x} - \mu)\} = 0$ (and hence, by (3.17), $\widehat{\mu}_1$ has the same asymptotic variance as $\widehat{\mu}$).
   (c) If ψ is bounded and $\psi'(x) > 0$ for all x, and the asymptotic BP of $\widehat{\mu}_0$ is 0.5, show that also $\widehat{\mu}_1$ has an asymptotic BP of 0.5.

4

Linear Regression 1

4.1 Introduction

In this chapter we begin the discussion on the estimation of the parameters of linear regression models, which will be pursued in the next chapter. M-estimates for regression are developed in the same way as for location. In this chapter we deal with fixed (nonrandom) predictors. Recall that our estimates of choice for location were redescending M-estimates using the median as starting point and the MAD as dispersion. Redescending estimates will also be our choice for regression. When the predictors are fixed and fulfill certain conditions that are satisfied in particular for analysis of variance models, monotone M-estimates—which are easy to compute—are robust, and can be used as starting points to compute a redescending estimate. When the predictors are random, or when they are fixed but in some sense "unbalanced", monotone estimates cease to be reliable, and the starting points for redescending estimates must be computed otherwise. This problem is treated in the next chapter.

We start with an example that shows the weakness of the least-squares estimate.

Example 4.1 *The data in Table 4.1 (Bond, 1979) correspond to an experiment on the speed of learning of rats. Times were recorded for a rat to go through a shuttlebox in successive attempts. If the time exceeded 5 seconds, the rat received an electric shock for the duration of the next attempt. The data are the number of shocks received and the average time for all attempts between shocks.*

Figure 4.1 shows the data and the straight line fitted by least squares (LS) to the linear regression model

$$y_i = \beta_0 + \beta_1 x_i + u_i.$$

The relationship between the variables is seen to be roughly linear except for the three upper left points. The LS line does not fit the bulk of the data, being a compromise

Robust Statistics – Theory and Methods Ricardo A. Maronna, R. Douglas Martin and Víctor J. Yohai

Table 4.1 Rats data

Shocks	Time	Shocks	Time
0	11.4	8	5.7
1	11.9	9	4.4
2	7.1	10	4.0
3	14.2	11	2.8
4	5.9	12	2.6
5	6.1	13	2.4
6	5.4	14	5.2
7	3.1	15	2.0

between those three points and the rest. The figure also shows the LS fit computed without using the three points. It gives a better representation of the majority of the data, while pointing out the exceptional character of points 1, 2 and 4. Code **shock** *is used for this data set.*

We aim at developing procedures that give a good fit to the bulk of the data without being perturbed by a small proportion of outliers, and that do not require deciding

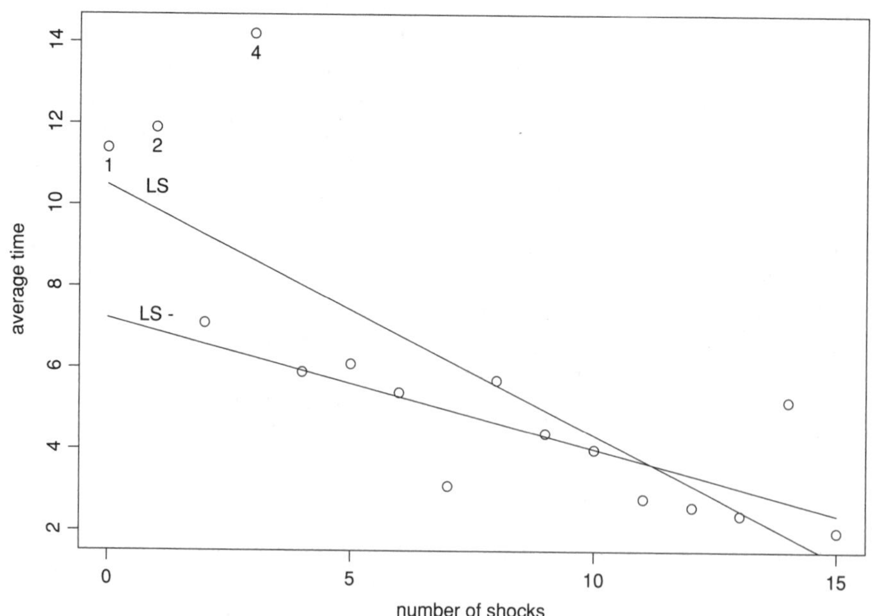

Figure 4.1 Shock data: LS fit with all data and omitting points 1, 2 and 4

Table 4.2 Regression estimates for rats data

	Int.	Slope
LS	10.48	−0.61
LS without 1,2,4	7.22	−0.32
L1	8.22	−0.42
Bisquare M-est.	7.83	−0.41

previously which observations are outliers. Table 4.2 gives the estimated parameters for LS with the complete data and with the three atypical points deleted, and also for two robust estimates (L1 and bisquare) to be defined later.

The LS fit of a straight line consists of finding $\hat{\beta}_0, \hat{\beta}_1$ such that the residuals

$$r_i = y_i - (\hat{\beta}_0 + \hat{\beta}_1 x_i)$$

satisfy

$$\sum_{i=1}^{n} r_i^2 = \min. \tag{4.1}$$

Recall that in the location case obtained by setting $\beta_1 = 0$ the solution of (2.15) is the sample mean, i.e., the LS estimate of location is the average of the data values. Since the median satisfies (2.17), the regression analog of the median, often called an L1 estimate (also called the least absolute deviation or LAD estimate), is defined by

$$\sum_{i=1}^{n} |r_i| = \min. \tag{4.2}$$

For our data the solution of (4.2) is given in Table 4.2, and one sees that its slope is smaller than that of the LS estimate, i.e., it is less affected by the outliers.

Now, consider the more general case of a data set of n observations $(x_{i1}, \ldots, x_{ip}, y_i)$ where $x_{i1}, \ldots x_{ip}$ are predictor variables (the *predictors* or *independent variables*) and y_i is a response variable (the *response* or *dependent variable*). The data are assumed to follow the *linear model*

$$y_i = \sum_{j=1}^{p} x_{ij}\beta_j + u_i, \quad i = 1, \ldots, n \tag{4.3}$$

where $\beta_1, \ldots, \beta_p$ are unknown parameters to be estimated, and the u_i's are random variables (the "errors"). In a designed experiment, the x_{ij}'s are nonrandom (or *fixed*), i.e., determined before the experiment. When the data are observational the x_{ij} are random variables. We sometimes have mixed situations with both fixed and random predictors.

Calling $\mathbf{x}_i$ and $\boldsymbol{\beta}$ the p-dimensional column vectors with coordinates $(x_{i1}, \ldots, x_{ip})$ and $(\beta_1, \ldots, \beta_p)$ respectively, the model can be more compactly written as

$$y_i = \mathbf{x}_i'\boldsymbol{\beta} + u_i \tag{4.4}$$

where $\mathbf{x}'$ is the transpose of $\mathbf{x}$. In the frequently occurring case where the model has a constant term, the first coordinate of each $\mathbf{x}_i$ is 1 and the model may be written as

$$y_i = \beta_0 + \underline{\mathbf{x}}_i'\boldsymbol{\beta}_1 + u_i \tag{4.5}$$

where $\underline{\mathbf{x}}_i = (x_{i1}, \ldots, x_{i(p-1)})'$ and $\boldsymbol{\beta}_1$ are in R^{p-1} and

$$\mathbf{x}_i = \begin{bmatrix} 1 \\ \underline{\mathbf{x}}_i \end{bmatrix}, \quad \boldsymbol{\beta} = \begin{bmatrix} \beta_0 \\ \boldsymbol{\beta}_1 \end{bmatrix}. \tag{4.6}$$

Here β_0 is called the *intercept* and the elements of $\boldsymbol{\beta}_1$ are the *slopes*. Call $\mathbf{X}$ the $n \times p$ matrix with elements x_{ij} and let $\mathbf{y}$ and $\mathbf{u}$ be the vectors with elements y_i and u_i respectively ($i = 1, \ldots, n$). Then the linear model (4.4) may be written

$$\mathbf{y} = \mathbf{X}\boldsymbol{\beta} + \mathbf{u}. \tag{4.7}$$

The *fitted values* $\widehat{y}_i$ and the *residuals* r_i corresponding to a vector $\boldsymbol{\beta}$ are defined respectively as

$$\widehat{y}_i(\boldsymbol{\beta}) = \mathbf{x}_i'\boldsymbol{\beta} \quad \text{and} \quad r_i(\boldsymbol{\beta}) = y_i - \widehat{y}_i(\boldsymbol{\beta}).$$

The dependence of the fitted values and residuals on $\boldsymbol{\beta}$ will be dropped when this does not cause confusion. In order to combine robustness and efficiency following the lines of Chapter 2, we shall discuss regression M-estimates $\widehat{\boldsymbol{\beta}}$ defined as solutions of equations of the form

$$\sum_{i=1}^{n} \rho\left(\frac{r_i(\widehat{\boldsymbol{\beta}})}{\widehat{\sigma}}\right) = \min. \tag{4.8}$$

Here ρ is a ρ-function (Definition 2.1 of Chapter 2), and $\widehat{\sigma}$ is an auxiliary scale estimate that is required to make $\widehat{\boldsymbol{\beta}}$ scale equivariant (see (2.4) and (4.16) below). The LS estimate and the L1 estimate correspond respectively to $\rho(t) = t^2$ and $\rho(t) = |t|$. In these two cases $\widehat{\sigma}$ becomes a constant factor outside the summation sign and minimizing (4.8) is equivalent to minimizing $\sum_{i=1}^{n} r_i^2$ or $\sum_{i=1}^{n} |r_i|$, respectively. Thus neither the LS nor the L1 estimates require a scale estimate.

In a designed experiment, the predictors x_{ij} are fixed. An important special case of fixed predictors is when they represent *categorical* predictors with values of either 0 or 1. The simplest situation is the comparison of several treatments, usually called a *one-way analysis of variance* (or "one-way ANOVA"). Here we have p samples y_{ik} ($i = 1, \ldots, n_k, k = 1, \ldots, p$) and the model

$$y_{ik} = \beta_k + u_{ik} \tag{4.9}$$

where the u_{ik}'s are i.i.d. Call $\mathbf{1}_m$ the column vector of m ones. Then the matrix $\mathbf{X}$ of predictors is

$$\mathbf{X} = \begin{bmatrix} \mathbf{1}_{n_1} & & & \\ & \mathbf{1}_{n_2} & & \\ & & \ddots & \\ & & & \mathbf{1}_{n_p} \end{bmatrix}$$

with the blank positions filled with zeros. The next level of model complexity is a factorial design with two factors which are represented by two categorical variables, usually called a *two-way analysis of variance*. In this case we have data y_{ijk}, $i = 1, \ldots, I$, $j = 1, \ldots, J$, $k = 1, \ldots, K_{ij}$, following an *additive model* usually written in the form

$$y_{ijk} = \mu + \alpha_i + \gamma_j + u_{ijk} \tag{4.10}$$

with "cells" i, j and K_{ij} observations per cell. Here $p = I + J + 1$ and $\boldsymbol{\beta}$ has coordinates $(\mu, \alpha_1, \ldots, \alpha_I, \gamma_1, \ldots, \gamma_J)$. The rank of $\mathbf{X}$ is $p^* = I + J - 1 < p$ and constraints on the parameters need to be added to make the estimates unique, typically

$$\sum_{i=1}^{I} \alpha_i = \sum_{j=1}^{J} \gamma_j = 0. \tag{4.11}$$

4.2 Review of the LS method

The LS method was proposed in 1805 by Legendre (for a fascinating account, see Stigler (1986)). The main reason for its immediate and lasting success was that it was the only method of estimation that could be effectively computed before the advent of electronic computers. We shall review the main properties of LS for multiple regression. (See any standard text on regression analysis, e.g., Weisberg (1985), Draper and Smith (2001), Montgomery et al. (2001) or Stapleton (1995).) The LS estimate of $\boldsymbol{\beta}$ is the $\widehat{\boldsymbol{\beta}}$ such that

$$\sum_{i=1}^{n} r_i^2(\widehat{\boldsymbol{\beta}}) = \min. \tag{4.12}$$

Differentiating with respect to $\boldsymbol{\beta}$ yields

$$\sum_{i=1}^{n} r_i(\widehat{\boldsymbol{\beta}})\mathbf{x}_i = \mathbf{0}, \tag{4.13}$$

which is equivalent to the linear equations

$$\mathbf{X}'\mathbf{X}\widehat{\boldsymbol{\beta}} = \mathbf{X}'\mathbf{y}$$

The above equations are usually called the "normal equations". If the model contains a constant term, it follows from (4.13) that the residuals have zero average.

The matrix of predictors $\mathbf{X}$ is said to have *full rank* if its columns are linearly independent. This is equivalent to

$$\mathbf{Xa} \neq \mathbf{0} \; \forall \; \mathbf{a} \neq \mathbf{0}$$

and also equivalent to the nonsingularity of $\mathbf{X}'\mathbf{X}$. If $\mathbf{X}$ has full rank then the solution of (4.13) is unique and is given by

$$\widehat{\boldsymbol{\beta}}_{LS} = \widehat{\boldsymbol{\beta}}_{LS}(\mathbf{X}, \mathbf{y}) = \left(\mathbf{X}'\mathbf{X}\right)^{-1} \mathbf{X}'\mathbf{y}. \tag{4.14}$$

If the model contains a constant term, then the first column of $\mathbf{X}$ is identically one, and the full rank condition implies that no other column is constant. If $\mathbf{X}$ is not of full rank, e.g., as in (4.10), then we have what is called *collinearity*. When there is collinearity the parameters are not *identifiable* in the sense that there exist $\boldsymbol{\beta}_1 \neq \boldsymbol{\beta}_2$ such that $\mathbf{X}\boldsymbol{\beta}_1 = \mathbf{X}\boldsymbol{\beta}_2$, which implies that (4.13) has infinite solutions, all yielding the same fitted values and hence the same residuals.

The LS estimate satisfies

$$\widehat{\boldsymbol{\beta}}_{LS}(\mathbf{X}, \mathbf{y} + \mathbf{X}\boldsymbol{\gamma}) = \widehat{\boldsymbol{\beta}}_{LS}(\mathbf{X}, \mathbf{y}) + \boldsymbol{\gamma} \quad \text{for all } \boldsymbol{\gamma} \in R^p \tag{4.15}$$

$$\widehat{\boldsymbol{\beta}}_{LS}(\mathbf{X}, \lambda\mathbf{y}) = \lambda\widehat{\boldsymbol{\beta}}_{LS}(\mathbf{X}, \mathbf{y}) \quad \text{for all } \lambda \in R \tag{4.16}$$

and for all nonsingular $p \times p$ matrices $\mathbf{A}$

$$\widehat{\boldsymbol{\beta}}_{LS}(\mathbf{XA}, \mathbf{y}) = \mathbf{A}^{-1}\widehat{\boldsymbol{\beta}}_{LS}(\mathbf{X}, \mathbf{y}). \tag{4.17}$$

The properties (4.15), (4.16) and (4.17) are called respectively *regression*, *scale* and *affine equivariance*. These are desirable properties, since they allow us to know how the estimate changes under these transformations of the data. A more precise justification is given in Section 4.9.1.

Assume now that the u_i's are i.i.d. with

$$\mathrm{E}u_i = 0 \quad \text{and} \quad \mathrm{Var}(u_i) = \sigma^2$$

and that $\mathbf{X}$ is fixed, i.e., nonrandom, and of full rank. Under the linear model (4.4) with $\mathbf{X}$ of full rank $\widehat{\boldsymbol{\beta}}_{LS}$ is unbiased and its mean and covariance matrix are given by

$$\mathrm{E}\widehat{\boldsymbol{\beta}}_{LS} = \boldsymbol{\beta}, \; \mathbf{Var}\left(\widehat{\boldsymbol{\beta}}_{LS}\right) = \sigma^2 \left(\mathbf{X}'\mathbf{X}\right)^{-1} \tag{4.18}$$

where henceforth $\mathbf{Var}(\mathbf{y})$ will denote the covariance matrix of the random vector $\mathbf{y}$.

Under model (4.5) we have the decomposition

$$\left(\mathbf{X}'\mathbf{X}\right)^{-1} = \begin{bmatrix} \overline{\mathbf{x}}'\mathbf{C}^{-1}\overline{\mathbf{x}} & -\left(\mathbf{C}^{-1}\overline{\mathbf{x}}\right)' \\ -\mathbf{C}^{-1}\overline{\mathbf{x}} & \mathbf{C}^{-1} \end{bmatrix}$$

where

$$\underline{\overline{\mathbf{x}}} = \text{ave}_i(\underline{\mathbf{x}}_i), \ \ \mathbf{C} = \sum_{i=1}^{n} (\underline{\mathbf{x}}_i - \underline{\overline{\mathbf{x}}})(\underline{\mathbf{x}}_i - \underline{\overline{\mathbf{x}}})' \tag{4.19}$$

and hence

$$\mathbf{Var}\left(\widehat{\beta}_{1,LS}\right) = \sigma^2 \mathbf{C}^{-1}. \tag{4.20}$$

If $\mathrm{E}u_i \neq 0$, then $\widehat{\boldsymbol{\beta}}_{LS}$ will be biased. However, if the model contains an intercept, the bias will only affect the intercept and *not* the slopes. More precisely, under (4.5)

$$\mathrm{E}\widehat{\boldsymbol{\beta}}_{1,LS} = \boldsymbol{\beta}_1 \tag{4.21}$$

although $\mathrm{E}\widehat{\beta}_{0,LS} \neq \beta_0$ (see Section 4.9.2 for details).

Let p^* be the rank of $\mathbf{X}$ and recall that if $p^* < p$, i.e., if $\mathbf{X}$ is collinear, then $\widehat{\boldsymbol{\beta}}_{LS}$ is not uniquely defined but all solutions of (4.13) yield the same residuals. Then an unbiased estimate of σ^2 is well defined by

$$s^2 = \frac{1}{n - p^*} \sum_{i=1}^{n} r_i^2, \tag{4.22}$$

whether or not $\mathbf{X}$ is of full rank.

If the u_i's are normal and $\mathbf{X}$ is of full rank, then $\widehat{\boldsymbol{\beta}}_{LS}$ is multivariate normal

$$\widehat{\boldsymbol{\beta}}_{LS} \sim \mathrm{N}_p\left(\boldsymbol{\beta}, \sigma^2 \left(\mathbf{X}'\mathbf{X}\right)^{-1}\right), \tag{4.23}$$

where $\mathrm{N}_p(\boldsymbol{\mu}, \boldsymbol{\Sigma})$ denotes the p-variate normal distribution with mean vector $\boldsymbol{\mu}$ and covariance matrix $\boldsymbol{\Sigma}$.

Let γ now be a linear combination of the parameters: $\gamma = \boldsymbol{\beta}'\mathbf{a}$ with $\mathbf{a}$ a constant vector. Then the natural estimate of γ is $\widehat{\gamma} = \widehat{\boldsymbol{\beta}}'\mathbf{a}$, which according to (4.23) is $\mathrm{N}(\gamma, \sigma_\gamma^2)$ with

$$\sigma_\gamma^2 = \sigma^2 \mathbf{a}' \left(\mathbf{X}'\mathbf{X}\right)^{-1} \mathbf{a}.$$

An unbiased estimate of σ_γ^2 is

$$\widehat{\sigma}_\gamma^2 = s^2 \mathbf{a}' \left(\mathbf{X}'\mathbf{X}\right)^{-1} \mathbf{a}. \tag{4.24}$$

Confidence intervals and tests for γ may be obtained from the fact that under normality the "t-statistic"

$$T = \frac{\widehat{\gamma} - \gamma}{\widehat{\sigma}_\gamma} \tag{4.25}$$

has a t-distribution with $n - p^*$ degrees of freedom, where $p^* = \text{rank}(\mathbf{X})$. In particular, a confidence upper bound and a two-sided confidence interval for γ with level $1 - \alpha$ are given by

$$\widehat{\gamma} + \widehat{\sigma}_\gamma t_{n-p^*,1-\alpha} \quad \text{and} \quad \left(\widehat{\gamma} - \widehat{\sigma}_\gamma t_{n-p^*,1-\alpha/2}, \widehat{\gamma} + \widehat{\sigma}_\gamma t_{n-p^*,1-\alpha/2}\right) \tag{4.26}$$

where $t_{n,\delta}$ is the δ-quantile of a t-distribution with n degrees of freedom. Similarly, the tests of level α for the null hypothesis $H_0 : \gamma = \gamma_0$ against the two-sided alternative $\gamma \neq \gamma_0$ and the one-sided alternative $\gamma > \gamma_0$ have the rejection regions

$$|\widehat{\gamma} - \gamma_0| > \widehat{\sigma}_\gamma t_{n-p^*,1-\alpha/2} \quad \text{and} \quad \widehat{\gamma} > \gamma_0 + \widehat{\sigma}_\gamma t_{n-p^*,1-\alpha}, \tag{4.27}$$

respectively.

If the u_i's are not normal but have a finite variance, then for large n it can be shown using the central limit theorem that $\widehat{\beta}_{LS}$ is approximately normal, with parameters given by (4.18), provided that

$$\text{none of the } \mathbf{x}_i \text{ is "much larger" than the rest.} \tag{4.28}$$

This condition is formalized in (10.33) in Section 10.9.2. Recall that for large n the quantiles $t_{n,\beta}$ of the t-distribution converge to the quantiles z_β of N(0, 1). For the large-sample theory of the LS estimate see Stapleton (1995) and Huber (1981, p. 157).

4.3 Classical methods for outlier detection

The most popular way to deal with regression outliers is to use LS and try to find the influential observations. After they are identified, some decision must be taken such as modifying or deleting them and applying LS to the modified data. Many numerical and/or graphical procedures called *regression diagnostics* are available for detecting influential observations based on an initial LS fit. They include the familiar Q–Q plots of residuals and plots of residuals vs. fitted values. See Weisberg (1985), Belsley, Kuh and Welsch (1980) or Chatterjee and Hadi (1988) for further details on these methods, as well as for proofs of the statements in this section.

The influence of one observation $\mathbf{z}_i = (\mathbf{x}_i, y_i)$ on the LS estimate depends both on y_i being too large or too small compared to y's from similar $\mathbf{x}$'s and on how "large" $\mathbf{x}_i$ is, i.e., how much *leverage* $\mathbf{x}_i$ has. Most popular diagnostics for measuring the influence of $\mathbf{z}_i = (\mathbf{x}_i, y_i)$ are based on comparing the LS estimate based on the full data with LS based on omitting $\mathbf{z}_i$. Call $\widehat{\beta}$ and $\widehat{\beta}_{(i)}$ the LS estimates based on the full data and on the data without $\mathbf{z}_i$, and let

$$\widehat{\mathbf{y}} = \mathbf{X}\widehat{\beta}, \widehat{\mathbf{y}}_{(i)} = \mathbf{X}\widehat{\beta}_{(i)}$$

where $r_i = r_i(\widehat{\beta})$. Note that if $p^* < p$, then $\widehat{\beta}_{(i)}$ is not unique, but $\widehat{\mathbf{y}}_{(i)}$ is unique. Then the *Cook distance* of $\mathbf{z}_i$ is

$$D_i = \frac{1}{p^* s^2} \left\| \widehat{\mathbf{y}}_{(i)} - \widehat{\mathbf{y}} \right\|^2$$

where $p^* = \text{rank}\,(\mathbf{X})$ and $\widehat{\sigma}$ is the residual standard deviation estimate

$$s^2 = \frac{1}{n - p^*} \sum_{i=1}^{n} r_i^2.$$

Call $\mathbf{H}$ the matrix of the orthogonal projection on the image of $\mathbf{X}$; that is, on the subspace $\{\mathbf{X}\boldsymbol{\beta} : \boldsymbol{\beta} \in R^p\}$. The matrix $\mathbf{H}$ is the so-called "hat matrix" and its diagonal elements $h_1, \ldots, h_n$ are the *leverages* of $\mathbf{x}_1, \ldots, \mathbf{x}_n$. If $p^* = p$, then $\mathbf{H}$ fulfills

$$\mathbf{H} = \mathbf{X}\left(\mathbf{X}'\mathbf{X}\right)^{-1}\mathbf{X}' \quad \text{and} \quad h_i = \mathbf{x}_i'\left(\mathbf{X}'\mathbf{X}\right)^{-1}\mathbf{x}_i. \tag{4.29}$$

The h_i's satisfy

$$\sum_{i=1}^{n} h_i = p^*, \, h_i \in [0, 1]. \tag{4.30}$$

It can be shown that the Cook distance is easily computed in terms of the h_i:

$$D_i = \frac{r_i^2}{s^2} \frac{h_i}{p^*(1-h_i)^2}. \tag{4.31}$$

It follows from (4.31) that observations with high leverage are more influential than observations with low leverage having the same residuals.

When the regression has an intercept,

$$h_i = \frac{1}{n} + (\underline{\mathbf{x}}_i - \overline{\underline{\mathbf{x}}})'\left(\mathbf{X}^{*\prime}\mathbf{X}^*\right)^{-1}(\mathbf{x}_i - \overline{\underline{\mathbf{x}}}) \tag{4.32}$$

where $\overline{\underline{\mathbf{x}}}$ is the average of the $\underline{\mathbf{x}}_i$'s and $\mathbf{X}^*$ is the $n \times (p-1)$ matrix whose i-th row is $(\underline{\mathbf{x}}_i - \overline{\underline{\mathbf{x}}})'$. In this case h_i is a measure of how far $\underline{\mathbf{x}}_i$ is from the average value $\overline{\underline{\mathbf{x}}}$.

Calculating h_i does not always require the explicit computation of $\mathbf{H}$. For example, in the case of the two-way design (4.10) it follows from the symmetry of the design that all the h_i's are equal, and then (4.30) yields

$$h_i = \frac{p^*}{n} = \frac{I+J-1}{IJ}.$$

While D_i can detect outliers in simple situations, it fails for more complex configurations and may even fail to recognize a single outlier. The reason is that r_i, h_i and s may be largely influenced by the outlier. It is safer to use statistics based on the "leave-one-out" approach , as follows. The leave-one-out residual $r_{(i)} = y_i - \widehat{\boldsymbol{\beta}}_{(i)}'\mathbf{x}_i$ is known to be expressible as

$$r_{(i)} = \frac{r_i}{1-h_i}. \tag{4.33}$$

and it is shown in the above references that

$$\text{Var}(r_{(i)}) = \frac{\sigma^2}{1-h_i}.$$

An estimate of σ^2 which is free of the influence of $\mathbf{x}_i$ is the quantity $s_{(i)}^2$ that is defined like s^2, but deleting the i-th observation from the sample. It is also shown in the above-mentioned references that

$$s_{(i)}^2 = \frac{1}{n-p^*-1}\left[(n-p^*)s^2 - \frac{r_i^2}{1-h_i}\right], \tag{4.34}$$

and a Studentized version of $r_{(i)}$ is given by

$$t_{(i)} = \sqrt{1-h_i}\frac{r_{(i)}}{s_{(i)}} = \frac{1}{\sqrt{1-h_i}}\frac{r_i}{s_{(i)}}. \tag{4.35}$$

Under the normal distribution model, $t_{(i)}$ has a t-distribution with $n-1$ degrees of freedom. Then a test of outlyingness with significance level α is to decide that the i-th observation is an outlier if $|t_{(i)}| > t_{n-1,(1-\alpha)/2}$. A graphical analysis is provided by the normal Q–Q plot of $t_{(i)}$.

While the above "complete" leave-one-out approach ensures the detection of an isolated outlier, it can still be fooled by the combined action of several outliers, an effect that is referred to as *masking*.

Example 4.2 *The data set* ***oats*** *in Table 4.3 (Scheffé, 1959, p. 138) lists the yield of grain for eight varieties of oats in five replications of a randomized-block experiment.*

Fitting (4.10) by LS (code **oats**) yields residuals with no noticeable structure, and the usual F-tests for row and column effects have highly significant p-values of 0.00002 and 0.001, respectively. To show the effect of outliers on the classical procedure, we have modified five data values. Table 4.4 shows the data with the five altered values in boldface.

Figure 4.2 shows the normal Q–Q plot of $t_{(i)}$ for the altered data. Again, nothing suspicious appears. But the p-values of the F-tests are now 0.13 and 0.04, the first of which is quite insignificant and the second of which is barely significant at the liberal 0.05 level. The diagnostics have thus failed to point out a departure from the model, with serious consequences.

Table 4.3 Oats data

	Block				
Variety	I	II	III	IV	V
1	296	357	340	331	348
2	402	390	431	340	320
3	437	334	426	320	296
4	303	319	310	260	242
5	469	405	442	487	394
6	345	342	358	300	308
7	324	339	357	352	230
8	488	374	401	338	320

Table 4.4 Modified oats data

	Block				
Variety	I	II	III	IV	V
1	**476**	357	340	331	348
2	402	390	431	340	320
3	437	334	426	320	296
4	303	319	310	260	**382**
5	469	405	442	**287**	394
6	345	342	358	300	308
7	324	339	357	352	**410**
8	**288**	374	401	338	320

There is a vast literature on regression diagnostics. A more complex but more reliable method of detecting influential groups of outliers may be found in Peña and Yohai (1999).

All these procedures are fast, and are much better than naively fitting LS without further care. But they are inferior to robust methods in several senses:

- they may fail in the presence of masking
- the distribution of the resulting estimate is unknown

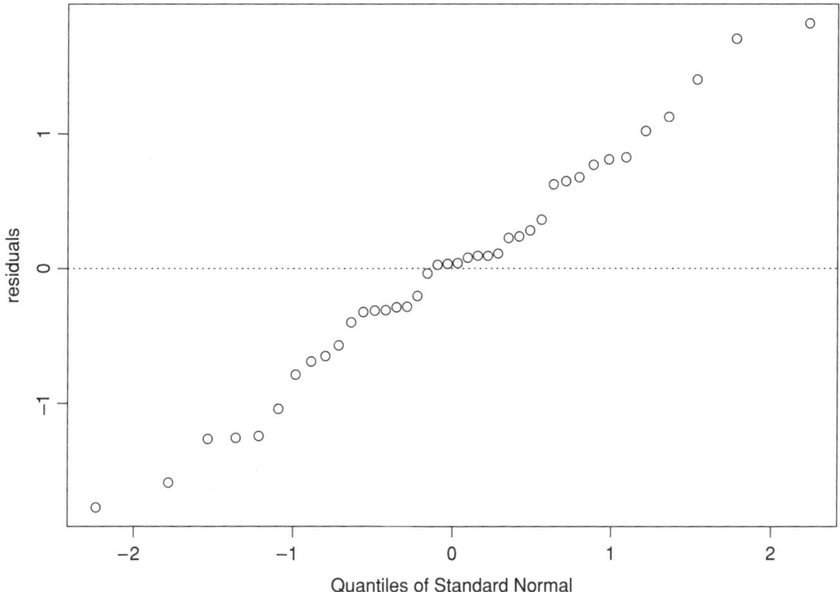

Figure 4.2 Altered oats data: Q–Q plot of LS residuals

- the variability may be underestimated
- once an outlier is found further ones may appear, and it is not clear when one should stop.

4.4 Regression M-estimates

As in Section 2.2 we shall now develop estimates combining robustness and efficiency. Assume model (4.4) with *fixed* $\mathbf{X}$ where u_i has a density

$$\frac{1}{\sigma} f_0\left(\frac{u}{\sigma}\right),$$

where σ is a scale parameter. For the linear model (4.4) the y_i's are independent but not identically distributed, y_i has density

$$\frac{1}{\sigma} f_0\left(\frac{y - \mathbf{x}_i'\boldsymbol{\beta}}{\sigma}\right)$$

and the likelihood function for $\boldsymbol{\beta}$ assuming a fixed value of σ is

$$L(\boldsymbol{\beta}) = \frac{1}{\sigma^n} \prod_{i=1}^{n} f_0\left(\frac{y_i - \mathbf{x}_i'\boldsymbol{\beta}}{\sigma}\right).$$

Calculating the MLE means maximizing $L(\boldsymbol{\beta})$, which is equivalent to finding $\widehat{\boldsymbol{\beta}}$ such that

$$\frac{1}{n} \sum_{i=1}^{n} \rho_0\left(\frac{r_i(\widehat{\boldsymbol{\beta}})}{\sigma}\right) + \log \sigma = \min, \tag{4.36}$$

where $\rho_0 = -\log f_0$ as in (2.13). We shall deal with estimates defined by (4.36). Continuing to assume σ is known and differentiating with respect to $\boldsymbol{\beta}$ we have the analog of the normal equations:

$$\sum_{i=1}^{n} \psi_0\left(\frac{r_i(\widehat{\boldsymbol{\beta}})}{\sigma}\right) \mathbf{x}_i = \mathbf{0}, \tag{4.37}$$

where $\psi_0 = \rho_0' = -f_0'/f_0$. If f_0 is the standard normal density then $\widehat{\boldsymbol{\beta}}$ is the LS estimate (4.12), and if f_0 is the double-exponential density then $\widehat{\boldsymbol{\beta}}$ satisfies

$$\sum_{i=1}^{n} \left| r_i(\widehat{\boldsymbol{\beta}}) \right| = \min$$

and $\widehat{\boldsymbol{\beta}}$ is called an *L1 estimate,* which is the regression equivalent of the median. It is remarkable that this estimate was studied before LS (by Boscovich in 1757 and Laplace in 1799). Differentiating the likelihood function in this case gives

$$\sum_{i=1}^{n} \operatorname{sgn}(r_i(\widehat{\boldsymbol{\beta}}))\mathbf{x}_i = \mathbf{0} \tag{4.38}$$

where "sgn" denotes the sign function (2.19). If the model contains an intercept term (4.38) implies that the residuals have zero median.

Unlike LS there are in general no explicit expressions for an L1 estimate. However, there exist very fast algorithms to compute it (Barrodale and Roberts, 1973; Portnoy and Koenker, 1997). We note also that an L1 estimate $\widehat{\boldsymbol{\beta}}$ may not be unique, and it has the property that at least p residuals are zero (Bloomfield and Staiger, 1983).

We define *regression M-estimates* as solutions $\widehat{\boldsymbol{\beta}}$ to

$$\sum_{i=1}^{n} \rho\left(\frac{r_i(\widehat{\boldsymbol{\beta}})}{\widehat{\sigma}}\right) = \min \tag{4.39}$$

where $\hat{\sigma}$ is an error scale estimate. Differentiating (4.39) yields the equation

$$\sum_{i=1}^{n} \psi\left(\frac{r_i(\widehat{\boldsymbol{\beta}})}{\widehat{\sigma}}\right) \mathbf{x}_i = \mathbf{0} \tag{4.40}$$

where $\psi = \rho'$. The last equation need not be the estimating equation of a MLE. In most situations considered in this chapter, $\widehat{\sigma}$ is computed previously, but it can also be computed simultaneously through a scale M-estimating equation.

It will henceforth be assumed that ρ and ψ are respectively a ρ- and a ψ-function in the sense of Definitions 2.1–2.2. The matrix $\mathbf{X}$ will be assumed to have full rank . In the special case where σ is assumed known, the reader may verify that the estimates are regression and affine equivariant (see Problem 4.1). The case of estimated σ is treated in Section 4.4.2.

Solutions to (4.40) with monotone (resp. redescending) ψ are called *monotone* (resp. *redescending) regression M-estimates*. The main advantage of monotone estimates is that all solutions of (4.40) are solutions of (4.39). Furthermore, if ψ is increasing then the solution is unique (see Theorem 10.15). The example in Section 2.7.1 showed that in the case of redescending location estimates, the estimating equation may have "bad" roots. This cannot happen with monotone estimates. On the other hand, we have seen in Section 3.4 that redescending M-estimates of location yield a better trade-off between robustness and efficiency, and the same can be shown to hold in the regression context. Computing redescending estimates requires a starting point, and this will be the main role of monotone estimates. This matter is pursued further in Section 4.4.2.

4.4.1 M-estimates with known scale

Assume model (4.4) with u such that

$$\mathrm{E}\psi\left(\frac{u}{\sigma}\right) = 0 \tag{4.41}$$

which holds in particular if u is symmetric. Then if (4.28) holds, $\widehat{\boldsymbol{\beta}}$ is consistent for $\boldsymbol{\beta}$ in the sense that

$$\widehat{\boldsymbol{\beta}} \to_p \boldsymbol{\beta} \tag{4.42}$$

when $n \to \infty$, and furthermore for large n

$$\mathcal{D}(\widehat{\boldsymbol{\beta}}) \approx \mathrm{N}_p(\boldsymbol{\beta}, v(\mathbf{X}'\mathbf{X})^{-1}) \tag{4.43}$$

where v is the same as in (2.63):

$$v = \sigma^2 \frac{\mathrm{E}\psi\,(u/\sigma)^2}{(\mathrm{E}\psi'\,(u/\sigma))^2}. \tag{4.44}$$

A general proof is given by Yohai and Maronna (1979).

Thus the approximate covariance matrix of an M- estimate differs only by a constant factor from that of the LS estimate. Hence its efficiency for normal u's does not depend on $\mathbf{X}$, i.e.,

$$\mathrm{Eff}\big(\widehat{\boldsymbol{\beta}}\big) = \frac{\sigma_0^2}{v} \tag{4.45}$$

where v is given by (4.44) with the expectations computed for $u \sim \mathrm{N}\big(0, \sigma_0^2\big)$. It is easy to see that the efficiency does not depend on σ_0.

It is important to note that if we have a model with intercept (4.5) and (4.41) does not hold, then the intercept is asymptotically biased, but the slope estimates are nonetheless consistent (see Section 4.9.2):

$$\widehat{\boldsymbol{\beta}}_1 \to_p \boldsymbol{\beta}_1. \tag{4.46}$$

4.4.2 M-estimates with preliminary scale

For estimating location with an M-estimate in Section 2.6.1 we estimated σ using the MAD. Here the equivalent procedure is first to compute the L1 fit and from it obtain the analog of the normalized MAD by taking the median of the nonnull absolute residuals:

$$\widehat{\sigma} = \frac{1}{0.675}\mathrm{Med}_i(|r_i| \mid r_i \neq 0). \tag{4.47}$$

The reason for using only *nonnull* residuals is that since at least p residuals are null, including all residuals when p is large could lead to underestimating σ. Recall that the L1 estimate does not require estimating a scale.

Write $\widehat{\sigma}$ in (4.47) as $\widehat{\sigma}(\mathbf{X}, \mathbf{y})$. Then since the L1 estimate is regression, scale and affine equivariant, it is easy to show that

$$\widehat{\sigma}(\mathbf{X}, \mathbf{y} + \mathbf{X}\boldsymbol{\gamma}) = \widehat{\sigma}(\mathbf{X}, \mathbf{y}), \widehat{\sigma}(\mathbf{X}\mathbf{A}, \mathbf{y}) = \widehat{\sigma}(\mathbf{X}, \mathbf{y}),\ \widehat{\sigma}(\mathbf{X}, \lambda\mathbf{y}) = |\lambda|\,\widehat{\sigma}(\mathbf{X}, \mathbf{y}) \tag{4.48}$$

for all $\boldsymbol{\gamma} \in R^p$, nonsingular $\mathbf{A} \in R^{p\times p}$ and $\lambda \in R$. We say that $\widehat{\sigma}$ is regression and affine *invariant* and scale equivariant.

We then obtain a regression M-estimate by solving (4.39) or (4.40) with $\widehat{\sigma}$ instead of σ. Then (4.48) implies that $\widehat{\boldsymbol{\beta}}$ is regression, affine and scale equivariant (Problem 4.2).

Assume that $\widehat{\sigma} \to_p \sigma$ and that (4.41) holds. Under (4.4) we would expect that for large n the distribution of $\widehat{\beta}$ is approximated by (4.43)–(4.44), i.e., that $\widehat{\sigma}$ can be replaced by σ. Since ψ is odd, this holds in general if the distribution of u_i is symmetric. Thus the efficiency of the estimate does not depend on **X**.

If the model contains an intercept the approximate distribution result holds for the slopes *without* any requirement on u_i. More precisely, assume model (4.5). Then $\widehat{\beta}_1$ is approximately normal with mean β_1 and covariance matrix $v\mathbf{C}^{-1}$ with v given by (4.44) and **C** defined in (4.19) (see Section 10.9.2 for a heuristic proof).

We can estimate v in (4.44) as

$$\widehat{v} = \widehat{\sigma}^2 \frac{\text{ave}_i \left\{\psi\left(r_i/\widehat{\sigma}\right)^2\right\}}{\left[\text{ave}_i \left\{\psi'\left(r_i/\widehat{\sigma}\right)\right\}\right]^2} \frac{n}{n-p} \tag{4.49}$$

where the denominator $n - p$ appears for the same reasons as in (4.22). Hence for large n we may treat $\widehat{\beta}$ as approximately normal:

$$\mathcal{D}(\widehat{\beta}) \approx \mathrm{N}_p\left(\beta, \widehat{v}\left(\mathbf{X}'\mathbf{X}\right)^{-1}\right). \tag{4.50}$$

Thus we can proceed as in (4.24), (4.25), (4.26) and (4.27), but replacing s^2 in (4.24) by the estimate $\widehat{v}$ above so that $\widehat{\sigma}_\gamma^2 = \widehat{v}\mathbf{a}'\left(\mathbf{X}'\mathbf{X}\right)^{-1}\mathbf{a}$, to obtain approximate confidence intervals and tests. In the case of intervals and tests for a single coefficient β_i we have

$$\widehat{\sigma}_{\beta_i}^2 = \widehat{v}\left(\mathbf{X}'\mathbf{X}\right)_{ii}^{-1}$$

where the subscripts ii mean the i-th diagonal element of matrix $\left(\mathbf{X}'\mathbf{X}\right)^{-1}$.

As we have seen in the location case, one important advantage of redescending estimates is that they give null weight to large residuals, which implies the possibility of a high efficiency for both normal and heavy-tailed data. This is valid also for regression since the efficiency depends only on v which is the same as for location. Therefore our recommended procedure is to use L1 as a basis for computing $\widehat{\sigma}$ and as a starting point for the iterative computing of a bisquare M-estimate.

Example 4.1 (continued) *The slope and intercept values for the bisquare M-estimate with 0.85 efficiency are shown in Table 4.2, along with those of the LS estimate using the full data, the LS estimate computed without the points labeled 1, 2 and 4, and the L1 estimate. The corresponding fitted lines are shown in Figure 4.3. The results are very similar to the LS estimate computed without the three atypical points.*

The estimated standard deviations of the slope are 0.122 for LS and 0.050 for the bisquare M-estimate, and the respective confidence intervals with level 0.95 are (−0.849, −0.371) and (−0.580, −0.384). It is seen that the outliers inflate the confidence interval based on the LS estimate relative to that based on the bisquare M-estimate.

Example 4.2 (continued) *Figure 4.4 shows the residual Q–Q plot based on the bisquare M-estimate (code* **oats***), and it is seen that the five modified values stand out from the rest.*

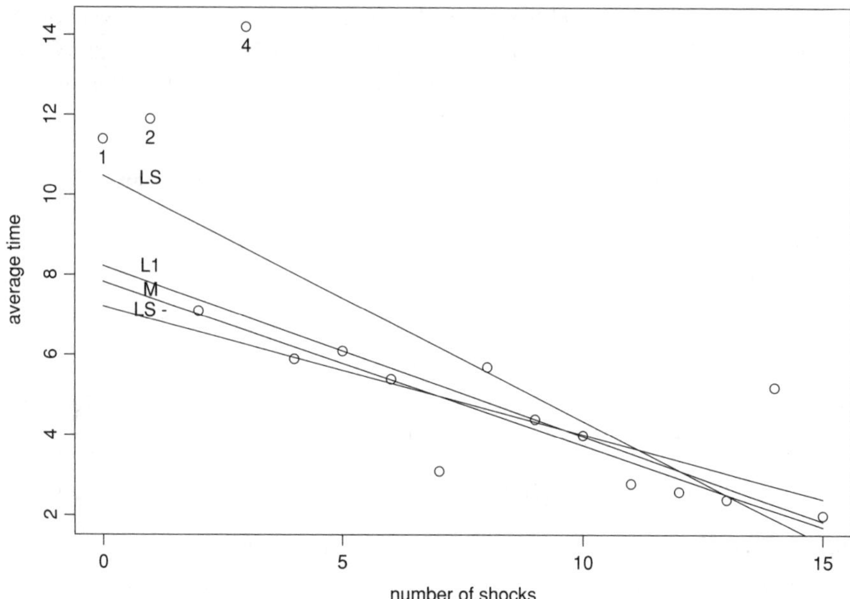

Figure 4.3 Rats data: fits by least squares (LS), L_1, bisquare M-estimate (M) and least squares with outliers omitted (LS-)

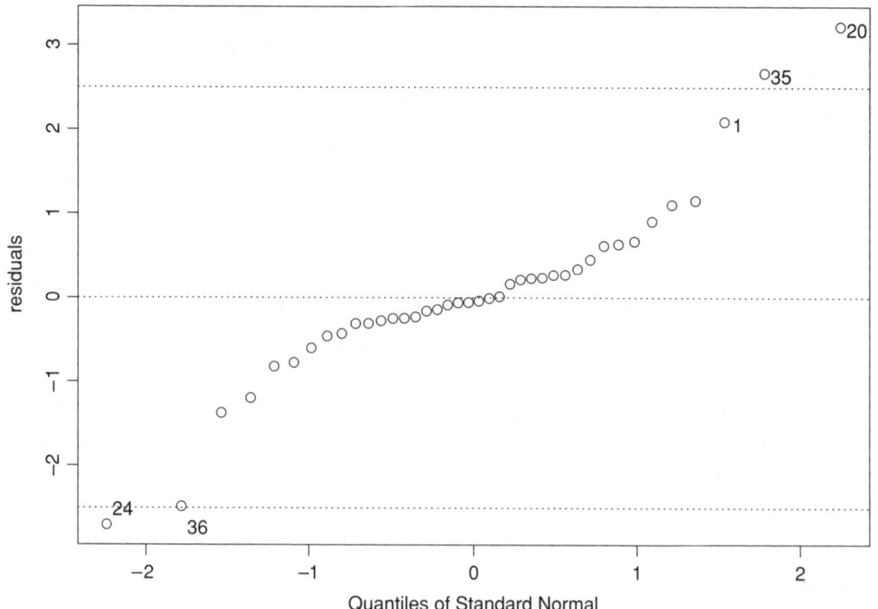

Figure 4.4 Altered oats data: normal Q–Q plot of residuals from M-estimate

Table 4.5 Oats data: p-values of tests

	Rows		Columns	
	F	Robust	F	Robust
Original	0.00002	3×10^{-6}	0.001	0.00008
Altered	0.13	0.0004	0.04	0.0015

Table 4.5 gives the p-values of the robust likelihood ratio-type test to be described in Section 4.7.2 for row and column effects based on the original and the altered data, together with those of the classical F-test already given.

We see that the M-estimate results for the altered data are quite close to those for the original data. Furthermore, for the altered data the robust test again gives strong evidence of row and column effects.

4.4.3 Simultaneous estimation of regression and scale

Another approach to deal with the estimation of σ is to proceed as in Section 2.6.2: that is, to add to the estimating equation (4.40) for $\boldsymbol{\beta}$ an M-estimating equation for σ, resulting in the system

$$\sum_{i=1}^{n} \psi\left(\frac{r_i(\widehat{\boldsymbol{\beta}})}{\widehat{\sigma}}\right) \mathbf{x}_i = \mathbf{0}, \tag{4.51}$$

$$\frac{1}{n}\sum_{i=1}^{n} \rho_{\text{scale}}\left(\frac{r_i(\boldsymbol{\beta})}{\widehat{\sigma}}\right) = \delta, \tag{4.52}$$

where ρ_{scale} is a ρ-function. Note that differentiating (4.36) with respect to $\boldsymbol{\beta}$ and σ yields a system of the form (4.51)–(4.52), with ρ_{scale} given in (2.71). Therefore this class of estimates includes the MLE.

Simultaneous estimates with monotonic ψ are less robust than those of the former section (recall Section 3.2.4 for the location case), but they will be used with redescending ψ in another context in Section 5.6.1.

4.5 Numerical computation of monotone M-estimates

4.5.1 The L1 estimate

As was mentioned above, computing the L1 estimate requires sophisticated algorithms like the one due to Barrodale and Roberts (1973). There are, however, some cases in which this estimate can be computed explicitly. For regression through the origin

$(y_i = \beta x_i + u_i)$, the reader can verify that $\widehat{\beta}$ is a "weighted median" (Problem 4.4). For one-way ANOVA (4.9) we immediately have that $\widehat{\beta}_k = \mathrm{Med}_i\,(y_{ik})$. And for two-way ANOVA with one observation per cell (i.e., (4.10)–(4.11) with $K_{ij} = 1$), there is a simple method that we now describe.

Let $y_{ij} = \mu + \alpha_i + \gamma_j + u_{ij}$. Then differentiating $\sum_i \sum_j |y_{ij} - \mu - \alpha_i - \gamma_j|$ with respect to μ, α_i and γ_j, and recalling that the derivative of $|x|$ is $\mathrm{sgn}\,(x)$, it follows that (4.38) is equivalent to

$$\mathrm{Med}_{i,j}\left(r_{ij}\right) = \mathrm{Med}_j\left(r_{ij}\right) = \mathrm{Med}_i\left(r_{ij}\right) = 0 \text{ for all } i,\, j \tag{4.53}$$

where $r_{ij} = y_{ij} - \widehat{\mu} - \widehat{\alpha}_i - \widehat{\gamma}_j$. These equations suggest an iterative procedure due to J.W. Tukey called "median polish" (Tukey, 1977), which goes as follows where "$a \longleftarrow b$" stands for "replace a by b":

1. Put $\widehat{\alpha}_i = \widehat{\gamma}_j = 0$ for $i = 1, \ldots, I$ and $j = 1, \ldots, J$, and $\widehat{\mu} = 0$, and hence $r_{ij} = y_{ij}$.
2. For $i = 1, \ldots, I$: let $\delta_i = \mathrm{Med}_j\left(r_{ij}\right)$. Update $\widehat{\alpha}_i \longleftarrow \widehat{\alpha}_i + \delta_i$ and $r_{ij} \longleftarrow r_{ij} - \delta_i$.
3. For $j = 1, \ldots, J$: let $\delta_j = \mathrm{Med}_i\left(r_{ij}\right)$. Update $\widehat{\gamma}_j \longleftarrow \widehat{\gamma}_j + \delta_j$ and $r_{ij} \longleftarrow r_{ij} - \delta_j$.
4. Repeat steps 2–3 until no more changes take place.
5. Put $a = \sum_i \widehat{\alpha}_i$ and $b = \sum_j \widehat{\gamma}_j$, and $\widehat{\alpha}_i \longleftarrow \widehat{\alpha}_i - a$, $\widehat{\gamma}_j \longleftarrow \widehat{\gamma}_j - b$, $\widehat{\mu} \longleftarrow a + b$.

If I or J is even, the median must be understood as the "high" or "low" median (Section 1.2), otherwise the procedure may oscillate indefinitely.

It can be shown (Problem 4.5) that the sum of absolute residuals

$$\sum_{i=1}^{I} \sum_{j=1}^{J} \left| y_{ij} - \widehat{\mu} - \widehat{\alpha}_i - \widehat{\gamma}_j \right|$$

decreases at each step of the algorithm. The result frequently coincides with an L1 estimate, and is otherwise generally close to it. Sposito (1987) gives conditions under which the median polish coincides with the L1 estimate.

4.5.2 M-estimates with smooth ψ-function

In the case of a smooth ψ-function one can solve (4.37) using an iterative reweighting method similar to that of Section 2.7. Define W as in (2.30), and then with σ replaced by $\widehat{\sigma}$ the M-estimate equation (4.37) for $\widehat{\beta}$ may be written as

$$\sum_{i=1}^{n} w_i r_i \mathbf{x}_i = \sum_{i=1}^{n} w_i \mathbf{x}_i \left(y_i - \mathbf{x}_i'\widehat{\beta}\right) = \mathbf{0} \tag{4.54}$$

with $w_i = W(r_i/\widehat{\sigma})$. These are "weighted normal equations", and if the w_i's were known, the equations could be solved by applying LS to $\sqrt{w_i}\, y_i$ and $\sqrt{w_i}\,\mathbf{x}_i$. But the

w_i's are not known and depend upon the data. So the procedure, which depends on a tolerance parameter ε, is

1. Compute an initial L1 estimate $\widehat{\boldsymbol{\beta}}_0$ and compute $\widehat{\sigma}$ from (4.47).
2. For $k = 0, 1, 2, \ldots$:
  (a) Given $\widehat{\boldsymbol{\beta}}_k$, for $i = 1, \ldots, n$ compute $r_{i,k} = y_i - \mathbf{x}_i'\widehat{\boldsymbol{\beta}}_k$ and $w_{i,k} = W(r_{i,k}/\widehat{\sigma})$.
  (b) Compute $\widehat{\boldsymbol{\beta}}_{k+1}$ by solving

$$\sum_{i=1}^{n} w_{i,k}\mathbf{x}_i\left(y_i - \mathbf{x}_i'\widehat{\boldsymbol{\beta}}\right) = \mathbf{0}.$$

3. Stop when $\max_i\left(\left|r_{i,k} - r_{i,k+1}\right|\right)/\widehat{\sigma} < \varepsilon$.

This algorithm converges if $W(x)$ is nonincreasing for $x > 0$ (Section 9.1). If ψ is monotone, since the solution is essentially unique, the choice of the starting point influences the number of iterations but not the final result. This procedure is called "iteratively reweighted least squares" (IRWLS).

For simultaneous estimation of $\boldsymbol{\beta}$ and σ the procedure is the same, except that at each iteration $\widehat{\sigma}$ is also updated as in (2.78).

4.6 Breakdown point of monotone regression estimates

In this section we discuss the breakdown point of monotone estimates for nonrandom predictors. Assume $\mathbf{X}$ is of full rank so that the estimates are well defined. Since $\mathbf{X}$ is fixed only $\mathbf{y}$ can be changed, and this requires a modification of the definition of the breakdown point (BP). The FBP for regression with fixed predictors is defined as

$$\varepsilon^* = \frac{m^*}{n},$$

with

$$m^* = \max\left\{m \geq 0 : \widehat{\boldsymbol{\beta}}(\mathbf{X}, \mathbf{y}_m) \text{ bounded } \forall\, \mathbf{y}_m \in \mathcal{Y}_m\right\} \tag{4.55}$$

where $\mathcal{Y}_m$ is the set of n-vectors with at least $n - m$ elements in common with $\mathbf{y}$. It is clear that the LS estimate has $\varepsilon^* = 0$.

Let $k^* = k^*(\mathbf{X})$ be the maximum number of $\mathbf{x}_i$ lying on the same subspace of dimension $< p$:

$$k^*(\mathbf{X}) = \max\left\{\#\left(\boldsymbol{\theta}'\mathbf{x}_i = 0\right) : \boldsymbol{\theta} \in R^p,\ \boldsymbol{\theta} \neq \mathbf{0}\right\} \tag{4.56}$$

where a subspace of dimension 0 is the set $\{0\}$. In the case of simple straight-line regression k^* is the maximum number of repeated x_i's. We have $k^* \geq p - 1$ always.

If $k^* = p - 1$ then $\mathbf{X}$ is said to be in *general position*. In the case of a model with intercept (4.6), $\mathbf{X}$ is in general position iff no more than $p - 1$ of the $\underline{\mathbf{x}}_i$ lie on a hyperplane.

It is shown in Section 4.9.3 that for all regression equivariant estimates

$$\varepsilon^* \leq \overset{*}{\varepsilon}_{\max} := \frac{\overset{*}{m}_{\max}}{n} \tag{4.57}$$

where

$$\overset{*}{m}_{\max} = \left[\frac{n - k^* - 1}{2}\right] \leq \left[\frac{n - p}{2}\right]. \tag{4.58}$$

In the location case, $k^* = 0$ and $\overset{*}{m}_{\max}/n$ becomes (3.25). The FBP of monotone M-estimates is given in Section 4.9.4. For the one-way design (4.9) and the two-way design (4.10) it can be shown that the FBP of monotone M-estimates attains the maximum (4.57) (see Section 4.9.3). In the first case

$$m^* = \overset{*}{m}_{\max} = \left[\frac{\min_j n_j - 1}{2}\right], \tag{4.59}$$

and so if at least half of the elements of the smallest sample are outliers then one of the $\widehat{\beta}_j$ is unbounded. In the second case

$$m^* = \overset{*}{m}_{\max} = \left[\frac{\min(I, J) - 1}{2}\right], \tag{4.60}$$

and so if at least half of the elements of a row or column are outliers then at least one of the estimates $\widehat{\mu}, \widehat{\alpha}_i$ or $\widehat{\gamma}_j$ breaks down. It is natural to conjecture that the FBP of monotone M-estimates attains the maximum (4.57) for all $\mathbf{X}$ such that x_{ij} is either 0 or 1, but no general proof is known.

For designs which are not zero–one designs, the FBP of M-estimates will in general be lower than $\overset{*}{\varepsilon}_{\max}$. This may happen even when there are no leverage points. For example, in the case of a uniform design $x_i = i,\ i = 1, \ldots, n$, for the fitting of a straight line through the origin, we have $k^* = 1$ and hence $\overset{*}{\varepsilon}_{\max} \approx 1/2$, while for large n it can be shown that $\varepsilon^* \approx 0.3$ (see Section 4.9.4). The situation is worse for fitting a polynomial (Problem 4.7). It is even worse when there are leverage points. Consider for instance the design

$$x_i = i \quad \text{for} \quad i = 1, \ldots, 10,\ x_{11} = 100. \tag{4.61}$$

Then it can be shown that $m^* = 0$ for a linear fit (Problem 4.8). The intuitive reason for this fact is that here the estimate is determined almost solely by y_{11}.

As a consequence, monotone M-estimates can be recommended as initial estimates for zero–one designs, and perhaps also for uniform designs, but not for designs where $\mathbf{X}$ has leverage points. The case of random $\mathbf{X}$ will be treated in the next chapter. The techniques discussed there will also be applicable to fixed designs with leverage points.

4.7 Robust tests for linear hypothesis

Regression M-estimates can be used to obtain robust approximate confidence intervals and tests for a single linear combination of the parameters. Define $\widehat{\sigma}_\gamma^2 = \widehat{\sigma}^2\mathbf{a}'\left(\mathbf{X}'\mathbf{X}\right)^{-1}\mathbf{a}$ as in (4.24) but with s^2 replaced by $\widehat{v}$ defined in (4.49). Then the tests and intervals are of the form (4.26)–(4.27). We shall now extend the theory to inference for several linear combinations of the β_j's represented by the vector $\gamma = \mathbf{A}\beta$ where $\mathbf{A}$ is a $q \times p$ matrix of rank q.

4.7.1 Review of the classical theory

To simplify the exposition it will be assumed that $\mathbf{X}$ has full rank, i.e., $p^* = p$, but the results can be shown to hold for general p^*. Assume normally distributed errors and let $\widehat{\gamma} = \mathbf{A}\widehat{\beta}$, where $\widehat{\beta}$ is the LS estimate. Then $\widehat{\gamma} \sim \mathrm{N}(\gamma, \mathbf{\Sigma}_\gamma)$ where

$$\mathbf{\Sigma}_\gamma = \sigma^2\mathbf{A}(\mathbf{X}'\mathbf{X})^{-1}\mathbf{A}'.$$

An estimate of $\mathbf{\Sigma}_\gamma$ is given by

$$\widehat{\mathbf{\Sigma}}_\gamma = s^2\mathbf{A}(\mathbf{X}'\mathbf{X})^{-1}\mathbf{A}'. \tag{4.62}$$

It is proved in standard regression textbooks that $(\widehat{\gamma}-\gamma)'\widehat{\mathbf{\Sigma}}_\gamma^{-1}(\widehat{\gamma}-\gamma)'/q$ has an F-distribution with q and $n - p^*$ degrees of freedom, and hence a confidence ellipsoid for γ of level $1 - \alpha$ is given by

$$\left\{\gamma : (\widehat{\gamma} - \gamma)'\widehat{\mathbf{\Sigma}}_\gamma^{-1}(\widehat{\gamma} - \gamma) \leq qF_{q,n-p^*}(1 - \alpha)\right\},$$

where $F_{n_1,n_2}(\delta)$ is the δ-quantile of an F-distribution with n_1 and n_2 degrees of freedom.

We consider testing the linear hypothesis $H_0 : \gamma = \gamma_0$ for a given γ_0, with level α. The so-called *Wald-type test* rejects H_0 when γ_0 does not belong to the confidence ellipsoid, and hence has rejection region

$$T > F_{q,n-p^*}(1 - \alpha) \tag{4.63}$$

with

$$T = \frac{1}{q}(\widehat{\gamma} - \gamma_0)'\widehat{\mathbf{\Sigma}}_\gamma^{-1}(\widehat{\gamma} - \gamma_0). \tag{4.64}$$

It is also shown in standard texts, e.g., Scheffé (1959), that the statistic T can be written in the form

$$T = \frac{(S_R - S)/q}{S/(n - p^*)} \tag{4.65}$$

where

$$S = \sum_{i=1}^{n} r_i^2\left(\widehat{\beta}\right), \quad S_R = \sum_{i=1}^{n} r_i^2\left(\widehat{\beta}_R\right),$$

where $\widehat{\boldsymbol{\beta}}_R$ is the LS estimate with the restriction $\gamma = \mathbf{A}\boldsymbol{\beta} = \gamma_0$, and that the test based on (4.65) coincides with the likelihood ratio test (LRT). We can also write the test statistic T (4.65) as

$$T = \frac{\left(S_R^* - S^*\right)}{q} \tag{4.66}$$

where

$$S^* = \sum_{i=1}^{n}\left(\frac{r_i\left(\widehat{\beta}\right)}{s}\right)^2, \quad S_R^* = \sum_{i=1}^{n}\left(\frac{r_i\left(\widehat{\beta}_R\right)}{s}\right)^2. \tag{4.67}$$

The most common application of these tests is when H_0 is the hypothesis that some of the coefficients β_i are zero. We may assume without loss of generality that the hypothesis is

$$H_0 = \left\{\beta_1 = \beta_2 = \ldots = \beta_q = 0\right\}$$

which can be written as $H_0 : \boldsymbol{\lambda} = \mathbf{A}\boldsymbol{\beta} = \mathbf{0}$ with $\mathbf{A} = (\mathbf{I}, \mathbf{0})$, where $\mathbf{I}$ is the $q \times q$ identity matrix and $\mathbf{0}$ is a $(p - q) \times p$ matrix with all its elements zero.

When $q = 1$ we have $\gamma = \mathbf{a}'\boldsymbol{\beta}$ with $\mathbf{a} \in R^p$ and then the variance of $\widehat{\gamma}$ is estimated by

$$\widehat{\sigma}_\gamma^2 = \widehat{\sigma}^2\mathbf{a}'(\mathbf{X}'\mathbf{X})^{-1}\mathbf{a}.$$

In this special case the Wald test (4.64) simplifies to

$$T = \left(\frac{\widehat{\gamma} - \gamma_0}{\widehat{\sigma}_\gamma}\right)^2$$

and is equivalent to the two-sided test in (4.27).

When the errors u_i are not normal, but the conditions for the asymptotic normality of $\widehat{\boldsymbol{\beta}}$ given at the end of Section 4.2 hold, the test and confidence regions given in this section will still be approximately valid for large n. For this case, recall that if T has an $F(q, m)$ distribution, then when $m \to \infty$, $qT \to_d \chi_q^2$.

4.7.2 Robust tests using M-estimates

Let $\widehat{\boldsymbol{\beta}}$ now be an M-estimate, and let $\widehat{\Sigma}_{\widehat{\beta}} = \widehat{v}\left(\mathbf{X}'\mathbf{X}\right)^{-1}$ be the estimate of its covariance matrix with $\widehat{v}$ defined in (4.49). Let

$$\widehat{\gamma} = \mathbf{A}\widehat{\boldsymbol{\beta}}, \qquad \widehat{\Sigma}_\gamma = \mathbf{A}\widehat{\Sigma}_{\widehat{\beta}}\mathbf{A}' = \widehat{v}\mathbf{A}(\mathbf{X}'\mathbf{X})^{-1}\mathbf{A}'.$$

Then a robust "Wald-type test" (WTT) is defined by the rejection region

$$\left\{T_W > F_{q,n-p^*}(1-\alpha)\right\}$$

with T_W equal to the right-hand side of (4.64), but the classical quantities there are replaced by the above robust estimates $\widehat{\gamma}$ and $\widehat{\Sigma}_\gamma$.

Let $\widehat{\beta}_R$ be the M-estimate computed with the restriction that $\gamma = \gamma_0$:

$$\widehat{\beta}_R = \arg\min_{\beta} \left\{ \sum_{i=1}^{n} \rho\left(\frac{r_i(\beta)}{\widehat{\sigma}}\right) : \mathbf{A}\beta = \gamma_0 \right\}.$$

A "likelihood ratio-type test" (LRTT) could be defined by the region

$$\left\{T > F_{q,n-p^*}(1-\alpha)\right\},$$

with T equal to the right-hand side of (4.66), but where the residuals in (4.67) correspond to an M-estimate $\widehat{\beta}$. But this test would not be robust, since outliers in the observations y_i would result in corresponding residual outliers and hence an overdue influence on the test statistic.

A robust LRTT can instead be defined by the statistic

$$T_L = \sum_{i=1}^{n} \rho\left(\frac{r_i(\widehat{\beta}_R)}{\widehat{\sigma}}\right) - \sum_{i=1}^{n} \rho\left(\frac{r_i(\widehat{\beta})}{\widehat{\sigma}}\right)$$

with a bounded ρ. Let

$$\xi = \frac{\mathrm{E}\psi'(u/\sigma)}{\mathrm{E}\psi(u/\sigma)^2}.$$

Then it can be shown (see Hampel et al., 1986) that under adequate regularity conditions, ξT_L converges in distribution under H_0 to a chi-squared distribution with q degrees of freedom. Since ξ can be estimated by

$$\widehat{\xi} = \frac{\mathrm{ave}_i\left\{\psi'(r_i(\widehat{\beta})/\widehat{\sigma})\right\}}{\mathrm{ave}_i\left\{\psi(r_i(\widehat{\beta})/\widehat{\sigma})^2\right\}},$$

an approximate LRTT for large n has rejection region

$$\widehat{\xi} T_L > \chi_q^2(1-\alpha),$$

where $\chi_n^2(\delta)$ denotes the δ-quantile of the chi-squared distribution with n degrees of freedom.

Wald-type tests have the drawback of being based on $\mathbf{X}'\mathbf{X}$, which may affect the robustness of the test when there are high leverage points. This makes LRTTs preferable. The influence of high leverage points on inference is discussed further in Section 5.8.

4.8 *Regression quantiles

Let for $\alpha \in (0, 1)$

$$\rho_\alpha(x) = \begin{cases} \alpha x & \text{if} \quad x \geq 0 \\ -(1-\alpha)x & \text{if} \quad x < 0. \end{cases}$$

Then it is easy to show (Problem 2.13) that the solution of

$$\sum_{i=1}^{n} \rho_\alpha(y_i - \mu) = \min$$

is the sample α-quantile. In the same way, the solution of

$$\mathrm{E}\rho_\alpha(y - \mu) = \min$$

is an α-quantile of the random variable y.

Koenker and Bassett (1978) extended this concept to regression, defining the *regression* α-quantile as the solution $\widehat{\boldsymbol{\beta}}$ of

$$\sum_{i=1}^{n} \rho_\alpha\left(y_i - \mathbf{x}_i'\widehat{\boldsymbol{\beta}}\right) = \min. \tag{4.68}$$

The case $\alpha = 0.5$ corresponds to the L1 estimate. Assume the model

$$y_i = \mathbf{x}_i'\boldsymbol{\beta}_\alpha + u_i,$$

where the $\mathbf{x}_i$ are fixed and the α-quantile of u_i is zero; this is equivalent to assuming that the α-quantile of y_i is $\mathbf{x}_i'\boldsymbol{\beta}_\alpha$. Then $\widehat{\boldsymbol{\beta}}$ is an estimate of $\boldsymbol{\beta}_\alpha$.

Regression quantiles are especially useful with heteroskedastic data. Assume the usual situation when the model contains a constant term. If the u_i's are identically distributed, then the $\boldsymbol{\beta}_\alpha$ for different α's differ only in the intercept, and hence regression quantiles do not give much useful information. But if the u_i's have different variability, then the $\boldsymbol{\beta}_\alpha$ will also have different slopes.

If the model is correct, one would like to have for $\alpha_1 < \alpha_2$ that $\mathbf{x}_0'\boldsymbol{\beta}_{\alpha_1} < \mathbf{x}_0'\boldsymbol{\beta}_{\alpha_2}$ for all $\mathbf{x}_0$ in the range of the data. But this cannot be mathematically insured. Although this fact may be taken as an indication of model failure, it is better to insure it from the start. Methods for avoiding the "crossing" of regression quantiles have been proposed by He (1997) and Zhao (2000).

There is a very large literature on regression quantiles; see Koenker, Hammond and Holly (2005) for references.

4.9 Appendix: proofs and complements

4.9.1 Why equivariance?

In this section we want to explain why equivariance is a desirable property for a regression estimate. Let $\mathbf{y}$ verify the model (4.7). Here $\boldsymbol{\beta}$ is the vector of model parameters. If we put for some vector γ

$$\mathbf{y}^* = \mathbf{y} + \mathbf{X}\gamma, \tag{4.69}$$

then $\mathbf{y}^* = \mathbf{X}(\beta + \gamma) + \mathbf{u}$, so that $\mathbf{y}^*$ verifies the model with parameter vector

$$\beta^* = \beta + \gamma. \tag{4.70}$$

If $\widehat{\beta} = \widehat{\beta}(\mathbf{X}, \mathbf{y})$ is an estimate, it would be desirable that if the data were transformed according to (4.69), the estimate would also transform according to (4.70), i.e., $\widehat{\beta}(\mathbf{X}, \mathbf{y}^*) = \widehat{\beta}(\mathbf{X}, \mathbf{y}) + \gamma$, which corresponds to regression equivariance (4.15).

Likewise, if $\mathbf{X}^* = \mathbf{XA}$ for some matrix $\mathbf{A}$, then $\mathbf{y}$ verifies the model

$$\mathbf{y} = (\mathbf{X}^*\mathbf{A}^{-1})\beta + \mathbf{u} = \mathbf{X}^*(\mathbf{A}^{-1}\beta) + \mathbf{u},$$

which is (4.7) with $\mathbf{X}$ replaced by $\mathbf{X}^*$ and β by $\mathbf{A}^{-1}\beta$. Again, it is desirable that estimates transform the same way, i.e., $\widehat{\beta}(\mathbf{X}^*, \mathbf{y}) = \mathbf{A}^{-1}\widehat{\beta}(\mathbf{X}, \mathbf{y})$, which corresponds to affine equivariance (4.17). Scale equivariance (4.16) is dealt with in the same manner.

It must be noted that although equivariance is desirable, it must sometimes be sacrificed for other properties such as a lower prediction error. In particular, the estimates resulting from a procedure for variable selection treated in Section 5.12 are neither regression nor affine equivariant. The same thing happens in general with procedures for dealing with a large number of variables like ridge regression or least-angle regression (Hastie, Tibshirani and Friedman, 2001).

4.9.2 Consistency of estimated slopes under asymmetric errors

We shall first prove (4.21). Let $\alpha = \mathrm{E}u_i$. Then (4.5) may be rewritten as

$$y_i = \beta_0^* + \underline{\mathbf{x}}_i'\beta_1 + u_i^*, \tag{4.71}$$

where

$$u_i^* = u_i - \alpha, \quad \beta_0^* = \beta_0 + \alpha. \tag{4.72}$$

Since $\mathrm{E}u_i^* = 0$, the LS estimate is unbiased for the parameters, which means that $\mathrm{E}(\widehat{\beta}_1) = \beta_1$ and $\mathrm{E}(\widehat{\beta}_0) = \beta_0^*$, so that only the intercept will be biased.

We now prove (4.46) along the same lines as above. Let α be such that

$$\mathrm{E}\psi\left(\frac{u_i - \alpha}{\sigma}\right) = 0.$$

Then reexpressing the model as (4.71)–(4.72), since $\mathrm{E}\psi(u_i^*/\sigma) = 0$, we may apply (4.42), and hence

$$\widehat{\beta}_0 \to_p \beta_0^*, \quad \widehat{\beta}_1 \to_p \beta_1,$$

which implies that the estimate of the slopes is consistent, although that of the intercept may be inconsistent.

4.9.3 Maximum FBP of equivariant estimates

The definition of the FBP in Section 4.6 can be modified to include the case of rank($\mathbf{X}$) $< p$. Since in this case there exists $\boldsymbol{\theta} \neq \mathbf{0}$ such that $\mathbf{X}\boldsymbol{\theta} = \mathbf{0}$, (4.56) is modified as

$$k^*(\mathbf{X}) = \max\left\{\#\left(\boldsymbol{\theta}'\mathbf{x}_i = 0\right) : \boldsymbol{\theta} \in R^p,\ \mathbf{X}\boldsymbol{\theta} \neq \mathbf{0}\right\}. \tag{4.73}$$

If rank($\mathbf{X}$) $< p$, there are infinite solutions to the equations, but all of them yield the same fit $\mathbf{X}\widehat{\boldsymbol{\beta}}$. We thus modify (4.55) with the requirement that the *fit* remains bounded:

$$m^* = \max\left\{m \geq 0 : \mathbf{X}\widehat{\boldsymbol{\beta}}(\mathbf{X}, \mathbf{y}_m) \text{ bounded } \forall\, \mathbf{y}_m \in \mathcal{Y}_m\right\}.$$

We now prove the bound (4.58).

Let $m = m^*_{\max} + 1$. We have to show that $\mathbf{X}\widehat{\boldsymbol{\beta}}(\mathbf{X}, \mathbf{y})$ is unbounded for $\mathbf{y} \in \mathcal{Y}_m$. By decomposing into the case of even and odd $n - k^*$, it follows that

$$2m \geq n - k^*. \tag{4.74}$$

In fact, if $n - k^*$ is even, $n - k^* = 2q$, hence

$$m^*_{\max} = \left[\frac{n - k^* - 1}{2}\right] = [q - 0.5] = q - 1,$$

which implies $m = q$ and hence $2m = n - k^*$; the other case follows similarly.

By the definition of k^*, there exists $\boldsymbol{\theta}$ such that $\mathbf{X}\boldsymbol{\theta} \neq \mathbf{0}$ and $\boldsymbol{\theta}'\mathbf{x}_i = 0$ for a set of size k^*. To simplify notation, we reorder the $\mathbf{x}_i$'s so that

$$\boldsymbol{\theta}'\mathbf{x}_i = 0 \text{ for } i = 1, \ldots, k^*. \tag{4.75}$$

Let for some $t \in R$

$$y_i^* = y_i + t\boldsymbol{\theta}'\mathbf{x}_i \quad \text{for} \quad i = k^* + 1, \ldots, k^* + m \tag{4.76}$$

$$y_i^* = y_i \qquad\qquad \text{otherwise.} \tag{4.77}$$

Then $\mathbf{y}^* \in \mathcal{Y}_m$. Now let $\mathbf{y}^{**} = \mathbf{y}^* - t\mathbf{X}\boldsymbol{\theta}$. Then $y_i^{**} = y_i$ for $1 \leq i \leq k^*$ by (4.75), and also for $k^* + 1 \leq i \leq k^* + m$ by (4.76). Then $\mathbf{y}^{**} \in \mathcal{Y}_m$, since

$$\#\left(i : y_i^{**} = y_i\right) \geq k^* + m$$

and $n - (k^* + m) \leq m$ by (4.74). Hence the equivariance (4.15) implies that

$$\mathbf{X}\widehat{\boldsymbol{\beta}}(\mathbf{X}, \mathbf{y}^*) - \mathbf{X}\widehat{\boldsymbol{\beta}}(\mathbf{X}, \mathbf{y}^{**}) = \mathbf{X}\left(\widehat{\boldsymbol{\beta}}(\mathbf{X}, \mathbf{y}^*) - \widehat{\boldsymbol{\beta}}(\mathbf{X}, \mathbf{y}^* - t\mathbf{X}\boldsymbol{\theta})\right) = t\mathbf{X}\boldsymbol{\theta},$$

which is unbounded for $t \in R$, and thus both $\mathbf{X}\widehat{\boldsymbol{\beta}}(\mathbf{X}, \mathbf{y}^*)$ and $\mathbf{X}\widehat{\boldsymbol{\beta}}(\mathbf{X}, \mathbf{y}^{**})$ cannot be bounded.

4.9.4 The FBP of monotone M-estimates

We now state the FBP of monotone M-estimates, which was derived by Ellis and Morgenthaler (1992) for the L1 estimate and generalized by Maronna and Yohai (2000).

Let ψ be nondecreasing and bounded. Call Ξ the image of $\mathbf{X} : \Xi = \{\mathbf{X}\boldsymbol{\theta} : \boldsymbol{\theta} \in R^p\}$. For each $\boldsymbol{\xi} = (\xi_1, \ldots, \xi_n)' \in R^n$ let $\{i_j : j = 1, \ldots, n\} = \{i_j(\boldsymbol{\xi})\}$ be a permutation that sorts the $|\xi_i|$'s in reverse order:

$$|\xi_{i_1}| \geq \ldots \geq |\xi_{i_n}|; \tag{4.78}$$

and let

$$m(\boldsymbol{\xi}) = \min\left\{m : \sum_{j=1}^{m+1} |\xi_{i_j}| \geq \sum_{j=m+2}^{n} |\xi_{i_j}|\right\}. \tag{4.79}$$

Then it is proved in Maronna and Yohai (1999) that

$$m^* = m^*(\mathbf{X}) = \min\{m(\boldsymbol{\xi}) : \boldsymbol{\xi} \in \Xi,\ \boldsymbol{\xi} \neq \mathbf{0}\}. \tag{4.80}$$

Ellis and Morgenthaler (1992) give a version of this result for the L1 estimate, and use the ratio of the sums on both sides of the inequality in (4.79) as a measure of leverage.

In the location case we have $x_i \equiv 1$, hence all ξ_i are equal, and the condition in (4.79) is equivalent to $m + 1 \geq n - m + 1$ which yields $m(\boldsymbol{\xi}) = [(n-1)/2]$ as in (3.25).

Consider now fitting a straight line through the origin with a uniform design $x_i = i$ $(i = 1, \ldots, n)$. Then for all $\boldsymbol{\xi} \neq \mathbf{0}$, ξ_i is proportional to $n - i + 1$, and hence

$$m(\boldsymbol{\xi}) = \min\left\{m : \sum_{j=1}^{m+1}(n - j + 1) \geq \sum_{j=m+2}^{n} (n - j + 1)\right\}.$$

The condition between braces is equivalent to

$$n(n+1) \geq 2(n-m)(n-m-1),$$

and for large n this is equivalent to $(1 - m/n)^2 \leq 1/2$, i.e.,

$$\frac{m}{n} = 1 - \sqrt{1/2} \approx 0.29.$$

The case of a general straight line is dealt with similarly. The proof of (4.59) is not difficult, but that of (4.60) is rather involved (see Maronna and Yohai, 1999).

If $\mathbf{X}$ is uniformly distributed on a p-dimensional spherical surface, it can be proved that $\varepsilon^* \approx \sqrt{0.5/p}$ for large p (Maronna, Bustos and Yohai, 1979) showing that even a fixed design without leverage points may yield a low BP if p is large.

Table 4.6 Hearing data

	Occupation						
Freq.	I	II	III	IV	V	VI	VII
500	2.1	6.8	8.4	1.4	14.6	7.9	4.8
1000	1.7	8.1	8.4	1.4	12.0	3.7	4.5
2000	14.4	14.8	27.0	30.9	36.5	36.4	31.4
3000	57.4	62.4	37.4	63.3	65.5	65.6	59.8
4000	66.2	81.7	53.3	80.7	79.7	80.8	82.4
6000	75.2	94.0	74.3	87.9	93.3	87.8	80.5
Normal	4.1	10.2	10.7	5.5	18.1	11.4	6.1

4.10 Problems

4.1. Let $\widehat{\beta}$ be a solution of (4.39) with *fixed* σ. Show that
(a) if y_i is replaced by $y_i + \mathbf{x}_i'\gamma$, then $\widehat{\beta} + \gamma$ is a solution
(b) if $\mathbf{x}_i$ is replaced by $\mathbf{A}\mathbf{x}_i$, then $\mathbf{A}'^{-1}\widehat{\beta}$ is a solution.

4.2. Let $\widehat{\beta}$ be a solution of (4.39) where $\widehat{\sigma}$ verifies (4.48). Show that $\widehat{\beta}$ is regression, affine and scale equivariant.

4.3. Show that the solution $\widehat{\beta}$ of (4.37) is the LS estimate of the regression of y_i^* on $\mathbf{x}_i$, where $y_i^* = \xi\left(y_i, \mathbf{x}_i'\widehat{\beta}, \widehat{\sigma}\right)$ with ξ defined in (2.92) are "pseudo-observations". Use this fact to define an iterative procedure to compute a regression M-estimate.

4.4. Show that the L1 estimate for the model of regression through the origin $y_i = \beta x_i + u_i$ is the median of $z_i = y_i/x_i$ where z_i has probability proportional to $|x_i|$.

4.5. Verify (4.53) and show that at each step of the median polish algorithm the sum $\sum_i \sum_j \left|y_{ij} - \widehat{\mu} - \widehat{\alpha}_i - \widehat{\gamma}_j\right|$ does not increase.

4.6. Write computer code for the median polish algorithm and apply it to the original and modified oats data of Example 4.2 and to the data of Problem 4.9.

4.7. Show that for large n the FBP given by (4.80) for fitting $y_i = \beta x_i^k + u_i$ with a uniform design of n points is approximately $1 - 0.5^{1/k}$.

4.8. Show that for the fit of $y_i = \beta x_i + u_i$ with the design (4.61), the FBP given by (4.80) is zero.

4.9. Table 4.6 (Roberts and Cohrssen, 1968) gives prevalence rates in percent for men aged 55–64 with hearing levels 16 decibels or more above the audiometric zero, at different frequencies (hertz) and for normal speech. The columns classify the data in seven occupational groups: professional–managerial, farm, clerical sales, craftsmen, operatives, service, laborers. (data set **hearing**). Fit an additive ANOVA model by LS and robustly. Compare the effect of the estimations. These data have also been analyzed by Daniel (1978).

5

Linear Regression 2

5.1 Introduction

The previous chapter concentrated on robust regression estimates for situations where the predictor matrix $\mathbf{X}$ contains no rows $\mathbf{x}_i$ with high leverage, and only the responses $\mathbf{y}$ may contain outliers. In that case a monotone M-estimate is a reliable starting point for computing a robust scale estimate and a redescending M-estimate. But when $\mathbf{X}$ is random, outliers in $\mathbf{X}$ operate as leverage points, and may completely distort the value of a monotone M-estimate when some pairs $(\mathbf{x}_i, y_i)$ are atypical. This chapter will deal with the case of random predictors and one of its main issues is how to obtain good initial values for redescending M-estimates.

The following example shows the failure of a monotone M-estimate when $\mathbf{X}$ is random and there is a single atypical observation.

Example 5.1 *Smith, Campbell and Lichfield (1984) measured the contents (in parts per million) of 22 chemical elements in 53 samples of rocks in Western Australia. The data are given in Table 5.1 (dataset* **miner95***).*

Figure 5.1 plots the zinc (Zn) vs. the copper (Cu) contents. Observation 15 stands out as clearly atypical. The LS fit is seen to be influenced more by this observation than by the rest. But the L1 fit shows the same drawback! Neither the LS nor the L1 fits (code **mineral***) represent the bulk of the data, since they are "attracted" by observation 15, which has a very large abscissa and too high an ordinate. By contrast, the LS fit omitting observation 15 gives a good fit to the rest of the data . Figures 5.2 and 5.3 show the Q–Q plot and the plot of residuals vs. fitted values for the LS estimate. Neither figure reveals the existence of an outlier as indicated by an exceptionally large residual. However, the second figure shows an approximate linear relationship between residuals and fitted values—excepting the point with largest fitted value—and this indicates that the fit is not correct.*

Table 5.1 Mineral data: copper (Cu) and zinc (Zn) contents

Obs.	Cu	Zn	Obs.	Cu	Zn	Obs.	Cu	Zn
1	102	4	2	96	56	3	265	2
4	185	8	5	229	26	6	20	1
7	49	9	8	28	9	9	128	28
10	83	16	11	126	8	12	79	22
13	116	12	14	34	14	15	633	140
16	258	46	17	264	32	18	189	19
19	70	19	20	71	19	21	121	22
22	60	19	23	37	11	24	60	17
25	23	40	26	19	17	27	35	19
28	45	27	29	52	24	30	44	35
31	24	24	32	48	24	33	42	27
34	46	11	35	99	10	36	17	15
37	33	33	38	78	12	39	201	6
40	89	14	41	4	4	42	18	10
43	43	13	44	29	18	45	26	12
46	33	10	47	24	10	48	12	3
49	14	10	50	179	25	51	68	17
52	66	22	53	102	19			

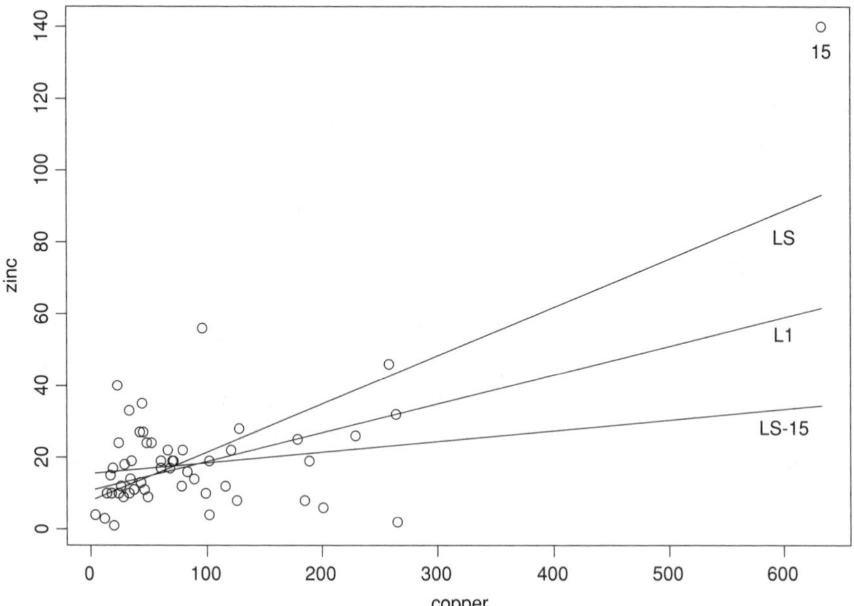

Figure 5.1 Mineral data: fits by LS, L_1, and LS without observation 15

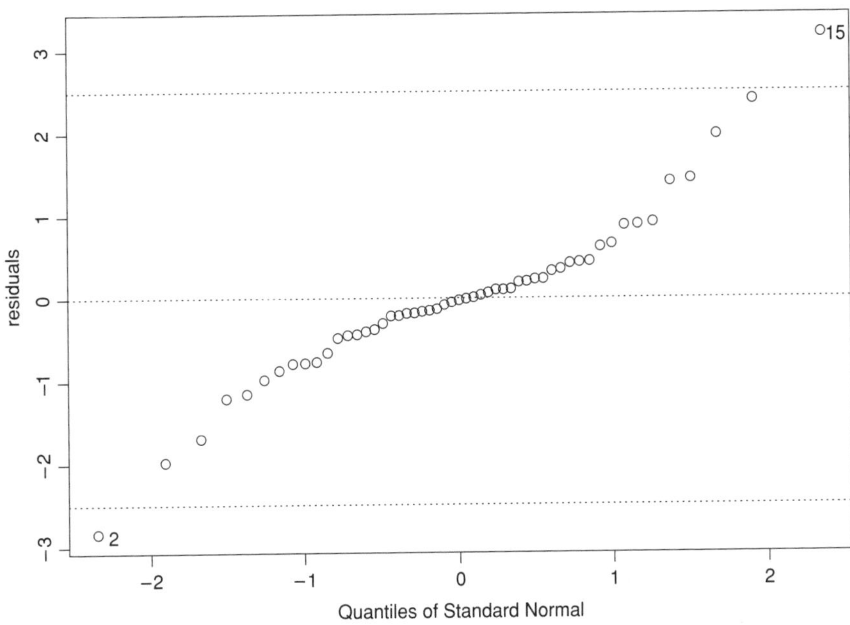

Figure 5.2 Mineral data: Q–Q plot of LS residuals

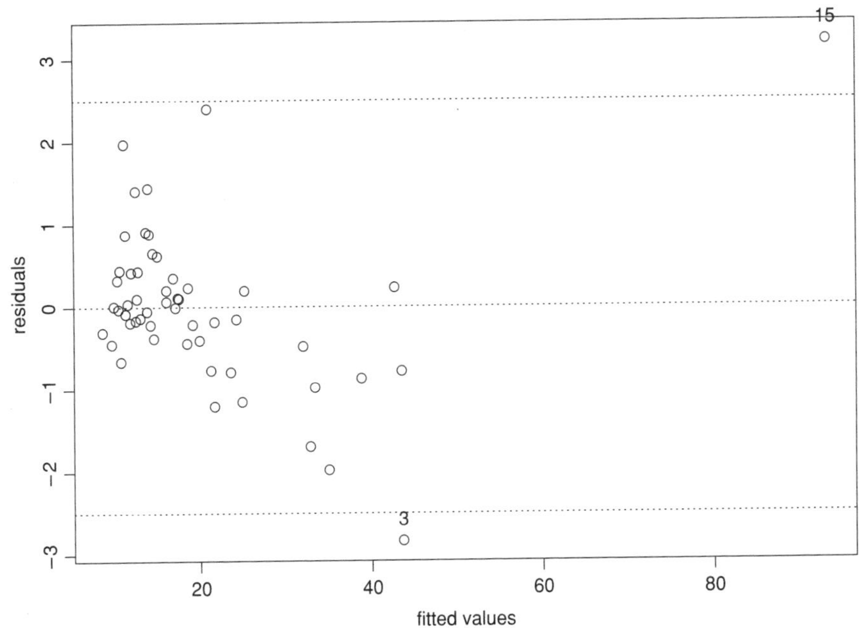

Figure 5.3 Mineral data: LS residuals vs. fit

Table 5.2 Regression coefficients for mineral data

	LS	L1	LS(−15)	Robust
Int.	7.96	10.41	15.49	14.05
Slope	0.134	0.080	0.030	0.022

Table 5.2 gives the estimated parameters for the LS and L1 fits, as well as for the LS fit computed without observation 15, and for a redescending regression M-estimate to be described shortly.

The intuitive reason for the failure of the L1 estimate (and of monotone M-estimates in general) in this situation is that the $\mathbf{x}_i$ outlier dominates the solution to (4.40) in the following sense. If for some i, $\mathbf{x}_i$ is "much larger than the rest", then in order to make the sum zero the residual $y_i - \mathbf{x}_i'\widehat{\beta}$ must be near zero and hence $\widehat{\beta}$ is essentially determined by $(\mathbf{x}_i, y_i)$. This does not happen with the redescending M-estimate.

5.2 The linear model with random predictors

Situations like the one in the previous example occur primarily when $\mathbf{x}_i$ are not fixed as in designed experiments, but instead are random variables observed together with y_i. We now briefly discuss the properties of a linear model with random $\mathbf{X}$. Our observations are now the i.i.d. $(p+1)$-dimensional random vectors $(\mathbf{x}_i, y_i)$ $(i = 1, \ldots, n)$ satisfying the linear model relation

$$y_i = \mathbf{x}_i'\beta + u_i. \tag{5.1}$$

In the case of fixed $\mathbf{X}$ we assumed that the distribution of u_i does not depend on $\mathbf{x}_i$. The analogous assumption here is that

$$\text{the } u_i\text{'s are i.i.d. and independent of the } \mathbf{x}_i\text{'s.} \tag{5.2}$$

The analog of assuming $\mathbf{X}$ is of full rank is to assume that the distribution of $\mathbf{x}$ is not concentrated on any subspace, i.e., $\mathrm{P}(\mathbf{a}'\mathbf{x} = 0) < 1$ for all $\mathbf{a} \neq \mathbf{0}$. This condition implies that the probability that $\mathbf{X}$ has full rank tends to one when $n \rightarrow \infty$, and holds in particular if the distribution of $\mathbf{x}$ has a density. Then the LS estimate is well defined, and (4.18) holds *conditionally* on $\mathbf{X}$:

$$\mathrm{E}\left(\widehat{\beta}_{LS}\middle|\,\mathbf{X}\right) = \beta, \ \mathbf{Var}\left(\widehat{\beta}_{LS}\middle|\,\mathbf{X}\right) = \sigma^2\left(\mathbf{X}'\mathbf{X}\right)^{-1},$$

where $\sigma^2 = \mathrm{Var}(u)$,

Also (4.23) holds conditionally: if the u_i's are normal then the conditional distribution of $\widehat{\beta}_{LS}$ given $\mathbf{X}$ is multivariate normal. If the u_i's are not normal, assume that

$$\mathbf{V}_{\mathbf{x}} = \mathrm{E}\mathbf{x}\mathbf{x}' \tag{5.3}$$

exists. It can be shown that

$$\mathcal{D}(\widehat{\beta}_{LS}) \approx \mathrm{N}_p\left(\beta, \frac{\mathbf{C}_{\widehat{\beta}}}{n}\right) \tag{5.4}$$

where

$$\mathbf{C}_{\widehat{\beta}} = \sigma^2 \mathbf{V}_{\mathbf{x}}^{-1} \tag{5.5}$$

is the asymptotic covariance matrix of $\widehat{\beta}$. See Section 10.9.3. The estimation of $\mathbf{C}_{\widehat{\beta}}$ is discussed in Section 5.8.

In the case (4.5) where the model has an intercept term, it follows from (5.5) that the asymptotic covariance matrix of (β_0, β_1) is

$$\sigma^2 \begin{bmatrix} 1 + \mu_{\mathbf{x}}' \mathbf{C}_{\mathbf{x}}^{-1} \mu_{\mathbf{x}} & \mu_{\mathbf{x}}' \\ \mu_{\mathbf{x}} & \mathbf{C}_{\mathbf{x}}^{-1} \end{bmatrix} \tag{5.6}$$

where

$$\mu_{\mathbf{x}} = \mathrm{E}\mathbf{x}, \quad \mathbf{C}_{\mathbf{x}} = \mathbf{Var}(\mathbf{x}).$$

5.3 M-estimates with a bounded ρ-function

Our approach to robust regression estimates where both $\mathbf{x}_i$ and the y_i may contain outliers is to use an M-estimate $\widehat{\beta}$ defined by

$$\sum_{i=1}^{n} \rho\left(\frac{r_i(\widehat{\beta})}{\widehat{\sigma}}\right) = \min \tag{5.7}$$

with a *bounded* ρ and a high breakdown point preliminary scale $\widehat{\sigma}$. The scale $\widehat{\sigma}$ will be required to fulfill certain requirements discussed later in Section 5.5. If ρ has a derivative ψ it follows that

$$\sum_{i=1}^{n} \psi\left(\frac{r_i}{\widehat{\sigma}}\right) \mathbf{x}_i = \mathbf{0} \tag{5.8}$$

where ψ is redescending (it is easy to verify that a function ρ with a monotonic derivative ψ cannot be bounded). Consequently the estimating equation (5.8) may have multiple solutions corresponding to multiple *local* minima of the function on the left-hand side of (5.7), and generally only one of them (the "good solution") corresponds to the global minimizer $\widehat{\beta}$ defined by (5.7). We shall see that ρ and $\widehat{\sigma}$ may be chosen in order to attain both a high breakdown point and a high efficiency.

In Section 5.5 we describe a particular computing method of approximating $\widehat{\beta}$ defined by (5.7). The method is called an "MM-estimate", and as motivation for its use we apply it to the data of Example 5.1. The results, displayed in Figure 5.4, show that the MM-estimate almost coincides with the LS estimate computed with the data point 15 deleted (code **mineral**). The MM-estimate intercept and slope parameters are now 14.05 and 0.02, respectively, as compared to 7.96 and 0.13 for the LS estimate (recall Table 5.2).

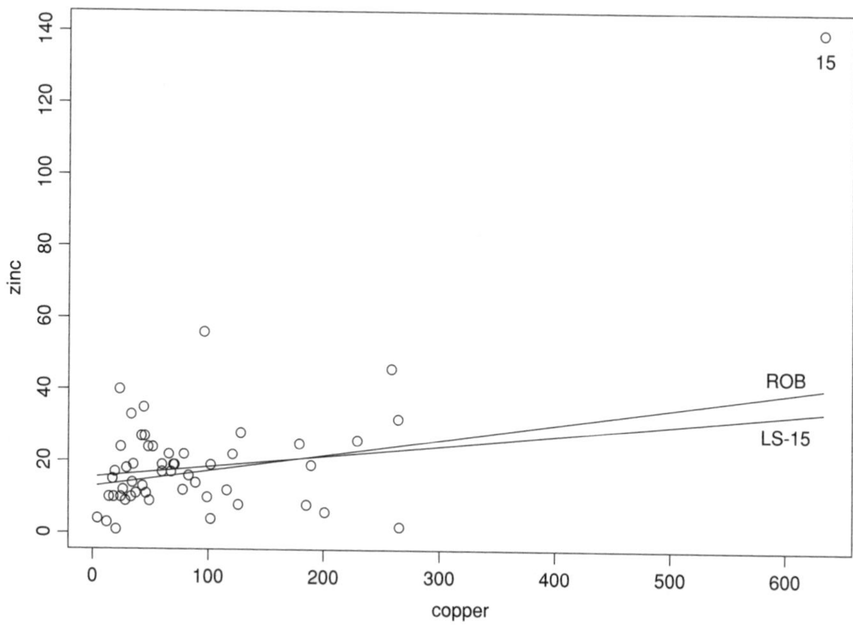

Figure 5.4 Mineral data: fits by MM-estimate ("ROB") and by LS without the outlier

Figure 5.5 shows the residuals vs. fitted values and Figure 5.6 the Q–Q plot of residuals. The former now lacks the suspicious structure of Figure 5.3 and point 15 is now revealed as a large outlier in the residuals as well as the fit, with a considerably reduced value of fit (roughly 40 instead of more than 90). And compared to Figure 5.2 the Q–Q plot now clearly reveals point 15 as an outlier. Figure 5.7 compares the sorted absolute values of residuals from the MM-estimate fit and the LS fit, with point 15 omitted for reasons of scale. It is seen that most points lie below the identity diagonal, showing that except for the outlier the sorted absolute MM-residuals are smaller than those from the LS estimate, and hence the MM-estimate fits the data better.

5.4 Properties of M-estimates with a bounded ρ-function

If $\widehat{\sigma}$ verifies (4.48), then the estimate $\widehat{\beta}$ defined by (4.39) is regression, scale and affine equivariant. We now discuss the breakdown point, influence function and asymptotic normality of such an estimate.

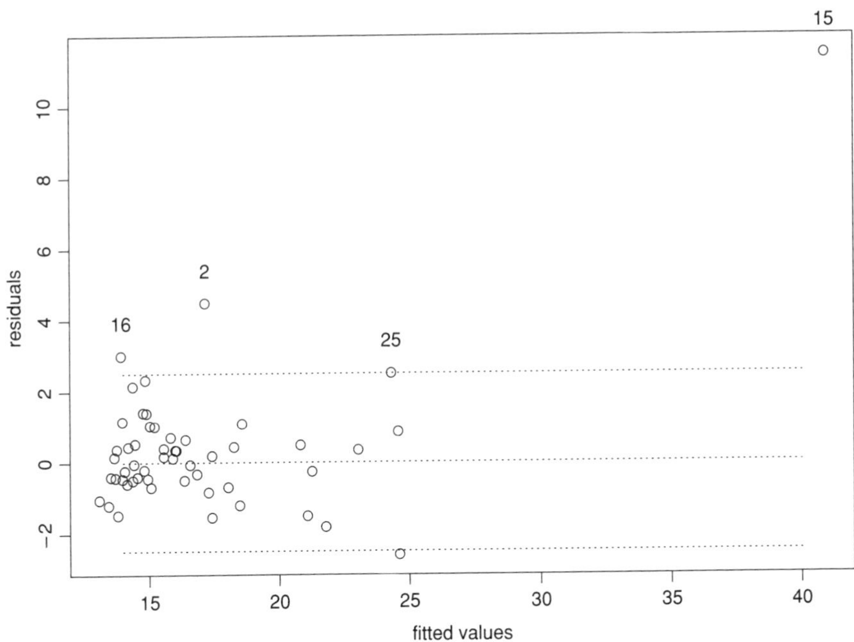

Figure 5.5 Mineral data: residuals vs. fitted values of MM-estimate

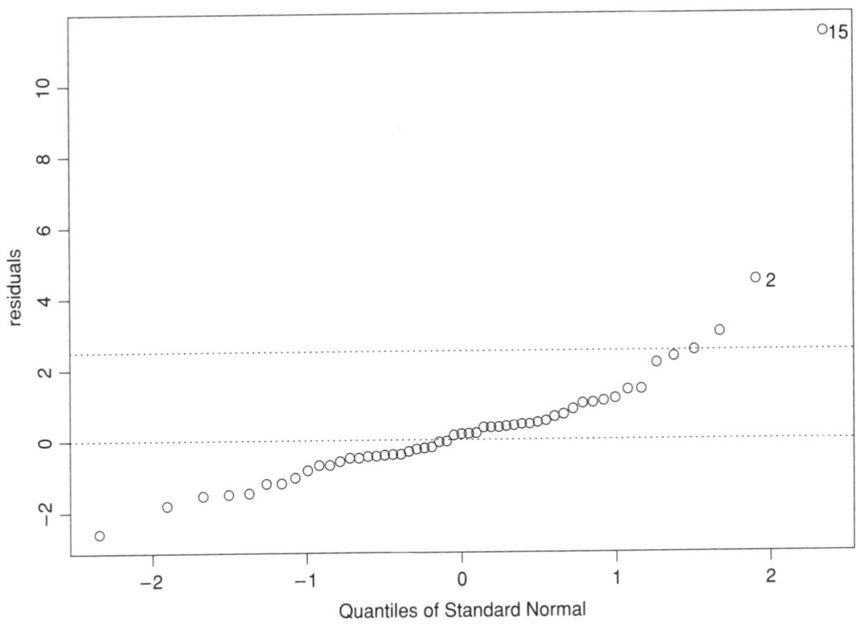

Figure 5.6 Mineral data: Q–Q plot of robust residuals

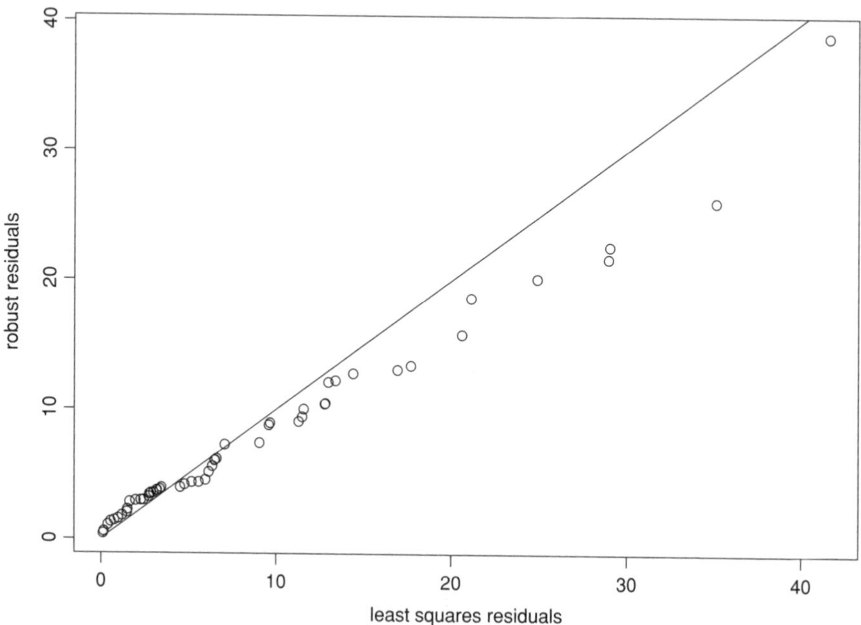

Figure 5.7 Mineral data: sorted absolute values of robust vs. LS residuals (point 15 omitted)

5.4.1 Breakdown point

We focus on the finite breakdown point (FBP) of $\widehat{\beta}$. Since the **x**'s are now random, we are in the situation of Section 3.2.5. Put $\mathbf{z}_i = (\mathbf{x}_i, y_i)$ and write the estimate as $\widehat{\beta}(\mathbf{Z})$ with $\mathbf{Z} = \{\mathbf{z}_1, \ldots, \mathbf{z}_n\}$. Then instead of (4.55) define $\varepsilon^* = m^*/n$ where

$$m^* = \max\left\{m \geq 0 : \widehat{\beta}(\mathbf{Z}_m) \text{ bounded } \forall\, \mathbf{Z}_m \in \mathcal{Z}_m\right\}, \tag{5.9}$$

and $\mathcal{Z}_m$ is the set of datasets with at least $n - m$ elements in common with **Z**. Note that since not only **y** but also **X** are variable here, the FBP given by (5.9) is less than or equal to that given earlier by (4.55).

Then it is easy to show that the FBP of monotone M-estimates is zero (Section 5.16.1). Intuitively this is due to the fact that a term with a "large" $\mathbf{x}_i$ "dominates" the sum in (4.40). Then the scale used in Section 4.4.2, which is based on the residuals from the L1 estimate, also has a zero BP.

On the other hand, it can be shown that the maximum FBP of any regression equivariant estimate is again the one given in Section 4.6

$$\varepsilon^* \leq \varepsilon^*_{\max} =: \frac{1}{n}\left[\frac{n - k^* - 1}{2}\right] \leq \frac{1}{n}\left[\frac{n - p}{2}\right], \tag{5.10}$$

with k^* as in (4.56):

$$k^*(\mathbf{X}) = \max\left\{\#\left(\boldsymbol{\theta}'\mathbf{x}_i = 0\right) : \boldsymbol{\theta} \in R^p,\ \boldsymbol{\theta} \neq \mathbf{0}\right\}. \tag{5.11}$$

The proof is similar to that of Section 4.9.3, and we shall see that this bound is attained by several types of estimates to be defined in this chapter. It can be shown in the same way that the maximum asymptotic BP for regression equivariant estimates is $(1-\alpha)/2$, where

$$\alpha = \max_{\boldsymbol{\theta}\neq\mathbf{0}} \mathrm{P}(\boldsymbol{\theta}'\mathbf{x} = 0). \tag{5.12}$$

In the previous chapter our method for developing robust estimates was to generalize the MLE, which leads to M-estimates with unbounded ρ. In the present setting, calculating the MLE again yields (4.36); in particular, LS is the MLE for normal u, for any $\mathbf{x}$. Thus no new class of estimates emerges from the ML approach.

5.4.2 Influence function

If the joint distribution F of $(\mathbf{x}, y)$ is given by the model (5.1)–(5.2), then it follows from (3.47) that the influence function (IF) of an M-estimate with known σ under the model is

$$\mathrm{IF}((\mathbf{x}_0, y_0), F) = \frac{\sigma}{b}\psi\left(\frac{y_0 - \mathbf{x}_0'\boldsymbol{\beta}}{\sigma}\right)\mathbf{V}_{\mathbf{x}}^{-1}\mathbf{x}_0 \text{ with } b = \mathrm{E}\psi'\left(\frac{u}{\sigma}\right) \tag{5.13}$$

and with $\mathbf{V}_{\mathbf{x}}$ defined by (5.3). The proof is similar to that of Section 3.8.1. It follows that the IF is unbounded. However, the IFs for the cases of monotone and of redescending ψ are rather different. If ψ is monotone, then the IF tends to infinity for any fixed $\mathbf{x}_0$ if y_0 tends to infinity. If ψ is redescending and is such that $\psi(x) = 0$ for $|x| \geq k$, then the IF will tend to infinity only when $\mathbf{x}_0$ tends to infinity and $\left|y_0 - \mathbf{x}_0'\boldsymbol{\beta}\right|/\sigma \leq k$, which means that large outliers have no influence on the estimate.

When σ is unknown and is estimated by $\widehat{\sigma}$, it can be shown that if the distribution of u_i is symmetric, then (5.13) also holds, with σ replaced by the asymptotic value of $\widehat{\sigma}$.

The fact that the IF is unbounded does not necessarily imply that the bias is unbounded for any positive contamination rate ε. In fact, while a monotone ψ implies $\mathrm{BP} = 0$, we shall see in Section 5.5 that with a bounded ρ it is possible to attain a high BP, and hence that the bias is bounded for large values of ε. On the other hand, in Section 5.11 we shall define a family of estimates with bounded IF but such that their BP may be very low for large p. These facts indicate that the IF need not yield a reliable approximation to the bias.

5.4.3 Asymptotic normality

Assume that the model (5.1)–(5.2) holds, that $\mathbf{x}$ has finite variances, and that $\widehat{\sigma}$ converges in probability to some σ. Then it can be proved under rather general conditions (see Section 10.9.3 for details) that the estimate $\widehat{\beta}$ defined by (4.39) is consistent and asymptotically normal. More precisely

$$\sqrt{n}(\widehat{\beta} - \beta) \to_d \mathrm{N}_p \left(0, v\mathbf{V}_{\mathbf{x}}^{-1}\right), \tag{5.14}$$

where $\mathbf{V}_{\mathbf{x}} = \mathrm{E}\mathbf{x}\mathbf{x}'$ and v is as in (4.44)

$$v = \sigma^2 \frac{\mathrm{E}\psi\,(u/\sigma)^2}{(\mathrm{E}\psi'\,(u/\sigma))^2}. \tag{5.15}$$

This result implies that as long as the $\mathbf{x}_i$'s have finite variance, the efficiency of $\widehat{\beta}$ does not depend on the distribution of $\mathbf{x}$.

We have seen in the previous chapter that a leverage point forces the fit of a monotone M-estimate to pass near the point, and this has a double-edged effect: if the point is a "typical" observation, the fit improves (although the normal approximation to the distribution of the estimate deteriorates); if it is "atypical", the overall fit worsens. The implications of these facts for the case of random $\mathbf{x}$ are as follows. Suppose that $\mathbf{x}$ is heavy tailed so that its variances do not exist. If the model (5.1)–(5.2) holds, then the normal ceases to be a good approximation to the distribution of $\widehat{\beta}$, but at the same time $\widehat{\beta}$ is "closer" to β than in the case of "typical" $\mathbf{x}$ (see Section 5.16.2 for details). But if the model does not hold, then $\widehat{\beta}$ may have a higher bias than in the case of "typical" $\mathbf{x}$.

5.5 MM-estimates

Computing an M-estimate requires finding the absolute minimum of

$$L(\beta) = \sum_{i=1}^{n} \rho\left(\frac{r_i\,(\beta)}{\widehat{\sigma}}\right). \tag{5.16}$$

When ρ is bounded, this is an exceedingly difficult task except for $p = 1$ or 2 where a grid search would work. However, we shall see that it suffices to find a "good" local minimum to achieve both a high BP and high efficiency at the normal distribution. This local minimum will be obtained by starting from a reliable starting point and applying the IRWLS algorithm of Section 4.5.2. This starting point will also be used to compute the robust residual scale $\widehat{\sigma}$ required to define the M-estimate, and hence it is necessary that it can be computed without requiring a previous scale.

The L1 estimate does not require a scale, but we have already seen that it is not a convenient estimate when $\mathbf{X}$ is random. Hence we need an initial estimate that is robust toward any kind of outliers and that does not require a previously computed scale. Such a class of estimates will be defined in Section 5.6

The steps of the proposed procedure are thus:

1. Compute an initial consistent estimate $\widehat{\beta}_0$ with high BP but possibly low normal efficiency.

2. Compute a robust scale $\widehat{\sigma}$ of the residuals $r_i(\widehat{\beta}_0)$.
3. Find a solution $\widehat{\beta}$ of (5.8) using an iterative procedure starting at $\widehat{\beta}_0$.

We shall show that in this way we can obtain $\widehat{\beta}$ having both a high BP and a prescribed high efficiency at the normal distribution.

Now we state the details of the above steps. The robust initial estimate $\widehat{\beta}_0$ must be regression, scale and affine equivariant, which ensures that $\widehat{\beta}$ inherits the same properties. For the purposes of the discussion at hand we assume that we have a good initial estimate $\widehat{\beta}_0$ available, and return to discuss such an estimate in Section 5.6. We shall use two different functions ρ and ρ_0, and each of these must be a *bounded* ρ-function in the sense of Definition 2.1 at the end of Section 2.2.4. The scale estimate $\widehat{\sigma}$ must be an M-scale estimate (2.54)

$$\frac{1}{n}\sum_{i=1}^{n}\rho_0\left(\frac{r_i}{\widehat{\sigma}}\right) = 0.5. \tag{5.17}$$

By (3.22) the asymptotic BP of $\widehat{\sigma}$ is 0.5. As was seen at the end of Section 2.5, we can always find c_0 such that using $\rho_0(r/c_0)$ ensures that the asymptotic value of σ coincides with the standard deviation when the u_i's are normal. For the bisquare scale given by (2.57) this value is $c_0 = 1.56$.

The key result is given by Yohai (1987), who called these estimates *MM-estimates*. Recall that all local minima of $L(\beta)$ satisfy (5.8). Let ρ satisfy

$$\rho_0 \geq \rho. \tag{5.18}$$

Yohai (1987) shows that if $\widehat{\beta}$ is such that

$$L(\widehat{\beta}) \leq L(\widehat{\beta}_0), \tag{5.19}$$

then $\widehat{\beta}$ is consistent. It can also be shown in the same way as the similar result for location in Section 3.2.3 that its BP is not less than that of $\widehat{\beta}_0$. If furthermore $\widehat{\beta}$ is any solution of (5.8), then it has the same efficiency as the global minimum. Thus it is not necessary to find the absolute minimum of (5.7) to ensure a high BP and high efficiency.

The numerical computation of the estimate follows the approach in Section 4.5: starting with $\widehat{\beta}_0$ we use the IRWLS algorithm to attain a solution of (5.8). It is shown in Section 9.1 that $L(\beta)$ given in (5.16) decreases at each iteration, which insures (5.19).

It remains to choose ρ in order to attain the desired normal efficiency, which is $1/v$, where v is the expression (5.15) computed at the standard normal. Let ρ_1 be the bisquare ρ-function given by (2.37) with $k = 1$, namely

$$\rho_1(t) = \min\left\{1, 1 - \left(1 - t^2\right)^3\right\}.$$

Let

$$\rho_0(r) = \rho_1\left(\frac{r}{c_0}\right) \quad \text{and} \quad \rho(r) = \rho_1\left(\frac{r}{c_1}\right),$$

with $c_0 = 1.56$ for consistency of the scale at the normal. In order that $\rho \leq \rho_0$ we must have $c_1 \geq c_0$: the larger c_1, the higher the efficiency at the normal distribution. The values of c_1 for prescribed efficiencies are the values of k in the table (2.39).

In Section 5.9 we shall demonstrate the basic trade-off between normal efficiency and bias under contamination: the larger the efficiency, the larger the bias. It is therefore important to choose the efficiency so as to maintain reasonable bias control. The results in Section 5.9 show that an efficiency of 0.95 yields too high a bias, and hence it is safer to choose an efficiency of 0.85 which gives a smaller bias while retaining a sufficiently high efficiency.

Note that M-estimates, and MM-estimates in particular, have an unbounded IF but a high BP. This seeming contradiction can be resolved by noting that an infinite gross-error sensitivity means only that the maximum bias $\mathrm{MB}(\varepsilon)$ is not $O(\varepsilon)$ for small ε, but does not imply that it is infinite! Actually, Yohai and Zamar (1997) have shown that $\mathrm{MB}(\varepsilon) = O(\sqrt{\varepsilon})$ for the estimates considered in this section. This implies that the bias induced by altering a single observation is bounded by $c/\sqrt{n}$ for some constant c, instead of the stronger bound c/n.

Example 5.2 *Maguna, Núñez, Okulik and Castro (2003) measured the aquatic toxicity of 38 carboxylic acids, together with nine molecular descriptors, in order to find a predicting equation for* $y = \log(\text{toxicity})$ *(dataset* ***toxicity****).*

Figures 5.8 and 5.9 respectively show the plot of the residuals vs. fit and the normal Q–Q plot for the LS estimate. No outliers are apparent. Figures 5.10 and 5.11 are the respective plots for the 85% normal efficiency MM-estimate (code **toxicity**), showing about 10 outliers. Figure 5.12 plots the ordered absolute residuals from LS as the abscissa and those from the MM-estimate as the ordinate, as compared to the identity line; the observations with the 10 largest absolute residuals from the MM-estimate were omitted for reasons of scale. The plot shows that the MM-residuals are in general smaller than the LS residuals, and hence MM gives a better fit to the bulk of the data.

5.6 Estimates based on a robust residual scale

In this section we shall present a family of regression estimates that do not depend on a preliminary scale and are thus useful as initial estimates for the MM-estimate. A particular member of this family provides a good initial estimate $\widehat{\boldsymbol{\beta}}_0$ and corresponding residuals to define the preliminary scale $\widehat{\sigma}$ in (5.7) in the MM-estimate method.

The LS and the L1 estimates minimize the averages of the squared and of the absolute residuals respectively, and therefore they minimize measures of residual largeness that can be seriously influenced by even a single residual outlier. A more robust alternative is to minimize a scale measure of residuals that is insensitive to large values, and one such possibility is the *median* of the absolute residuals. This is the basis of the least median of squares (LMS) estimate, introduced as the first estimate of this kind by Hampel (1975) and by Rousseeuw (1984) who also proposed

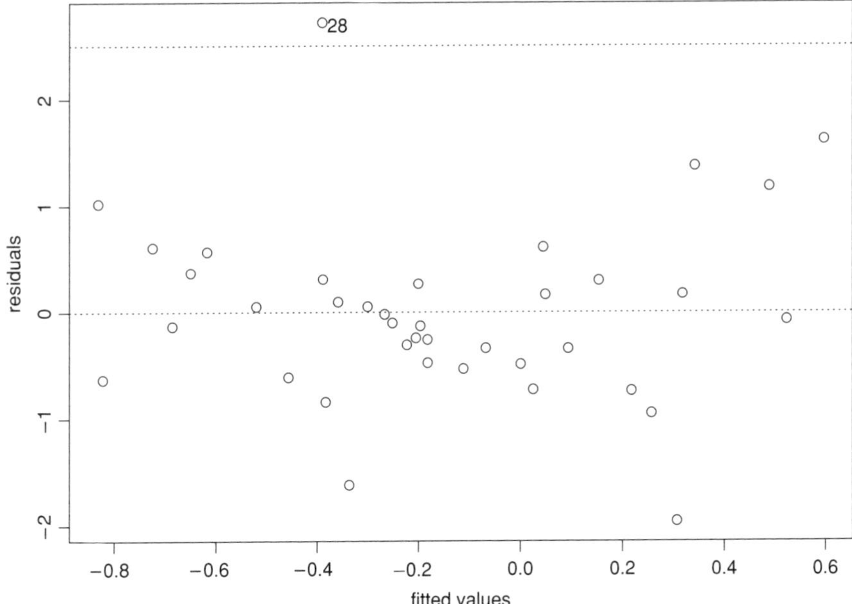

Figure 5.8 Toxicity data: LS residuals vs. fit

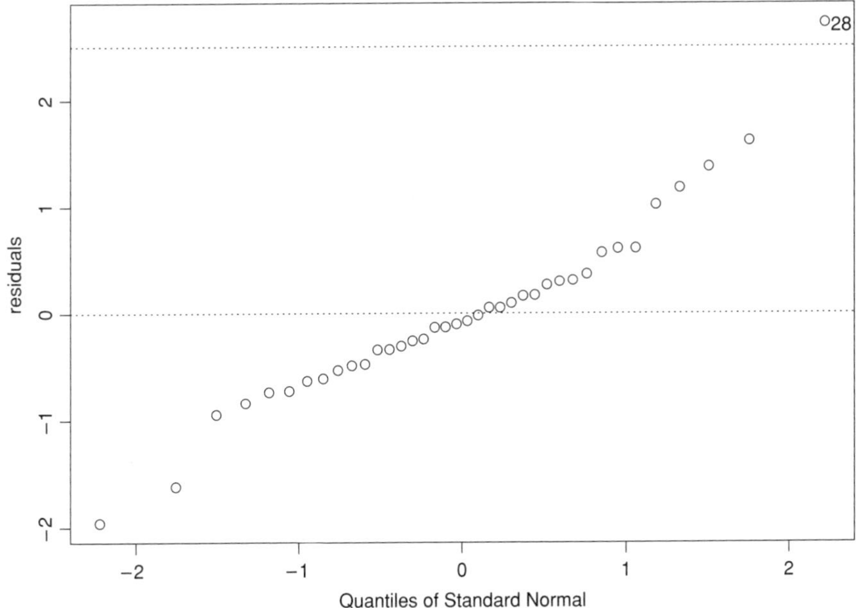

Figure 5.9 Toxicity data: Q–Q plot of LS residuals

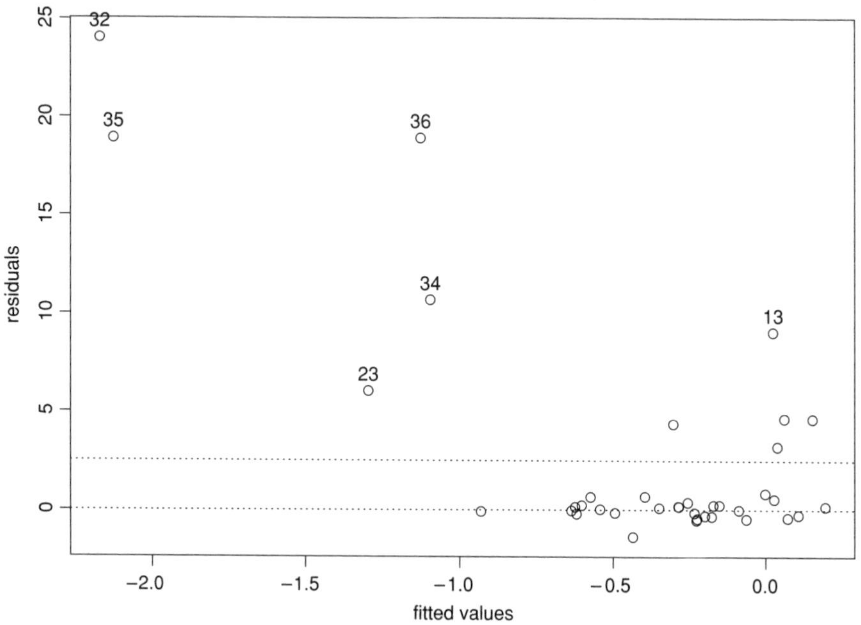

Figure 5.10 Toxicity data: MM-residuals vs. fit

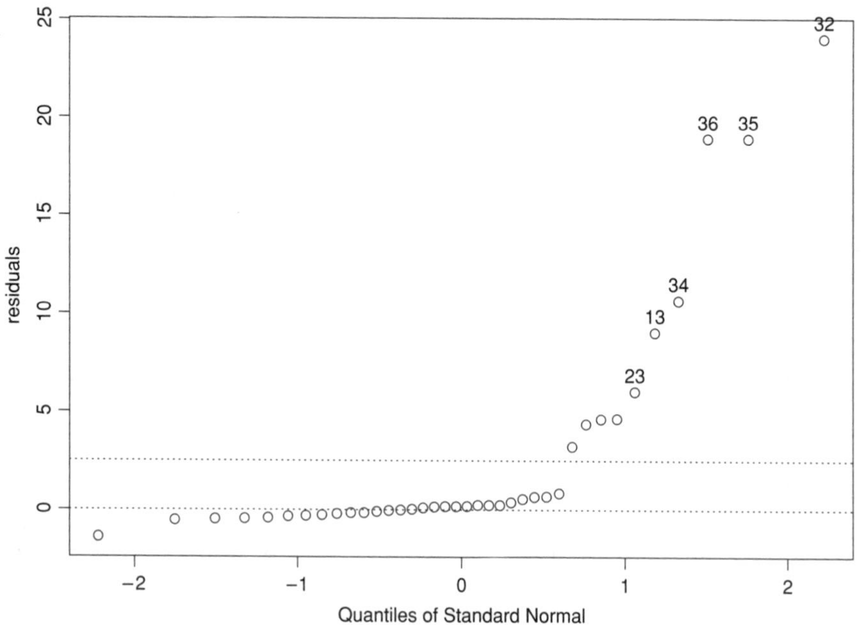

Figure 5.11 Toxicity data: Q–Q plot of robust residuals

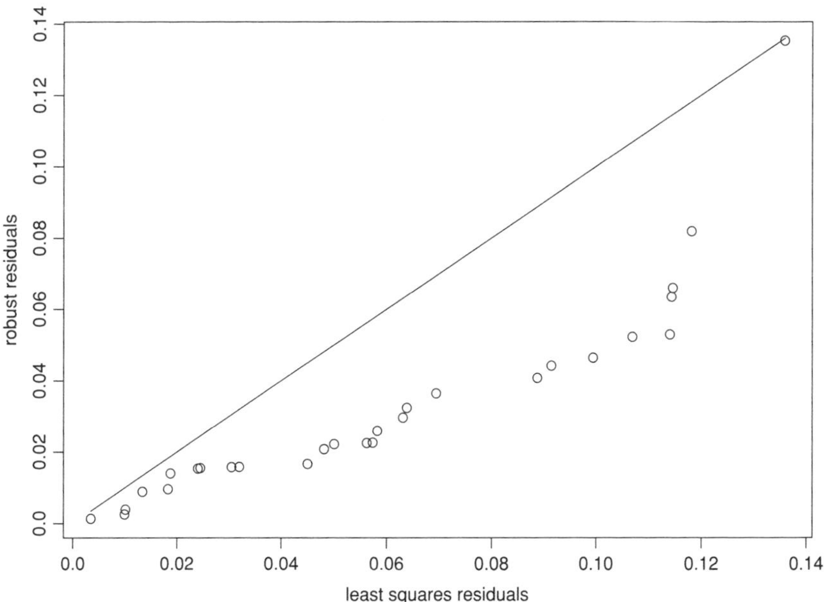

Figure 5.12 Toxicity data: ordered absolute residuals from MM and from LS (largest residuals omitted)

a computational algorithm. In the location case the LMS estimate is equivalent to the Shorth estimate defined as the midpoint of the shortest half of the data (see Problem 2.16a). For fitting a linear model the LMS estimate has the intuitive property of generating a strip of minimum width that contains half of the observations (Problem 5.9).

Let $\widehat{\sigma} = \widehat{\sigma}(\mathbf{r})$ be a scale equivariant robust scale estimate based on a vector of residuals

$$\mathbf{r}(\boldsymbol{\beta}) = (r_1(\boldsymbol{\beta}), \ldots, r_n(\boldsymbol{\beta})). \tag{5.20}$$

Then a regression estimate can be defined as

$$\widehat{\boldsymbol{\beta}} = \arg\min_{\boldsymbol{\beta}} \widehat{\sigma}(\mathbf{r}(\boldsymbol{\beta})). \tag{5.21}$$

Such estimates are regression, scale and affine equivariant (Problem 5.1).

5.6.1 S-estimates

A very important case of (5.21) is when $\widehat{\sigma}(\mathbf{r})$ is a scale M-estimate defined for each $\mathbf{r}$ by

$$\frac{1}{n}\sum_{i=1}^{n} \rho\left(\frac{r_i}{\widehat{\sigma}}\right) = \delta, \tag{5.22}$$

where ρ is a bounded ρ-function. By (3.22) the asymptotic BP of $\widehat{\sigma}$ is $\min(\delta, 1-\delta)$. The resulting estimate (5.21) is called an S-estimate (Rousseeuw and Yohai, 1984). When computing an initial estimate $\widehat{\boldsymbol{\beta}}_0$ for the MM-estimate, the bisquare ρ works quite well and we recommend its use for this purpose.

We now consider the BP of S-estimates. Proofs of all results on the BP are given in Section 5.16.4. The maximum FBP of an S-estimate with a bounded ρ-function is

$$\varepsilon^*_{\max} = \frac{m^*_{\max}}{n} \tag{5.23}$$

where $m^*_{\max}$ is the same as in (4.58), namely

$$m^*_{\max} = \left[\frac{n-k^*-1}{2}\right], \tag{5.24}$$

where k^* is defined in (5.11). Hence $\varepsilon^*_{\max}$ coincides with the maximum BP for equivariant estimates given in (5.10). This BP is attained by taking any δ of the form

$$\delta = \frac{m^*_{\max}+\gamma}{n} \quad \text{with } \gamma \in (0,1). \tag{5.25}$$

Recall that $k^* \geq p-1$ and if $k^* = p-1$ we say that $\mathbf{X}$ is in general position. When $\mathbf{X}$ is in general position, the maximum FBP is

$$\varepsilon^*_{\max} = \frac{1}{n}\left[\frac{n-p}{2}\right]$$

which is approximately 0.5 for large n.

Similarly, the maximum asymptotic BP of a regression S-estimate with a bounded ρ is

$$\varepsilon^* = \frac{1-\alpha}{2}, \tag{5.26}$$

with α defined in (5.12), and thus coincides with the maximum asymptotic BP for equivariant estimates given in Section 5.4.1 This maximum is attained by taking $\delta = (1-\alpha)/2$. If $\mathbf{x}$ has a density then $\alpha = 0$ and hence $\delta = 0.5$ yields BP $= 0.5$.

Since the median of absolute values is a scale M-estimate, the LMS estimate may be written as the estimate minimizing the scale $\widehat{\sigma}$ given by (5.22) with $\rho(t) = \mathrm{I}(|t| < 1)$ and $\delta = 0.5$. For a general δ, a solution $\widehat{\sigma}$ of (5.22) is the h-th order statistics of $|r_i|$, with $h = n - [n\delta]$ (Problem 2.14). The regression estimate defined by minimizing $\widehat{\sigma}$ is called the *least α-quantile estimate* with $\alpha = h/n$. Although it has a discontinuous ρ-function, the proof of the preceding results (5.24)–(5.25) can be shown to imply that the maximum BP is again (5.23) and that it can be attained by choosing

$$h = n - m^*_{\max} = \left[\frac{n+k^*+2}{2}\right], \tag{5.27}$$

which is slightly larger than $n/2$. See the end of Section 5.16.4.

We now deal with the efficiency of S-estimates. Since an S-estimate $\widehat{\boldsymbol{\beta}}$ minimizes $\widehat{\sigma} = \widehat{\sigma}(\mathbf{r}(\boldsymbol{\beta}))$ it follows that $\widehat{\boldsymbol{\beta}}$ is also an M-estimate (4.39) in that

$$\sum_{i=1}^{n} \rho\left(\frac{r_i(\widehat{\boldsymbol{\beta}})}{\widehat{\sigma}}\right) \leq \sum_{i=1}^{n} \rho\left(\frac{r_i(\widetilde{\boldsymbol{\beta}})}{\widehat{\sigma}}\right) \quad \text{for all } \widetilde{\boldsymbol{\beta}}, \tag{5.28}$$

where $\widehat{\sigma} = \widehat{\sigma}(\mathbf{r}(\widehat{\boldsymbol{\beta}}))$ is the same in the denominator on both sides of the equation. To see that this is indeed the case, suppose that for some $\widetilde{\boldsymbol{\beta}}$ we had

$$\sum_{i=1}^{n} \rho\left(\frac{r_i(\widetilde{\boldsymbol{\beta}})}{\widehat{\sigma}}\right) < \sum_{i=1}^{n} \rho\left(\frac{r_i(\widehat{\boldsymbol{\beta}})}{\widehat{\sigma}}\right).$$

Then since ρ is monotone, there would exist $\widetilde{\sigma} < \widehat{\sigma}$ such that

$$\sum_{i=1}^{n} \rho\left(\frac{r_i(\widetilde{\boldsymbol{\beta}})}{\widetilde{\sigma}}\right) = n\delta$$

and hence $\widehat{\sigma}$ would not be the minimum scale.

If ρ has a derivative ψ it follows that $\widehat{\boldsymbol{\beta}}$ is also an M-estimate in the sense of (5.8), but with the condition that the scale $\widehat{\sigma} = \widehat{\sigma}(\mathbf{r}(\widehat{\boldsymbol{\beta}}))$ is estimated simultaneously with $\widehat{\boldsymbol{\beta}}$.

Because an S-estimate is an M-estimate, it follows that the asymptotic distribution of an S-estimate with a smooth ρ under the model (5.1)–(5.2) is given by (4.43)–(4.44). See Davies (1990) and Kim and Pollard (1990) for a rigorous proof. For the LMS estimate, which has a discontinuous ρ, Davies (1990) shows that $\widehat{\boldsymbol{\beta}} - \boldsymbol{\beta}$ has the slow convergence rate $n^{-1/3}$, while estimates based on a smooth ρ-function have the usual convergence rate $n^{-1/2}$. Thus the LMS estimate is very inefficient for large n.

Unfortunately S-estimates with a smooth ρ cannot simultaneously have a high BP and high efficiency. In particular it was shown by Hössjer (1992) that an S-estimate with BP $= 0.5$ has an asymptotic efficiency under normally distributed errors that is not larger than 0.33. Numerical computation yields that the normal distribution efficiency of the S-estimate based on the bisquare scale is 0.29, which is adequately close to the upper bound.

Since an S-estimate with a differentiable ρ-function satisfies (5.8), its IF is given by (5.13) and hence is unbounded. See, however, the comments on page 126. Note also that S-estimates are "redescending" in the sense that if some of the y_i's are "too large" the estimate is completely unaffected by these observations, and coincides with an M-estimate computed after deleting such outliers. A precise statement is given in Problem 5.10.

Numerical computation of S-estimates is discussed in Section 5.7.

5.6.2 L-estimates of scale and the LTS estimate

An alternative to using an M-scale is to use an L-estimate of scale. Call $|r|_{(1)} \leq \ldots \leq |r|_{(n)}$ the ordered absolute values of residuals. Then we can define scale estimates as linear combinations of $|r|_{(i)}$ in one of the two following forms:

$$\widehat{\sigma} = \sum_{i=1}^{n} a_i \, |r|_{(i)} \quad \text{or} \quad \widehat{\sigma} = \left(\sum_{i=1}^{n} a_i \, |r|_{(i)}^2 \right)^{1/2}$$

where the a_i's are nonnegative constants.

A particular version of the second form is the α-trimmed squares scale where $\alpha \in (0, 1)$, and $n - h = [n\alpha]$ of the largest absolute residuals are trimmed:

$$\widehat{\sigma} = \left(\sum_{i=1}^{h} |r|_{(i)}^2 \right)^{1/2}. \tag{5.29}$$

The corresponding regression estimate is called the least trimmed squares (LTS) estimate (Rousseeuw, 1984). The FBP of the LTS estimate depends on h in the same way as that of the LMS estimate, so that for the LTS estimate to attain the maximum BP one must choose h in (5.29) as in (5.27). In particular, when $\mathbf{X}$ is in general position one must choose

$$h = n - \left[\frac{n - p - 2}{2} \right] = \left[\frac{n + p + 1}{2} \right],$$

which is approximately $n/2$ for large n. The asymptotic behavior of the LTS estimate is more complicated than that of smooth S-estimates. However, it is known that they have the standard convergence rate of $n^{-1/2}$, and it can be shown that the asymptotic efficiency of the LTS estimate at the normal distribution has the exceedingly low value of about 7% (see page 180 of Rousseeuw and Leroy (1987)).

5.6.3 Improving efficiency with one-step reweighting

We have seen that estimates based on a robust scale cannot have both a high BP and high normal efficiency. As we have already discussed, one can attain a desired normal efficiency by using an S-estimate as the starting point for an iterative procedure leading to an MM-estimate. In this section we consider a simpler alternative procedure proposed by Rousseeuw and Leroy (1987) to increase the efficiency of an estimate $\widehat{\boldsymbol{\beta}}_0$ without decreasing its BP.

Let $\widetilde{\sigma}$ be a robust scale of $\mathbf{r}(\widehat{\boldsymbol{\beta}}_0)$, e.g., the normalized median of absolute values (4.47). Then compute a new estimate $\widehat{\boldsymbol{\beta}}$, defined as the weighted LS estimate of the dataset with weights $w_i = W(r_i(\widehat{\boldsymbol{\beta}}_0)/\widetilde{\sigma})$ where $W(t)$ is a decreasing function of $|t|$. Rousseeuw and Leroy proposed the "weight function" W to be chosen as the "hard rejection" function $W(t) = \mathrm{I}(|t| \le k)$ with k equal to a γ-quantile of the distribution of $|x|$ where x has a standard normal distribution, e.g., $\gamma = 0.975$. Under normality this amounts to discarding a proportion of about $1 - \gamma$ of the points with largest absolute residuals.

He and Portnoy (1992) show that in general such reweighting methods preserve the order of consistency of $\widehat{\boldsymbol{\beta}}_0$, so in the standard situation where $\widehat{\boldsymbol{\beta}}_0$ is

$\sqrt{n}$-consistent then so is $\widehat{\beta}$. Unfortunately this means that because the LMS estimate is $n^{1/3}$-consistent, so is the reweighted LMS estimate.

In general the reweighted estimate $\widehat{\beta}$ is more efficient than $\widehat{\beta}_0$, but its asymptotic distribution is complicated (more so when W is discontinuous) and this makes it difficult to tune it for a given efficiency; in particular, it has to be noted than choosing $\gamma = 0.95$ for hard rejection does not make the asymptotic efficiency of $\widehat{\beta}$ equal to 0.95: it continues to be zero. A better approach for increasing the efficiency is described in Section 5.6.4.

5.6.4 A fully efficient one-step procedure

None of the estimators discussed so far can achieve full efficiency at the normal distribution and at the same time have a high BP and small maximum bias. We now discuss an adaptive one-step estimation method due to Gervini and Yohai (2002) that attains full asymptotic efficiency at the normal error distribution and at the same time has a high BP and small maximum bias. It is a weighted LS estimate computed from an initial estimate $\widehat{\beta}_0$ with high BP, but rather than deleting the values larger than a fixed k the procedure will keep a number N of observations $(\mathbf{x}_i, y_i)$ corresponding to the smallest values of $t_i = \left|r_i(\widehat{\beta}_0)\right| / \widehat{\sigma}$, $i = 1, \ldots, n$, where N depends on the data as will be described below. This N has the property that in large samples under normality it will have $N/n \to 1$, so that a vanishing fraction of data values will be deleted and full efficiency will be obtained.

Call G the distribution function of the absolute errors $|u_i| / \sigma$ under the normal model; that is,

$$G(t) = 2\Phi(t) - 1 = \mathrm{P}(|x| \leq t),$$

with $x \sim \Phi$ which is the standard normal distribution function. Let $t_{(1)} \leq \ldots \leq t_{(n)}$ denote the order statistics of the t_i. Let $\eta = G^{-1}(\gamma)$ where γ is a large value such as $\gamma = 0.95$. Define

$$i_0 = \min\left\{i : t_{(i)} \geq \eta\right\}, \qquad q = \min_{i \geq i_0} \left(\frac{i-1}{G\left(t_{(i)}\right)} \right), \tag{5.30}$$

and

$$N = [q] \tag{5.31}$$

where [.] denotes the integer part. The one-step estimate is the LS estimate of the observations corresponding to $t_{(i)}$ for $i \leq N$.

We now justify this procedure. The intuitive idea is to consider as potential outliers only those observations whose $t_{(i)}$ are not only greater than a given value, but also sufficiently larger than the corresponding order statistic of a sample from G. Note that if the data contain one or more outliers, then in a normal Q–Q plot of the $t_{(i)}$ against the respective quantiles of G, some large $t_{(i)}$ will appear well above the identity line, and we would delete it and all larger ones. The idea of the proposed procedure is to delete observations with large $t_{(i)}$ until the Q–Q plot of the remaining ones remains below the identity line at least for large values of $\left|t_{(i)}\right|$. Since we are interested only in the tails of the distribution, we consider only values larger than some given η.

More precisely, for $N \leq n$ call G_N the empirical distribution function of $t_{(1)} \leq \ldots \leq t_{(N)}$:

$$G_N(t) = \frac{1}{N}\#\left\{t_{(i)} \leq t\right\}.$$

It follows that

$$G_N(t) = \frac{i-1}{N} \text{ for } t_{(i-1)} \leq t < t_{(i)}$$

and hence each t in the half-open interval $[t_{(i-1)}, t_{(i)})$ is an α_i-quantile of G_N with

$$\alpha_i = \frac{i-1}{N}.$$

The α_i-quantile of G is $G^{-1}(\alpha_i)$. Then we look for N such that for $i_0 \leq i \leq N$ the α_i-quantile of G_N is not larger than that of G: that is,

$$\text{for } i \in [i_0, N]: \; t_{(i-1)} \leq t < t_{(i)} \Longrightarrow t \leq G^{-1}\left(\frac{i-1}{N}\right) \Longleftrightarrow G(t) \leq \frac{i-1}{N}. \tag{5.32}$$

Since G is continuous, (5.32) implies that

$$G\left(t_{(i)}\right) \leq \frac{i-1}{N} \text{ for } i_0 \leq i \leq N. \tag{5.33}$$

Also since

$$i > N \Longrightarrow \frac{i-1}{N} \geq 1 > G(t) \;\forall t,$$

the restriction $i \leq N$ may be dropped in (5.33), which is seen to be equivalent to

$$N \leq \frac{i-1}{G\left(t_{(i)}\right)} \text{ for } i \geq i_0 \Longleftrightarrow N \leq q \tag{5.34}$$

with q defined in (5.30). We want the largest $N \leq q$ and since N is an integer, we finally have (5.31).

Gervini and Yohai show that under very general assumptions on $\widehat{\beta}$ and $\widehat{\sigma}$ and regardless of the consistency rate of $\widehat{\beta}$, these estimates attain the maximum BP and full asymptotic efficiency for normally distributed errors.

5.7 Numerical computation of estimates based on robust scales

Minimizing the function $\widehat{\sigma}(\mathbf{r}(\beta))$ is difficult not only because of the existence of several local minima, but also because for some proposed robust estimates it is very

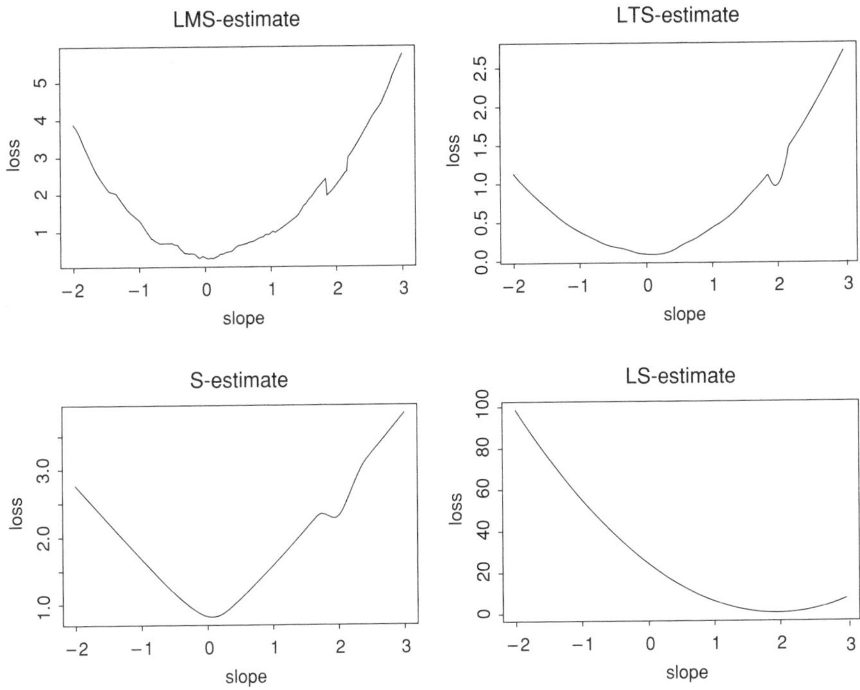

Figure 5.13 Loss functions for different regression estimates

nonsmooth. For example, consider fitting a straight line through the origin, i.e., the model is $y_i = \beta x_i + u_i$, for an artificial dataset with $n = 50$, x_i and u_i i.i.d. $N(0, 1)$, and three outliers at (10,20). Figure 5.13 shows the "loss" functions to be minimized in computing the LMS estimate, the LTS estimate with $\alpha = 0.5$, the S-estimate with bisquare ρ, and the LS estimate.

It is seen that the LMS loss function is very nonsmooth, and that all estimates except LS exhibit a local minimum at about 2 which is a "bad solution". The global minima of the loss functions for these four estimates are attained at the values of 0.06, 0.14, 0.07 and 1.72, respectively.

The loss functions for the LMS and LTS estimates are not differentiable, and hence gradient methods cannot be applied to them. Stromberg (1993a, 1993b) gives an exact algorithm for computing the LMS estimate, but the number of operations it requires is of order $\binom{n}{p+1}$ which is only practical for very small values of n and p. For other approaches see Agulló (1997, 2001) and Hawkins (1994). The loss function for the bisquare S-estimate is differentiable, but since gradient methods ensure only the attainment of a local minimum, a "good" starting point is needed.

In Section 5.7.1 we describe iterative procedures for computing S-estimates and LTS estimates, such that the objective function decreases at each iteration and thus leads to local minima. Since there are usually several such minima, attaining or at least approximating the global minimum depends on the starting point. Section 5.7.2 presents a general stochastic method for generating candidate solutions that can be used to obtain a good initial estimate for the LTS or S-estimates, or as approximations to the LMS estimate. Since this method is computer intensive, care is needed to reduce the computing time and strategies for this goal are presented in Section 5.7.3.

5.7.1 Finding local minima

S-estimates

For an S-estimate the simplest procedure to find a local minimum that satisfies (5.8) is the IRWLS algorithm described in Section 4.5, in which σ is updated at each step. That is, if $\widehat{\beta}_k$ is the estimate at the k-th iteration, the scale estimate $\widehat{\sigma}_k$ is obtained by solving

$$\sum_{i=1}^{n} \rho\left(\frac{r_i\left(\widehat{\beta}_k\right)}{\sigma_k}\right) = \delta \tag{5.35}$$

with the method of Section 2.7.2. Then $\widehat{\beta}_{k+1}$ is obtained by weighted least squares with weights $w_i = W(r_i/\widehat{\sigma}_k)$. It can be shown that if W is decreasing then $\widehat{\sigma}_k$ decreases at each step (Section 9.2). Since computing $\widehat{\sigma}$ consumes an important proportion of the time, it is important to do it economically, and in this regard it is best to start the iteration for $\widehat{\sigma}_k$ from the previous value $\widehat{\sigma}_{k-1}$.

The LTS estimate

A local minimum of (5.29) can be attained iteratively using the "concentration step" (C-step) of Rousseeuw and van Driessen (2000). Given a candidate $\widehat{\beta}_1$, let $\widehat{\beta}_2$ be the LS estimate based on the data corresponding to the h smallest absolute residuals. It is proved in Section 9.3 that the trimmed L-scale $\widehat{\sigma}$ given by (5.29) is not larger for $\widehat{\beta}_2$ than for $\widehat{\beta}_1$. This procedure is exact in the sense that after a finite number of steps it attains a value of $\widehat{\beta}$ such that further steps do not decrease the values of $\widehat{\sigma}$. It can be shown that this $\widehat{\beta}$ is a local minimum, but not necessarily a global one.

5.7.2 The subsampling algorithm

To find an approximate solution to (5.21) we shall compute a "large" finite set of candidate solutions, and replace the minimization over $\beta \in R^p$ in (5.21) by minimizing $\widehat{\sigma}(\mathbf{r}(\beta))$ over that finite set. To compute the candidate solutions we take subsamples of size p

$$\{(\mathbf{x}_i, y_i) : i \in J\}, \quad J \subset \{1, \ldots, n\}, \ \#(J) = p.$$

For each J find β_J that satisfies the exact fit $\mathbf{x}_i'\beta_J = y_i$ for $i \in J$. If a subsample is collinear, it is replaced by another. Then the problem of minimizing $\widehat{\sigma}(\mathbf{r}(\beta))$

for $\beta \in R^p$ is replaced by the finite problem of minimizing $\widehat{\sigma}(\mathbf{r}(\beta_J))$ over J. Since choosing all $\binom{n}{p}$ subsamples would be prohibitive unless both n and p are rather small, we choose N of them at random: $\{J_k : k = 1, \ldots, N\}$ and the estimate $\widehat{\beta}_{J_{k^*}}$ is defined by

$$k^* = \arg\min \left\{ \widehat{\sigma}\left(\mathbf{r}\left(\beta_{J_k}\right)\right) : k = 1, \ldots, N \right\}. \tag{5.36}$$

Suppose the sample contains a proportion ε of outliers. The probability of an outlier-free subsample is $\alpha = (1-\varepsilon)^p$, and the probability of at least one "good" subsample is $1-(1-\alpha)^N$. If we want this probability to be larger than $1-\gamma$, we must have

$$\log \gamma \geq N \log(1-\alpha) \approx -N\alpha$$

and hence

$$N \geq \frac{|\log \gamma|}{\left|\log\left(1-(1-\varepsilon)^p\right)\right|} \approx \frac{|\log \gamma|}{(1-\varepsilon)^p} \tag{5.37}$$

for p not too small. Therefore N must grow exponentially with p. Table 5.3 gives the minimum N for $\gamma = 0.01$. Since the number of "good" (outlier-free) subsamples is binomial, the expected number of good samples is $N\alpha$, and so for $\gamma = 0.01$ the expected number of "good" subsamples is $|\log 0.01| = 4.6$.

A shortcut saves much computing time. Suppose we have examined $M-1$ subsamples and $\widehat{\sigma}_{M-1}$ is the current minimum. Now we draw the M-th subsample which yields the candidate estimate $\widehat{\beta}_M$. We may spare the effort of computing the new scale estimate $\widehat{\sigma}_M$ in those cases where it will turn out not to be smaller than $\widehat{\sigma}_{M-1}$. The reason is as follows. If $\widehat{\sigma}_M < \widehat{\sigma}_{M-1}$, then since ρ is monotonic

$$n\delta = \sum_{i=1}^{n} \rho\left(\frac{r_i\left(\widehat{\beta}_M\right)}{\widehat{\sigma}_M}\right) \geq \sum_{i=1}^{n} \rho\left(\frac{r_i\left(\widehat{\beta}_M\right)}{\widehat{\sigma}_{M-1}}\right).$$

Table 5.3 Minimum N for $\gamma = 0.01$

p	$\varepsilon = 0.10$	0.15	0.20	0.25	0.50
5	6	8	12	17	146
10	11	22	41	80	4714
15	20	51	129	343	150900
20	36	117	398	1450	4.83×10^6
30	107	602	3718	25786	4.94×10^9
40	310	3064	34644	457924	5.06×10^{12}
50	892	15569	322659	8.13×10^7	5.18×10^{15}

Thus if

$$n\delta \leq \sum_{i=1}^{n} \rho\left(\frac{r_i\left(\widehat{\beta}_M\right)}{\widehat{\sigma}_{M-1}}\right) \tag{5.38}$$

we may discard $\widehat{\beta}_M$ since $\widehat{\sigma}_M \geq \widehat{\sigma}_{M-1}$. Therefore $\widehat{\sigma}$ is computed only for those subsamples that do not verify the condition (5.38).

Although the N given by (5.37) ensures that the approximate algorithm has the desired BP in a probabilistic sense, it does not imply that it is a good approximation to the exact estimate. Furthermore, because of the randomness of the subsampling procedure the resulting estimate is *stochastic*, i.e., repeating the computation may lead to another local minimum and hence to another $\widehat{\beta}$, with the unpleasant consequence that repeating the computation may yield different results. In our experience a carefully designed algorithm usually gives good results, and the above infrequent but unpleasant effects can be mitigated by increasing N as much as the available computing power will allow.

The subsampling procedure may be used to compute an approximate LMS estimate. Since total lack of smoothness precludes any kind of iterative improvement, the result is usually followed by one-step reweighting (Section 5.6.3), which, besides improving the efficiency of the estimate, makes it more stable with respect to the randomness of subsampling. It must be recalled, however, that the resulting estimate is not asymptotically normal, and hence it is not possible to use it as a basis for approximate inference on the parameters. Since the resulting estimate is a weighted LS estimate, it would be intuitively attractive to apply classical LS inference as if these weights were constant, but this procedure is not valid.

5.7.3 A strategy for fast iterative estimates

For the LTS or S-estimates, which admit iterative improvement steps as described in Section 5.7.1 for an S-estimate, it is possible to dramatically speed up the search for a global minimum. In the discussion below "iteration" refers to one of the two iterative procedures described in that section (though the method to be described could be applied to any estimate that admits iterative improvement steps). Consider the following two extreme strategies for combining the subsampling and the iterative parts of the minimization:

A use the "best" result (5.36) of the subsampling as a starting point from which to iterate until convergence to a local minimum, or

B iterate to convergence from each of the N candidates $\widehat{\beta}_{J_k}$ and keep the result with smallest $\widehat{\sigma}$.

Clearly strategy B would yield a better approximation of the absolute minimum than A, but is also much more expensive. An intermediate strategy, which depends on two parameters K_{iter} and K_{keep}, consists of the following steps:

1. For $k = 1, \ldots, N$, compute $\widehat{\beta}_{J_k}$ and perform K_{iter} iterations, which yields the candidates $\widetilde{\beta}_k$ with residual scale estimates $\widetilde{\sigma}_k$.

2. Only the K_{keep} candidates with smallest K_{keep} estimates $\widetilde{\sigma}_k$ are kept in storage, which only needs to be updated when the current $\widetilde{\sigma}_k$ is lower than at least one of the current best K_{iter} values. Call these estimates $\widetilde{\boldsymbol{\beta}}_{(k)}$, $k = 1, \ldots, K_{keep}$.
3. For $k = 1, \ldots, K_{keep}$, iterate to convergence starting from $\widetilde{\boldsymbol{\beta}}_{(k)}$, obtaining the candidate $\widehat{\boldsymbol{\beta}}_k$ with residual scale estimate $\widehat{\sigma}_k$.
4. The final result is the candidate $\widehat{\boldsymbol{\beta}}_k$ with minimum $\widehat{\sigma}_k$.

Option A above corresponds to $K_{iter} = 0$ and $K_{keep} = 1$, while B corresponds to $K_{iter} = \infty$ and $K_{keep} = 1$.

This strategy was first proposed by Rousseeuw and van Driessen (2000) for the LTS estimate (the "Fast LTS") with $K_{iter} = 2$ and $K_{keep} = 10$, and as mentioned above the general method can be used with any estimate that can be improved iteratively.

Salibian-Barrera and Yohai (2005) proposed a "Fast S-estimate", based on this strategy. A theoretical study of the properties of this procedure seems impossible, but their simulations show that it is not worthwhile to increase K_{iter} and K_{keep} beyond the values 1 and 10, respectively. They also show that $N = 500$ gives reliable results at least for $p \leq 40$ and contamination fraction up to 10%. Their simulation also shows that the "Fast S" is better than the "Fast LTS" with respect to both mean squared errors and the probability of converging to a "wrong" solution. The simulation in Salibian-Barrera and Yohai (2005) also indicates that the Fast LTS works better with $K_{iter} = 1$ than with $K_{iter} = 2$.

A further saving in time is obtained by replacing $\widehat{\sigma}$ in step 1 of the procedure with an approximation obtained by one step of the Newton–Raphson algorithm starting from the normalized median of absolute residuals.

Ruppert (1992) proposes a more complex random search method. However, the simulations by Salibian-Barrera and Yohai (2005) show that its behavior is worse than that of both the Fast S- and the Fast LTS estimates.

5.8 Robust confidence intervals and tests for M-estimates

In general, estimates that fulfill an M-estimating equation like (5.8) are asymptotically normal, and hence approximate confidence intervals and tests can be obtained as in Sections 4.4.2 and 4.7.2. Recall that according to (5.14), $\widehat{\boldsymbol{\beta}}$ has an approximately normal distribution with covariance matrix given $vn^{-1}\mathbf{V}_{\mathbf{x}}^{-1}$. For the purposes of inference, $\mathbf{V}_{\mathbf{x}}$ and v can be estimated by

$$\widehat{\mathbf{V}}_{\mathbf{x}} = \frac{1}{n}\sum_{i=1}^{n}\mathbf{x}_i\mathbf{x}_i' = \frac{1}{n}\mathbf{X}'\mathbf{X}, \quad \widehat{v} = \widehat{\sigma}^2 \frac{\text{ave}_i\left\{\psi\left(r_i/\widehat{\sigma}\right)^2\right\}}{\left[\text{ave}_i\left\{\psi'\left(r_i/\widehat{\sigma}\right)\right\}\right]^2}\frac{n}{n-p}, \tag{5.39}$$

and hence the resulting confidence intervals and tests are the same as those for fixed $\mathbf{X}$.

Actually, this estimate of $\mathbf{V_x}$ has the drawback of not being robust. In fact just one large $\mathbf{x}_i$ corresponding to an outlying observation with a large residual may have a large distorting influence on $\mathbf{X'X}$ with diagonal elements typically inflated. Since $\widehat{v}$ is stable with respect to outlier influence, the confidence intervals based on $\widehat{v}(\mathbf{X'X})^{-1}$ may be too small and hence the coverage probabilities may be much smaller than the nominal.

Yohai, Stahel and Zamar (1991) proposed a more robust estimate of the matrix $\mathbf{V_x}$, defined as

$$\widetilde{\mathbf{V}}_{\mathbf{x}} = \frac{1}{\sum_{i=1}^{n} w_i} \sum_{i=1}^{n} w_i \mathbf{x}_i \mathbf{x}_i',$$

with $w_i = W(r_i/\widehat{\sigma})$, where W is the weight function (2.30). Under the model with n large, the residual r_i is close to the error u_i, and since u_i is independent of $\mathbf{x}_i$ we have as $n \to \infty$

$$\frac{1}{n} \sum_{i=1}^{n} w_i \, \mathbf{x}_i \mathbf{x}_i' \to_p \mathrm{E}W\left(\frac{u}{\sigma}\right)(\mathrm{E}\mathbf{x}\mathbf{x}')$$

and

$$\frac{1}{n} \sum_{i=1}^{n} w_i \to_p \mathrm{E}W\left(\frac{u}{\sigma}\right),$$

and thus

$$\widetilde{\mathbf{V}}_{\mathbf{x}} \to_p \mathrm{E}\mathbf{x}\mathbf{x}'.$$

Assume that $\psi(t) = 0$ if $|t| > k$ for some k, as happens with the bisquare. Then if $|r_i|/\widehat{\sigma} > k$, the weight w_i is zero. If observation i has high leverage (i.e., $\mathbf{x}_i$ is "large") and is an outlier, then since the estimate is robust, $|r_i|/\widehat{\sigma}$ is also "large", and this observation will have null weight and hence will not influence $\widetilde{\mathbf{V}}_{\mathbf{x}}$. On the other hand, if $\mathbf{x}_i$ is "large" but r_i is small or moderate, then w_i will be nonnull, and $\mathbf{x}_i$ will still have a beneficial influence on $\widetilde{\mathbf{V}}_{\mathbf{x}}$ by virtue of reducing the variance of $\widehat{\boldsymbol{\beta}}$. Hence the advantage of $\widetilde{\mathbf{V}}_{\mathbf{x}}$ is that it downweights high-leverage observations only when they are outlying. Therefore we recommend the routine use of $\widetilde{\mathbf{V}}_{\mathbf{x}}$ instead of $\mathbf{V_x}$ for all instances of inference, in particular the Wald-type tests defined in Section 4.7.2 which also require estimating the covariance matrix of the estimates.

The following example shows how different the inference can be when using an MM-estimate compared to using the LS estimate.

For the straight-line regression of the mineral data in Example 5.1, the slope given by LS and its estimated SD are 0.135 and 0.020 respectively, while the corresponding values for the MM-estimate are 0.044 and 0.021; hence the classical and robust two-sided intervals with level 0.95 are (0.0958, 0.1742) and (0.00284, 0.08516), which are disjoint, showing the influence of the outlier.

5.8.1 Bootstrap robust confidence intervals and tests

Since the confidence intervals and tests for robust estimates are asymptotic, their actual level may be a poor approximation to the desired one if n is small. This occurs especially when the error distribution is very heavy tailed or asymmetric (see the end of Section 10.3). Better results can be obtained by using the bootstrap method, which estimates the distribution of an estimate by generating a large number of samples *with* replacement ("bootstrap samples") from the sample and recomputing the estimate for each of them. See for example Efron and Tibshirani (1993) and Davison and Hinkley (1997).

While the bootstrap approach has proved successful in many situations, its application to robust estimates presents special problems. One is that in principle the estimate should be recomputed for each bootstrap sample, which may require impractical computing times. Another is that the proportion of outliers in some of the bootstrap samples might be much higher than in the original one, leading to quite incorrect values of the recomputed estimate. Salibian-Barrera and Zamar (2002) proposed a method which avoids both pitfalls, and consequently is faster and more robust than the naive application of the bootstrap approach.

5.9 Balancing robustness and efficiency

Defining the asymptotic bias of regression estimates requires a measure of the "size" of the difference between the value of an estimate, which for practical purposes we take to be the asymptotic value $\widehat{\boldsymbol{\beta}}_\infty$, and the true parameter value $\boldsymbol{\beta}$. We shall use an approach based on prediction. Consider an observation $(\mathbf{x}, y)$ from the model (5.1)–(5.2):

$$y = \mathbf{x}'\boldsymbol{\beta} + u, \quad \mathbf{x} \text{ and } u \text{ independent.}$$

The prediction error corresponding to $\widehat{\boldsymbol{\beta}}_\infty$ is

$$e = y - \mathbf{x}'\widehat{\boldsymbol{\beta}}_\infty = u - \mathbf{x}'(\widehat{\boldsymbol{\beta}}_\infty - \boldsymbol{\beta}).$$

Let $\mathrm{E}u^2 = \sigma^2 < \infty$, $\mathrm{E}u = 0$ and $\mathbf{V_x} = \mathrm{E}\mathbf{x}\mathbf{x}'$. Then the mean squared prediction error is

$$\mathrm{E}e^2 = \sigma^2 + (\widehat{\boldsymbol{\beta}}_\infty - \boldsymbol{\beta})'\mathbf{V_x}(\widehat{\boldsymbol{\beta}}_\infty - \boldsymbol{\beta}).$$

The second term is measure of the increase in the prediction error due to the parameter estimation bias, and so we define the bias as

$$\mathrm{b}(\widehat{\boldsymbol{\beta}}_\infty) = \sqrt{(\widehat{\boldsymbol{\beta}}_\infty - \boldsymbol{\beta})'\mathbf{V_x}(\widehat{\boldsymbol{\beta}}_\infty - \boldsymbol{\beta})}. \tag{5.40}$$

Note that if $\widehat{\boldsymbol{\beta}}$ is regression, scale and affine equivariant, this measure is *invariant* under the respective transformations, i.e., $\mathrm{b}(\widehat{\boldsymbol{\beta}}_\infty)$ does not change when any of those transformations is applied to $(\mathbf{x}, y)$. If $\mathbf{V_x}$ is a multiple of the identity—such as when the elements of $\mathbf{x}$ are i.i.d. zero mean normal—then (5.40) is a multiple of $\|\widehat{\boldsymbol{\beta}}_\infty - \boldsymbol{\beta}\|$, so in this special case the Euclidean norm is an adequate bias measure.

Now consider a model with intercept, i.e.,

$$\mathbf{x} = \begin{bmatrix} 1 \\ \underline{\mathbf{x}} \end{bmatrix}, \qquad \boldsymbol{\beta} = \begin{bmatrix} \beta_0 \\ \boldsymbol{\beta}_1 \end{bmatrix}$$

and let $\boldsymbol{\mu} = \mathrm{E}\underline{\mathbf{x}}$ and $\mathbf{U} = \mathrm{E}\underline{\mathbf{x}}\underline{\mathbf{x}}'$, so that

$$\mathbf{V}_\mathbf{x} = \begin{bmatrix} 1 & \boldsymbol{\mu}' \\ \boldsymbol{\mu} & \mathbf{U} \end{bmatrix}.$$

For a regression and affine equivariant estimate there is no loss of generality in assuming $\boldsymbol{\mu} = \mathbf{0}$ and $\beta_0 = 0$, and in this case

$$\mathrm{b}(\widehat{\boldsymbol{\beta}}_\infty)^2 = \widehat{\beta}^2_{0,\infty} + \widehat{\boldsymbol{\beta}}'_{1,\infty}\mathbf{U}\widehat{\boldsymbol{\beta}}_{1,\infty},$$

with the first term representing the contribution to bias of the intercept and the second that of the slopes.

A frequently used benchmark for comparing estimates is to assume that the joint distribution of $(\underline{\mathbf{x}}, y)$ belongs to a contamination neighborhood of a multivariate normal. By the affine and regression equivariance of the estimates, there is no loss of generality in assuming that this central normal distribution is $\mathrm{N}_{p+1}(\mathbf{0}, \mathbf{I})$. In this case it can be shown that the maximum biases of M-estimates do not depend on p. A proof is outlined in Section 5.16.5. The same is true of the other estimates treated in this chapter except GM-estimates in Section 5.11.

The maximum asymptotic bias of S-estimates can be derived from the results of Martin, Yohai and Zamar (1989), and those of the LTS and LMS estimates from Berrendero and Zamar (2001). Table 5.4 compares the maximum asymptotic biases of LTS, LMS and the S-estimate with bisquare scale and three MM-estimates with bisquare ρ, in all cases with asymptotic BP equal to 0.5, when the joint distribution of $\mathbf{x}$ and y is in an ε-neighborhood of the multivariate normal $\mathrm{N}_{p+1}(\mathbf{0}, \mathbf{I})$. One MM-estimate is given by the global minimum of (5.16) with normal distribution

Table 5.4 Maximum bias of regression estimates for contamination ε

	ε				
	0.05	0.10	0.15	0.20	Eff.
LTS	0.63	1.02	1.46	2.02	0.07
S-E	0.56	0.88	1.23	1.65	0.29
LMS	0.53	0.83	1.13	1.52	0.0
MM (global)	0.78	1.24	1.77	2.42	0.95
MM (local)	0.56	0.88	1.23	1.65	0.95
MM (local)	0.56	0.88	1.23	1.65	0.85

efficiency 0.95. The other two MM-estimates correspond to local minima of (5.16) obtained using the IRWLS algorithm starting from the S-estimate, with efficiencies 0.85 and 0.95. The LMS estimate has the smallest bias for all the values of ε considered, but also has zero asymptotic efficiency. It is remarkable that the maximum biases of both "local" MM-estimates are much lower than those of the "global" MM-estimate, and close to the maximum biases of the LMS estimate. This shows the importance of a good starting point. The fact that an estimate obtained as a local minimum starting from a very robust estimate may have a lower bias than one defined by the absolute minimum was pointed out by Hennig (1995), who also gave bounds for the bias of MM-estimates with general ρ-functions in contamination neighborhoods.

It is also curious that the two local MM-estimates with different efficiencies have the same maximum biases. To understand this phenomenon, we show in Figure 5.14 the asymptotic biases of the S- and MM-estimates for contamination fraction $\varepsilon = 0.2$ and point contamination located at (x_0, Kx_0) with $x_0 = 2.5$, as a function of the contamination slope K. It is seen that the bias of each estimate is worse than that of the LS estimate up to a certain value of K and then drops to zero. But the range of values where the MM-estimate with efficiency 0.95 has a larger bias than the LS estimate is greater than those for the 0.85 efficient MM-estimate and the S-estimate. This is the price paid for a higher normal efficiency. The MM-estimate with efficiency 0.85 is closer in behavior to the S-estimate than the one with efficiency 0.95. If one

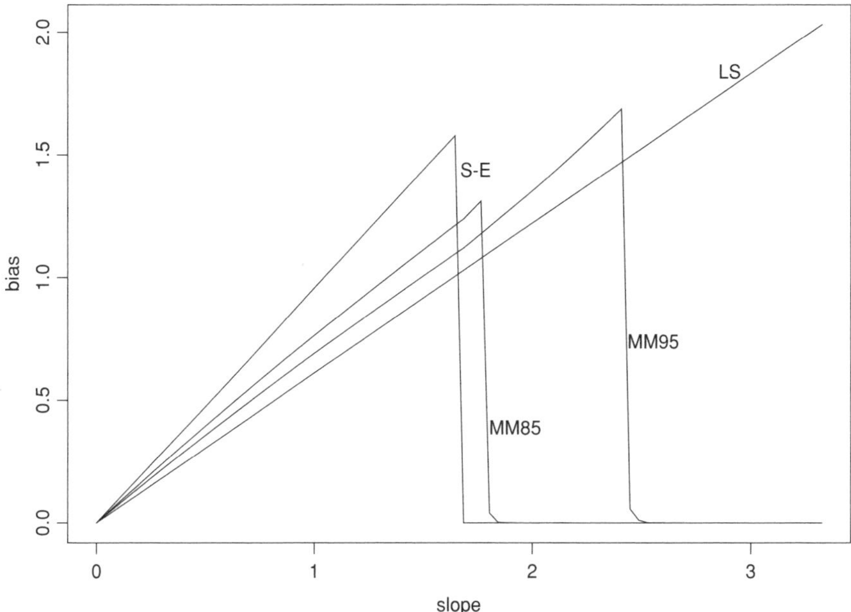

Figure 5.14 Biases of LS, S-estimate and MM-estimates with efficiencies 0.85 and 0.95, as a function of the contamination slope, for $\varepsilon = 0.2$, when $x_0 = 2.5$

makes similar plots for other values of x_0, one finds that for $x_0 \geq 5$ the curves for the S-estimate and the two MM-estimates are very similar.

For these reasons we recommend an MM-estimate with bisquare function and efficiency 0.85, computed starting from a bisquare S-estimate. One could also compute such MM-estimates with several different efficiencies, say between 0.7 and 0.95, and compare the results. Yohai et al. (1991) give a test to compare the biases of estimates with different efficiencies.

The former results on MM-estimates suggest a general approach for the choice of robust estimates. Consider in general an estimate $\widehat{\boldsymbol{\theta}}$ with high efficiency, defined by the absolute minimum of a target function. We have an algorithm that starting from any initial value $\boldsymbol{\theta}_0$ yields a local minimum of the target function, that we shall call $A(\theta_0)$. Assume also that we have an estimate $\boldsymbol{\theta}^*$ with lower bias, although possibly with low efficiency. Define a new estimate $\widetilde{\boldsymbol{\theta}}$ (call it the "approximate estimate") by the local minimum of the target function obtained by applying the algorithm starting from $\boldsymbol{\theta}^*$, i.e. $\widetilde{\boldsymbol{\theta}} = A(\boldsymbol{\theta}^*)$. Then in general $\widetilde{\boldsymbol{\theta}}$ has the same efficiency as $\widehat{\boldsymbol{\theta}}$ under the model, while it has a lower bias than $\widehat{\boldsymbol{\theta}}$ in contamination neighborhoods. If furthermore $\boldsymbol{\theta}^*$ is fast to compute, then also $\widetilde{\boldsymbol{\theta}}$ will be faster than $\widehat{\boldsymbol{\theta}}$. An instance of this approach in multivariate analysis will be presented in Section 6.7.5.

5.9.1 "Optimal" redescending M-estimates

In Section 3.5.4 we gave the solution to the Hampel dual problems for a one-dimensional parametric model, namely: (1) finding the M-estimate minimizing the asymptotic variance subject to an upper bound on the gross-error sensitivity (GES), and (2) minimizing the GES subject to an upper bound on the asymptotic variance. This approach cannot be taken with regression M-estimates with random predictors since (5.13) implies that the GES is infinite. However, it can be suitably modified, as we now show.

Consider a regression estimate $\widehat{\boldsymbol{\beta}} = (\widehat{\beta}_0, \widehat{\boldsymbol{\beta}}_1)$ where $\widehat{\beta}_0$ corresponds to the intercept and $\widehat{\boldsymbol{\beta}}_1$ to the slopes. Yohai and Zamar (1997) showed that for an M-estimate $\widehat{\boldsymbol{\beta}}$ with bounded ρ, the maximum biases $\mathrm{MB}(\varepsilon, \widehat{\beta}_0)$ and $\mathrm{MB}(\varepsilon, \widehat{\boldsymbol{\beta}}_1)$ in an ε-contamination neighborhood are of order $\sqrt{\varepsilon}$. Therefore the biases of these estimates are continuous at zero, which means that a small amount of contamination produces only a small change in the estimate. Because of this the approach in Section 3.5.4 can then be adapted to the present situation by replacing the GES with a different measure called the *contamination sensitivity (CS)*, which is defined as

$$\mathrm{CS}(\widehat{\boldsymbol{\beta}}_j) = \lim_{\varepsilon \to 0} \frac{\mathrm{MB}(\varepsilon, \widehat{\boldsymbol{\beta}}_j)}{\sqrt{\varepsilon}} \quad (j = 0, 1).$$

Recall that the asymptotic covariance matrix of a regression M-estimate depends on ρ only through

$$v(\psi, F) = \frac{\mathrm{E}_F\left(\psi(u)^2\right)}{(\mathrm{E}_F \psi'(u))^2}$$

Table 5.5 Constant for optimal estimate

Efficiency	0.80	0.85	0.90	0.95
c	0.060	0.044	0.028	0.013

where $\psi = \rho'$ and F is the error distribution. We consider only the slopes $\boldsymbol{\beta}_1$ which are usually more important than the intercept. The analogs of the direct and dual Hampel problems can now be stated as finding the function ψ that

- minimizes $v(\psi, F)$ subject to the constraint $\mathrm{CS}(\widehat{\boldsymbol{\beta}}_1) \leq k_1$

or

- minimizes $\mathrm{CS}(\widehat{\boldsymbol{\beta}}_1)$ subject to $v(\psi, F) \leq k_2$

where k_1 and k_2 are given constants.

Yohai and Zamar (1997) found that the optimal ψ for both problems has the form

$$\psi_c(u) = \mathrm{sgn}(u)\left(-\frac{\varphi'(|u|) + c}{\varphi(|u|)}\right)^+ \tag{5.41}$$

where φ is the standard normal density, c is a constant and $t^+ = \max(t, 0)$ denotes the positive part of t. For $c = 0$ we have the LS estimate: $\psi(u) = u$.

Table 5.5 gives the values of c corresponding to different efficiencies, and Figure 5.15 shows the bisquare and optimal ψ-functions with efficiency 0.95. We observe that the optimal ψ increases almost linearly and then redescends much faster than the bisquare ψ. This optimal ψ-function is a smoothed, differentiable version of the hard-rejection function $\psi(u) = u\mathrm{I}(|u| \leq a)$ for some constant a. As such it not only is good from the numerical optimization perspective, but also has the intuitive feature of making a rather rapid transition from its maximum absolute values to zero in the "flanks" of the nominal normal distribution. The latter is a region in which it is most difficult to tell whether a data point is an outlier or not, while outside that transition region outliers are clearly identified and rejected, and inside the region data values are left essentially unaltered. As a minor point, the reader should note that (5.41) implies that the optimal ψ has the curious feature of vanishing exactly in a small interval around zero. For example, if $c = 0.013$ the interval is $(-0.032, 0.032)$, which is so small it is not visible in the figure.

Svarc, Yohai and Zamar (2002) considered the two optimization problems stated above, but used the actual maximum bias $\mathrm{MB}(\varepsilon, \widehat{\boldsymbol{\beta}}_1)$ for a range of positive values of ε instead of the approximation given by the contamination sensitivity $\mathrm{CS}(\widehat{\boldsymbol{\beta}}_1)$. They calculated the optimal ψ and showed numerically that for $\varepsilon \leq 0.20$ it is almost identical to the one based on the contamination sensitivity. Therefore the optimal solution corresponding to an infinitesimal contamination is a good approximation to the one corresponding to $\varepsilon > 0$, at least for $\varepsilon \leq 0.20$.

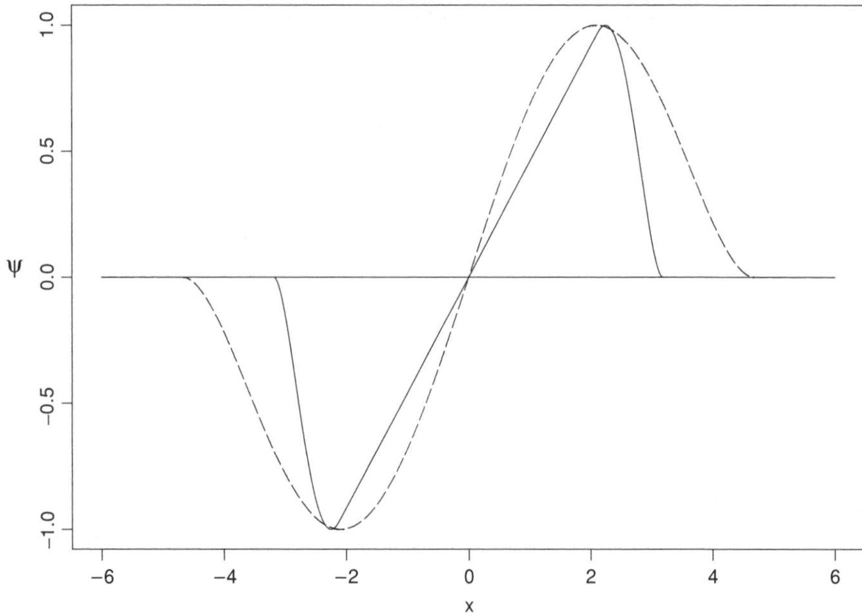

Figure 5.15 Optimal (—) and bisquare (- - -) ψ-functions with efficiency 0.95

5.10 The exact fit property

The so-called *exact fit property* states essentially that if a proportion α of observations lies exactly on a subspace, and $1 - \alpha$ is less than the BP of a regression and scale equivariant estimate, then the fit given by the estimate coincides with the subspace. More precisely, let the FBP of $\widehat{\boldsymbol{\beta}}$ be $\varepsilon^* = m^*/n$, and let the dataset contain q points such that $y_i = \mathbf{x}_i'\boldsymbol{\gamma}$ for some $\boldsymbol{\gamma}$. We prove in Section 5.16.3 that if $q \geq n - m^*$ then $\widehat{\boldsymbol{\beta}} = \boldsymbol{\gamma}$. For example, in the location case if more than half the sample points are concentrated at x_0, then the median coincides with x_0. In practice if a sufficiently large number q of observations satisfy an approximate linear fit $y_i \approx \mathbf{x}_i'\boldsymbol{\gamma}$ for some $\boldsymbol{\gamma}$, then the estimate coincides approximately with that fit: $\widehat{\boldsymbol{\beta}} \approx \boldsymbol{\gamma}$.

The exact fit property implies that if a dataset is composed of two linear substructures, an estimate with a high BP will choose to fit one of them, and this will allow the other to be discovered through the analysis of the residuals. A nonrobust estimate such as LS will instead try to make a compromise fit, with the undesirable result that the existence of two structures passes unnoticed.

Example 5.3 *Jalali-Heravi and Knouz (2002) measured for 32 chemical compounds a physical property called the Krafft point, together with several molecular descriptors, in order to find a predictive equation.*

Table 5.6 Krafft point data

Heat	Krafft pt.	Heat	Krafft pt.
296.1	7.0	261	48.0
303.0	16.0	267	50.0
314.3	11.0	274	62.0
309.8	20.8	281	57.0
316.7	21.0	307	20.2
335.5	31.5	306	0.0
330.4	31.0	320	8.1
328.0	25.0	334	24.2
337.2	38.2	347	36.2
344.1	40.5	307	0.0
341.7	30.0	321	12.5
289.3	8.0	335	26.5
226.8	22.0	349	39.0
240.5	33.0	378	36.0
247.4	35.5	425	24.0
254.2	42.0	471	19.0

Table 5.6 gives the Krafft points and one of the descriptors, called heat of formation. Figure 5.16 shows the data with the LS and the MM fits. It is seen that there are two linear structures, and that the LS estimate fits neither of them, while MM fits the majority of the observation. The points in the smaller group correspond to compounds called sulfonates (code **krafft**).

5.11 Generalized M-estimates

In this section we treat a family of estimates which is of historical importance. The simplest way to robustify a monotone M-estimate is to downweight the influential $\mathbf{x}_i$'s to prevent them from dominating the estimating equations. Hence we may define an estimate by

$$\sum_{i=1}^{n} \psi\left(\frac{r_i(\boldsymbol{\beta})}{\widehat{\sigma}}\right) \mathbf{x}_i W(d(\mathbf{x}_i)) = \mathbf{0} \tag{5.42}$$

where W is a weight function and $d(\mathbf{x})$ is some measure of the "largeness" of $\mathbf{x}$. Here ψ is monotone and $\widehat{\sigma}$ is simultaneously estimated by an M-estimating equation of the form (5.22). For instance, to fit a straight line $y_i = \beta_0 + \beta_1 x_i + \varepsilon_i$, we may choose

$$d(x_i) = \frac{\left|x_i - \widehat{\mu}_x\right|}{\widehat{\sigma}_x} \tag{5.43}$$

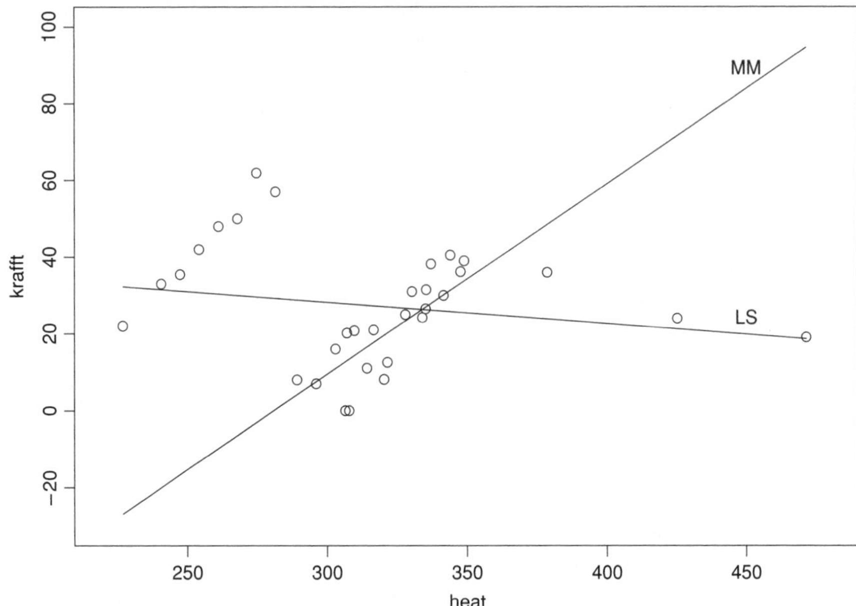

Figure 5.16 Krafft point data: consequences of the exact fit property

where $\widehat{\mu}_x$ and $\widehat{\sigma}_x$ are respectively robust location and dispersion statistics of the x_i's, such as the median and MAD. In order to bound the effect of influential points W must be such that $W(t)t$ is bounded.

More generally, we may let the weights depend on the residuals as well as the predictor variables and use a *generalized* M-estimate (a "GM-estimate") $\widehat{\boldsymbol{\beta}}$ defined by

$$\sum_{i=1}^{n} \eta\left(d(\mathbf{x}_i), \frac{r_i\left(\widehat{\boldsymbol{\beta}}\right)}{\widehat{\sigma}}\right)\mathbf{x}_i = \mathbf{0} \tag{5.44}$$

where for each s, $\eta(s,r)$ is a nondecreasing and bounded ψ-function of r, and $\widehat{\sigma}$ is obtained by a simultaneous M-scale estimate equation of the form

$$\frac{1}{n}\sum_{i=1}^{n} \rho_{\text{scale}}\left(\frac{r_i\left(\boldsymbol{\beta}\right)}{\widehat{\sigma}}\right) = \delta.$$

Two particular forms of GM-estimate have been of primary interest in the literature. The first is the estimate (5.42), which corresponds to the choice $\eta(s,r) = W(s)\psi(r)$ and is called a "Mallows estimate" (Mallows, 1975). The second form is the choice

$$\eta(s,r) = \frac{\psi(sr)}{s}, \tag{5.45}$$

which was first proposed by Schweppe, Wildes and Rom (1970) in the context of electric power systems.

The GM-estimate with the Schweppe function (5.45) is also called the "Hampel–Krasker–Welsch" estimate (Krasker and Welsch, 1982). When ψ is Huber's ψ_k, it is a solution to Hampel's problem (Section 3.5.4). See Section 5.16.6 for details.

Note that the function $d(\mathbf{x})$ in (5.43) depends on the data, and for this reason it will be better to denote it by $d_n(\mathbf{x})$. The most usual way to measure largeness is as a generalization of (5.43). Let $\widehat{\boldsymbol{\mu}}_n$ and $\widehat{\boldsymbol{\Sigma}}_n$ be a robust location vector and robust dispersion matrix, to be further studied in Chapter 6. Then d_n is defined as

$$d_n(\mathbf{x}) = \left(\mathbf{x} - \widehat{\boldsymbol{\mu}}_n\right)' \widehat{\Sigma}_n^{-1} \left(\mathbf{x} - \widehat{\boldsymbol{\mu}}_n\right). \tag{5.46}$$

In the case where $\widehat{\boldsymbol{\mu}}_n$ and $\widehat{\boldsymbol{\Sigma}}_n$ are the sample mean and covariance matrix, $\sqrt{d_n(\mathbf{x})}$ is known as the Mahalanobis distance.

Assume that $\widehat{\boldsymbol{\mu}}_n$ and $\widehat{\boldsymbol{\Sigma}}_n$ converge in probability to $\boldsymbol{\mu}$ and Σ respectively. With this assumption it can be shown that if the errors are symmetric, then the IF of a GM-estimate in the model (5.1)–(5.2) is

$$\mathrm{IF}((\mathbf{x}_0, y_0), F) = \sigma\, \eta\left(d(\mathbf{x}_0), \frac{y_0 - \mathbf{x}_0'\boldsymbol{\beta}}{\sigma}\right) \mathbf{B}^{-1}\mathbf{x}_0 \tag{5.47}$$

with

$$\mathbf{B} = -\mathrm{E}\dot{\eta}\left(d(\mathbf{x}), \frac{u}{\sigma}\right)\mathbf{x}\mathbf{x}', \quad \dot{\eta}(s, r) = \frac{\partial \eta(s, r)}{\partial r} \tag{5.48}$$

and

$$d(\mathbf{x}) = (\mathbf{x} - \boldsymbol{\mu})' \Sigma^{-1} (\mathbf{x} - \boldsymbol{\mu}).$$

Hence the IF is the same as would be obtained from (3.47) using d instead of d_n. It can be shown that $\widehat{\boldsymbol{\beta}}$ is asymptotically normal, and as a consequence of (5.47), the asymptotic covariance matrix of $\widehat{\boldsymbol{\beta}}$ is

$$\sigma^2 \mathbf{B}^{-1\prime}\mathbf{C}\mathbf{B}^{-1} \tag{5.49}$$

with

$$\mathbf{C} = \mathrm{E}\eta\left(d(\mathbf{x}), \frac{y - \mathbf{x}'\boldsymbol{\beta}}{\sigma}\right)^2 \mathbf{x}\mathbf{x}'.$$

It follows from (5.47) that GM-estimates have several attractive properties:

- If $\eta(s, r)s$ is bounded, then their IF is bounded.
- The same condition ensures a positive BP (Maronna et al. 1979).
- They are defined by estimating equations, and hence easy to compute like ordinary monotone M-estimates.

However, GM-estimates also have several drawbacks:

- Their efficiency depends on the distribution of $\mathbf{x}$: if $\mathbf{x}$ is heavy tailed they cannot be simultaneously very efficient and very robust.

- Their BP is less than 0.5 and is quite low for large p. For example, if $\mathbf{x}$ is multivariate normal, the BP is $O(p^{-1/2})$ (Maronna et al., 1979).
- To obtain affine equivariance it is necessary that $\widehat{\boldsymbol{\mu}}_n$ and $\widehat{\Sigma}_n$ used in (5.46) are affine equivariant (see Chapter 6). It will be seen in Chapter 6 that computing robust affine equivariant multivariate estimates presents the same computational difficulties we have seen for redescending M-estimates, and hence combining robustness and equivariance entails losing computational simplicity, which is one important feature of GM-estimates.
- A further drawback is that the simultaneous estimation of σ reduces the BP, especially for large p (Maronna and Yohai, 1991).

For these reasons GM-estimates, although much treated in the literature, are not a good choice except perhaps for small p. However, their computational simplicity is attractive, and they are much used in power systems. See for example Mili, Cheniae, Vichare and Rousseeuw (1996) and Pires, Simões Costa and Mili (1999).

5.12 Selection of variables

In many situations the main purpose of fitting a regression equation is to predict the response variable. If the number of predictor variables is large and the number of observations relatively small, fitting the model using all the predictors will yield poorly estimated coefficients, especially when predictors are highly correlated. More precisely, the variances of the estimated coefficients will be high and therefore the forecasts made with the estimated model will have a large variance too. A common practice to overcome this difficulty is to fit a model using only a subset of variables selected according to some statistical criterion.

Consider evaluating a model using the mean squared error (MSE) of the forecast. This MSE is composed of the variance plus the squared bias. Deleting some predictors may cause an increase in the bias and a reduction of the variance. Hence the problem of finding the best subset of predictors can be viewed as that of finding the best trade-off between bias and variance. There is a very large literature on the subset selection problem, when the LS estimate is used as an estimation procedure. See for example Miller (1990), Seber (1984) and Hastie et al. (2001).

Let the sample be $(\mathbf{x}_i, y_i)$, $i = 1, \ldots, n$, where $\mathbf{x}_i = (x_{i1}, \ldots, x_{ip})$. The predictors are assumed to be random but the case of fixed predictors is treated in a similar manner. For each set $C \subset \{1, 2, \ldots, p\}$ let $q = \#(C)$ and $\mathbf{x}_{iC} = (x_{ij})_{j \in C} \in R^q$. Akaike's (1970) *Final Prediction Error* (FPE) criterion based on the LS estimate is defined as

$$\mathrm{FPE}(C) = \mathrm{E}\left(y_0 - \mathbf{x}_{0C}'\widehat{\boldsymbol{\beta}}_C\right)^2 \tag{5.50}$$

where $\widehat{\boldsymbol{\beta}}_C$ is the estimate based on the set C and $(\mathbf{x}_0, y_0)$ have the same joint distribution as $(\mathbf{x}_i, y_i)$ and are independent of the sample. The expectation on the right hand side of (5.50) is with respect to both $(\mathbf{x}_0, y_0)$ and $\widehat{\boldsymbol{\beta}}_C$. Then it is shown that an approximately unbiased estimate of FPE is

$$\mathrm{FPE}^*(C) = \frac{1}{n}\sum_{i=1}^{n} r_{iC}^2 \left(1 + 2\frac{q}{n}\right), \tag{5.51}$$

where

$$r_{iC} = y_i - \mathbf{x}_{iC}'\widehat{\boldsymbol{\beta}}_C.$$

The first term of (5.51) evaluates the goodness of the fit when the estimate is $\widehat{\boldsymbol{\beta}}_C$ and the second term penalizes the use of a large number of explanatory variables. The best subset C is chosen as the one minimizing $\mathrm{FPE}^*(C)$.

It is clear, however, that a few outliers may distort the value of $\mathrm{FPE}^*(C)$, so that the choice of the predictors may be determined by a few atypical observations. To robustify FPE we must note that not only must the regression estimate be robust, but also the value of the criterion should not be sensitive to a few residuals.

We shall therefore robustify the FPE criterion by using for $\widehat{\boldsymbol{\beta}}$ a robust M-estimate (5.7) along with a robust error scale estimate $\widehat{\sigma}$. In addition we shall bound the influence of large residuals by replacing the square in (5.50) with a bounded ρ-function, namely the same ρ as in (5.7). To make the procedure invariant under scale changes the error must be divided by a scale σ, and to make consistent comparisons among different subsets of the predictor variable σ must remain the same for all C. Thus the proposed criterion, which will be called the *robust final prediction error* (RFPE), is defined as

$$\mathrm{RFPE}(C) = \mathrm{E}\rho\left(\frac{y_0 - \mathbf{x}_{0C}'\widehat{\boldsymbol{\beta}}_C}{\sigma}\right) \tag{5.52}$$

where σ is the asymptotic value of $\widehat{\sigma}$.

To estimate RFPE for each subset C we first compute

$$\widehat{\boldsymbol{\beta}}_C = \arg\min_{\boldsymbol{\beta}\in R^q} \sum_{i=1}^{n} \rho\left(\frac{y_i - \mathbf{x}_{iC}'\boldsymbol{\beta}}{\widehat{\sigma}}\right)$$

where the scale estimate $\widehat{\sigma}$ is based on the full set of variables, and define the estimate by

$$\mathrm{RFPE}^*(C) = \frac{1}{n}\sum_{i=1}^{n} \rho\left(\frac{r_{iC}}{\widehat{\sigma}}\right) + \frac{q}{n}\frac{\widehat{A}}{\widehat{B}} \tag{5.53}$$

where

$$\widehat{A} = \frac{1}{n}\sum_{i=1}^{n} \psi\left(\frac{r_{iC}}{\widehat{\sigma}}\right)^2, \quad \widehat{B} = \frac{1}{n}\sum_{i=1}^{n} \psi'\left(\frac{r_{iC}}{\widehat{\sigma}}\right). \tag{5.54}$$

Note that if $\rho(r) = r^2$ then $\psi(r) = 2r$, and the result is equivalent to (5.51) since $\widehat{\sigma}$ cancels out. The criterion (5.53) is justified in Section 5.16.7.

When p is large, finding the optimal subset may be very costly in terms of computation time and therefore strategies to find suboptimal sets can be used. Two problems arise:

- Searching over all subsets may be impractical because of the extremely large number of subsets.
- Each computation of RFPE* requires recomputing a robust estimate $\widehat{\beta}_C$ for each C, which can be very time consuming when performed a large number of times.

In the case of the LS estimate there exist very efficient algorithms to compute the classical FPE (5.51) for all subsets (see the references above) and so the first problem above is tractable for the classical approach if p is not too large. But computing a robust estimate $\widehat{\beta}_C$ for all subsets C would be infeasible unless p were small. A simple but frequently effective suboptimal strategy is *stepwise regression*: add or remove one variable at a time ("forward" or "backward" regression), choosing the one whose inclusion or deletion yields the lowest value of the criterion. Various simulation studies indicate that the backward procedure is better. Starting with $C = \{1, \ldots, p\}$ we remove one variable at a time. At step k $(= 1, \ldots, p-1)$ we have a subset C with $\#(C) = p - k + 1$, and the next predictor to be deleted is found by searching over all subsets of C of size $p - k$ to find the one with smallest RFPE*.

The second problem above arises because robust estimates are computationally intensive, the more so when there are a large number of predictors. A simple way to reduce the computational burden is to avoid repeating the subsampling for each subset C by computing $\widehat{\beta}_C$, starting from the approximation given by the weighted LS estimate with weights w_i obtained from the estimate corresponding to the full model.

To demonstrate the advantages of using a robust model selection approach based on RFPE*, we shall use a simulated dataset from a known model for which the "correct solution" is clear. We generated $n = 50$ observations from the model $y_i = \beta_0 + \mathbf{x}_i'\boldsymbol{\beta}_1 + u_i$ with $\beta_0 = 1$ and $\boldsymbol{\beta}_1' = (1, 1, 1, 0, 0, 0)$, so that $p = 7$. The u_i's and x_{ij}'s are i.i.d. standard normal. Here a perfect model selection method will select the variables $\{1, 2, 3\}$. To this dataset we added six outliers $(\mathbf{x}_0, y_0)$ with $y_0 = 20$ and $\mathbf{x}_0 = \mathbf{0}$ for three of them and $\mathbf{x}_0' = (0, 0, 0, 1, 1, 1)$ for the other three, which should have the effect of decreasing the estimates of the first three slopes and increasing those of the last three. We then applied the backward stepwise procedure using both the classical LS estimate with the C_p criterion, and our proposed robust RFPE* criterion. Table 5.7 shows the results (code **modelselection**). The minima of the criteria are attained by the sets $\{1, 2, 3\}$ for the robust criterion and by $\{4, 5, 6\}$ for the classical one.

Other approaches to robust model selection were given by Qian and Künsch (1998), Ronchetti and Staudte (1994) and Ronchetti, Field and Blanchard (1997).

Table 5.7 Variable selection for simulated data

LS		Robust	
Variable	C_p	Variable	RFPE*
1 2 4 5 6	269	1 2 3 5 6	15.37
2 4 5 6	261	1 2 3 6	15.14
4 5 6	260	1 2 3	15.11
5 6	263	2 3	18.19

5.13 Heteroskedastic errors

The asymptotic theory for M-estimates, which includes S- and MM-estimates, has been derived under the assumption that the errors are i.i.d. and hence homoskedastic.

These assumptions do not always hold in practice. When the y_i's are time series or spatial variables the errors may be correlated. And in many cases the variability of the error may depend on the explanatory variables, in particular the conditional variance of y given $\mathbf{x}$ may depend on $\boldsymbol{\beta}'\mathbf{x} = \mathrm{E}(y|\mathbf{x})$.

Actually the assumptions of independent and homoskesdastic errors are not necessary for the consistency and asymptotic normality of M-estimates. In fact, it can be shown that these properties hold under much weaker conditions. Nevertheless we can mention two problems:

- The estimates may have lower efficiency than others which take into account the correlation or heteroskedasticity of the errors.
- The asymptotic covariance matrix of M-estimates may be different from $v\mathbf{V}_{\mathbf{x}}^{-1}$ given in (5.14), which was derived assuming i.i.d. errors. Therefore the estimate $\widetilde{\mathbf{V}}_{\mathbf{x}}$ given in Section 5.8 would not converge to the true asymptotic covariance matrix of $\widehat{\boldsymbol{\beta}}$.

We deal with these problems in the next two subsections.

5.13.1 Improving the efficiency of M-estimates

To improve the efficiency of M-estimates under heteroskedasticity, the dependence of the error scale on $\mathbf{x}$ should be included in the model. For instance, we can replace model (5.1) by

$$y_i = \boldsymbol{\beta}'\mathbf{x}_i + h(\boldsymbol{\lambda}, \boldsymbol{\beta}'\mathbf{x})u_i,$$

where the u_i's are i.i.d. and independent of $\mathbf{x}_i$, and $\boldsymbol{\lambda}$ is an additional vector parameter. In this case the error scales are proportional to $h^2(\lambda, \boldsymbol{\beta}'\mathbf{x})$.

Observe that if we knew $h(\boldsymbol{\lambda}, \boldsymbol{\beta}'\mathbf{x})$, then the transformed variables

$$y_i^* = \frac{y_i}{h(\boldsymbol{\lambda}, \boldsymbol{\beta}'\mathbf{x})}, \quad \mathbf{x}_i^* = \frac{\mathbf{x}_i}{h(\boldsymbol{\lambda}, \boldsymbol{\beta}'\mathbf{x})}$$

would follow the homoskedastic regression model

$$y_i^* = \boldsymbol{\beta}'\mathbf{x}_i^* + u_i.$$

This suggests the following procedure to obtain robust estimates of $\boldsymbol{\beta}$ and $\boldsymbol{\lambda}$:

1. Compute an initial robust estimate $\widehat{\boldsymbol{\beta}}_0$ for homoskedastic regression, e.g., $\widehat{\boldsymbol{\beta}}_0$ may be an MM-estimate.
2. Compute the residuals $r_i(\widehat{\boldsymbol{\beta}}_0)$.
3. Use these residuals to obtain an estimate $\widehat{\boldsymbol{\lambda}}$ of $\boldsymbol{\lambda}$. For example, if

$$h(\boldsymbol{\lambda}, t) = \exp(\lambda_1 + \lambda_2 |t|),$$

   then λ can be estimated by a robust linear fit of $\log(|r_i(\widehat{\boldsymbol{\beta}}_0)|)$ on $|\widehat{\boldsymbol{\beta}}_0'\mathbf{x}_i|$.
4. Compute a robust estimate for homoskedastic regression based on the transformed variables

$$y_i^* = \frac{y_i}{h(\widehat{\boldsymbol{\lambda}}, \widehat{\boldsymbol{\beta}}_0'\mathbf{x})}, \mathbf{x}_i^* = \frac{\mathbf{x}_i}{h(\widehat{\boldsymbol{\lambda}}, \widehat{\boldsymbol{\beta}}_0'\mathbf{x})}.$$

Steps 1–4 may be iterated.

Robust methods for heteroskedastic regression have been proposed by Carroll and Ruppert (1982) who used monotone M-estimates; by Giltinan, Carroll and Ruppert (1986) who employed GM-estimates, and by Bianco, Boente and Di Rienzo (2000) and Bianco and Boente (2002) who defined estimates with high BP and bounded influence starting with an initial MM-estimate followed by one Newton–Raphson step of a GM-estimate.

5.13.2 Estimating the asymptotic covariance matrix under heteroskedastic errors

Simpson, Carroll and Ruppert (1992) proposed an estimate for the asymptotic covariance matrix of regression GM-estimates which does not require homoskedasticity but requires symmetry of the error distribution. Croux, Dhaene and Hoorelbeke (2003) proposed a method to estimate the asymptotic covariance matrix of a regression M-estimate which requires neither homoskedasticity nor symmetry. This method can also be applied to simultaneous M-estimates of regression of scale (which includes S-estimates) and to MM-estimates.

We shall give some details of the method for the case of MM-estimates. Let $\widehat{\gamma}$ and $\widehat{\sigma}$ be the initial S-estimate used to compute the MM-estimate and the corresponding scale estimate, respectively. Since $\widehat{\gamma}$ and $\widehat{\sigma}$ are M-estimates of regression and of scale, they satisfy equations of the form

$$\sum_{i=1}^{n} \psi_1\left(\frac{r_i(\widehat{\gamma})}{\widehat{\sigma}}\right)\mathbf{x}_i = \mathbf{0}, \tag{5.55}$$

$$\frac{1}{n}\sum_{i=1}^{n}\left[\rho_1\left(\frac{r_i(\widehat{\gamma})}{\widehat{\sigma}}\right) - \delta\right] = 0, \tag{5.56}$$

with $\psi_1 = \rho_1'$. The final MM-estimate $\widehat{\boldsymbol{\beta}}$ is a solution of

$$\sum_{i=1}^{n} \psi_2\left(\frac{r_i(\widehat{\boldsymbol{\beta}})}{\widehat{\sigma}}\right)\mathbf{x}_i = \mathbf{0}. \tag{5.57}$$

To explain the proposed method we need to express the system (5.55)–(5.56)–(5.57) as a unique set of M-estimating equations. To this end let the vector γ represent the values taken on by $\widehat{\gamma}$. Let $\mathbf{z} = (\mathbf{x}, y)$. For $\gamma, \boldsymbol{\beta} \in R^p$ and $\sigma \in R$ put $\boldsymbol{\alpha} = (\gamma', \sigma, \boldsymbol{\beta}')' \in R^{2p+1}$, and define the function

$$\Psi(\mathbf{z}, \boldsymbol{\alpha}) = (\Psi_1(\mathbf{z}, \boldsymbol{\alpha}), \Psi_2(\mathbf{z}, \boldsymbol{\alpha}), \Psi_3(\mathbf{z}, \boldsymbol{\alpha}))$$

where

$$\Psi_1(\mathbf{z}, \boldsymbol{\alpha}) = \psi_1\left(\frac{y - \gamma'\mathbf{x}}{\sigma}\right)\mathbf{x}$$

$$\Psi_2(\mathbf{z}, \boldsymbol{\alpha}) = \rho_1\left(\frac{y - \gamma'\mathbf{x}}{\sigma}\right) - \delta$$

$$\Psi_3(\mathbf{z}, \boldsymbol{\alpha}) = \psi_2\left(\frac{y - \boldsymbol{\beta}'\mathbf{x}}{\sigma}\right)\mathbf{x}.$$

Then $\widehat{\boldsymbol{\alpha}} = (\widehat{\gamma}, \widehat{\sigma}, \widehat{\boldsymbol{\beta}})$ is an M-estimate satisfying

$$\sum_{i=1}^{n} \Psi(\mathbf{z}_i, \widehat{\alpha}) = 0,$$

and therefore according to (3.48), its asymptotic covariance matrix is

$$\mathbf{V} = \mathbf{A}^{-1}\mathbf{B}\mathbf{A}'^{-1}$$

with

$$\mathbf{A} = \mathrm{E}\left[\Psi(\mathbf{z}, \alpha)\Psi(\mathbf{z}, \alpha)'\right]$$

and

$$\mathbf{B} = \mathrm{E}\left(\frac{\partial \Psi(\mathbf{z}, \alpha)}{\partial \boldsymbol{\alpha}}\right),$$

where the expectation is calculated under the model $y = \mathbf{x}'\boldsymbol{\beta} + u$ and taking $\gamma = \boldsymbol{\beta}$.

Then $\mathbf{V}$ can be estimated by

$$\widehat{\mathbf{V}} = \widehat{\mathbf{A}}^{-1}\widehat{\mathbf{B}}\widehat{\mathbf{A}}'^{-1}$$

where $\widehat{\mathbf{A}}$ and $\widehat{\mathbf{B}}$ are the empirical versions of $\mathbf{A}$ and $\mathbf{B}$ obtained by replacing α with $\widehat{\alpha}$.

Observe that the only requirement for $\widehat{\mathbf{V}}$ to be a consistent estimate of $\mathbf{V}$ is that the observations $(\mathbf{x}_1, y_1), \ldots, (\mathbf{x}_n, y_n)$ be i.i.d., but this does not require any condition on the conditional distribution of y given $\mathbf{x}$, e.g., homoskedasticity.

Croux et al. (2003) also consider estimates of $\mathbf{V}$ when the errors are not independent.

5.14 *Other estimates

5.14.1 τ-estimates

These estimates were proposed by Yohai and Zamar (1988). They have a high BP and a controllable normal distribution efficiency, but unlike MM-estimates they do not require a preliminary scale estimate. Let $\widehat{\sigma}(\mathbf{r})$ be a robust M-scale based on $\mathbf{r} = (r_1, \ldots, r_n)$, namely the solution of

$$\frac{1}{n}\sum_{i=1}^{n} \rho_0\left(\frac{r_i}{\widehat{\sigma}}\right) = \delta, \tag{5.58}$$

and define the scale τ as

$$\tau(\mathbf{r})^2 = \widehat{\sigma}(\mathbf{r})^2 \frac{1}{n}\sum_{i=1}^{n} \rho\left(\frac{r_i}{\widehat{\sigma}(\mathbf{r})}\right) \tag{5.59}$$

where ρ_0 and ρ are bounded ρ-functions. Put $\mathbf{r}(\boldsymbol{\beta}) = (r_1(\boldsymbol{\beta}), \ldots, r_n(\boldsymbol{\beta}))$ with $r_i(\boldsymbol{\beta}) = y_i - \mathbf{x}_i'\boldsymbol{\beta}$ $(i = 1, \ldots, n)$. Then a regression τ-estimate is defined by

$$\widehat{\boldsymbol{\beta}} = \arg\min_{\boldsymbol{\beta}} \tau(\mathbf{r}(\boldsymbol{\beta})). \tag{5.60}$$

A τ-estimate minimizes a robust scale of the residuals, but unlike the S-estimates of Section 5.6.1 it has a controllable efficiency. The intuitive reason is that the LS estimate is obtained as a special case of (5.60) when $\rho(r) = r^2$ so that $\tau(\mathbf{r})^2 = \text{ave}\left(\mathbf{r}^2\right)$, and hence by an adequate choice of ρ the estimate can be made arbitrarily close to the LS estimate, and so arbitrarily efficient at the normal distribution.

Yohai and Zamar (1988) showed that $\widehat{\boldsymbol{\beta}}$ satisfies an M-estimating equation (5.8) where ψ is a linear combination of ρ' and ρ_0' with coefficients depending on the data. From this property it is shown that $\widehat{\boldsymbol{\beta}}$ is asymptotically normal. Its asymptotic efficiency at the normal distribution can be adjusted to be arbitrarily close to one, just as in the case of MM-estimates. It is also shown that its BP is the same as that of an S-estimate based on ρ_0, and so by adequately choosing ρ_0 the estimate can attain the maximum BP for regression estimates.

Numerical computation of these estimates follows the same lines as for S-estimates. See Yohai and Zamar (1988).

5.14.2 Projection estimates

Note that the residuals of the LS estimate are uncorrelated with any linear combination of the predictors. In fact the normal equations (4.13) imply that for any $\boldsymbol{\lambda} \in R^p$ the LS regression of the residuals r_i on the projections $\boldsymbol{\lambda}'\mathbf{x}_i$ is zero, since $\sum_{i=1}^n r_i \boldsymbol{\lambda}'\mathbf{x}_i = 0$. The LS estimate of regression through the origin is defined for $\mathbf{z} = (z_1, \ldots, z_n)'$ and $\mathbf{y} = (y_1, \ldots, y_n)'$ as

$$b(\mathbf{z}, \mathbf{y}) = \frac{\sum_{i=1}^n z_i y_i}{\sum_{i=1}^n z_i^2},$$

and it follows that the LS estimate $\widehat{\boldsymbol{\beta}}$ of $\mathbf{y}$ on $\mathbf{x}$ satisfies

$$b(\mathbf{X}\boldsymbol{\lambda}, \mathbf{r}(\widehat{\boldsymbol{\beta}})) = 0 \; \forall \; \boldsymbol{\lambda} \in R^p, \; \boldsymbol{\lambda} \neq \mathbf{0}. \tag{5.61}$$

A robust regression estimate could be obtained by (5.61) using for b a *robust* estimate of regression through the origin. But in general it is not possible to obtain equality in (5.61). Hence we must content ourselves with making b "as small as possible". Let $\widehat{\sigma}$ be a robust scale estimate, such as $\widehat{\sigma}(\mathbf{z}) = \mathrm{Med}(|\mathbf{z}|)$. Then the projection estimates for regression ("P-estimates") proposed by Maronna and Yohai (1993) are defined as

$$\widehat{\boldsymbol{\beta}} = \arg\min_{\boldsymbol{\beta}} \left(\max_{\boldsymbol{\lambda} \neq \mathbf{0}} \left| b(\mathbf{X}\boldsymbol{\lambda}, \mathbf{r}(\widehat{\boldsymbol{\beta}})) \right| \widehat{\sigma}(\mathbf{X}\boldsymbol{\lambda}) \right) \tag{5.62}$$

which means that the residuals are "as uncorrelated as possible" with all projections. Note that the condition $\boldsymbol{\lambda} \neq \mathbf{0}$ can be replaced by $\|\boldsymbol{\lambda}\| = 1$. The factor $\widehat{\sigma}(\mathbf{X}\boldsymbol{\lambda})$ is needed to make the regression estimate scale equivariant.

The "median of slopes" estimate for regression through the origin is defined as the conditional median

$$b(\mathbf{x}, \mathbf{y}) = \mathrm{Med}\left(\frac{y_i}{x_i} \middle| x_i \neq 0 \right). \tag{5.63}$$

Martin et al. (1989) extended Huber´s minimax result for the median (Section 3.8.5) showing that (5.63) minimizes asymptotic bias among regression invariant estimates. Maronna and Yohai (1993) studied P-estimates with b given by (5.63), which they called MP estimates, and found that their maximum asymptotic bias is lower than that of MM- and S-estimates. They have $n^{-1/2}$ consistency rate, but are not asymptotically normal, which makes their use difficult for inference.

Maronna and Yohai (1993) show that if the $\mathbf{x}_i$ are multivariate normal, then the maximum asymptotic bias of P-estimates does not depend on p, and is not larger than twice the minimax asymptotic bias for all regression equivariant estimates.

Numerical computation of P-estimates is difficult because of the nested optimization in (5.62). An approximate solution can be found by reducing the searches over

$\boldsymbol{\beta}$ and $\boldsymbol{\lambda}$ to finite sets. A set of N candidate estimates $\boldsymbol{\beta}_k$, $k = 1, \ldots, N$, is obtained by subsampling as in Section 5.7.2, and from them the candidate directions are computed as

$$\boldsymbol{\lambda}_{jk} = \frac{\boldsymbol{\beta}_j - \boldsymbol{\beta}_k}{\|\boldsymbol{\beta}_j - \boldsymbol{\beta}_k\|}, \quad j \neq k.$$

Then (5.62) is replaced by

$$\widehat{\boldsymbol{\beta}} = \arg\min_k \left(\max_{j \neq k} \left| b(\mathbf{X}\boldsymbol{\lambda}_j, \mathbf{r}(\widehat{\boldsymbol{\beta}}_k)) \right| \widehat{\sigma}\left(\mathbf{X}\boldsymbol{\lambda}_j\right) \right).$$

The resulting approximate estimate is regression and affine equivariant. In principle the procedure requires $N(N-1)$ evaluations, but this can be reduced to $O(N \log N)$ by a suitable trick.

5.14.3 Constrained M-estimates

Mendes and Tyler (1996) define *constrained M-estimates* (CM-estimates for short) as in (4.36)

$$(\widehat{\boldsymbol{\beta}}, \widehat{\sigma}) = \arg\min_{\boldsymbol{\beta},\sigma} \left\{ \frac{1}{n} \sum_{i=1}^{n} \rho\left(\frac{r_i(\boldsymbol{\beta})}{\sigma} \right) + \log \sigma \right\} \tag{5.64}$$

with the restriction

$$\frac{1}{n} \sum_{i=1}^{n} \rho\left(\frac{r_i(\boldsymbol{\beta})}{\sigma} \right) \leq \varepsilon, \tag{5.65}$$

where ρ is a bounded ρ-function and $\varepsilon \in (0, 1)$. Note that if ρ is bounded (5.64) cannot be handled without restrictions, for then $\sigma \to 0$ would yield a trivial solution.

Mendes and Tyler show that CM-estimates are M-estimates with the same ρ. Thus they are asymptotically normal with a normal distribution efficiency that depends only on ρ (but not on ε), and hence the efficiency can be made arbitrarily high. Mendes and Tyler also show that for a continuous distribution the solution asymptotically attains the bound (5.65), so that $\widehat{\sigma}$ is an M-scale of the residuals. It follows that the estimate has an asymptotic BP equal to $\min(\varepsilon, 1-\varepsilon)$, and taking $\varepsilon = 0.5$ yields the maximum BP.

5.14.4 Maximum depth estimates

Maximum regression depth estimates were introduced by Rousseeuw and Hubert (1999). Define the *regression depth* of $\boldsymbol{\beta} \in R^p$ with respect to a sample $(\mathbf{x}_i, y_i)$ as

$$d(\boldsymbol{\beta}) = \frac{1}{n} \min_{\boldsymbol{\lambda} \neq \mathbf{0}} \# \left\{ \frac{r_i(\boldsymbol{\beta})}{\boldsymbol{\lambda}'\mathbf{x}_i} < 0, \boldsymbol{\lambda}'\mathbf{x}_i \neq 0 \right\}, \tag{5.66}$$

where $\boldsymbol{\lambda} \in R^p$. Then the *maximum depth regression estimate* is defined as

$$\widehat{\boldsymbol{\beta}} = \arg\max_{\boldsymbol{\beta}} d(\boldsymbol{\beta}). \tag{5.67}$$

The solution need not be unique. Since only the direction matters, the infimum in (5.66) may be restricted to $\{\|\boldsymbol{\lambda}\| = 1\}$. Like the P-estimates of Section 5.14.2, maximum depth estimates are based on the univariate projections $\boldsymbol{\lambda}'\mathbf{x}_i$ of the predictors. In the case of regression through the origin, $\widehat{\boldsymbol{\beta}}$ coincides with the median of slopes given by (5.63). But when $p > 1$, the maximum asymptotic BP of these estimates at the linear model (5.1) is 1/3, and for an arbitrary joint distribution of $(\mathbf{x}, y)$ it can only be asserted to be $\geq 1/(p+1)$.

Adrover, Maronna and Yohai (2002) discuss the relationships between maximum depth and P-estimates. They derive the asymptotic bias of the former and compare it to that of the MP-estimates defined in Section 5.14.2. Both biases turn out to be similar for moderate contamination (in particular, the GESs are equal), while the MP-estimate is better for large contamination. They define an approximate algorithm for computing the maximum depth estimate, based on an analogous idea already studied for the MP-estimate.

5.15 Models with numeric and categorical predictors

Consider a linear model of the form

$$y_i = \mathbf{x}_{1i}'\boldsymbol{\beta}_1 + \mathbf{x}_{2i}'\boldsymbol{\beta}_2 + u_i, \quad i = 1, \ldots n, \tag{5.68}$$

where the $\mathbf{x}_{1i} \in R^{p_1}$ are fixed 0–1 vectors, such as a model with some categorical variables as in the case of the example in Section 1.4.2, and the $\mathbf{x}_{2i} \in R^{p_2}$ are continuous random variables. In the model (5.68) the presence of the continuous variables means that a monotone M-estimate would not be robust. On the other hand an S-estimate will often be too expensive since a subsampling procedure would require at least $O(2^{p_1+p_2})$ evaluations; for models with categorical variables the number of parameters p_1 is often beyond the reach of reasonable computing times. See the number of subsamples required as a function of the number of parameters in Section 5.7.2, and note for example that the model in Section 1.4.2 has $p_1 + p_2 = 22 + 5 = 27$ parameters. In any event the sub-sampling will be a waste if $p_2 \ll p_1$. Besides, in an unbalanced structured design there is a high probability that a subsampling algorithm yields collinear samples. For example, if there are five independent explanatory dummy variables that take the value 1 with probability 0.1, then the probability of selecting a noncollinear sample of size 5 is only 0.011!

Maronna and Yohai (2000) proposed an estimate based on the idea that if one knew $\boldsymbol{\beta}_2$ (respectively $\boldsymbol{\beta}_1$) in (5.68), it would be natural to use a monotone M-estimate (S-estimate) for the parameter $\boldsymbol{\beta}_1$ (the parameter $\boldsymbol{\beta}_2$). Let $M(\mathbf{X}, \mathbf{y})$ be a monotone M-estimate such as L1. Then, for each $\boldsymbol{\beta}_2 \in R^{p_2}$ define

$$\boldsymbol{\beta}_1^*(\boldsymbol{\beta}_2) = M(\mathbf{X}_1, \mathbf{y} - \mathbf{X}_2\boldsymbol{\beta}_2), \tag{5.69}$$

where $\mathbf{X}_1$ and $\mathbf{X}_2$ are the matrices with the rows $\mathbf{x}'_{i1}$ and $\mathbf{x}'_{i2}$, respectively. Denote the residuals of (5.68) by $\mathbf{r}(\widehat{\boldsymbol{\beta}}_1, \widehat{\boldsymbol{\beta}}_2) = \mathbf{y} - \mathbf{X}_1\widehat{\boldsymbol{\beta}}_1 - \mathbf{X}_2\widehat{\boldsymbol{\beta}}_2$, and let $S(\mathbf{r})$ be an M-scale estimate of the residuals $\mathbf{r} = (r_1, \ldots, r_n)$. An *MS-estimate* $(\widehat{\boldsymbol{\beta}}_1, \widehat{\boldsymbol{\beta}}_2)$ is defined by

$$\widehat{\boldsymbol{\beta}}_2 = \arg\min_{\boldsymbol{\beta}_2} \; S(\mathbf{r}(\boldsymbol{\beta}_1^*(\boldsymbol{\beta}_2), \boldsymbol{\beta}_2)) \tag{5.70}$$

and $\widehat{\boldsymbol{\beta}}_1 = \boldsymbol{\beta}_1^*(\widehat{\boldsymbol{\beta}}_2)$. This estimate is regression and affine equivariant. For example, if (5.68) consists of p_2 continuous predictors and an intercept, i.e., $p_1 = 1$ and $\mathbf{x}_{1i} \equiv 1$ and $M(\mathbf{X}, \mathbf{y})$ is the L1 estimate, then $\boldsymbol{\beta}_1^*(\boldsymbol{\beta}_2) = \mathrm{Med}(\mathbf{y} - \mathbf{X}_2\boldsymbol{\beta}_2)$. An MS-estimate was used in the example of Section 1.4.2 and for the comparison of classical and robust t-statistics and p-values in Section 5.8.

Rousseeuw and Wagner (1994), and Hubert and Rousseeuw (1996, 1997), have proposed other approaches to this problem.

Example 5.4 *Each row of the dataset* **algae** *(from Hettich and Bay, 1999) is a set of 90 measurements at a river in some place in Europe. There are 11 predictors. The first three are categorical: the season of the year, river size (small, medium and large) and fluid velocity (low, medium and high). The other eight are the concentrations of several chemical substances. The response is the logarithm of the abundance of a certain class of algae.*

Figures 5.17 and 5.18 (code **algae**) are the normal Q–Q plots of the residuals corresponding to the LS estimate and to the MS-estimate described above. The first gives the impression of short-tailed residuals, while the residuals from the robust fit indicate the existence of least two outliers.

Example 1.3 (continued) In the multiple linear regression Section 1.4.2 the response variable was rate of unemployment and the predictor variables were PA, GPA, HS, GHS, Region and Period. The last two are categorical variables with 22 and 2 parameters respectively, while the other predictors are continuous variables. The estimator used for that example was the MS-estimate. Figures 1.4 and 1.5 revealed that for these data the LS estimate found no outliers at all, while the MS-estimate found a number of large outliers. In this example three of the LS and MS-estimate t-statistics and p-values give opposite results using 0.05 as the level of the test:

Estimate	Variable	t-value	p-value
MS	Region 20	−1.0944	0.2811
LS	Region 20	−3.0033	0.0048
MS	HS	1.3855	0.1744
LS	HS	2.4157	0.0209
MS	Period2	2.1313	0.0400
LS	Period2	0.9930	0.3273

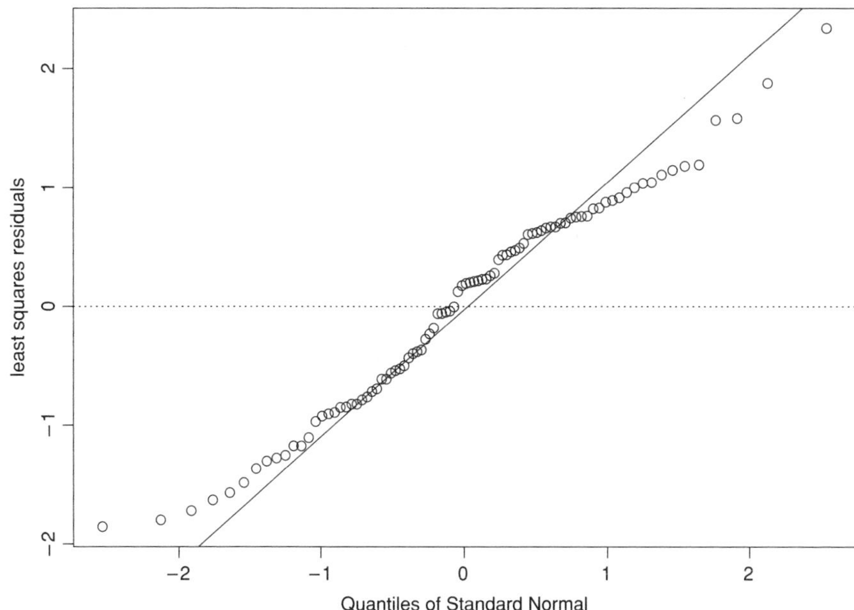

Figure 5.17 Algae data: normal Q–Q plot of LS residuals

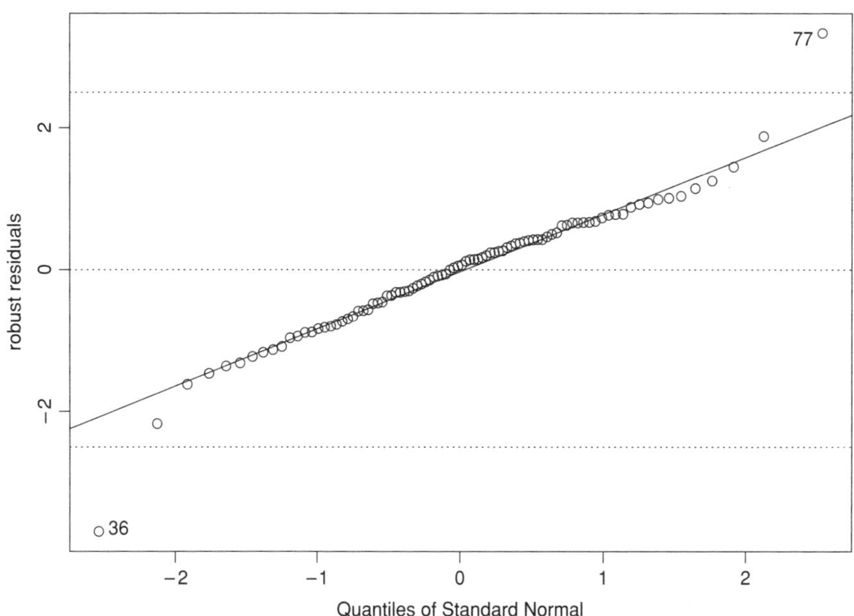

Figure 5.18 Algae data: normal Q–Q plot of robust residuals

For the "Region 20" level of the "Region" categorical variable and the HS variables the LS fit declares these variables as significant while the robust fit declares them insignificant. The opposite is the case for the Period2 level of the Period categorical variable. This shows that outliers can have a large influence on the classical test statistics of a LS fit.

5.16 *Appendix: proofs and complements

5.16.1 The BP of monotone M-estimates with random X

We assume σ is known and equal to one. The estimate verifies

$$\psi\left(y_1 - \mathbf{x}_1'\widehat{\beta}\right)\mathbf{x}_1 + \sum_{i=2}^{n} \psi\left(y_i - \mathbf{x}_i'\widehat{\beta}\right)\mathbf{x}_i = \mathbf{0}. \tag{5.71}$$

Let y_1 and $\mathbf{x}_1$ tend to infinity in such a way that $y_1/\|\mathbf{x}_1\| \to \infty$. If $\widehat{\beta}$ remained bounded, we would have

$$y_1 - \mathbf{x}_1'\widehat{\beta} \geq y_1 - \|\mathbf{x}_1\| \left\|\widehat{\beta}\right\| = \|\mathbf{x}_1\| \left(\frac{y_1}{\|\mathbf{x}_1\|} - \left\|\widehat{\beta}\right\|\right) \to \infty.$$

Since ψ is nondecreasing, $\psi\left(y_1 - \mathbf{x}_1'\widehat{\beta}\right)$ would tend to $\sup \psi > 0$ and hence the first term in (5.71) would tend to infinity, while the sum would remain bounded.

5.16.2 Heavy-tailed x

The behavior of the estimates under heavy-tailed $\mathbf{x}$ is most easily understood when the estimate is the LS estimate and $p = 1$, i.e.,

$$y_i = \beta x_i + u_i,$$

where $\{x_i\}$ and $\{u_i\}$ are independent i.i.d. sequences. Then

$$\widehat{\beta}_n = \frac{\sum_{i=1}^{n} x_i y_i}{T_n} \quad \text{with } T_n = \sum_{i=1}^{n} x_i^2.$$

Assume $\mathrm{E}u_i = 0$ and $\mathrm{Var}(u_i) = 1$. Then

$$\mathrm{Var}\left(\widehat{\beta}_n \middle| \mathbf{X}\right) = \frac{1}{T_n} \quad \text{and} \quad \mathrm{E}\left(\widehat{\beta}_n \middle| \mathbf{X}\right) = \beta,$$

and hence, by a well-known property of the variance (see, e.g., Feller, 1971),

$$\mathrm{Var}(\sqrt{n}\widehat{\beta}_n) = n\left\{\mathrm{E}\left[\mathrm{Var}\left(\widehat{\beta}_n \middle| \mathbf{X}\right)\right] + \mathrm{Var}\left[\mathrm{E}\left(\widehat{\beta}_n \middle| \mathbf{X}\right)\right]\right\} = \mathrm{E}\frac{1}{T_n/n}.$$

If $a = \mathrm{E}x_i^2 < \infty$, the law of large numbers implies that $T_n/n \to_p a$, and under suitable conditions on x_i this implies that

$$\mathrm{E}\frac{1}{T_n/n} \to \frac{1}{a}, \tag{5.72}$$

hence

$$\mathrm{Var}(\sqrt{n}\widehat{\beta}_n) \to \frac{1}{a};$$

that is, $\widehat{\beta}$ is $n^{-1/2}$-consistent.

If instead $\mathrm{E}x_i^2 = \infty$, then $T_n/n \to_p \infty$, which implies that

$$\mathrm{Var}(\sqrt{n}\widehat{\beta}_n) \to 0,$$

and hence $\widehat{\beta}$ tends to β at a higher rate than $n^{-1/2}$

A simple sufficient condition for (5.72) is that $x_i \geq \alpha$ for some $\alpha > 0$, for then $n/T_n \leq 1/\alpha^2$ and (5.72) holds by the bounded convergence theorem (Theorem 10.6). But the result can be shown to hold under more general assumptions.

5.16.3 Proof of the exact fit property

Define for $t \in R$

$$\mathbf{y}^* = \mathbf{y} + t(\mathbf{y} - \mathbf{X}\gamma).$$

Then the regression and scale equivariance of $\widehat{\beta}$ implies

$$\widehat{\beta}(\mathbf{X}, \mathbf{y}^*) = \widehat{\beta}(\mathbf{X}, \mathbf{y}) + t\left(\widehat{\beta}(\mathbf{X}, \mathbf{y}) - \gamma\right).$$

Since for all t, $\mathbf{y}^*$ has at least $q \geq n - m^*$ values in common with $\mathbf{y}$, the above expression must remain bounded, and this requires $\widehat{\beta}(\mathbf{X}, \mathbf{y}) - \gamma = \mathbf{0}$.

5.16.4 The BP of S-estimates

It will be shown that the finite BP of an S-estimate defined in Section 5.6.1 does not depend on $\mathbf{y}$, and that its maximum is given by (5.23)–(5.24).

This result has been proved by Rousseeuw and Leroy (1987) and Mili and Coakley (1996) under slightly more restricted conditions. The main result of this section is the following.

Theorem 5.1 *Let m^* be as in (5.9) and $m^*_{\max}$ as in (5.24). Call $m(\delta)$ the largest integer $< n\delta$. Then:*

(a) $m^* \leq n\delta$,
(b) if $[n\delta] \leq m^*_{\max}$, then $m^* \geq m(\delta)$.

It follows from this result that if $n\delta$ is not an integer and $\delta \le m^*_{\max}/n$, then $m^* = [n\delta]$, and hence the δ given by (5.25) yields $m^* = m^*_{\max}$.

To prove the theorem we first need an auxiliary result.

Lemma 5.2 *Consider any sequence* $\mathbf{r}_N = \left(r_{N,1}, \ldots, r_{N,n}\right)$ *with* $\sigma_N = \widehat{\sigma}(\mathbf{r}_N)$. *Then*

(i) Let $C = \left\{i : |r_{N,i}| \to \infty\right\}$. *If* $\#(C) > n\delta$, *then* $\sigma_N \to \infty$.
(ii) Let $D = \left\{i : |r_{N,i}| \text{ is bounded}\right\}$. *If* $\#(D) > n - n\delta$, *then* σ_N *is bounded.*

Proof of lemma:

(i) Assume σ_N is bounded. Then the definition of σ_N implies

$$n\delta \ge \lim_{N\to\infty} \sum_{i\in C} \rho\left(\frac{r_{N,i}}{\sigma_N}\right) = \#(C) > n\delta,$$

which is a contradiction.

(ii) To show that σ_N remains bounded, assume that $\sigma_N \to \infty$. Then $r_{N,i}/\sigma_N \to 0$ for $i \in D$, which implies

$$n\delta = \lim_{N\to\infty} \sum_{i=1}^{n} \rho\left(\frac{r_{N,i}}{\sigma_N}\right) = \lim_{N\to\infty} \sum_{i\notin D} \rho\left(\frac{r_{N,i}}{\sigma_N}\right) \le n - \#(D) < n\delta,$$

which is a contradiction.

Proof of (a): It will be shown that $m^* \le n\delta$. Let $m > n\delta$. Take $C \subset \{1, \ldots, n\}$ with $\#(C) = m$. Let $\mathbf{x}_0 \in R^p$ with $\|\mathbf{x}_0\| = 1$. Given a sequence $(\mathbf{X}_N, \mathbf{y}_N)$, define for $\boldsymbol{\beta} \in R^p$

$$\mathbf{r}_N(\boldsymbol{\beta}) = \mathbf{y}_N - \mathbf{X}_N\boldsymbol{\beta}.$$

Take $(\mathbf{X}_N, \mathbf{y}_N)$ such that

$$(\mathbf{x}_{N,i}, y_{N,i}) = \begin{cases} (N\mathbf{x}_0, N^2) & \text{if } \ i \in C \\ (\mathbf{x}_i, y_i) & \text{otherwise} \end{cases} \tag{5.73}$$

It will be shown that the estimate $\widehat{\boldsymbol{\beta}}_N$ based on $(\mathbf{X}_N, \mathbf{Y}_N)$ cannot be bounded.

Assume first that $\widehat{\boldsymbol{\beta}}_N$ is bounded, which implies that $|r_{N,i}| \to \infty$ for $i \in C$. Then part (i) of the lemma implies that $\widehat{\sigma}(\mathbf{r}_N(\widehat{\boldsymbol{\beta}}_N)) \to \infty$. Since $n\delta/m < 1 = \rho(\infty)$, condition R3 of Definition 2.1 implies that there is a single value γ such that

$$\rho\left(\frac{1}{\gamma}\right) = \frac{n\delta}{m}. \tag{5.74}$$

It will be shown that

$$\frac{1}{N^2}\widehat{\sigma}(\mathbf{r}_N(\widehat{\boldsymbol{\beta}}_N)) \to \gamma. \tag{5.75}$$

In fact,

$$n\delta = \sum_{i \notin C} \rho\left(\frac{y_i - \mathbf{x}_i'\widehat{\boldsymbol{\beta}}_N}{\widehat{\sigma}_N}\right) + \sum_{i \in C} \rho\left(\frac{N^2 - N\mathbf{x}_0'\widehat{\boldsymbol{\beta}}_N}{\widehat{\sigma}_N}\right).$$

The first sum tends to zero. The second one is

$$m\rho\left(\frac{1 - N^{-1}\mathbf{x}_0'\widehat{\boldsymbol{\beta}}_N}{N^{-2}\widehat{\sigma}_N}\right).$$

The numerator of the fraction tends to one. If a subsequence $\{N_j^{-2}\widehat{\sigma}_{N_j}\}$ has a (possibly infinite) limit t, then it must fulfill $n\delta = m\underset{\sim}{\rho}(1/t)$, which proves (5.75).

Now define $\widetilde{\boldsymbol{\beta}}_N = \mathbf{x}_0 N/2$, so that $\mathbf{r}_N(\boldsymbol{\beta}_N)$ has elements

$$\widetilde{r}_{N,i} = \frac{N^2}{2} \text{ for } i \in C, \quad \widetilde{r}_{N,i} = y_i - \mathbf{x}_0'\mathbf{x}_i \frac{N}{2} \text{ otherwise.}$$

Since $\#\left\{i : \left|\widetilde{r}_{N,i}\right| \to \infty\right\} = n$, part (i) of the lemma implies that $\widehat{\sigma}(\mathbf{r}_N(\widetilde{\boldsymbol{\beta}}_N)) \to \infty$, and proceeding as in (5.75) yields

$$\frac{1}{N^2}\widehat{\sigma}(\mathbf{r}_N(\widetilde{\boldsymbol{\beta}}_N)) \to \frac{\gamma}{2},$$

and hence

$$\widehat{\sigma}(\mathbf{r}_N(\widetilde{\boldsymbol{\beta}}_N)) < \widehat{\sigma}(\mathbf{r}_N(\widehat{\boldsymbol{\beta}}_N))$$

for large N, so that $\widehat{\boldsymbol{\beta}}_N$ cannot minimize σ.

Proof of (b): Let $m \leq m(\delta) < n\delta$, and consider a contamination sequence in a set C of size m. It will be shown that the corresponding estimate $\widehat{\boldsymbol{\beta}}_N$ is bounded. Assume first that $\widehat{\boldsymbol{\beta}}_N \to \infty$. Then

$$i \notin C,\ |r_{N,i}(\widehat{\boldsymbol{\beta}}_N)| \to \infty \implies \widehat{\boldsymbol{\beta}}_N'\mathbf{x}_{N,i} \neq 0,$$

and hence

$$\#\{i : |r_{N,i}(\widehat{\boldsymbol{\beta}}_N)| \to \infty\} \geq \#\left\{\widehat{\boldsymbol{\beta}}_N'\mathbf{x}_{N,i} \neq 0, i \notin C\right\}$$
$$= n - \#\left(\left\{i : \widehat{\boldsymbol{\beta}}_N'\mathbf{x}_{N,i} = 0\right\} \cup C\right).$$

The Bonferroni inequality implies that

$$\#\left(\left\{i : \widehat{\boldsymbol{\beta}}_N'\mathbf{x}_{N,i} = 0\right\} \cup C\right) \leq \#\left\{i : \widehat{\boldsymbol{\beta}}_N'\mathbf{x}_{N,i} = 0\right\} + \#(C),$$

and $\#\left\{i : \widehat{\boldsymbol{\beta}}_N'\mathbf{x}_{N,i} = 0\right\} \leq k^*(\mathbf{X})$ by (4.56). Hence

$$\#\{i : |r_{N,i}(\widehat{\boldsymbol{\beta}}_N)| \to \infty\} \geq n - k^*(\mathbf{X}) - m.$$

Now (4.58) implies that $n - k^*(\mathbf{X}) \geq 2m^*_{\max} + 1$, and since

$$m \leq m(\delta) \leq [n\delta] \leq m^*_{\max},$$

we have

$$\#\{i : |r_{N,i}(\widehat{\boldsymbol{\beta}}_N)| \to \infty\} \geq 1 + m^*_{\max} \geq 1 + [n\delta] > n\delta,$$

which by part (i) of the lemma implies $\widehat{\sigma}(\mathbf{r}_N(\widehat{\boldsymbol{\beta}}_N)) \to \infty$.

Assume now that $\widehat{\boldsymbol{\beta}}_N$ is bounded. Then

$$\#\{i : |r_{N,i}(\widehat{\boldsymbol{\beta}}_N)| \to \infty\} \leq m < n\delta,$$

which by part (ii) of the lemma implies $\widehat{\sigma}\left(\mathbf{r}_N(\widehat{\boldsymbol{\beta}}_N)\right)$ is bounded. Hence $\widehat{\boldsymbol{\beta}}_N$ cannot be unbounded. This completes the proof of the finite BP.

The least quantile estimate corresponds to the scale given by $\rho(t) = \mathrm{I}(|t| > 1)$, and according to Problem 2.14 it has $\widehat{\sigma} = |r|_{(h)}$ where $|r|_{(i)}$ are the ordered absolute residuals, and $h = n - [n\delta]$. The optimal choice of δ in (5.25) yields $h = n - m^*_{\max}$, and formal application of the theorem would imply that this h yields the maximum FBP. Actually the proof of the theorem does not hold because ρ is discontinuous and hence does not fulfill (5.74), but the proof can be reworked for $\widehat{\sigma} = |r|_{(h)}$ to show that the result also holds in this case.

The asymptotic BP A proof similar to but much simpler than that of Theorem 5.1, with averages replaced by expectations, shows that in the asymptotic case $\varepsilon^* \leq \delta$ and if $\delta \leq (1-\alpha)/2$ then $\varepsilon^* \geq \delta$. It follows that $\varepsilon^* = \delta$ for $\delta \leq (1-\alpha)/2$, and this proves that the maximum asymptotic BP is (5.26).

5.16.5 Asymptotic bias of M-estimates

Let $F = \mathcal{D}(\mathbf{x}, y)$ be $\mathrm{N}_{p+1}(\mathbf{0}, \mathbf{I})$. We shall first show that the asymptotic bias under point mass contamination of M-estimates and of estimates which minimize a robust scale does not depend on the dimension p.

To simplify the exposition we consider only the case of an M-estimate with known scale $\sigma = 1$. Call $(\mathbf{x}_0, y_0)$ the contamination location. The asymptotic value of the estimate is given by

$$\widehat{\boldsymbol{\beta}}_\infty = \arg\min_{\boldsymbol{\beta}} L(\boldsymbol{\beta}),$$

where

$$L(\boldsymbol{\beta}) = (1-\varepsilon)\mathrm{E}_F\rho(y - \mathbf{x}'\boldsymbol{\beta}) + \varepsilon\rho(y_0 - \mathbf{x}_0'\boldsymbol{\beta}). \tag{5.76}$$

Since $\mathcal{D}(y - \mathbf{x}'\boldsymbol{\beta}) = \mathrm{N}(0, 1 + \|\boldsymbol{\beta}\|^2)$ under F, we have

$$\mathrm{E}_F\rho(y - \mathbf{x}'\boldsymbol{\beta}) = g(\|\boldsymbol{\beta}\|),$$

where

$$g(t) = \mathrm{E}\rho\left(z\sqrt{1+t^2}\right), \quad z \sim \mathrm{N}(0, 1).$$

It is easy to show that g is an increasing function.

By the affine equivariance of the estimate, $\left\|\widehat{\boldsymbol{\beta}}_\infty\right\|$ does not change if we take $\mathbf{x}_0$ along the first coordinate axis, i.e., of the form $\mathbf{x}_0 = (x_0, 0, \ldots, 0)$, and thus

$$L(\boldsymbol{\beta}) = (1-\varepsilon)g\left(\|\boldsymbol{\beta}\|\right) + \varepsilon\rho(y_0 - x_0\beta_1),$$

where β_1 is the first coordinate of $\boldsymbol{\beta}$.

Given $\boldsymbol{\beta} = (\beta_1, \beta_2, \ldots, \beta_p)$, with $\beta_j \neq 0$ for some $j \geq 2$, the vector $\widetilde{\boldsymbol{\beta}} = (\beta_1, 0, \ldots, 0)$ has $\left\|\widetilde{\boldsymbol{\beta}}\right\| < \|\boldsymbol{\beta}\|$, which implies $g(\left\|\widetilde{\boldsymbol{\beta}}\right\|) < g(\|\boldsymbol{\beta}\|)$ and $L(\widetilde{\boldsymbol{\beta}}) < L(\boldsymbol{\beta})$. Then, we may restrict the search to the vectors of the form $(\beta_1, 0, \ldots, 0)$, for which

$$L(\boldsymbol{\beta}) = L_1(\beta_1) = (1-\varepsilon)g(\beta_1) + \varepsilon\rho(y_0 - x_0\beta_1),$$

and therefore the value minimizing $L_1(\beta_1)$ depends only on x_0 and y_0, and not on p, which proves the initial assertion.

It follows that the maximum asymptotic bias for point mass contamination does not depend on p. Actually, it can be shown that the maximum asymptotic bias for unrestricted contamination coincides with the former, and hence does not depend on p either.

The same results hold for M-estimates with the previous scale, and for S-estimates, but the details are more involved.

5.16.6 Hampel optimality for GM-estimates

We now deal with general M-estimates for regression through the origin $y = \beta x + u$, defined by

$$\sum_{i=1}^{n} \Psi(x_i, y_i; \beta) = 0,$$

with Ψ of the form

$$\Psi(x, y; \beta) = \eta\,(x, y - x\beta)\,x.$$

Assume σ is known and equal to one. It follows that the influence function is

$$\mathrm{IF}((x_0, y_0), F) = \frac{1}{b}\eta\,(x_0, y_0 - x_0\beta)\,x_0,$$

where

$$b = -\mathrm{E}\dot{\eta}(x, y - \beta x)x^2,$$

with $\dot{\eta}$ defined in (5.48), and hence the GES is

$$\gamma^* = \sup_{x_0, y_0} |\mathrm{IF}((x_0, y_0), F)| = \frac{1}{b}\sup_{s>0} K(s), \quad \text{with } K(s) = \sup_r |\eta(s, r)|\,.$$

The asymptotic variance is

$$v = \frac{1}{b^2}\mathrm{E}\eta\,(x, y - x\beta)^2\,x^2.$$

The direct and dual Hampel problems can now be stated as minimizing v subject to a bound on γ^*, and minimizing γ^* subject to a bound on v, respectively.

Let F correspond to the model (5.1)–(5.2), with normal u. The MLE corresponds to

$$\Psi_0(x, y; \beta) = (y - x\beta)\,x.$$

Since the estimates are equivariant, it suffices to treat the case of $\beta = 0$. Proceeding as in Section 3.8.7, it follows that the solutions to both problems,

$$\widetilde{\Psi}(x, y; \beta) = \widetilde{\eta}\,(x, y - x\beta)\,x,$$

have the form $\widetilde{\Psi}(x, y; \beta) = \psi_k\,(\Psi_0(x, y; \beta))$ for some $k > 0$, where ψ_k is Huber's ψ, which implies that $\widetilde{\eta}$ has the form (5.45).

The case $p > 1$ is more difficult to deal with, since $\boldsymbol{\beta}$—and hence the IF—are multidimensional. But the present reasoning gives some justification for the use of (5.45).

5.16.7 Justification of RFPE*

We are going to give a heuristic justification of (5.53). Let $(\mathbf{x}_i, y_i)$, $i = 0, 1, \ldots, n$, be i.i.d. and satisfy the model

$$y_i = \mathbf{x}_i'\boldsymbol{\beta} + u_i \quad (i = 0, \ldots, n), \tag{5.77}$$

where u_i and $\mathbf{x}_i$ are independent, and

$$\mathrm{E}\psi\left(\frac{u_i}{\sigma}\right) = 0. \tag{5.78}$$

Call $C_0 = \left\{j : \beta_j \neq 0\right\}$ the set of variables that actually have some predictive power. Given $C \subseteq \{1, \ldots, p\}$ let

$$\boldsymbol{\beta}_C = \left(\beta_j, j \in C\right), \quad \mathbf{x}_{iC} = \left(x_{ij} : j \in C\right), \ i = 0, \ldots, n.$$

Put $q = \#(C)$ and call $\widehat{\boldsymbol{\beta}}_C \in R^q$ the estimate based on $\{(\mathbf{x}_{iC}, y_i), i = 1, \ldots, n\}$. Then the residuals are $r_i = y_i - \widehat{\boldsymbol{\beta}}_C'\mathbf{x}_{iC}$ for $i = 1, \ldots, n$.

Assume that $C \supseteq C_0$. Then $\mathbf{x}_i'\boldsymbol{\beta} = \mathbf{x}_{iC}'\boldsymbol{\beta}_C$ and hence the model (5.77) can be rewritten as

$$y_i = \mathbf{x}_{iC}'\boldsymbol{\beta}_C + u_i, \quad i = 0, \ldots, n. \tag{5.79}$$

Put $\boldsymbol{\Delta} = \widehat{\boldsymbol{\beta}}_C - \boldsymbol{\beta}_C$. A second-order Taylor expansion yields

$$\rho\left(\frac{y_0 - \widehat{\boldsymbol{\beta}}_C'\mathbf{x}_{0C}}{\sigma}\right) = \rho\left(\frac{u_0 - \mathbf{x}_{0C}'\boldsymbol{\Delta}}{\sigma}\right)$$

$$\approx \rho\left(\frac{u_0}{\sigma}\right) - \psi\left(\frac{u_0}{\sigma}\right)\frac{\mathbf{x}_{0C}'\boldsymbol{\Delta}}{\sigma} + \frac{1}{2}\,\psi'\left(\frac{u_0}{\sigma}\right)\left(\frac{\mathbf{x}_{0C}'\boldsymbol{\Delta}}{\sigma}\right)^2. \tag{5.80}$$

The independence of u_0 and $\widehat{\beta}_C$, and (5.78), yield

$$\mathrm{E}\psi\left(\frac{u_0}{\sigma}\right)\mathbf{x}_{0C}'\boldsymbol{\Delta} = \mathrm{E}\psi\left(\frac{u_0}{\sigma}\right)\mathrm{E}(\mathbf{x}_{0C}'\boldsymbol{\Delta}) = 0. \tag{5.81}$$

According to (5.14), we have for large n

$$\mathcal{D}\left(\sqrt{n}\boldsymbol{\Delta}\right) \approx \mathrm{N}\left(0, \frac{\sigma^2 A}{B^2}\mathbf{V}^{-1}\right),$$

where

$$A = \mathrm{E}\psi^2\left(\frac{u}{\sigma}\right), \quad B = \mathrm{E}\psi'\left(\frac{u}{\sigma}\right), \quad \mathbf{V} = \mathrm{E}(\mathbf{x}_{0C}\mathbf{x}_{0C}').$$

Since u_0, Δ and $\mathbf{x}_0$ are independent we have

$$\begin{aligned}\mathrm{E}\psi'\left(\frac{u_0}{\sigma}\right)\left(\frac{\Delta'\mathbf{x}_{0C}}{\sigma}\right)^2 &= \mathrm{E}\psi'\left(\frac{u_0}{\sigma}\right)\mathrm{E}\left(\frac{\Delta'\mathbf{x}_{0C}}{\sigma}\right)^2 \\ &\approx B\frac{A}{nB^2}\mathrm{E}\mathbf{x}_{0C}'\mathbf{V}^{-1}\mathbf{x}_{0C}. \end{aligned} \tag{5.82}$$

Let $\mathbf{U}$ be any matrix such that $\mathbf{V} = \mathbf{U}\mathbf{U}'$, and hence such that

$$\mathrm{E}(\mathbf{U}^{-1}\mathbf{x}_{0C})(\mathbf{U}^{-1}\mathbf{x}_{0C})' = \mathbf{I}_q,$$

where $\mathbf{I}_q$ is the $q \times q$ identity matrix. Then

$$\mathrm{E}\mathbf{x}_{0C}'\mathbf{V}^{-1}\mathbf{x}_{0C} = \mathrm{E}\left\|\mathbf{U}^{-1}\mathbf{x}_{0C}\right\|^2 = \mathrm{trace}(\mathbf{I}_q) = q, \tag{5.83}$$

and hence (5.80), (5.81) and (5.82) yield

$$\mathrm{RFPE}(C) \approx \mathrm{E}\rho\left(\frac{u_0}{\sigma}\right) + \frac{q}{2n}\frac{A}{B}. \tag{5.84}$$

To estimate RFPE(C) using (5.84) we need to estimate $\mathrm{E}\rho(u_0/\sigma)$. A second-order Taylor expansion yields

$$\begin{aligned}\frac{1}{n}\sum_{i=1}^{n}\rho\left(\frac{r_i}{\widehat{\sigma}}\right) &= \frac{1}{n}\sum_{i=1}^{n}\rho\left(\frac{u_i - \mathbf{x}_{iC}'\Delta}{\widehat{\sigma}}\right) \\ &\approx \frac{1}{n}\sum_{i=1}^{n}\rho\left(\frac{u_i}{\widehat{\sigma}}\right) - \frac{1}{n\widehat{\sigma}}\sum_{i=1}^{n}\psi\left(\frac{u_i}{\widehat{\sigma}}\right)\mathbf{x}_{iC}'\Delta \\ &\quad + \frac{1}{2n\widehat{\sigma}^2}\sum_{i=1}^{n}\psi'\left(\frac{u_i}{\widehat{\sigma}}\right)\left(\mathbf{x}_{iC}'\Delta\right)^2. \end{aligned} \tag{5.85}$$

The estimate $\widehat{\beta}_C$ satisfies the equation

$$\sum_{i}^{n}\psi\left(\frac{r_{iC}}{\widehat{\sigma}}\right)\mathbf{x}_{iC} = \mathbf{0}, \tag{5.86}$$

and a first-order Taylor expansion of (5.86) yields

$$\mathbf{0} = \sum_i^n \psi\left(\frac{r_{iC}}{\widehat{\sigma}}\right)\mathbf{x}_{iC} = \sum_i^n \psi\left(\frac{u_i - \mathbf{x}_{iC}'\Delta}{\widehat{\sigma}}\right)\mathbf{x}_{iC}$$

$$\approx \sum_i^n \psi\left(\frac{u_i}{\widehat{\sigma}}\right)\mathbf{x}_{iC} - \frac{1}{\widehat{\sigma}}\sum_{i=1}^n \psi'\left(\frac{u_i}{\widehat{\sigma}}\right)(\mathbf{x}_{iC}'\Delta)\mathbf{x}_{iC},$$

and hence

$$\sum_i^n \psi\left(\frac{u_i}{\widehat{\sigma}}\right)\mathbf{x}_{iC} \approx \frac{1}{\widehat{\sigma}}\sum_{i=1}^n \psi'\left(\frac{u_i}{\widehat{\sigma}}\right)(\mathbf{x}_{iC}'\Delta)\mathbf{x}_{iC}. \tag{5.87}$$

Replacing (5.87) in (5.85) yields

$$\frac{1}{n}\sum_{i=1}^n \rho\left(\frac{r_i}{\widehat{\sigma}}\right) \approx \frac{1}{n}\sum_{i=1}^n \rho\left(\frac{u_i}{\widehat{\sigma}}\right) - \frac{1}{2n\widehat{\sigma}^2}\sum_{i=1}^n \psi'\left(\frac{u_i}{\widehat{\sigma}}\right)(\mathbf{x}_{iC}'\Delta)^2.$$

Since

$$\frac{1}{n}\sum_{i=1}^n \psi'\left(\frac{u_i}{\widehat{\sigma}}\right)\mathbf{x}_{iC}\mathbf{x}_{iC}' \to_p \mathrm{E}\psi'\left(\frac{u_0}{\sigma}\right)\mathrm{E}(\mathbf{x}_{0C}\mathbf{x}_{0C}') = B\mathbf{V},$$

we obtain using (5.83)

$$\frac{1}{n}\sum_{i=1}^n \rho\left(\frac{r_i}{\widehat{\sigma}}\right) \approx \frac{1}{n}\sum_{i=1}^n \rho\left(\frac{u_i}{\widehat{\sigma}}\right) - \frac{B}{2\widehat{\sigma}^2}\Delta'\mathbf{V}\Delta$$

$$= \frac{1}{n}\sum_{i=1}^n \rho\left(\frac{u_i}{\widehat{\sigma}}\right) - \frac{A}{2Bn}q,$$

Hence by the law of large numbers and the consistency of $\widehat{\sigma}$

$$\mathrm{E}\rho\left(\frac{u_0}{\sigma}\right) \approx \frac{1}{n}\sum_{i=1}^n \rho\left(\frac{u_i}{\widehat{\sigma}}\right) \approx \frac{1}{n}\sum_{i=1}^n \rho\left(\frac{r_i}{\widehat{\sigma}}\right) + \frac{Aq}{2Bn} \tag{5.88}$$

and finally inserting (5.88) in (5.84) yields

$$\mathrm{RFPE}(C) \approx \frac{1}{n}\sum_{i=1}^n \rho\left(\frac{r_i}{\widehat{\sigma}}\right) + \frac{Aq}{Bn} \approx \frac{1}{n}\sum_{i=1}^n \rho\left(\frac{r_i}{\widehat{\sigma}}\right) + \frac{\widehat{A}q}{\widehat{B}n} = \mathrm{RFPE}^*(C).$$

When C does not contain C_0, it can be shown that the use of RFPE* continues to be asymptotically valid.

5.16.8 A robust multiple correlation coefficient

In a multiple linear regression model, the R^2 statistic measures the proportion of the variation in the dependent variable accounted for by the explanatory variables. It is defined for a model with intercept (4.5) as

$$R^2 = \frac{S_0^2 - S^2}{S_0^2}$$

with

$$S^2 = \sum_{i=1}^{n} r_i^2, \quad S_0^2 = \sum_{i=1}^{n} (y_i - \overline{y})^2, \tag{5.89}$$

where r_i are the LS residuals.

Note that $\overline{y}$ is the LS estimate of the regression coefficients under model (4.5) with the restriction $\boldsymbol{\beta}_1 = 0$.

Recall that $S^2/(n - p^*)$ and $S_0^2/(n - 1)$ are unbiased estimates of the error variance for the complete model, and for the model with $\boldsymbol{\beta}_1 = 0$, respectively. To take the degrees of freedom into account, an adjusted R^2 is defined by

$$R_a^2 = \frac{S_0^2/(n-1) - S^2/(n-p^*)}{S_0^2/(n-p)}. \tag{5.90}$$

If instead of the LS estimate we use an M-estimate with general scale defined as in (5.7), a robust R^2 statistic and adjusted robust R^2 statistic can be defined by (5.89) and (5.90) respectively but replacing S^2 and S_0^2 with

$$S^2 = \min_{\boldsymbol{\beta} \in R^p} \sum \rho\left(\frac{r_i(\boldsymbol{\beta})}{\widehat{\sigma}}\right), \quad S_0^2 = \min_{\beta_0 \in R} \sum_{i=1}^{p} \rho\left(\frac{y_i - \beta_0}{\widehat{\sigma}}\right).$$

Croux and Dehon (2003) have considered alternative definitions of robust R^2.

5.17 Problems

5.1. Show that S-estimates are regression, affine and scale equivariant.

5.2. The **stack loss** dataset (Brownlee, 1965, p.454) given in Table 5.8 contains observations from 21 days' operation of a plant for the oxidation of ammonia as a stage in the production of nitric acid. The predictors X_1, X_2, X_3 are respectively the air flow, the cooling water inlet temperature, and the acid concentration, and the response Y is the stack loss. Fit a linear model to these data using the LS estimate, and the MM-estimates with efficiencies 0.95 and 0.85, and compare the results. Fit the residuals vs. the day. Is there a pattern?.

5.3. The dataset **alcohol** (Romanelli, Martino and Castro, 2001) gives for 44 aliphatic alcohols the logarithm of their solubility together with six physicochemical characteristics. The interest is in predicting the solubility. Compare the results of using the LS and MM-estimates to fit the log-solubility as a function of the characteristics.

5.4. The dataset **waste** (from Chatterjee and Hadi, 1988) contains for 40 regions the solid waste and five variables on land use. Fit a linear model to these data using the LS, L1 and MM-estimates. Draw the respective Q–Q plots of residuals, and the plots of residuals vs. fitted values, and compare the estimates and the plots.

Table 5.8 Stack loss data

Day	X_1	X_2	X_3	Y
1	80	27	58.9	4.2
2	80	27	58.8	3.7
3	75	25	59.0	3.7
4	62	24	58.7	2.8
5	62	22	58.7	1.8
6	62	23	58.7	1.8
7	62	24	59.3	1.9
8	62	24	59.3	2.0
9	58	23	58.7	1.5
10	58	18	58.0	1.4
11	58	18	58.9	1.4
12	58	17	58.8	1.3
13	58	18	58.2	1.1
14	58	19	59.3	1.2
15	50	18	58.9	0.8
16	50	18	58.6	0.7
17	50	19	57.2	0.8
18	50	19	57.9	0.8
19	50	20	58.0	0.9
20	56	20	58.2	1.5
21	70	20	59.1	1.5

5.5. Show that the "median of slopes" estimate (5.63) is a GM-estimate (5.44).

5.6. For the "median of slopes" estimate and the model $y_i = \beta x_i + u_i$, calculate the following, assuming that $\mathrm{P}(x = 0) = 0$:
(a) the asymptotic breakdown point
(b) the influence function and the gross-error sensitivity
(c) the maximum asymptotic bias [hint: use (3.67)].

5.7. Show that when using the shortcut (5.38), the number of times that the M-scale is computed has expectation $\sum_{i=1}^{N}(1/i) \leq \log N$, where N is the number of subsamples.

5.8. The minimum α-quantile regression estimate is defined for $\alpha \in (0, 1)$ as the value of β minimizing the α-quantile of $\left|y - \mathbf{x}'\beta\right|$. Show that this estimate is an S-estimate for the scale given by $\rho(u) = \mathrm{I}(|u| > 1)$ and $\delta = 1 - \alpha$. Find its asymptotic breakdown point.

5.9. For each β let $c(\beta)$ be the minimum c such that

$$\#\{i : \beta'\mathbf{x}_i - c \leq y_i \leq \beta'\mathbf{x}_i + c\} \geq n/2.$$

Show that the LMS estimate minimizes $c(\beta)$.

5.10. Let $\{(\mathbf{x}_1, y_1), \ldots, (\mathbf{x}_n, y_n)\}$ be a regression data set, and $\widehat{\boldsymbol{\beta}}$ an S-estimate with finite BP equal to ε^*. Let $D \subset (1, \ldots, n)$ with $\#(D) < n\varepsilon^*$.

(a) Show that there exists K such that

(i) $\widehat{\boldsymbol{\beta}}$ as a function of the y_i's is constant if the y_i's with $i \notin D$ remain fixed and those with $i \in D$ are changed in any way such that $|y_i| \geq K$.

(ii) There exists $\widehat{\sigma}$ depending only on D such that $\widehat{\boldsymbol{\beta}}$ verifies

$$\sum_{i \notin D} \rho\left(\frac{r_i(\widehat{\boldsymbol{\beta}})}{\widehat{\sigma}}\right) = \min.$$

(b) Discuss why the former property does not mean that the value of the estimate is the same if we omit the points $(\mathbf{x}_i, y_i)$ with $i \in D$.

(c) Show that property (a) holds also for MM-estimates.

6

Multivariate Analysis

6.1 Introduction

Multivariate analysis deals with situations in which several variables are measured on each experimental unit. In most cases of interest it is known or assumed that some form of relationship exists among the variables, and hence that considering each of them separately would entail a loss of information. Some possible goals of the analysis are: reduction of dimensionality (principal components, factor analysis, canonical correlation); identification (discriminant analysis); explanatory models (multivariate linear model). The reader is referred to Seber (1984) and Johnson and Wichern (1998) for further details.

A p-variate observation is now a vector $\mathbf{x} = (x_1, \ldots, x_p)' \in R^p$ and a distribution F now means a distribution on R^p. In the classical approach, location of a p-variate random variable $\mathbf{x}$ is described by the expectation $\boldsymbol{\mu} = \mathrm{E}\mathbf{x} = (\mathrm{E}x_1, \ldots, \mathrm{E}x_n)'$ and dispersion is described by the covariance matrix

$$\mathbf{Var}(\mathbf{x}) = \mathrm{E}((\mathbf{x} - \boldsymbol{\mu})(\mathbf{x} - \boldsymbol{\mu})').$$

It is well known that $\mathbf{Var}(\mathbf{x})$ is symmetric and positive semidefinite, and that for each constant vector $\mathbf{a}$ and matrix $\mathbf{A}$

$$\mathrm{E}(\mathbf{Ax} + \mathbf{a}) = \mathbf{A}\ \mathrm{E}\mathbf{x} + \mathbf{a}, \quad \mathbf{Var}(\mathbf{Ax} + \mathbf{a}) = \mathbf{A}\mathbf{Var}(\mathbf{x})\mathbf{A}'. \tag{6.1}$$

Classical multivariate methods of estimation are based on the assumption of an i.i.d. sample of observations $X = \{\mathbf{x}_1, \ldots, \mathbf{x}_n\}$ with each $\mathbf{x}_i$ having a p-variate normal $\mathrm{N}_p(\boldsymbol{\mu}, \boldsymbol{\Sigma})$ distribution with density

$$f(\mathbf{x}) = \frac{1}{(2\pi)^{p/2}\sqrt{|\boldsymbol{\Sigma}|}} \exp\left(-\frac{1}{2}(\mathbf{x} - \boldsymbol{\mu})'\Sigma^{-1}(\mathbf{x} - \boldsymbol{\mu})\right), \tag{6.2}$$

where $\boldsymbol{\Sigma} = \mathbf{Var}(\mathbf{x})$ and $|\boldsymbol{\Sigma}|$ stands for the determinant of $\boldsymbol{\Sigma}$. The contours of constant density are the elliptical surfaces

$$\{\mathbf{z} : (\mathbf{z} - \boldsymbol{\mu})'\boldsymbol{\Sigma}^{-1}(\mathbf{z} - \boldsymbol{\mu}) = c\}.$$

Assuming $\mathbf{x}$ is multivariate normal implies that for any constant vector $\mathbf{a}$, all linear combinations $\mathbf{a}'\mathbf{x}$ are normally distributed. It also implies that since the conditional expectation of one coordinate with respect to any group of coordinates is a linear function of the latter, the type of dependence among variables is *linear*. Thus methods based on multivariate normality will yield information only about linear relationships among coordinates. As in the univariate case, the main reason for assuming normality is simplicity.

It is known that under the normal distribution (6.2), the MLEs of $\boldsymbol{\mu}$ and $\boldsymbol{\Sigma}$ for a sample $\mathbf{x}$ are respectively the sample mean and sample covariance matrix

$$\overline{\mathbf{x}} = \text{ave}(X) = \frac{1}{n}\sum_{i=1}^{n}\mathbf{x}_i, \quad \mathbf{Var}(X) = \text{ave}\{(X - \overline{\mathbf{x}})(X - \overline{\mathbf{x}})'\}.$$

The sample mean and sample covariance matrix share the behavior of the distribution mean and covariance matrix under affine transformations, namely (6.1) for each vector $\mathbf{a}$ and matrix $\mathbf{A}$

$$\text{ave}(\mathbf{A}X + \mathbf{a}) = \mathbf{A}\text{ave}(X) + \mathbf{a}, \quad \mathbf{Var}(\mathbf{A}X + \mathbf{a}) = \mathbf{A}\mathbf{Var}(X)\mathbf{A}',$$

where $\mathbf{A}X + \mathbf{a}$ is the data set $\{\mathbf{A}\mathbf{x}_i + \mathbf{a}, i = 1, \ldots, n\}$. This property is known as the *affine equivariance* of the sample mean and covariances.

Just as in the univariate case, a few atypical observations may completely alter the sample means and/or covariances. Worse still, a multivariate outlier need not be an outlier in any of the coordinates considered separately.

Example 6.1 *Table 6.1 (from Seber, 1984, Table 9.12) contains measurements of phosphate and chloride in the urine of 12 men with similar weights. The data are plotted in Figure 6.1.*

We see in Figure 6.1 that observation 3, which has the lowest phosphate value, stands out clearly from the rest. However, Figure 6.2, which shows the normal Q–Q plot of phosphate, does not reveal any atypical value, and the same occurs in the Q–Q plot of chloride (not shown). Thus the atypical character of observation 3 is visible only when considering both variables simultaneously.

The table below shows that omitting this observation has no important effect on means or variances, but the correlation almost doubles in magnitude, i.e., the influence of the outlier has been to decrease the correlation by a factor of two relative to that without the outlier:

	Means		Vars.		Correl.
Complete data	1.79	6.01	0.26	3.66	−0.49
Without obs. 3	1.87	6.16	0.20	3.73	−0.80

Table 6.1 Biochemical data

Phosphate	Chloride
1.50	5.15
1.65	5.75
0.90	4.35
1.75	7.55
1.40	8.50
1.20	10.25
1.90	5.95
1.65	6.30
2.30	5.45
2.35	3.75
2.35	5.10
2.50	4.05

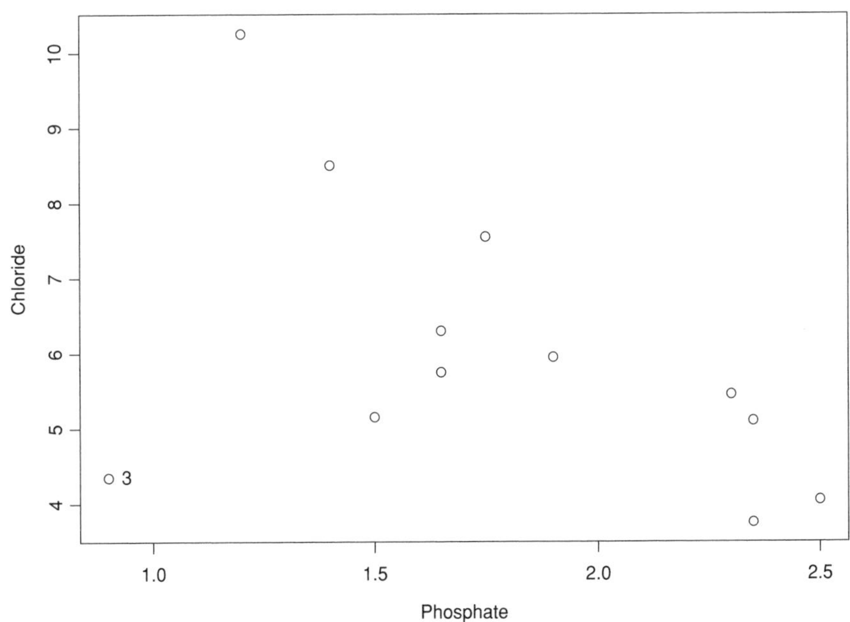

Figure 6.1 Biochemical data

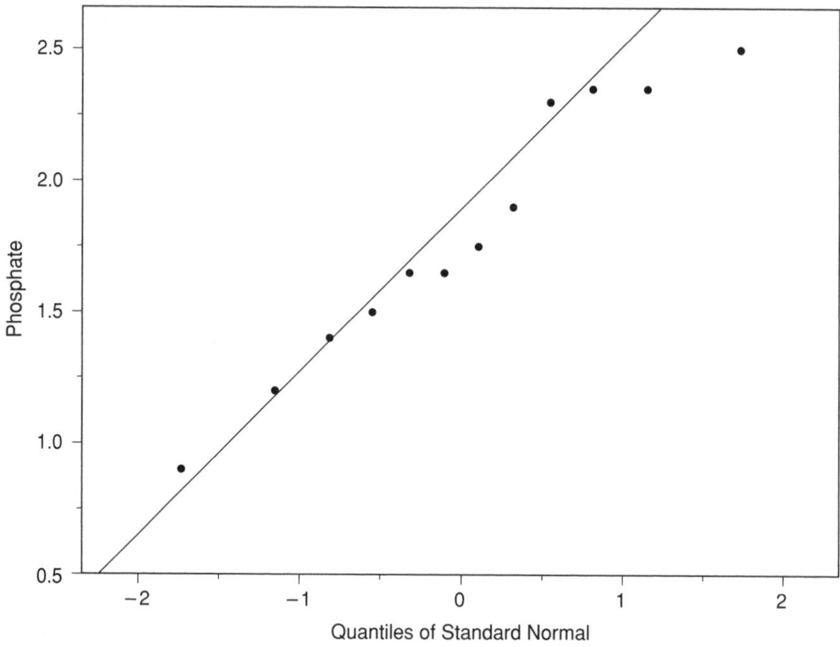

Figure 6.2 Normal Q–Q plot of phosphate

Here we have an example of an observation which is not a one-dimensional outlier in either coordinate but strongly affects the results of the analysis. This example shows the need for robust substitutes of the mean vector and covariance matrix, which will be the main theme of this chapter.

Some methods in multivariate analysis make no use of means or covariances, such as Breiman, Friedman, Olshen and Stone's (1984) nonparametric "CART" (Classification And Regression Trees) methods. To some extent such (nonequivariant) methods have a certain built-in robustness. But if we want to retain the simplicity of the normal distribution as the "nominal" model, with corresponding linear relationships, elliptical distributional shapes and affine equivariance for the bulk of the data, then the appropriate approach is to consider slight or moderate departures from normality.

Let $(\widehat{\boldsymbol{\mu}}(X), \widehat{\boldsymbol{\Sigma}}(X))$ be location and dispersion estimates corresponding to a sample $X = \{\mathbf{x}_1, \ldots, \mathbf{x}_n\}$. Then the estimates are affine equivariant if

$$\widehat{\boldsymbol{\mu}}(\mathbf{A}X + \mathbf{b}) = \mathbf{A}\widehat{\boldsymbol{\mu}}(X) + \mathbf{b}, \quad \widehat{\boldsymbol{\Sigma}}(\mathbf{A}X + \mathbf{a}) = \mathbf{A}\widehat{\boldsymbol{\Sigma}}\mathbf{A}'. \tag{6.3}$$

Affine equivariance is a desirable property of an estimate. The reasons are given in Section 6.12.1. This is, however, not a mandatory property, and may in some cases be sacrificed for other properties such as computational speed; an instance of this trade-off is given in Section 6.9.1.

As in the univariate case, one may consider the approach of outlier detection. The squared *Mahalanobis distance* between the vectors $\mathbf{x}$ and $\boldsymbol{\mu}$ with respect to the matrix

$\mathbf{\Sigma}$ is defined as

$$d(\mathbf{x}, \boldsymbol{\mu}, \mathbf{\Sigma}) = (\mathbf{x} - \boldsymbol{\mu})'\mathbf{\Sigma}^{-1}(\mathbf{x} - \boldsymbol{\mu}). \tag{6.4}$$

For simplicity d will be sometimes referred to as "distance", although it should be kept in mind that it is actually a *squared* distance. Then the multivariate analog of t_i^2, where $t_i = (x_i - \overline{x})/s$ is the univariate outlyingness measure in (1.3), is $D_i = d(\mathbf{x}_i, \overline{\mathbf{x}}, \mathbf{C})$ with $\mathbf{C} = \mathbf{Var}(X)$. When $p = 1$ we have $D_i = t_i^2 n/(n-1)$. It is known (Seber, 1984) that if $\mathbf{x} \sim \mathrm{N}_p(\boldsymbol{\mu}, \mathbf{\Sigma})$ then $d(\mathbf{x}, \boldsymbol{\mu}, \mathbf{\Sigma}) \sim \chi_p^2$. Thus, assuming the estimates $\overline{\mathbf{x}}$ and $\mathbf{C}$ are close to their true values, we may examine the Q–Q plot of D_i vs. the quantiles of a χ_p^2 distribution and delete observations for which D_i is "too high". This approach may be effective when there is a single outlier, but as in the case of location it can be useless when n is small (recall Section 1.3) and, as in regression, several outliers may mask one another.

Example 6.2 *The data set* **wine** *is a part of one given in Hettich and Bay (1999). It contains, for each of 59 wines grown in the same region in Italy, the quantities of 13 constituents. The original purpose of the analysis (de Vel, Aeberhard and Coomans, 1993) was to classify wines from different cultivars by means of these measurements. In this example we treat cultivar 1.*

The upper row of Figure 6.3 shows the plots of the classical squared distances as a function of observation number, and their Q–Q plots with respect to the χ_p^2 distribution (code **wine**). No clear outliers stand out.

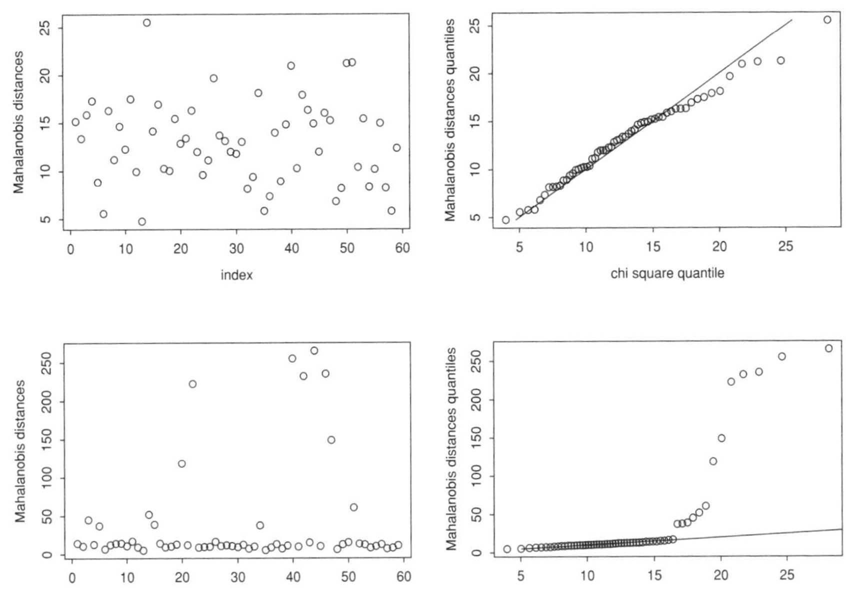

Figure 6.3 Wine example: Mahalanobis distances vs. index number for classical and SR-05 estimates (left column), and Q–Q plots of distances (right column)

The lower row shows the results of using a robust estimate to be defined in Section 6.4.4 (called SR-05 there). At least seven points stand out clearly. The failure of the classical analysis in the upper row of Figure 6.3 shows that several outliers may "mask" one another. These seven outliers have a strong influence on the results of the analysis.

Simple robust estimates of multivariate location can be obtained by applying a robust univariate location estimate to each coordinate, but this lacks affine equivariance. For dispersion, there exist simple robust estimates of the covariance between two variables ("pairwise covariances") which could be used to construct a "robust covariance" matrix (see Devlin, Gnanadesikan and Kettenring, 1981; Huber, 1981). Apart from not being equivariant, the resulting matrix may not be positive semidefinite. See, however, Section 6.9 for an approach that ensures positive definiteness and "approximate" equivariance. Nonequivariant procedures may also lack robustness when the data are very collinear (Section 6.6). In subsequent sections we shall discuss a number of equivariant estimates that are robust analogs of the mean and covariance matrix. They will be generally called *location vectors* and *dispersion matrices*. The latter are also called *robust covariance matrices* in the literature.

Note that if the matrix $\widehat{\Sigma}$ with elements $\sigma_{jk},\ j,k=1,\ldots,p$, is a "robust covariance matrix", then the matrix $\mathbf{R}$ with elements

$$r_{jk}=\frac{\sigma_{jk}}{\sqrt{\sigma_{jj}\sigma_{kk}}} \tag{6.5}$$

is a robust analog of the correlation matrix.

6.2 Breakdown and efficiency of multivariate estimates

The concepts of breakdown point and efficiency will be necessary to understand the advantages and drawbacks of the different families of estimates discussed in this chapter.

6.2.1 Breakdown point

To define the breakdown point of $(\widehat{\mu},\widehat{\Sigma})$ based on the ideas in Section 3.2 we must establish the meaning of "bounded, and also bounded away from the boundary of the parameter space". For the location vector the parameter space is a finite-dimensional Euclidean space, and so the statement means simply that $\widehat{\mu}$ remains in a bounded set. However, the dispersion matrix has a more complex parameter space consisting of the set of symmetric nonnegative definite matrices. Each such matrix is characterized by the matrix of its eigenvectors and associated nonnegative eigenvalues. Thus "$\widehat{\Sigma}$ bounded, and also bounded away from the boundary" is equivalent to the eigenvalues being bounded away from zero and infinity.

From a more intuitive point of view, recall that if $\Sigma=\mathbf{Var}(\mathbf{x})$ and $\mathbf{a}$ is a constant vector then $\mathrm{Var}(\mathbf{a}'\mathbf{x})=\mathbf{a}'\Sigma\mathbf{a}$. Hence if Σ is any robust dispersion matrix then $\sqrt{\mathbf{a}'\Sigma\mathbf{a}}$

can be considered as a robust measure of dispersion of the linear combination $\mathbf{a}'\mathbf{x}$. Let $\lambda_1(\mathbf{\Sigma}) \geq \ldots \geq \lambda_p(\mathbf{\Sigma})$ be the eigenvalues of $\mathbf{\Sigma}$ in descending order, and $\mathbf{e}_1, \ldots, \mathbf{e}_p$ the corresponding eigenvectors. It is a fact of linear algebra that for any symmetric matrix $\mathbf{\Sigma}$, the minimum (resp. maximum) of $\mathbf{a}'\mathbf{\Sigma}\boldsymbol{a}$ over $\|\mathbf{a}\| = 1$ is equal to λ_p (λ_1) and this minimum is attained for $\mathbf{a} = \mathbf{e}_p$ ($\mathbf{e}_1$). If we are interested in linear relationships among the variables, then it is dangerous not only that the largest eigenvalue becomes too large ("explosion") but also that the smallest one becomes too small ("implosion"). The first case is caused by outliers (observations far away from the bulk of the data), the second by "inliers" (observations concentrated at some point or in general on a region of lower dimensionality).

For $0 \leq m \leq n$ call $\mathcal{Z}_m$ the set of "samples" $Z = \{\mathbf{z}_1, \ldots, \mathbf{z}_n\}$ such that $\#\{\mathbf{z}_i = \mathbf{x}_i\} = m$, and call $\widehat{\boldsymbol{\mu}}(Z)$ and $\widehat{\mathbf{\Sigma}}(Z)$ the mean and dispersion matrix estimates based on the sample Z. The finite breakdown point of $(\widehat{\boldsymbol{\mu}}, \widehat{\mathbf{\Sigma}})$ is defined as $\varepsilon^* = m^*/n$ where m^* is the largest m such that there exist finite positive a, b, c such that

$$\|\widehat{\boldsymbol{\mu}}(Z)\| \leq a \text{ and } b \leq \lambda_p(\widehat{\mathbf{\Sigma}}(Z)) \leq \lambda_1(\widehat{\mathbf{\Sigma}}(Z)) \leq c$$

for all $Z \in \mathcal{Z}_m$.

For theoretical purposes it may be simpler to work with the asymptotic BP. An ε-contamination neighborhood $\mathcal{F}(F, \varepsilon)$ of a multivariate distribution F is defined as in (3.2). Applying Definition 3.1 and (3.19) to the present context we have that the asymptotic BP of $(\widehat{\boldsymbol{\mu}}, \widehat{\mathbf{\Sigma}})$ is the largest $\varepsilon^* \in (0, 1)$ for which there exist finite positive a, b, c such that the following holds for all G:

$$\|\widehat{\boldsymbol{\mu}}_\infty((1-\varepsilon)F + \varepsilon G)\| \leq a,$$

$$b \leq \lambda_p(\widehat{\mathbf{\Sigma}}((1-\varepsilon)F + \varepsilon G)) \leq \lambda_1(\widehat{\mathbf{\Sigma}}((1-\varepsilon)F + \varepsilon G)) \leq c.$$

In some cases we may restrict G to range over point mass distributions, and in that case we use the terms "point mass contamination neighborhoods" and "point mass breakdown point".

6.2.2 The multivariate exact fit property

A result analogous to that of Section 5.10 holds for multivariate location and dispersion estimation. Let the FBP of the affine equivariant estimate $(\widehat{\boldsymbol{\mu}}, \widehat{\mathbf{\Sigma}})$ be $\varepsilon^* = m^*/n$. Let the data set contain q points on a hyperplane $H = \left\{\mathbf{x} : \boldsymbol{\beta}'\mathbf{x} = \gamma\right\}$ for some $\boldsymbol{\beta} \in R^p$ and $\gamma \in R$. If $q \geq n - m^*$ then $\widehat{\boldsymbol{\mu}} \in H$, and $\widehat{\mathbf{\Sigma}}\boldsymbol{\beta} = \mathbf{0}$. The proof is given in Section 6.12.8.

6.2.3 Efficiency

The asymptotic efficiency of $(\widehat{\boldsymbol{\mu}}, \widehat{\mathbf{\Sigma}})$ is defined as in (3.45). Call $(\widehat{\boldsymbol{\mu}}_n, \widehat{\mathbf{\Sigma}}_n)$ the estimates for a sample of size n, and let $(\widehat{\boldsymbol{\mu}}_\infty, \widehat{\mathbf{\Sigma}}_\infty)$ be their asymptotic values. All estimates considered in this chapter are consistent at the normal distribution in the following sense: if $\mathbf{x}_i \sim \mathrm{N}_p(\boldsymbol{\mu}, \Sigma)$ then $\widehat{\boldsymbol{\mu}}_\infty = \boldsymbol{\mu}$ and $\widehat{\mathbf{\Sigma}}_\infty = c\mathbf{\Sigma}$ where c is a constant (if $c = 1$

we have the usual definition of consistency). This result will be seen to hold for the larger family of *elliptical* distributions, to be defined later. Most estimates defined in this chapter are also asymptotically normal:

$$\sqrt{n}\left(\widehat{\mu}_n - \widehat{\mu}_\infty\right) \to_d \mathrm{N}_p(\mathbf{0}, \mathbf{V}_\mu), \quad \sqrt{n}\mathrm{vec}\left(\widehat{\Sigma}_n - \widehat{\Sigma}_\infty\right) \to_d \mathrm{N}_q(\mathbf{0}, \mathbf{V}_\Sigma),$$

where $q = p(p+1)/2$ and for a symmetric matrix $\boldsymbol{\Sigma}$, vec($\boldsymbol{\Sigma}$) is the vector containing the q elements of the upper triangle of $\boldsymbol{\Sigma}$. The matrices $\mathbf{V}_\mu$ and $\mathbf{V}_\Sigma$ are the asymptotic covariance matrices of $\widehat{\boldsymbol{\mu}}$ and $\widehat{\boldsymbol{\Sigma}}$. In general the estimate can be defined in such a way that $c = 1$ for a given model, e.g., the multivariate normal.

We consider the efficiency of $\widehat{\boldsymbol{\mu}}$ when the data have a $\mathrm{N}_p(\boldsymbol{\mu}, \Sigma)$ distribution. In Section 6.12.2 it is shown that an affine equivariant location estimate $\widehat{\boldsymbol{\mu}}$ has an asymptotic covariance matrix of the form

$$\mathbf{V}_\mu = v\boldsymbol{\Sigma}, \tag{6.6}$$

where v is a constant depending on the estimate. In the case of the normal distribution MLE $\overline{\mathbf{x}}$ we have $v = 1$ and the matrix $\mathbf{V}_0$ in (3.45) is simply $\boldsymbol{\Sigma}$, which results in $\mathbf{V}_\mu^{-1}\mathbf{V}_0 = v^{-1}\mathbf{I}$ and eff($\widehat{\boldsymbol{\mu}}$) $= 1/v$. Thus the normal distribution efficiency of an affine equivariant location estimate is independent of $\boldsymbol{\mu}$ and $\boldsymbol{\Sigma}$. With one exception treated in Section 6.9.1, the location estimates considered in this chapter are affine equivariant.

The efficiency of $\widehat{\boldsymbol{\Sigma}}$ is much more complicated and will not be discussed here. It has been dealt with by Tyler (1983) in the case of the class of M-estimates to be defined in the next section.

6.3 M-estimates

Multivariate M-estimates will now be defined as in Section 2.2 by generalizing MLEs. Recall that in the univariate case it was possible to define separate robust equivariant estimates of location and of dispersion. This is more complicated to do in the multivariate case, and if we want equivariant estimates it is better to estimate location and dispersion *simultaneously*. We shall develop the multivariate analog of simultaneous M-estimates (2.69)–(2.70). Recall that a multivariate normal density has the form

$$f(\mathbf{x}, \boldsymbol{\mu}, \boldsymbol{\Sigma}) = \frac{1}{\sqrt{|\boldsymbol{\Sigma}|}} h(d(\mathbf{x}, \boldsymbol{\mu}, \boldsymbol{\Sigma})) \tag{6.7}$$

where $h(s) = c\exp(-s/2)$ with $c = (2\pi)^{-p/2}$ and $d(\mathbf{x}, \boldsymbol{\mu}, \boldsymbol{\Sigma}) = (\mathbf{x} - \boldsymbol{\mu})'\boldsymbol{\Sigma}^{-1}(\mathbf{x} - \boldsymbol{\mu})$. We note that the level sets of f are ellipsoidal surfaces. In fact for any choice of positive h such that f integrates to one, the level sets of f are ellipsoids, and so any density of this form is called *elliptically symmetric* (henceforth "elliptical" for short). In the special case where $\boldsymbol{\mu} = 0$ and $\boldsymbol{\Sigma} = c\mathbf{I}$ a density of the form (6.7) is called *spherically symmetric* or *radial* (henceforth "spherical" for short). It is easy to verify that the distribution $\mathcal{D}(\mathbf{x})$ is elliptical if and only if for some constant vector $\mathbf{a}$ and matrix $\mathbf{A}$, $\mathcal{D}(\mathbf{A}(\mathbf{x} - \mathbf{a}))$ is spherical. An important example of a nonnormal elliptical distribution is the p-variate *Student distribution* with ν degrees of freedom ($0 < \nu < \infty$), which will be called $T_{p,\nu}$, and is obtained by the choice

$$h(s) = \frac{c}{(s+\nu)^{(p+\nu)/2}} \tag{6.8}$$

where c is a constant. The case $\nu = 1$ is called the multivariate Cauchy density, and the limiting case $\nu \to \infty$ yields the normal distribution. If the mean (resp. the dispersion matrix) of an elliptical distribution exists, then it is equal to $\boldsymbol{\mu}$ (to a multiple of $\boldsymbol{\Sigma}$) (Problem 6.1). More details on elliptical distributions are given in Section 6.12.9.

Let $\mathbf{x}_1, \ldots, \mathbf{x}_n$ be an i.i.d. sample from an f of the form (6.7) in which h is assumed everywhere positive. To calculate the MLE of $\boldsymbol{\mu}$ and $\boldsymbol{\Sigma}$, note that the likelihood function is

$$L(\boldsymbol{\mu}, \boldsymbol{\Sigma}) = \frac{1}{|\boldsymbol{\Sigma}|^{n/2}} \prod_{i=1}^{n} h(d(\mathbf{x}_i, \boldsymbol{\mu}, \boldsymbol{\Sigma})),$$

and maximizing $L(\boldsymbol{\mu}, \boldsymbol{\Sigma})$ is equivalent to

$$-2\log L(\boldsymbol{\mu}, \boldsymbol{\Sigma}) = n\log|\widehat{\boldsymbol{\Sigma}}| + \sum_{i=1}^{n} \rho(d_i) = \min, \tag{6.9}$$

where

$$\rho(s) = -2\log h(s) \text{ and } d_i = d(\mathbf{x}_i, \widehat{\boldsymbol{\mu}}, \widehat{\boldsymbol{\Sigma}}). \tag{6.10}$$

Differentiating with respect to $\boldsymbol{\mu}$ and $\boldsymbol{\Sigma}$ yields the system of estimating equations (see Section 6.12.3 for details)

$$\sum_{i=1}^{n} W(d_i)(\mathbf{x}_i - \widehat{\boldsymbol{\mu}}) = \mathbf{0} \tag{6.11}$$

$$\frac{1}{n}\sum_{i=1}^{n} W(d_i)(\mathbf{x}_i - \widehat{\boldsymbol{\mu}})(\mathbf{x}_i - \widehat{\boldsymbol{\mu}})' = \widehat{\boldsymbol{\Sigma}} \tag{6.12}$$

with $W = \rho'$. For the normal distribution we have $W \equiv 1$ which yields the sample mean and sample covariance matrix for $\widehat{\boldsymbol{\mu}}$ and $\widehat{\boldsymbol{\Sigma}}$. For the multivariate Student distribution (6.8) we have

$$W(d) = \frac{p+\nu}{d+\nu}. \tag{6.13}$$

In general, we define M-estimates as solutions of

$$\sum_{i=1}^{n} W_1(d_i)(\mathbf{x}_i - \widehat{\boldsymbol{\mu}}) = \mathbf{0} \tag{6.14}$$

$$\frac{1}{n}\sum_{i=1}^{n} W_2(d_i)(\mathbf{x}_i - \widehat{\boldsymbol{\mu}})(\mathbf{x}_i - \widehat{\boldsymbol{\mu}})' = \widehat{\boldsymbol{\Sigma}} \tag{6.15}$$

where the functions W_1 and W_2 need not be equal. Note that by (6.15) we may interpret $\widehat{\boldsymbol{\Sigma}}$ as a weighted covariance matrix, and by (6.14) we can express $\widehat{\boldsymbol{\mu}}$ as the weighted mean

$$\widehat{\boldsymbol{\mu}} = \frac{\sum_{i=1}^{n} W_1(d_i)\mathbf{x}_i}{\sum_{i=1}^{n} W_1(d_i)} \tag{6.16}$$

with weights depending on an outlyingness measure d_i. This is similar to (2.31) in that with $w_i = W_1(d_i)$ we can express $\widehat{\boldsymbol{\mu}}$ as a weighted mean with data-dependent weights.

Existence and uniqueness of solutions were treated by Maronna (1976) and more generally by Tatsuoka and Tyler (2000). Uniqueness of solutions of (6.14)–(6.15) requires that $dW_2(d)$ be a nondecreasing function of d. To understand the reason for this condition, note that an M-scale estimate of a univariate sample $\mathbf{z}$ may be written as the solution of

$$\delta = \text{ave}\left(\rho\left(\frac{\mathbf{z}}{\widehat{\sigma}}\right)\right) = \text{ave}\left(\left(\frac{\mathbf{z}}{\widehat{\sigma}}\right) W\left(\frac{\mathbf{z}}{\widehat{\sigma}}\right)\right),$$

where $W(t) = \rho(t)/t$. Thus the condition on the monotonicity of $dW_2(d)$ is the multivariate version of the requirement that the ρ-function of a univariate M-scale be monotone.

We shall call an M-estimate of location and dispersion *monotone* if $dW_2(d)$ is nondecreasing, and *redescending* otherwise. Monotone M-estimates are defined as solutions to the estimating equations (6.14)–(6.15), while redescending ones must be defined by the minimization of some objective function, as happens with S-estimates or CM-estimates to be defined in Sections 6.4 and 6.11.2 respectively. Huber (1981) treats a slightly more general definition of monotone M-estimates. For practical purposes monotone estimates are essentially unique, in the sense that *all* solutions to the M-estimating equations are consistent estimates.

It is proved in Chapter 8 of Huber (1981) that if the $\mathbf{x}_i$ are i.i.d. with distribution F then under general assumptions when $n \to \infty$, monotone M-estimates defined as any solution $\widehat{\boldsymbol{\mu}}$ and $\widehat{\boldsymbol{\Sigma}}$ of (6.14) and (6.15) converge in probability to the solution $(\widehat{\boldsymbol{\mu}}_\infty, \widehat{\boldsymbol{\Sigma}}_\infty)$ of

$$\mathrm{E}W_1(d)(\mathbf{x} - \widehat{\boldsymbol{\mu}}_\infty) = \mathbf{0}, \tag{6.17}$$

$$\mathrm{E}W_2(d)(\mathbf{x} - \widehat{\boldsymbol{\mu}}_\infty)(\mathbf{x} - \widehat{\boldsymbol{\mu}}_\infty)' = \widehat{\boldsymbol{\Sigma}}_\infty \tag{6.18}$$

where $d = d(\mathbf{x}, \widehat{\boldsymbol{\mu}}_\infty, \widehat{\boldsymbol{\Sigma}}_\infty)$. Huber also proves that $\sqrt{n}\left(\widehat{\boldsymbol{\mu}}-\widehat{\boldsymbol{\mu}}_\infty, \widehat{\boldsymbol{\Sigma}}-\widehat{\boldsymbol{\Sigma}}_\infty\right)$ tends to a multivariate normal distribution. It is easy to show that M-estimates are affine equivariant (Problem 2) and so if $\mathbf{x}$ has an elliptical distribution (6.7) the asymptotic covariance matrix of $\widehat{\boldsymbol{\mu}}$ has the form (6.6) (see Sections 6.12.1 and 6.12.7).

6.3.1 Collinearity

If the data are collinear, i.e., all points lie on a hyperplane H, the sample covariance matrix is singular and $\overline{\mathbf{x}} \in H$. It follows from (6.16) that since $\widehat{\boldsymbol{\mu}}$ is a linear combination of elements of H, it lies in H. Furthermore (6.15) shows that $\widehat{\Sigma}$ must be singular. In fact, if a sufficiently large proportion of the observations lie on a hyperplane, $\widehat{\Sigma}$ must be singular (Section 6.2.2). But in this case $\widehat{\Sigma}^{-1}$, and hence the d_i's, do not exist and the M-estimate is not defined.

To make the estimate well defined in all cases, it suffices to extend the definition (6.4) as follows. Let $\lambda_1 \geq \lambda_2 \geq \ldots \geq \lambda_p$ and $\mathbf{b}_j$ $(j = 1, \ldots, p)$ be the eigenvalues and eigenvectors of $\widehat{\Sigma}$. For a given $\mathbf{x}$ let $z_j = \mathbf{b}_j'(\mathbf{x}-\widehat{\mu})$. Since $\mathbf{b}_1, \ldots, \mathbf{b}_p$ are an orthonormal basis, we have

$$\mathbf{x} - \widehat{\mu} = \sum_{j=1}^{p} z_j \mathbf{b}_j.$$

Then if $\widehat{\Sigma}$ is not singular, we have (Problem 6.12)

$$d(\mathbf{x}, \widehat{\mu}, \widehat{\Sigma}) = \sum_{j=1}^{p} \frac{z_j^2}{\lambda_j}. \tag{6.19}$$

On the other hand if $\widehat{\Sigma}$ is singular, its smallest q eigenvalues are zero and in this case we define

$$d(\mathbf{x}, \widehat{\mu}, \widehat{\Sigma}) = \begin{cases} \sum_{j=1}^{p-q} z_j^2/\lambda_j & \text{if} \quad z_{p-q+1} = \ldots = z_p = 0 \\ \infty & \text{otherwise} \end{cases} \tag{6.20}$$

which may be seen as the limit case of (6.19) when $\lambda_j \downarrow 0$ for $j > p - q$.

Note that d_i enters (6.14)–(6.15) only through the functions W_1 and W_2, which usually tend to zero at infinity, so this extended definition simply excludes those points which do not belong to the hyperplane spanned by the eigenvectors corresponding to the positive eigenvalues of $\widehat{\Sigma}$.

6.3.2 Size and shape

If one dispersion matrix is a scalar multiple of another, i.e., $\Sigma_2 = k\Sigma_1$, we say that they have the same *shape*, but different *sizes*. Several important features of the distribution, such as correlations, principal components and linear discriminant functions, depend only on shape.

Let $\widehat{\mu}_\infty$ and $\widehat{\Sigma}_\infty$ be the asymptotic values of location and dispersion estimates at an elliptical distribution F defined in (6.7). It is shown in Section 6.12.2 that in this case $\widehat{\mu}_\infty$ is equal to the center of symmetry μ, and $\widehat{\Sigma}_\infty$ is a constant multiple of Σ, with the proportionality constant depending on F and on the estimator. This situation is similar to the scaling problem in (2.50) and at the end of Section 2.5. Consider in particular an M-estimate at the distribution $F = \mathrm{N}_p(\mu, \Sigma)$. By the equivariance of the estimate we may assume that $\mu = \mathbf{0}$ and $\Sigma = \mathbf{I}$. Then $\widehat{\mu}_\infty = \mathbf{0}$ and $\widehat{\Sigma}_\infty = c\mathbf{I}$, and hence $d\left(\mathbf{x}, \widehat{\mu}_\infty, \widehat{\Sigma}_\infty\right) = \|\mathbf{x}\|^2/c$. Taking the trace in (6.18) yields

$$\mathrm{E}W_2\left(\frac{\|\mathbf{x}\|^2}{c}\right)\|\mathbf{x}\|^2 = pc.$$

Since $\|\mathbf{x}\|^2$ has a χ_p^2 distribution, we obtain a consistent estimate of the covariance matrix $\mathbf{\Sigma}$ in the normal case by replacing $\widehat{\mathbf{\Sigma}}$ by $\widehat{\mathbf{\Sigma}}/c$, with c defined as the solution of

$$\int_0^\infty W_2\left(\frac{z}{c}\right)\frac{z}{c}g(z)dz = p, \tag{6.21}$$

where g is the density of the χ_p^2 distribution.

Another approach to estimating the size of $\mathbf{\Sigma}$ is based on noting that if $\mathbf{x} \sim \mathrm{N}(\boldsymbol{\mu}, \mathbf{\Sigma})$, then $d(\mathbf{x}, \boldsymbol{\mu}, \mathbf{\Sigma}) \sim \chi_p^2$, and the fact that $\mathbf{\Sigma} = c\widehat{\mathbf{\Sigma}}_\infty$ implies

$$cd(\mathbf{x}, \boldsymbol{\mu}, \mathbf{\Sigma}) = d(\mathbf{x}, \boldsymbol{\mu}, \widehat{\mathbf{\Sigma}}_\infty).$$

Hence the empirical distribution of

$$\left\{d(\mathbf{x}_1, \widehat{\boldsymbol{\mu}}, \widehat{\mathbf{\Sigma}}), \ldots, d(\mathbf{x}_n, \widehat{\boldsymbol{\mu}}, \widehat{\mathbf{\Sigma}})\right\}$$

will resemble that of $d(\mathbf{x}, \widehat{\boldsymbol{\mu}}_\infty, \widehat{\mathbf{\Sigma}}_\infty)$ which is $c\chi_p^2$, and so we may estimate c robustly with

$$\widehat{c} = \frac{\mathrm{Med}\left\{d(\mathbf{x}_1, \widehat{\boldsymbol{\mu}}, \widehat{\mathbf{\Sigma}}), \ldots, d(\mathbf{x}_n, \widehat{\boldsymbol{\mu}}, \widehat{\mathbf{\Sigma}})\right\}}{\chi_p^2(0.5)} \tag{6.22}$$

where $\chi_p^2(\alpha)$ denotes the α-quantile of the χ_p^2 distribution.

6.3.3 Breakdown point

It is intuitively clear that robustness of the estimates requires that no term dominates the sums in (6.14)–(6.15), and to achieve this we assume

$$W_1(d)\sqrt{d} \text{ and } W_2(d)d \text{ are bounded for } d \geq 0. \tag{6.23}$$

Let

$$K = \sup_d W_2(d)d. \tag{6.24}$$

We first consider the asymptotic BP, which is easier to deal with. The "weak part" of joint M-estimates of $\boldsymbol{\mu}$ and $\mathbf{\Sigma}$ is the estimate $\widehat{\mathbf{\Sigma}}$, for if we take $\mathbf{\Sigma}$ as known, then it is not difficult to prove that the asymptotic BP of $\widehat{\boldsymbol{\mu}}$ is 1/2 (see Section 6.12.4). On the other hand, in the case where $\boldsymbol{\mu}$ is known the following result was obtained by Maronna (1976). If the underlying distribution F_0 attributes zero mass to any hyperplane, then the asymptotic BP of a *monotone* M-estimate of $\mathbf{\Sigma}$ with W_2 satisfying (6.23) is

$$\varepsilon^* = \min\left(\frac{1}{K}, 1 - \frac{p}{K}\right). \tag{6.25}$$

See Section 6.12.4 for a simplified proof. The above expression has a maximum value of $1/(p+1)$, attained at $K = p + 1$, and hence

$$\varepsilon^* \leq \frac{1}{p+1}. \tag{6.26}$$

Tyler (1987) proposed a monotone M-estimate with $W_2(d) = p/d$, which corresponds to the multivariate t-distribution MLE weights (6.13) with degrees of freedom $\nu \downarrow 0$. Tyler showed that the BP of this estimate is $\varepsilon^* = 1/p$, which is slightly larger than the bound (6.26). This result is not a contradiction with (6.26) since W_2 is not defined at zero and hence does not satisfy (6.23). Unfortunately this unboundedness may make the estimate unstable.

It is useful to understand the form of the breakdown under the assumptions (6.23). Take $F = (1 - \varepsilon) F_0 + \varepsilon G$ where G is any contaminating distribution. First let G be concentrated at $\mathbf{x}_0$. Then the term $1/K$ in (6.25) is obtained by letting $\mathbf{x}_0 \to \infty$, and the term $1 - p/K$ is obtained by letting $\mathbf{x}_0 \to \boldsymbol{\mu}$. Now consider a general G. For the joint estimation of $\boldsymbol{\mu}$ and $\boldsymbol{\Sigma}$, Tyler shows that if $\varepsilon > \varepsilon^*$ and one lets G tend to $\delta_{\mathbf{x}_0}$ then $\boldsymbol{\mu} \to \mathbf{x}_0$ and $\lambda_p(\boldsymbol{\Sigma}) \to 0$, i.e., inliers can make $\boldsymbol{\Sigma}$ nearly singular.

The FBP is similar but the details are more involved (Tyler, 1990). Define a sample to be in *general position* if no hyperplane contains more than p points. Davies (1987) showed that the maximum FBP of any equivariant estimate for a sample in general position is $m^*_{\max}/n$ with

$$m^*_{\max} = \left[\frac{n-p}{2}\right]. \tag{6.27}$$

It is therefore natural to search for estimates whose BP is nearer to this maximum BP than that of monotone M-estimates.

6.4 Estimates based on a robust scale

Just as with the regression estimates of Section 5.6 where we aimed at making the residuals "small", we shall define multivariate estimates of location and dispersion that make the distances d_i "small". To this end we look for $\widehat{\boldsymbol{\mu}}$ and $\widehat{\boldsymbol{\Sigma}}$ minimizing some measure of "largeness" of $d(\mathbf{x}, \widehat{\boldsymbol{\mu}}, \widehat{\boldsymbol{\Sigma}})$. If follows from (6.4) that this can be trivially attained by letting the smallest eigenvalue of $\widehat{\boldsymbol{\Sigma}}$ tend to zero. To prevent this we impose the constraint $\left|\widehat{\boldsymbol{\Sigma}}\right| = 1$. Call $\mathcal{S}_p$ the set of symmetric positive definite $p \times p$ matrices. For a data set X call $\mathbf{d}\left(X, \widehat{\boldsymbol{\mu}}, \widehat{\boldsymbol{\Sigma}}\right)$ the vector with elements $d(\mathbf{x}_i, \widehat{\boldsymbol{\mu}}, \widehat{\boldsymbol{\Sigma}})$, $i = 1, \ldots, n$, and let $\widehat{\sigma}$ be a robust scale estimate. Then we define the estimates $\widehat{\boldsymbol{\mu}}$ and $\widehat{\boldsymbol{\Sigma}}$ by

$$\widehat{\sigma}\left(\mathbf{d}\left(X, \widehat{\boldsymbol{\mu}}, \widehat{\boldsymbol{\Sigma}}\right)\right) = \min \text{ with } \widehat{\boldsymbol{\mu}} \in R^p, \ \widehat{\boldsymbol{\Sigma}} \in \mathcal{S}_p, \ \left|\widehat{\boldsymbol{\Sigma}}\right| = 1. \tag{6.28}$$

It is easy to show that the estimates defined by (6.28) are equivariant. An equivalent formulation of the above goal is to minimize $\left|\widehat{\boldsymbol{\Sigma}}\right|$ subject to a bound on $\widehat{\sigma}$ (Problems 6.7, 6.8, 6.9).

6.4.1 The minimum volume ellipsoid estimate

The simplest case of (6.28) is to mimic the approach that results in the LMS in Section 5.6, and let $\widehat{\sigma}$ be the sample median. The resulting location and dispersion matrix estimate is called the minimum volume ellipsoid (MVE) estimate. The name stems from the fact that among all ellipsoids $\{\mathbf{x} : d(\mathbf{x}, \boldsymbol{\mu}, \boldsymbol{\Sigma}) \leq 1\}$ containing at least half of the data points, the one given by the MVE estimate has minimum volume, i.e., the minimum $|\boldsymbol{\Sigma}|$. The consistency rate of the MVE is the same slow rate as the LMS, namely only $n^{-1/3}$, and hence is very inefficient (Davies, 1992).

6.4.2 S-estimates

To overcome the inefficiency of the MVE we consider a more general class of estimates called S-estimates (Davies, 1987), defined by (6.28) taking for $\widehat{\sigma}$ an M-scale estimate that satisfies

$$\frac{1}{n}\sum_{i=1}^{n}\rho\left(\frac{d_i}{\widehat{\sigma}}\right)=\delta \tag{6.29}$$

where ρ is a smooth bounded ρ-function. The same reasoning as in (5.28) shows that an S-estimate $(\widehat{\boldsymbol{\mu}}, \widehat{\boldsymbol{\Sigma}})$ is an M-estimate in the sense that for any $\widetilde{\boldsymbol{\mu}}, \widetilde{\boldsymbol{\Sigma}}$ with $|\widetilde{\boldsymbol{\Sigma}}| = 1$ and $\widehat{\sigma} = \widehat{\sigma}\left(\mathbf{d}\left(X, \widehat{\boldsymbol{\mu}}, \widehat{\boldsymbol{\Sigma}}\right)\right)$

$$\sum_{i=1}^{n}\rho\left(\frac{d\left(\mathbf{x}_i, \widehat{\boldsymbol{\mu}}, \widehat{\boldsymbol{\Sigma}}\right)}{\widehat{\sigma}}\right) \leq \sum_{i=1}^{n}\rho\left(\frac{d\left(\mathbf{x}_i, \widetilde{\boldsymbol{\mu}}, \widetilde{\boldsymbol{\Sigma}}\right)}{\widehat{\sigma}}\right). \tag{6.30}$$

If ρ is differentiable, it can be shown (Section 6.12.5) that the solution to (6.28) must satisfy estimating equations of the form (6.14)–(6.15), i.e.,

$$\sum_{i=1}^{n}W\left(\frac{d_i}{\widehat{\sigma}}\right)(\mathbf{x}_i-\widehat{\boldsymbol{\mu}})=\mathbf{0}, \tag{6.31}$$

$$\frac{1}{n}\sum_{i=1}^{n}W\left(\frac{d_i}{\widehat{\sigma}}\right)(\mathbf{x}_i-\widehat{\boldsymbol{\mu}})(\mathbf{x}_i-\widehat{\boldsymbol{\mu}})'=c\widehat{\boldsymbol{\Sigma}}, \tag{6.32}$$

where

$$W=\rho' \text{ and } \widehat{\sigma}=\widehat{\sigma}(d_1,\ldots,d_n), \tag{6.33}$$

and c is a scalar such that $|\widehat{\boldsymbol{\Sigma}}| = 1$. Note, however, that if ρ is bounded (as is the usual case), $dW(d)$ cannot be monotone (Problem 6.5); actually for the estimates usually employed $W(d)$ vanishes for large d. Hence the estimate is not a monotone M-estimate, and therefore the estimating equations yield only *local* minima of $\widehat{\sigma}$.

The choice $\rho(d) = d$ yields the average of the d_i's as a scale estimate. In this case $W \equiv 1$ and hence

$$\widehat{\boldsymbol{\mu}}=\overline{\mathbf{x}}, \qquad \widehat{\boldsymbol{\Sigma}}=\frac{\mathbf{C}}{|\mathbf{C}|^{1/p}} \tag{6.34}$$

where $\mathbf{C}$ is the sample covariance matrix. For this choice of scale estimate it follows that

$$\sum_{i=1}^{n}(\mathbf{x}_i-\overline{\mathbf{x}})'\,\boldsymbol{\Sigma}^{-1}\,(\mathbf{x}_i-\overline{\mathbf{x}})\leq\sum_{i=1}^{n}(\mathbf{x}_i-\boldsymbol{\nu})'\,\mathbf{V}^{-1}\,(\mathbf{x}_i-\boldsymbol{\nu}) \tag{6.35}$$

for all $\boldsymbol{\nu}$ and $\mathbf{V}$ with $|\mathbf{V}|=1$.

It can be shown (Davies, 1987) that if ρ is differentiable, then for S-estimates the distribution of $\sqrt{n}\left(\widehat{\boldsymbol{\mu}}-\widehat{\boldsymbol{\mu}}_\infty,\widehat{\boldsymbol{\Sigma}}-\widehat{\boldsymbol{\Sigma}}_\infty\right)$ tends to a multivariate normal.

Similarly to Section 5.6, it can be shown that the maximum FBP (6.27) is attained for S-estimates by taking in (6.29)

$$n\delta=m^*_{\max}=\left[\frac{n-p}{2}\right].$$

We define the bisquare multivariate S-estimate as the one with scale given by (6.29) with

$$\rho(t)=\min\left\{1,1-(1-t)^3\right\}, \tag{6.36}$$

which has weight function

$$W(t)=3\,(1-t)^2\,\mathrm{I}(t\leq1). \tag{6.37}$$

The reason for this definition is that in the univariate case the bisquare scale estimate—call it $\widehat{\eta}$ for notational convenience—based on centered data x_i with location $\widehat{\mu}$ is the solution of

$$\frac{1}{n}\sum_{i=1}^{n}\rho_{bisq}\left(\frac{x_i-\widehat{\mu}}{\widehat{\eta}}\right)=\delta \tag{6.38}$$

where $\rho_{bisq}(t)=\min\left\{1,1-\left(1-t^2\right)^3\right\}$. Since $\rho_{bisq}(t)=\rho\left(t^2\right)$ for the ρ defined in (6.36), it follows that (6.38) is equivalent to

$$\frac{1}{n}\sum_{i=1}^{n}\rho\left(\frac{(x_i-\widehat{\mu})^2}{\widehat{\sigma}}\right)=\delta$$

with $\widehat{\sigma}=\widehat{\eta}^2$. Now $d\,(\mathbf{x},\boldsymbol{\mu},\boldsymbol{\Sigma})$ is the normalized *squared* distance between $\mathbf{x}$ and $\boldsymbol{\mu}$, which explains the use of ρ.

6.4.3 The minimum covariance determinant estimate

Another possibility is to use a trimmed scale for $\widehat{\sigma}$ instead of an M-scale, as was done to define the LTS estimate in Section 5.6.2. Let $d_{(1)}\leq\ldots\leq d_{(n)}$ be the ordered values of the squared distances $d_i=d\,(\mathbf{x}_i,\boldsymbol{\mu},\boldsymbol{\Sigma})$, and for $1\leq h<n$ define the trimmed scale of the squared distances as

$$\widehat{\sigma}=\sum_{i=1}^{h}d_{(i)}. \tag{6.39}$$

An estimate $(\widehat{\mu}, \widehat{\Sigma})$ defined by (6.28) with this trimmed scale is called a *minimum covariance determinant* (MCD) estimate. The reason for the name is the following: for each ellipsoid $\{\mathbf{x} : d(\mathbf{x}, \mathbf{t}, \mathbf{V}) \leq 1\}$ containing at least h data points, compute the covariance matrix $\mathbf{C}$ of the data points in the ellipsoid. If $(\widehat{\mu}, \widehat{\Sigma})$ is an MCD estimate, then the ellipsoid with $\mathbf{t} = \widehat{\mu}$ and $\mathbf{V}$ equal to a scalar multiple of $\widehat{\Sigma}$ minimizes $|\mathbf{C}|$.

As in the case of the LTS estimate in Section 5.6, the maximum BP of the MCD estimate is attained by taking $h = n - m^*_{\max}$ with $m^*_{\max}$ defined in (6.27).

6.4.4 S-estimates for high dimension

Consider the multivariate S-estimate with a bisquare ρ-function. The following table gives the asymptotic efficiencies of the location estimate vector under normality for different dimensions p,

p	5	10	20	50	100
Eff.	0.845	0.932	0.969	0.989	0.994

It is seen that the efficiency approaches one for large p. The same thing happens with the dispersion matrix estimate. It is shown in Section 6.12.6 that this behavior holds for any S-estimate with a continuous weight function $W = \rho'$. This may seem like good news. However, the proof shows that for large p all observations, except those that are extremely far away from the bulk of the data, have approximately the same weight, and hence the estimate is approximately equal to the sample mean and sample covariance matrix. Thus observations outlying enough to be dangerous may also have nearly maximum weight, and as a result, the bias can be very large (bias is defined in Section 6.6). It will be seen later that this increase in efficiency and decrease in robustness with large p does not occur with the MVE.

Rocke (1996) pointed out the problem just described and proposed that the ρ-function change with dimension to prevent both the efficiency from increasing to values arbitrarily close to one, and correspondingly the bias becoming arbitrarily large. He proposed a family of ρ-functions with the property that when $p \to \infty$ the function ρ approaches the step function $\rho(d) = \mathrm{I}(d > 1)$. The latter corresponds to the scale estimate $\widehat{\sigma} = \mathrm{Med}\,(\mathbf{d})$ and so the limiting form of the estimate for large dimensions is the MVE estimate.

Put for brevity $d = d(\mathbf{x}, \widehat{\mu}_\infty, \widehat{\Sigma}_\infty)$. It is shown in Section 6.12.6 that if $\mathbf{x}$ is normal, then for large p

$$\mathcal{D}\left(\frac{d}{\sigma}\right) \approx \mathcal{D}\left(\frac{z}{p}\right) \quad \text{with } z \sim \chi^2_p$$

and hence that d/σ is increasingly concentrated around one. To have a high enough (but not too high) efficiency, we should give a high weight to the values of d/σ near one and downweight the extreme ones. A simple way to do this is to have $W(t) = 0$ for t between the α- and the $(1-\alpha)$-quantiles of d/σ. Call $\chi^2_p(\alpha)$ the α-quantile of χ^2_p. For large p the χ^2_p distribution is approximately symmetric, with $\chi^2_p(0.5) \approx \mathrm{E}z = p$

and $\chi^2_p(1-\alpha) - p \approx p - \chi^2_p(\alpha)$. Let

$$\gamma = \min\left(\frac{\chi^2_p(1-\alpha)}{p} - 1, 1\right). \tag{6.40}$$

We now define a smooth ρ-function such that the resulting weight function $W(t)$ vanishes for $t \notin [1-\gamma, 1+\gamma]$. Let

$$\rho(t) = \begin{cases} 0 & \text{for} \quad 0 \le t \le 1-\gamma \\ \left(\frac{t-1}{4\gamma}\right)\left[3 - \left(\frac{t-1}{\gamma}\right)^2\right] + \frac{1}{2} & \text{for} \quad 1-\gamma < t < 1+\gamma \\ 1 & \text{for} \quad t \ge 1+\gamma \end{cases} \tag{6.41}$$

which has as derivative the weight function

$$W(t) = \frac{3}{4\gamma}\left[1 - \left(\frac{t-1}{\gamma}\right)^2\right] \mathrm{I}(1-\gamma \le t \le 1+\gamma). \tag{6.42}$$

Figures 6.4 and 6.5 show the plots of ρ and of W for $\alpha = 0.05$ and the values $p = 10$ and 100. The corresponding functions for the bisquare (6.36) and (6.37) are also plotted for comparison. The weight functions are scaled so that $W(0) = 1$ to simplify viewing. Figure 6.6 shows the density of d/σ for $p = 10$ and 100. Note that

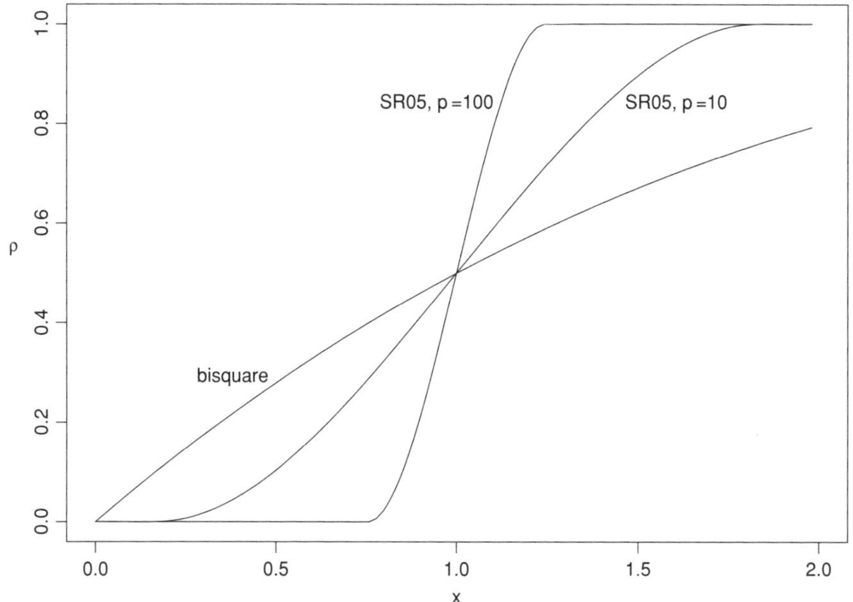

Figure 6.4 ρ-functions for S estimates: bisquare and Rocke-type estimates with $\alpha = 0.05$ and $p = 10$ and 100

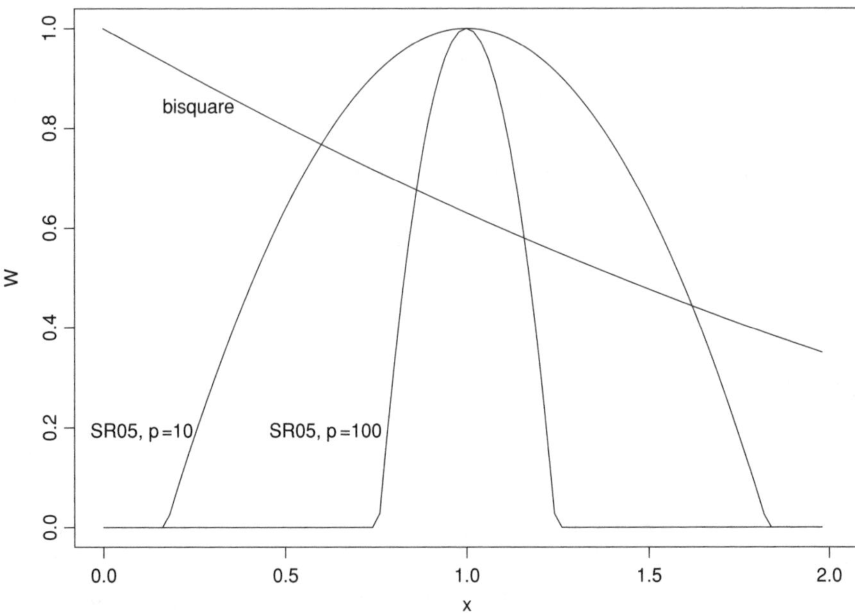

Figure 6.5 Weight functions of the bisquare and of SR-05 for $p = 10$ and 100

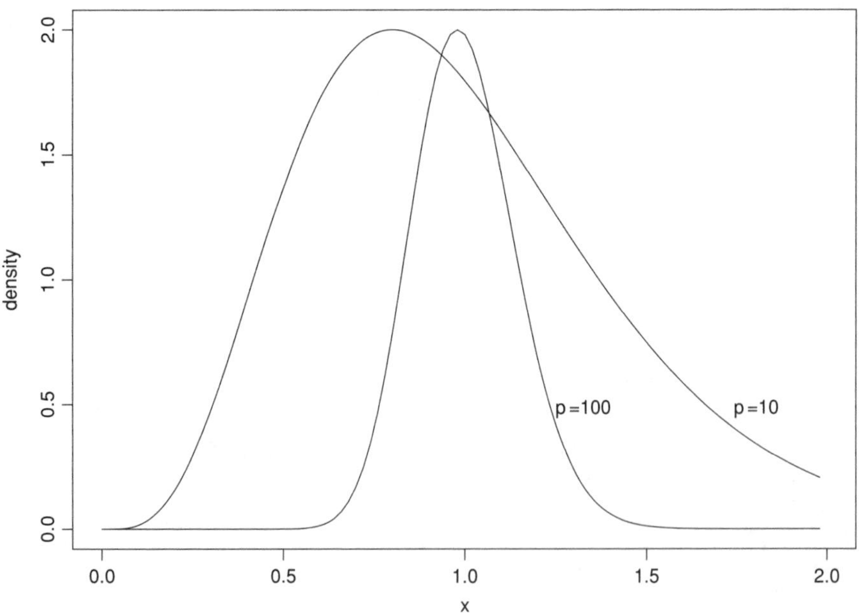

Figure 6.6 Densities of d/σ for normal data with $p = 10$ and 100

when p increases the density becomes more concentrated around one, the interval on which W is positive shrinks, and ρ tends to the step function corresponding to the MVE. One also sees that the bisquare weight function is quite inappropriate for assigning high weights to the bulk of the data and rapidly downweighting data far from the bulk of the data, except possibly for very small values of p.

Rocke's "biflat" family of weight functions is the squared values of the W in (6.42). It is smoother at the endpoints but gives less weights to inner points.

The asymptotic efficiency of the estimates based on the weight function (6.42) at the normal model can be computed. The table below gives the efficiency of the location estimate for a wide range of values of p at $\alpha = 0.05$ and 0.10. It is seen that the efficiency is almost constant for large p.

p	5	10	20	50	100
$\alpha = 0.05$	0.69	0.82	0.84	0.84	0.84
$\alpha = 0.10$	0.64	0.72	0.74	0.73	0.73

6.4.5 One-step reweighting

A one-step reweighting procedure that can be used with any pair of estimates $\widehat{\mu}$ and $\widehat{\Sigma}$ is similar to the one defined in Section 5.6.3. Let W be a weight function. Given the estimates $\widehat{\mu}$ and $\widehat{\Sigma}$, define new estimates $\widetilde{\mu}$, $\widetilde{\Sigma}$ as a weighted mean vector and weighted covariance matrix with weights $W(d_i)$, where the d_i's are the squared distances corresponding to $\widehat{\mu}$ and $\widehat{\Sigma}$. The most popular function is hard rejection, corresponding to $W(t) = \mathrm{I}(t \leq k)$ where k is chosen with the same criterion as in Section 5.6.3. For $\widehat{c}$ defined in (6.22) the distribution of $d_i/\widehat{c}$ is approximately χ^2_p under normality, and hence choosing $k = \widehat{c}\chi^2_{p,\beta}$ will reject approximately a fraction $1 - \beta$ of the "good" data if there are no outliers. It is customary to take $\beta = 0.95$ or 0.975. If the dispersion matrix estimate is singular we proceed as in Section 6.3.1.

Although no theoretical results are known, simulations have showed this procedure improves the bias and efficiency of the MVE and MCD estimates. But it cannot be asserted that such improvement happens with *any* estimate.

6.5 The Stahel–Donoho estimate

Recall that the simplest approach to the detection of outliers in a univariate sample is the one given in Section 1.3: for each data point compute an "outlyingness measure" (1.4) and identify those points having a "large" value of this measure. The key idea for the extension of this approach to the multivariate case is that a multivariate outlier should be an outlier in *some* univariate projection. More precisely, given a direction $\mathbf{a} \in R^p$ with $\|\mathbf{a}\| = 1$, denote by $\mathbf{a}'X=\{\mathbf{a}'\mathbf{x}_1, \ldots, \mathbf{a}'\mathbf{x}_n\}$ the projection of the data set X along $\mathbf{a}$. Let $\widehat{\mu}$ and $\widehat{\sigma}$ be robust univariate location and dispersion statistics, e.g., the median and MAD respectively. The outlyingness with respect to X of a point $\mathbf{x} \in R^p$

along $\mathbf{a}$ is defined as in (1.4) by

$$t(\mathbf{x}, \mathbf{a}) = \frac{\mathbf{x}'\mathbf{a} - \widehat{\mu}\left(\mathbf{a}'X\right)}{\widehat{\sigma}\left(\mathbf{a}'X\right)}.$$

The outlyingness of $\mathbf{x}$ is then defined by

$$t(\mathbf{x}) = \max_{\mathbf{a}} t(\mathbf{x}, \mathbf{a}). \tag{6.43}$$

In the above maximum $\mathbf{a}$ ranges over the set $\{\|\mathbf{a}\| = 1\}$, but in view of the equivariance of $\widehat{\mu}$ and $\widehat{\sigma}$, it is equivalent to take the set $\{\mathbf{a} \neq \mathbf{0}\}$.

The Stahel–Donoho estimate, proposed by Stahel (1981) and Donoho (1982), is a weighted mean and covariance matrix where the weight of $\mathbf{x}_i$ is a nonincreasing function of $t(\mathbf{x}_i)$. More precisely, let W_1 and W_2 be two weight functions, and define

$$\widehat{\boldsymbol{\mu}} = \frac{1}{\sum_{i=1}^n w_{i1}} \sum_{i=1}^n w_{i1}\mathbf{x}_i, \tag{6.44}$$

$$\widehat{\boldsymbol{\Sigma}} = \frac{1}{\sum_{i=1}^n w_{i2}} \sum_{i=1}^n w_{i2}\left(\mathbf{x}_i - \widehat{\boldsymbol{\mu}}\right)\left(\mathbf{x}_i - \widehat{\boldsymbol{\mu}}\right)' \tag{6.45}$$

with

$$w_{ij} = W_j(t(\mathbf{x}_i)), \;\; j = 1, 2. \tag{6.46}$$

If $\mathbf{y}_i = \mathbf{A}\mathbf{x}_i + \mathbf{b}$, then it is easy to show that $t(\mathbf{y}_i) = t(\mathbf{x}_i)$ (t is *invariant*) and hence the estimates are equivariant.

In order that no term dominates in (6.44)–(6.45) it is clear that the weight functions must satisfy the conditions

$$tW_1(t) \text{ and } t^2 W_2(t) \text{ are bounded for } t \geq 0. \tag{6.47}$$

It can be shown (see Maronna and Yohai, 1995) that under (6.47) the asymptotic BP is 1/2. For the FBP, Tyler (1994) and Gather and Hilker (1997) show that the estimate attains the maximum BP given by (6.27) if $\widehat{\boldsymbol{\mu}}$ is the sample median and the scale is

$$\widehat{\sigma}(z) = \frac{1}{2}\left(\widetilde{z}_k + \widetilde{z}_{k+1}\right)$$

where $\widetilde{z}_i$ denotes the ordered values of $|z_i - \mathrm{Med}(\mathbf{z})|$ and $k = [(n + p)/2]$.

The asymptotic normality of the estimate was shown by Zuo, Cui and He (2004), and its influence function and maximum asymptotic bias were derived by Zuo, Cui and Young (2004).

The choice of the weight functions in (6.44)–(6.45) is important for combining robustness and efficiency. A family of weight functions used in the literature is the "Huber weights"

$$W^H_{c,k}(t) = \min\left(1, \left(\frac{c}{t}\right)^k\right) \tag{6.48}$$

where $k \geq 2$ in order to satisfy (6.47). Maronna and Yohai (1995) used

$$W_1 = W_2 = W^H_{c,2} \text{ with } c = \sqrt{\chi^2_{p,\beta}} \tag{6.49}$$

in (6.46) with $\beta = 0.95$. For large p, Maronna and Zamar (2002) preferred $c = \min(4, \sqrt{\chi^2_{p,0.5}})$. to avoid c becoming too large and hence the estimate losing robustness.

Zuo et al. (2004a) proposed the family of weights

$$W^Z_{c,k}(t) = \min\left\{1, 1 - \frac{1}{b}\exp\left[-k\left(1 - \frac{1}{c(1+t)}\right)^2\right]\right\} \tag{6.50}$$

where $c = \text{Med}(1/(1 + t(\mathbf{x})))$, k is a tuning parameter and $b = 1 - e^{-k}$.

Simulations show that one-step reweighting does not improve the Stahel–Donoho estimate.

6.6 Asymptotic bias

We now deal with data from a contaminated distribution $F = (1 - \varepsilon)F_0 + \varepsilon G$, where F_0 describes the "typical" data. In order to define bias, we have to define which are the "true" parameters to estimate. To fix ideas assume $F_0 = \text{N}_p(\boldsymbol{\mu}_0, \boldsymbol{\Sigma}_0)$, but note that the following discussion applies to any other elliptical distribution. Let $\widehat{\boldsymbol{\mu}}_\infty$ and $\widehat{\boldsymbol{\Sigma}}_\infty$ be the asymptotic values of location and dispersion estimates.

Defining a single measure of bias for a multidimensional estimate is more complicated than in Section 3.3. Assume first that $\boldsymbol{\Sigma}_0 = \mathbf{I}$. In this case the symmetry of the situation makes it natural to choose the Euclidean norm $\|\widehat{\boldsymbol{\mu}}_\infty - \boldsymbol{\mu}_0\|$ as a reasonable bias measure for location. For the dispersion matrix size is relatively easy to adjust, by means of (6.21) or (6.22), and it will be most useful to focus on shape. Thus we want to measure the discrepancy between $\widehat{\boldsymbol{\Sigma}}_\infty$ and scalar multiples of $\mathbf{I}$. The simplest way to do so is with the *condition number*, which is defined as the ratio of the largest to the smallest eigenvalue,

$$\text{cond}(\widehat{\boldsymbol{\Sigma}}_\infty) = \frac{\lambda_1(\widehat{\boldsymbol{\Sigma}}_\infty)}{\lambda_p(\widehat{\boldsymbol{\Sigma}}_\infty)}.$$

The condition number equals one if and only if $\widehat{\boldsymbol{\Sigma}}_\infty = c\mathbf{I}$ for some $c \in R$. Other functions of the eigenvalues may be used for measuring shape discrepancies, such as the likelihood ratio test statistic for testing sphericity (Seber, 1984), which is the ratio of the arithmetic to the geometric mean of the eigenvalues:

$$\frac{\text{trace}(\widehat{\boldsymbol{\Sigma}}_\infty)}{|\widehat{\boldsymbol{\Sigma}}_\infty|^{1/p}}.$$

It is easy to show that in the special case of a spherical distribution the asymptotic value of an equivariant $\widehat{\boldsymbol{\Sigma}}$ is a scalar multiple of $\mathbf{I}$ (Problem 6.3), and so in this case there is no shape discrepancy.

For the case of an equivariant estimate and a general $\boldsymbol{\Sigma}_0$ we want to define bias so that it is invariant under affine transformations, i.e., the bias does not change if $\mathbf{x}$ is replaced by $\mathbf{Ax}+\mathbf{b}$. To this end we "normalize" the data so that they have an identity dispersion matrix. Let $\mathbf{A}$ be any matrix such that $\mathbf{A}'\mathbf{A}=\boldsymbol{\Sigma}_0^{-1}$ and define $\mathbf{y}=\mathbf{Ax}$. Then $\mathbf{y}$ has mean $\mathbf{A}\boldsymbol{\mu}_0$ and the identity dispersion matrix, and if the estimates $\widehat{\boldsymbol{\mu}}$ and $\widehat{\boldsymbol{\Sigma}}$ are equivariant then their asymptotic values based on data $\mathbf{y}_i$ are $\mathbf{A}\widehat{\boldsymbol{\mu}}_\infty$ and $\mathbf{A}\widehat{\boldsymbol{\Sigma}}_\infty\mathbf{A}'$ respectively, where $\widehat{\boldsymbol{\mu}}_\infty$ and $\widehat{\boldsymbol{\Sigma}}_\infty$ are their values based on data $\mathbf{x}_i$. Since their respective discrepancies are given by

$$\|\mathbf{A}\widehat{\boldsymbol{\mu}}_\infty-\mathbf{A}\boldsymbol{\mu}_0\|^2=\left(\widehat{\boldsymbol{\mu}}_\infty-\boldsymbol{\mu}_0\right)'\boldsymbol{\Sigma}_0^{-1}\left(\widehat{\boldsymbol{\mu}}_\infty-\boldsymbol{\mu}_0\right)\quad\text{and}\quad\operatorname{cond}(\mathbf{A}\widehat{\boldsymbol{\Sigma}}_\infty\mathbf{A}'),$$

and noting that $\mathbf{A}\widehat{\boldsymbol{\Sigma}}_\infty\mathbf{A}'$ has the same eigenvalues as $\boldsymbol{\Sigma}_0^{-1}\widehat{\boldsymbol{\Sigma}}_\infty$, it is natural to define

$$\operatorname{bias}(\widehat{\boldsymbol{\mu}})=\sqrt{\left(\widehat{\boldsymbol{\mu}}_\infty-\boldsymbol{\mu}_0\right)'\boldsymbol{\Sigma}_0^{-1}\left(\widehat{\boldsymbol{\mu}}_\infty-\boldsymbol{\mu}_0\right)}\text{ and }\operatorname{bias}(\widehat{\boldsymbol{\Sigma}})=\operatorname{cond}(\boldsymbol{\Sigma}_0^{-1}\widehat{\boldsymbol{\Sigma}}_\infty).\tag{6.51}$$

It is easy to show that if the estimates are equivariant then (6.51) does not depend upon either $\boldsymbol{\mu}_0$ or $\boldsymbol{\Sigma}_0$. Hence to evaluate equivariant estimates we may without loss of generality take $\boldsymbol{\mu}_0=\mathbf{0}$ and $\boldsymbol{\Sigma}_0=\mathbf{I}$.

Table 6.2 gives the maximum asymptotic biases (6.51) of several multivariate estimates at the multivariate normal distribution, for $\varepsilon=0.1$ and $p=5$, 10 and 20. "S-D9" is the Stahel–Donoho estimate using weights (6.49) with $\beta=0.9$; "Bisq." is the S-estimate with bisquare ρ and BP $=0.5$; "SR05" is the SR-α estimate based on (6.41) with $\alpha=0.05$. The results for MVE, S-D9 and MCD are from Adrover and Yohai (2002). No results on the condition numbers of S-D9 and MCD are available. The results show that SR05 and SR10 are the best competitors relative to the MVE, with somewhat smaller maximum biases than the MVE estimate for $p=5$, and somewhat higher relative biases for $p=10$ and 20 (overall SR10 is preferred).

Table 6.2 Maximum biases of multivariate estimates for contamination rate 0.10

	p	MVE	Bisq.	SR05	SR10	S-D9	MCD
Location	5	0.73	0.46	0.63	0.66	0.52	0.94
	10	0.75	1.40	0.92	0.90	1.07	1.97
	20	0.77	6.90	1.24	1.07	2.47	7.00
Dispersion	5	6.9	4.05	4.77	5.06		
	10	9.4	19.31	9.52	9.42		
	20	15.0	357.42	23.90	20.10		

6.7 Numerical computation of multivariate estimates

6.7.1 Monotone M-estimates

Equations (6.15) and (6.16) yield an iterative algorithm similar to the one for regression in Section 4.5. Start with initial estimates $\widehat{\boldsymbol{\mu}}_0$ and $\widehat{\boldsymbol{\Sigma}}_0$, e.g., the vector of coordinate-wise medians and the diagonal matrix with the squared normalized MADs of the variables in the diagonal. At iteration k let $d_{ki} = d(\mathbf{x}_i, \widehat{\boldsymbol{\mu}}_k, \widehat{\boldsymbol{\Sigma}}_k)$ and compute

$$\widehat{\boldsymbol{\mu}}_{k+1} = \frac{\sum_{i=1}^n W_1(d_{ki})\mathbf{x}_i}{\sum_{i=1}^n W_1(d_{ki})}, \quad \widehat{\boldsymbol{\Sigma}}_{k+1} = \frac{1}{n}\sum_{i=1}^n W_2(d_{ki})(\mathbf{x}_i - \widehat{\boldsymbol{\mu}}_{k+1})(\mathbf{x}_i - \widehat{\boldsymbol{\mu}}_{k+1})'. \quad (6.52)$$

If at some iteration $\widehat{\boldsymbol{\Sigma}}_k$ becomes singular, it suffices to compute the d_i's through (6.20). The convergence of the procedure is established in Section 9.5. Since the solution is unique for monotone M-estimates, the starting values influence only the number of iterations but not the end result.

6.7.2 Local solutions for S-estimates

Since local minima of $\widehat{\sigma}$ are solutions of the M-estimating equations (6.31)–(6.32), a natural procedure to minimize $\widehat{\sigma}$ is to use the iterative procedure (6.52) to solve the equations, with $W_1 = W_2$ equal to $W = \rho'$ as stated in (6.33). It must be recalled that since $tW(t)$ is redescending, this pair of equations yields only a *local* minimum of σ, and hence the starting values are essential. Assume for the moment that we have the initial $\widehat{\boldsymbol{\mu}}_0$ and $\widehat{\boldsymbol{\Sigma}}_0$ (their computation is treated below in Section 6.7.5).

At iteration k, call $\widehat{\boldsymbol{\mu}}_k$ and $\widehat{\boldsymbol{\Sigma}}_k$ the current values and compute

$$d_{ki} = d(\mathbf{x}_i, \widehat{\boldsymbol{\mu}}_k, \widehat{\boldsymbol{\Sigma}}_k), \widehat{\sigma}_k = \widehat{\sigma}(d_{k1}, \ldots, d_{kn}), \ w_{ki} = W\left(\frac{d_{ki}}{\widehat{\sigma}_k}\right). \quad (6.53)$$

Then compute

$$\widehat{\boldsymbol{\mu}}_{k+1} = \frac{\sum_{i=1}^n w_{ki}\mathbf{x}_i}{\sum_{i=1}^n w_{ki}}, \ \widehat{\mathbf{C}}_k = \sum_{i=1}^n w_{ki}(\mathbf{x}_i - \widehat{\boldsymbol{\mu}}_{k+1})(\mathbf{x}_i - \widehat{\boldsymbol{\mu}}_{k+1})', \ \widehat{\boldsymbol{\Sigma}}_{k+1} = \frac{\widehat{\mathbf{C}}_k}{\left|\widehat{\mathbf{C}}_k\right|^{1/p}}. \quad (6.54)$$

It is shown in Section 9.6 that if the weight function W is nonincreasing, then $\widehat{\sigma}_k$ decreases at each step. One can then stop the iteration when the relative change $(\widehat{\sigma}_k - \widehat{\sigma}_{k+1})/\widehat{\sigma}_k$ is below a given tolerance. Experience shows that since the decrease of $\widehat{\sigma}_k$ is generally slow, it is not necessary to recompute it at each step, but at, say, every 10th iteration.

If W is not monotonic, the iteration steps (6.53)–(6.54) are not guaranteed to cause a decrease in $\widehat{\sigma}_k$ at each step. However, the algorithm can be modified to insure a decrease at each iteration. Since the details are involved, they are deferred to Section 9.6.1.

6.7.3 Subsampling for estimates based on a robust scale

The obvious procedure to generate an initial approximation to an estimate defined by (6.28) is follow the general approach for regression described in Section 5.7.2, in which the minimization problem is replaced by a finite one where the candidate estimates are sample means and covariance matrices of subsamples. To obtain a finite set of candidate solutions take a subsample of size $p+1$, $\{\mathbf{x}_i : i \in J\}$ where the set $J \subset \{1, \ldots, n\}$ has $p+1$ elements, and compute

$$\widehat{\boldsymbol{\mu}}_J = \mathrm{ave}_{i\in J}(\mathbf{x}_i) \text{ and } \widehat{\boldsymbol{\Sigma}}_J = \frac{\widehat{\mathbf{C}}_J}{\left|\widehat{\mathbf{C}}_J\right|^{1/p}}, \tag{6.55}$$

where $\widehat{\mathbf{C}}_J$ is the covariance matrix of the subsample; and let

$$\mathbf{d}_J = \{d_{Ji} : i = 1, \ldots, n\}, \quad d_{Ji} = d\left(\mathbf{x}_i, \widehat{\boldsymbol{\mu}}_J, \widehat{\boldsymbol{\Sigma}}_J\right). \tag{6.56}$$

Then the problem of minimizing $\widehat{\sigma}$ is replaced by the finite problem of minimizing $\widehat{\sigma}(\mathbf{d}_J)$ over J. Since choosing all $\binom{n}{p+1}$ subsamples is prohibitive unless both n and p are rather small, we choose N of them at random, $\{J_k : k = 1, \ldots, N\}$, and the estimates are $\widehat{\boldsymbol{\mu}}_{J_{k^*}}, \widehat{\boldsymbol{\Sigma}}_{J_{k^*}}$ with

$$k^* = \arg\min_{k=1,\ldots,N} \widehat{\sigma}(\mathbf{d}_{J_k}). \tag{6.57}$$

If the sample contains a proportion ε of outliers, the probability of at least one "good" subsample is $1-(1-\alpha)^N$ where $\alpha = (1-\varepsilon)^{p+1}$. If we want this probability to be larger than $1-\delta$ we must have

$$N \geq \frac{|\log \delta|}{|\log(1-\alpha)|} \approx \frac{|\log \delta|}{(1-\varepsilon)^{p+1}}. \tag{6.58}$$

See Table 5.3 in Chapter 5 for the values of N required as a function of p and ε.

A seemingly trivial but important detail is in order. It would seem that the more candidates, the better. Adding the "subsample" consisting of the whole data set, resulting in the usual sample mean and covariance estimates, would decrease the scale at practically no cost. But the simulations described in Section 6.8 show that while this addition improves the efficiency at the normal model, it greatly increases the bias due to outliers. The reason is that for contaminated data it may happen that the chosen subsample is the whole data set, and hence the outcome may be the sample mean and covariance matrix.

A simple but effective improvement of the subsampling procedure is as follows. For subsample J with distances $\mathbf{d}_J$ defined in (6.56), let $\tau = \mathrm{Med}(\mathbf{d}_J)$ and compute

$$\boldsymbol{\mu}_J^* = \mathrm{ave}\{\mathbf{x}_i : d_{Ji} \leq \tau\}, \ \mathbf{C}_J^* = \mathbf{Var}\{\mathbf{x}_i : d_{Ji} \leq \tau\}, \ \boldsymbol{\Sigma}_J^* = \frac{\mathbf{C}_J^*}{\left|\mathbf{C}_J^*\right|^{1/p}}. \tag{6.59}$$

Then use $\boldsymbol{\mu}_J^*$ and $\mathbf{C}_J^*$ instead of $\widehat{\boldsymbol{\mu}}_J$ and $\widehat{\mathbf{C}}_J$. The motivation for this idea is that a subsample of $p+1$ points is too small to yield reliable means and covariances, and so it is desirable to enlarge the subsample in a suitable way. This is attained by selecting the half-sample with smallest distances. Although no theoretical results are known for this method, our simulations show that this extra effort yields a remarkable improvement in the behavior of the estimates with only a small increase in computational time. In particular, for the MVE estimate the minimum scale obtained this way is always much smaller than that obtained with the original subsamples. For example, the ratio is about 0.3 for $p = 40$ and 500 subsamples.

6.7.4 The MVE

We have seen that the objective function of S-estimates can be decreased by iteration steps, and the same thing happens with the MCD (Section 6.7.6). However, no such improvements are known for the MVE, which makes the outcome of the subsampling procedure the only available approximation to the estimate.

The simplest approach to this problem is to use directly the "best" subsample given by (6.57). However, in view of the success of the improved subsampling method given by (6.59) we make it our method of choice to compute the MVE.

An exact method for the MVE was proposed by Agulló (1996), but since it is not feasible except for small n and p we do not describe it here.

6.7.5 Computation of S-estimates

Once we have initial values $\widehat{\boldsymbol{\mu}}_0$ and $\widehat{\boldsymbol{\Sigma}}_0$, an S-estimate is computed by means of the iterative procedures described in Section 6.7.2. We present two approaches to compute $\widehat{\boldsymbol{\mu}}_0$ and $\widehat{\boldsymbol{\Sigma}}_0$.

The simplest approach is to obtain initial values of $\boldsymbol{\mu}_0$, $\boldsymbol{\Sigma}_0$ through subsampling and then apply the iterative algorithm. Much better results are obtained by following the same principles as the strategy described for regression in Section 5.7.3. This approach was first employed for multivariate estimation by Rousseeuw and van Driessen (1999) (see Section 6.7.6).

Our preferred approach, however, proceeds as was done in Section 5.5 to compute the MM-estimates of regression; that is, start the iterative algorithm from a bias-robust but possibly inefficient estimate, which is computed through subsampling. Since Table 6.2 provides evidence that the MVE estimate has the smallest maximum bias, it is natural to think of using it as an initial estimate.

It is important to note that although the MVE estimate has the unattractive feature of a slow $n^{-1/3}$ rate of consistency, this feature does not affect the efficiency of the local minimum which is the outcome of the iterative algorithm, since it satisfies the M-estimating equations (6.31)–(6.32); if equations (6.17)–(6.18) for the asymptotic values $\widehat{\boldsymbol{\mu}}_\infty$, $\widehat{\boldsymbol{\Sigma}}_\infty$ have a unique solution, then *all* solutions of (6.31)–(6.32) converge to $\left(\widehat{\boldsymbol{\mu}}_\infty, \widehat{\boldsymbol{\Sigma}}_\infty\right)$ with a rate of order $n^{-1/2}$.

Therefore we recommend use of the MVE as an initial estimate based on a subsampling approach, but using the improved method (6.59). Simulation results in

Section 6.8 show that this greatly improves the behavior of the estimates as compared to the simpler subsampling described at the beginning of the section.

Other numerical algorithms have been proposed by Ruppert (1992) and Woodruff and Rocke (1994).

6.7.6 The MCD

Rousseeuw and van Driessen (1999) found an iterative algorithm for the MCD, based on the following fact. Given any $\widehat{\boldsymbol{\mu}}_1$ and $\widehat{\boldsymbol{\Sigma}}_1$, let d_i be the corresponding squared distances. Then compute $\widehat{\boldsymbol{\mu}}_2$ and $\widehat{\mathbf{C}}$ as the sample mean and covariance matrix of the data with h smallest of the d_i's, and set $\widehat{\boldsymbol{\Sigma}}_2 = \widehat{\mathbf{C}}/\left|\widehat{\mathbf{C}}\right|^{1/p}$. Then $\widehat{\boldsymbol{\mu}}_2$ and $\widehat{\boldsymbol{\Sigma}}_2$ yield a lower value of $\widehat{\sigma}$ in (6.39) than $\widehat{\boldsymbol{\mu}}_1$ and $\widehat{\boldsymbol{\Sigma}}_1$. This is called the *concentration step* ("C-step" in the above authors' paper), and a proof of the above reduction in $\widehat{\sigma}$ is given in Section 9.6.2. In this case the modification (6.59) is not necessary, since the concentration steps are already of this sort. The overall strategy then is: for each of N candidate solutions obtained by subsampling, perform, say, two of the above steps, keep the 10 out of N that yield the smallest values of the criterion, and starting from each of them iterate the C-steps to convergence.

6.7.7 The Stahel–Donoho estimate

No exact algorithm for the Stahel–Donoho estimate is known. To approximate the estimate we need a large number of directions, and these can be obtained by subsampling. For each subsample $J = \left\{\mathbf{x}_{i_1}, \ldots, \mathbf{x}_{i_p}\right\}$ of size p, let $\mathbf{a}_J$ be a vector of norm 1 orthogonal to the hyperplane spanned by the subsample. The unit length vector $\mathbf{a}_J$ can be obtained by applying the QR orthogonalization procedure (see for example Chambers (1977) to $\left\{\mathbf{x}_{i_1} - \overline{\mathbf{x}}_J, \ldots, \mathbf{x}_{i_{p-1}} - \overline{\mathbf{x}}_J, \mathbf{b}\right\}$, where $\overline{\mathbf{x}}_J$ is the average of the subsample and $\mathbf{b}$ is any vector not collinear with $\overline{\mathbf{x}}_J$. Then we generate N subsamples $J_1, \ldots, J_N$ and replace (6.43) by

$$\widehat{t}(\mathbf{x}) = \max_k t(\mathbf{x}, \mathbf{a}_{J_k}).$$

It is easy to show that $\widehat{t}$ is invariant under affine transformations, and hence the approximate estimate is equivariant.

6.8 Comparing estimates

Until now we have relied on asymptotic results, and we need some sense of the behavior of the estimates defined in the previous sections in finite-sample sizes. Since the behavior of robust estimates for finite n is in general analytically intractable we must resort to simulation. Recall that the performance of an estimate is a combination of its bias and variability. The relative performances of two estimates depend on the underlying distribution and also on the sample size. Since the variability tends to zero when $n \to \infty$ while the bias remains essentially constant, an estimate with low bias but high variability may be very good for very large n but bad for moderate n. Hence no estimate can be "best" under all circumstances. Computing time is also an element to be taken into account.

The simulations presented in this section are all based on an underlying p-variate normal distribution $\mathrm{N}_p(\mathbf{0}, \mathbf{I})$ with 10% point mass contamination located at a distance k from the origin, with k ranging over a suitable interval. The criterion we use to compare the estimates is the root mean square error (RMSE) of the location estimate. The relative performances of the dispersion matrices are similar to those of the location vectors, and are hence not shown here. Details of the simulation are given at the end of this section.

We first compare the two versions of MVE mentioned in Section 6.7.4: simple subsampling (labeled "MVE-1"), and subsampling with the improvement (6.59) (labeled "MVE-2"). Figure 6.7 displays the results for $p = 10$ with 500 subsamples; the advantages of MVE-2 are clear. Hence in the remainder of the discussion only the improved version MVE-2 will be employed.

Now we compare two ways of approximating the bisquare S-estimate as described in Section 6.7.5: starting from the older "naive" subsampling-based estimate (labeled "Bisq-1"), and starting from the improved MVE-2 estimate (labeled "Bisq-2"). Figure 6.8 shows for $p = 10$ the advantages of a good starting point. All S-estimates considered henceforth are computed with the MVE-2 as starting point.

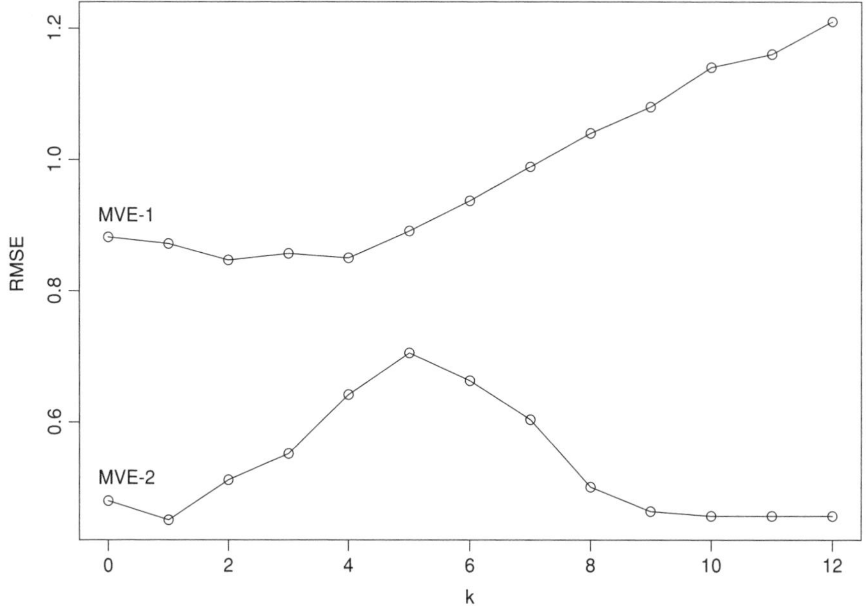

Figure 6.7 RMSE of location estimate from MVE with simple (MVE-1) and improved (MVE-2), subsampling for $p = 10$

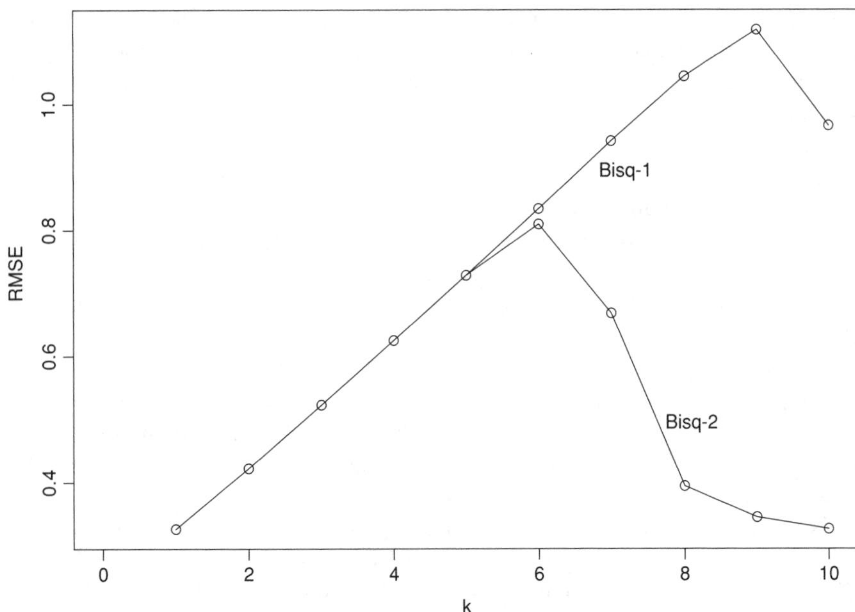

Figure 6.8 RMSE of bisquare location estimate from pure subsampling (Bisq-1) and starting from fast MVE (Bisq-2) for $p = 10$

To compare the behaviors of several estimates we performed a simulation experiment, which included

- the classical sample mean and covariance matrix
- the MVE-2 and MCD estimates
- the MVE-2 and MCD both with one-step reweighting (Section 6.4.5) with $\beta = 0.975$
- the S-estimate with bisquare function
- the Rocke-type S-estimate SR-α with $\alpha = 0.05$ and 0.10.

The sampling situations were ε-contaminated p-variate normal distributions $\mathrm{N}_p(\mathbf{0}, \mathbf{I})$, with $\varepsilon = 0.10$, $p = 5, 10, 20, 30$, and with $n = 5p$ and $10p$. We show only a part of the results. In each plot "$k = 0$" corresponds to the case $\varepsilon = 0$ so that we can compare normal distribution efficiencies. For the MVE and MCD we show only the results corresponding to the reweighted versions. None of the two versions of SR-α was systematically better than the other, but their relative performance depends on p. We show the results for $\alpha = 0.05$. Figure 6.9 shows the results for $p = 5$ and $n = 50$. The bisquare estimate is clearly more robust and more efficient than the others.

Figure 6.10 shows the results for $p = 20$ and $n = 200$. Here SR-05 is clearly better than its competitors. The values for the MCD eventually drop for large k (the maximum asymptotic bias of the MCD for $p = 20$ is attained at $k = 17$).

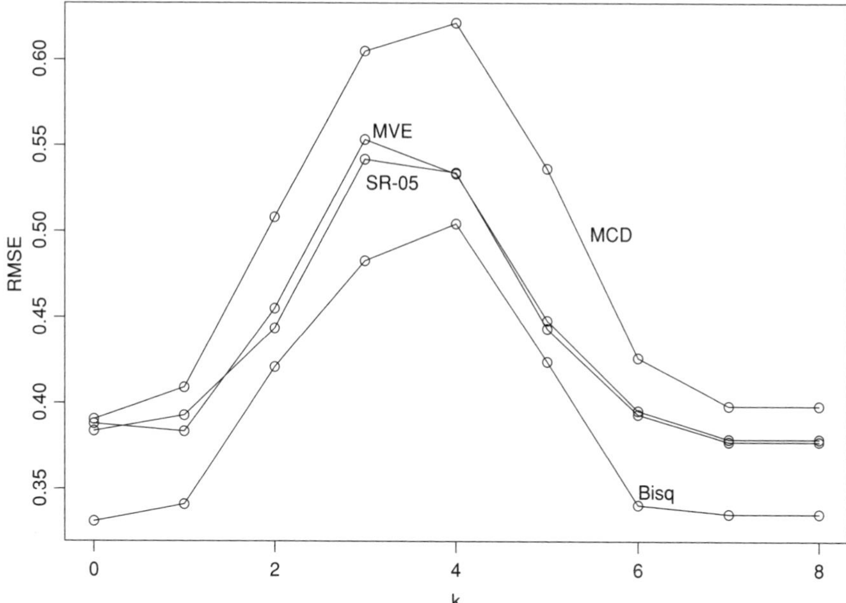

Figure 6.9 Simulation RMSEs of location estimates for $p = 5$ and $n = 50$

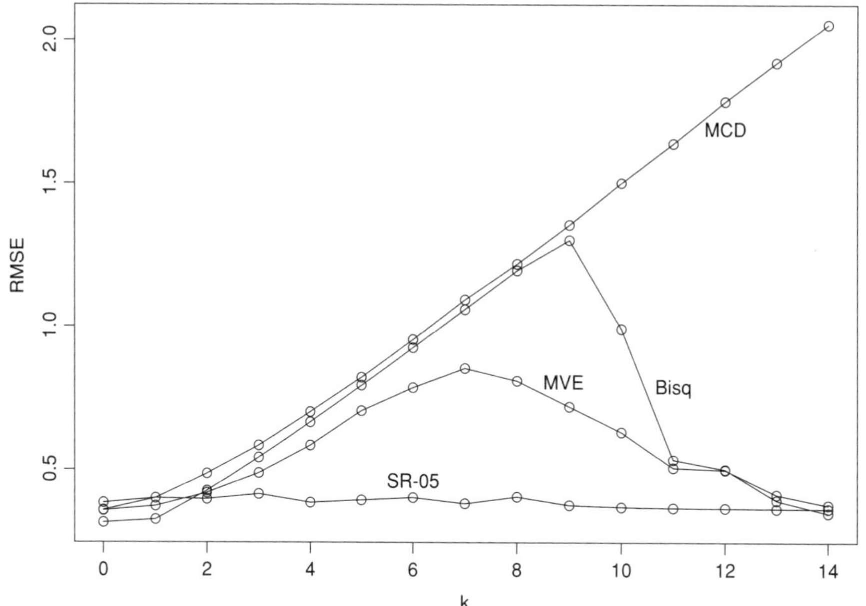

Figure 6.10 Simulation RMSEs for multivariate location estimates with $p = 20$ and $n = 200$

As a consequence of these simulations and of the former asymptotic results we propose the use of S-estimates, since they give an adequate balance between bias and variability and can be computed in feasible times. As a simple rule, we recommend the bisquare for $p < 10$ and SR-05 for $p \geq 10$.

Remark: S-PLUS includes Rocke's "translated bisquare" estimate, which has a monotonic weight function. Although it is better than the bisquare, it is not as robust for large p as the estimates described in Section 6.4.4. Besides, the MCD is used as the starting estimate in the S-PLUS implementation, and hence inherits the bad behavior of the MCD for large p. For these reasons, this version cannot be recommended for high dimensional data.

Details of the simulation

The sampling situations used were a version of an ε-contaminated $\mathrm{N}_p(\mathbf{0}, \mathbf{I})$ with point mass contamination. Let $m = [n\varepsilon]$. Then for each replication, a sample $\{\mathbf{x}_i : i = 1, \ldots, n\}$ is generated, with $\mathbf{x}_i \sim \mathrm{N}_p(\mathbf{0}, \mathbf{I})$ for $i = 1, \ldots, n$. Then for $i \leq m$, $\mathbf{x}_i$ is replaced by $\delta \mathbf{x}_i + k\mathbf{b}_1$ where $\mathbf{b}_1 = (1, 0, \ldots, 0)'$ and $\delta = 0.001$ (rather than $\delta = 0$, which may cause collinear subsamples). Note that this is not exactly a contaminated sample: in a sample from a true ε-contaminated distribution, the number of outliers would be binomial, $\mathrm{Bi}(n, \varepsilon)$, rather than the fixed number $n\varepsilon$.

The number of replications was $N_{rep} = 500$. For replication j ($j = 1, \ldots, N_{rep}$) we obtained the estimates $\left(\widehat{\boldsymbol{\mu}}_j, \widehat{\boldsymbol{\Sigma}}_j\right)$, which were evaluated by $\left\|\widehat{\boldsymbol{\mu}}_j\right\|$ and $\mathrm{cond}(\widehat{\boldsymbol{\Sigma}}_j)$. To summarize these N_{rep} numbers for each estimate, the "average errors" of $\widehat{\boldsymbol{\mu}}$ and $\widehat{\boldsymbol{\Sigma}}$ were used: namely, $\sqrt{\mathrm{ave}_j(\left\|\widehat{\boldsymbol{\mu}}_j\right\|^2)}$ (the RMSE) and $\mathrm{ave}_j\left(\log \mathrm{cond}(\widehat{\boldsymbol{\Sigma}}_j)\right)$. Using the median instead of the mean yields similar results. Logs were taken because of the large difference in orders of magnitude of condition numbers.

6.9 Faster robust dispersion matrix estimates

Estimates based on a subsampling approach will be too slow when p is large, e.g., when p is on the order of a few hundred. We now present two deterministic methods for high-dimensional data based on projections, the first based on pairwise robust covariances and the second based on the search of univariate projections with extreme values of the kurtosis.

6.9.1 Using pairwise robust covariances

Much faster estimators can be obtained if equivariance is given up. The simplest approaches for location and dispersion are respectively to apply a robust location estimate to each coordinate and a robust estimate of covariance to each pair of variables. Such pairwise robust covariance estimates are easy to compute, but unfortunately the resulting dispersion matrix lacks affine equivariance and positive definiteness. Besides, such estimates for location and dispersion may lack both bias robustness and high normal efficiency if the data are very correlated. This is because the coordinate-wise location estimates need to incorporate the correlation structure for full efficiency at the normal distribution, and because the pairwise covariance estimates may fail to downweight higher-dimensional outliers.

A simple way to define a robust covariance between two random variables x, y is by truncation or rejection. Let ψ be a bounded monotone or redescending ψ-function, and $\mu(.)$ and $\sigma(.)$ robust location and dispersion statistics. Then robust correlations and covariances can be defined as

$$\mathrm{RCov}(x, y) = \sigma(x)\sigma(y)\mathrm{E}\left[\psi\left(\frac{x-\mu(x)}{\sigma(x)}\right)\psi\left(\frac{y-\mu(y)}{\sigma(y)}\right)\right], \tag{6.60}$$

$$\mathrm{RCorr}(x, y) = \frac{\mathrm{RCov}(x, y)}{[\mathrm{RCov}(x, x)\mathrm{RCov}(y, y)]^{1/2}}. \tag{6.61}$$

See Sections 8.2–8.3 of Huber (1981). This definition satisfies $\mathrm{RCorr}(x, x) = 1$. When $\psi(x) = \mathrm{sgn}(x)$ and μ is the median, (6.61) and (6.64) are called the *quadrant* correlation and covariance estimates. The sample versions of (6.60) and (6.61) are obtained by replacing the expectation by the average and μ and σ by their estimates $\widehat{\mu}$ and $\widehat{\sigma}$.

These estimates are not consistent under a given model. In particular, if $\mathcal{D}(x, y)$ is bivariate normal with correlation ρ and ψ is monotone, then the value ρ_R of $\mathrm{RCorr}(x, y)$ is an increasing function $\rho_R = g(\rho)$ of ρ which can be computed (Problem 6.11). Then, the estimate $\widehat{\rho}_R$ of ρ_R can be corrected to ensure consistency at the normal model by using the inverse transformation $\widehat{\rho} = g^{-1}(\widehat{\rho}_R)$.

Another robust pairwise covariance initially proposed by Gnanadesikan and Kettenring (1972) and studied by Devlin et al. (1981) is based on the identity

$$\mathrm{Cov}(x, y) = \frac{1}{4}\left(\mathrm{SD}\,(x+y)^2 - \mathrm{SD}\,(x-y)^2\right). \tag{6.62}$$

They proposed to define a robust correlation by replacing the standard deviation by a robust dispersion σ (they chose a trimmed standard deviation):

$$\mathrm{RCorr}(x, y) = \frac{1}{4}\left(\sigma\left(\frac{x}{\sigma(x)} + \frac{y}{\sigma(y)}\right)^2 - \sigma\left(\frac{x}{\sigma(x)} - \frac{y}{\sigma(y)}\right)^2\right) \tag{6.63}$$

and a robust covariance defined by

$$\mathrm{RCov}(x, y) = \sigma(x)\sigma(y)\mathrm{RCorr}(x, y). \tag{6.64}$$

The latter satisfies

$$\mathrm{RCov}(t_1 x, t_2 y) = t_1 t_2 \mathrm{RCov}(x, y) \quad \text{for all } t_1, t_2 \in R \tag{6.65}$$

and

$$\mathrm{RCov}(x, x) = \sigma\,(x)^2.$$

Note that dividing x and y by their σ's in (6.63) is required for (6.65) to hold.

The above pairwise robust covariances can be used in the obvious way to define a "robust correlation (or covariance) matrix" of a random vector $\mathbf{x} = (x_1, \ldots, x_p)'$. The resulting dispersion matrix is symmetric but not necessarily positive semidefinite, and is not affine equivariant. Genton and Ma (1999) calculated the influence function and asymptotic efficiency of the estimates of such matrices. It can be shown that the above correlation matrix estimate is consistent if $\mathcal{D}(\mathbf{x})$ is an elliptical distribution, and a proof is given in Section 6.12.10.

Maronna and Zamar (2002) show that a simple modification of Gnanadesikan and Kettenring's approach yields a positive definite matrix and "approximately equivariant" estimates of location and dispersion. Recall that if $\boldsymbol{\Sigma}$ is the covariance matrix of the p-dimensional random vector $\mathbf{x}$ and σ denotes the standard deviation, then

$$\sigma(\mathbf{a}'\mathbf{x})^2 = \mathbf{a}'\boldsymbol{\Sigma}\boldsymbol{a} \tag{6.66}$$

for all $\mathbf{a} \in R^p$. The lack of positive semidefiniteness of the Gnanadesikan–Kettenring matrix is overcome by a modification that forces (6.66) for a robust σ and a set of "principal directions", and is based on the observation that the eigenvalues of the covariance matrix are the variances along the directions given by the respective eigenvectors.

Let $\mathbf{X} = [x_{ij}]$ be an $n \times p$ data matrix with rows $\mathbf{x}_i'$, $i = 1, \ldots, n$, and columns $\mathbf{x}^j$, $j = 1, \ldots, p$. Let $\widehat{\sigma}(.)$ and $\widehat{\mu}(.)$ be robust univariate location and dispersion statistics. For a data matrix $\mathbf{X}$ we shall define a robust dispersion matrix estimate $\widehat{\boldsymbol{\Sigma}}(\mathbf{X})$ and a robust location vector estimate $\widehat{\boldsymbol{\mu}}(\mathbf{X})$ by the following computational steps;

1. First compute a normalized data matrix $\mathbf{Y}$ with columns $\mathbf{y}^j = \mathbf{x}^j/\widehat{\sigma}(\mathbf{x}^j)$, and hence with rows
$$\mathbf{y}_i = \mathbf{D}^{-1}\mathbf{x}_i \ (i = 1, \ldots, n) \quad \text{where } \mathbf{D} = \operatorname{diag}(\widehat{\sigma}(\mathbf{x}^1), \ldots, \widehat{\sigma}(\mathbf{x}^p)). \tag{6.67}$$
2. Compute a robust "correlation matrix" $\mathbf{U} = [U_{jk}]$ of $\mathbf{X}$ as the "covariance matrix" of $\mathbf{Y}$ by applying (6.63) to the columns of $\mathbf{Y}$, i.e.,
$$U_{jj} = 1, \ U_{jk} = \frac{1}{4}\left[\widehat{\sigma}\left(\mathbf{y}^j + \mathbf{y}^k\right)^2 - \widehat{\sigma}\left(\mathbf{y}^j - \mathbf{y}^k\right)^2\right] \quad (j \neq k).$$
3. Compute the eigenvalues λ_j and eigenvectors $\mathbf{e}_j$ of $\mathbf{U}$ $(j = 1, \ldots, p)$, and let $\mathbf{E}$ be the matrix whose columns are the $\mathbf{e}_j$'s. It follows that $\mathbf{U} = \mathbf{E}\Lambda\mathbf{E}'$ where $\Lambda = \operatorname{diag}(\lambda_1, \ldots, \lambda_p)$. Here the λ_i's need not be nonnegative. This is the "principal component decomposition" of $\mathbf{Y}$.
4. Compute the matrix $\mathbf{Z}$ with
$$\mathbf{z}_i = \mathbf{E}'\mathbf{y}_i = \mathbf{E}'\mathbf{D}^{-1}\mathbf{x}_i \quad (i = 1, \ldots, n) \tag{6.68}$$
so that $(\mathbf{z}^1, \ldots, \mathbf{z}^p)$ are the "principal components" of $\mathbf{Y}$.
5. Compute $\widehat{\sigma}(\mathbf{z}^j)$ and $\widehat{\mu}(\mathbf{z}^j)$ for $j = 1, \ldots, p$, and set
$$\Gamma = \operatorname{diag}\left(\widehat{\sigma}(\mathbf{z}^1)^2, \ldots, \widehat{\sigma}(\mathbf{z}^p)^2\right), \quad \boldsymbol{\nu} = (\widehat{\mu}(\mathbf{z}^1), \ldots, \widehat{\mu}(\mathbf{z}^p))'.$$

Here the elements of Γ are nonnegative. Being "principal components" of $\mathbf{Y}$, the $\mathbf{z}^j$'s should be approximately uncorrelated with covariance matrix $\mathbf{\Gamma}$.

6. Now transform back to $\mathbf{X}$ with

$$\mathbf{x}_i = \mathbf{A}\mathbf{z}_i, \quad \text{with } \mathbf{A} = \mathbf{DE}, \tag{6.69}$$

and finally define

$$\widehat{\boldsymbol{\Sigma}}(\mathbf{X}) = \mathbf{A}\Gamma\mathbf{A}', \ \widehat{\boldsymbol{\mu}}(\mathbf{X}) = \mathbf{A}\boldsymbol{\nu}. \tag{6.70}$$

The justification for the last equation is that, if $\boldsymbol{\nu}$ and Γ were the mean and covariance matrix of $\mathbf{Z}$, since $\mathbf{x}_i = \mathbf{A}\mathbf{z}_i$ the mean and covariance matrix of $\mathbf{X}$ would be given by (6.70).

Note that (6.67) makes the estimate scale equivariant, and that (6.70) replaces the λ_i's, which may be negative, by the "robust variances" $\sigma(\mathbf{z}^j)^2$ of the corresponding directions. The reason for defining $\widehat{\boldsymbol{\mu}}$ as in (6.70) is that it is better to apply a coordinate-wise location estimate to the approximately uncorrelated $\mathbf{z}^j$ and then transform back to the $\mathbf{X}$-coordinates, than to apply a coordinate-wise location estimate directly to the $\mathbf{x}^j$'s.

The procedure can be iterated in the following way. Put $\widehat{\boldsymbol{\mu}}_{(0)} = \widehat{\boldsymbol{\mu}}(\mathbf{X})$ and $\widehat{\boldsymbol{\Sigma}}_{(0)} = \widehat{\boldsymbol{\Sigma}}(\mathbf{X})$. At iteration k we have $\widehat{\boldsymbol{\mu}}_{(k)}$ and $\widehat{\boldsymbol{\Sigma}}_{(k)}$, whose computation has required computing a matrix $\mathbf{A}$ as in (6.69). Call $\mathbf{Z}_{(k)}$ the matrix with rows $\mathbf{z}_i = \mathbf{A}^{-1}\mathbf{x}_i$. Then $\widehat{\boldsymbol{\mu}}_{(k+1)}$ and $\widehat{\boldsymbol{\Sigma}}_{(k+1)}$ are obtained by computing $\widehat{\boldsymbol{\Sigma}}$ and $\widehat{\boldsymbol{\mu}}$ for $\mathbf{Z}_{(k)}$ and then expressing them back in the original coordinate system. More precisely, we define

$$\widehat{\boldsymbol{\Sigma}}_{(k+1)}(\mathbf{X}) = \mathbf{A}\widehat{\boldsymbol{\Sigma}}(\mathbf{Z}_{(k)})\mathbf{A}', \ \widehat{\boldsymbol{\mu}}_{(k+1)}(\mathbf{X}) = \mathbf{A}\widehat{\boldsymbol{\mu}}(\mathbf{Z}_{(k)}). \tag{6.71}$$

The reason for iterating is that the first step works very well when the data have low correlations; and the $\mathbf{z}^j$'s are (hopefully) less correlated than the original variables. The resulting estimate will be called the "orthogonalized Gnanadesikan–Kettenring estimate" (OGK).

A final step is convenient both to increase the estimate's efficiency and to make it "more equivariant". The simplest and fastest option is the reweighting procedure in Section 6.4.5. But it is much better to use this estimate as the starting point for the iterations of an S-estimate.

Since a large part of the computing effort is consumed by the univariate estimates $\widehat{\mu}$ and $\widehat{\sigma}$, they must be fast. The experiments by Maronna and Zamar (2002) showed that it is desirable that $\widehat{\mu}$ and $\widehat{\sigma}$ be both bias robust and efficient at the normal distribution in order for $\widehat{\boldsymbol{\Sigma}}$ and $\widehat{\boldsymbol{\mu}}$ to perform satisfactorily. To this end, the dispersion estimate $\widehat{\sigma}$ is defined in a way similar to the "τ-scale" estimate (5.59), which is a truncated standard deviation, and the location estimate $\widehat{\mu}$ is a weighted mean. More precisely, let $\widehat{\mu}_0$ and $\widehat{\sigma}_0$ be the median and MAD. Let W be a weight function and ρ a ρ-function. Let $w_i = W\left(\left(x_i - \widehat{\mu}_0\right)/\widehat{\sigma}_0\right)$ and

$$\widehat{\mu} = \frac{\sum_{i=1}^n x_i w_i}{\sum_{i=1}^n w_i}, \quad \widehat{\sigma}^2 = \frac{\sigma_0^2}{n}\sum_{i=1}^n \rho\left(\frac{x_i - \widehat{\mu}}{\widehat{\sigma}_0}\right).$$

An adequate balance of robustness and efficiency is obtained with W, the bisquare weight function (2.62), with $k = 4.5$, and ρ as the bisquare ρ (2.37) with $k = 3$.

It is shown by Maronna and Zamar (2002) that if the BPs, of $\widehat{\mu}$ and $\widehat{\sigma}$ are not less than ε then so is the BP of $(\widehat{\mu}, \widehat{\Sigma})$, as long as the data are not collinear. Simulations in their paper show that two is an adequate number of iterations (6.71), and that further iterations do not seem to converge and yield no improvement.

An implementation of the OGK estimator for applications to data mining was discussed by Alqallaf, Konis, Martin and Zamar (2002), using the quadrant correlation estimator. A reason for focusing on the quadrant correlation was the desire to operate on huge data sets that are too large to fit in computer memory, and a fast bucketing algorithm can be used to compute this estimate on "streaming" input data (data read into the computer sequentially from a database). The median and MAD estimates were used for robust location and dispersion because there are algorithms for the approximate computation of order statistics from a single pass on large streaming data sets (Manku, Rajagopalan and Lindsay, 1999).

6.9.2 Using kurtosis

The *kurtosis* of a random variable x is defined as

$$\mathrm{Kurt}\,(x) = \frac{\mathrm{E}\,(x - \mathrm{E}x)^4}{\mathrm{SD}\,(x)^4}.$$

Peña and Prieto (2001) propose an equivariant procedure based on the following observation. A distribution is called *unimodal* if its density has a maximum at some point x_0, and is increasing for $x < x_0$ and decreasing for $x > x_0$. Then it can be shown that the kurtosis is a measure of both heavy-tailedness and unimodality. It follows that, roughly speaking, for univariate data a small proportion of outliers increases the kurtosis, since it makes the data tails heavier, and a large proportion decreases the kurtosis, since it makes the data more bimodal.

Hence Peña and Prieto look for projections which either maximize or minimize the kurtosis, and use them in a way similar to the Stahel–Donoho estimate. The procedure is complex, but it may be summarized as follows for the case of p-dimensional data:

1. Two sets of p directions **a** are found, one corresponding to local maxima and the other to local minima of the kurtosis.
2. The outlyingness of each data point is measured through (6.43), with the vector **a** ranging only over the $2p$ directions found in the previous step.
3. Points with outlyingness above a given threshold are transitorily deleted, and steps 1–2 are iterated on the remaining points until no more deletions take place.
4. The sample mean and covariance matrix of the remaining points are computed.
5. Deleted points whose Mahalanobis distances are below a threshold are again included.
6. Steps 4–5 are repeated until no more inclusions take place.

Step 1 is performed through an efficient iterative algorithm. The procedure is very fast for high dimensions, and the simulations by Peña and Prieto (2001) suggest that it has a promising performance.

6.10 Robust principal components

Principal components analysis (PCA) is a widely used method for dimensionality reduction. Let $\mathbf{x}$ be a p-dimensional random vector with mean $\boldsymbol{\mu}$ and covariance matrix $\boldsymbol{\Sigma}$. The first principal component is the univariate projection of maximum variance; more precisely, it is the linear combination $\mathbf{x}'\mathbf{b}_1$ where $\mathbf{b}_1$ (called the first *principal direction*) is the vector $\mathbf{b}$ such that

$$\mathrm{Var}(\mathbf{b}'\mathbf{x}) = \max \quad \text{subject to} \quad \|\mathbf{b}\| = 1. \tag{6.72}$$

The second principal component is $\mathbf{x}'\mathbf{b}_2$ where $\mathbf{b}_2$ (the second principal direction) satisfies (6.72) with $\mathbf{b}_2'\mathbf{b}_1 = 0$, and so on. Call $\lambda_1 \geq \lambda_2 \geq \ldots \geq \lambda_p$ the eigenvalues of $\boldsymbol{\Sigma}$. Then $\mathbf{b}_1, \ldots, \mathbf{b}_p$ are the respective eigenvectors and $\mathrm{Var}(\mathbf{b}_j'\mathbf{x}) = \lambda_j$. The number q of components can be chosen on the basis of the "proportion of unexplained variance"

$$\frac{\sum_{j=q+1}^{p} \lambda_j}{\sum_{j=1}^{p} \lambda_j}. \tag{6.73}$$

PCA can be viewed in an alternative geometric form in the spirit of regression modeling. Consider finding a q-dimensional hyperplane H such the orthogonal distance of $\mathbf{x}$ to H is "smallest", in the following sense. Call $\widehat{\mathbf{x}}_H$ the point of H closest in Euclidean distance to $\mathbf{x}$, i.e., such that

$$\widehat{\mathbf{x}}_H = \arg\min_{\mathbf{z}\in H} ||\mathbf{x} - \mathbf{z}||.$$

Then we look for H^* such that

$$\mathrm{E}\left\|\mathbf{x} - \widehat{\mathbf{x}}_{H^*}\right\|^2 = \min. \tag{6.74}$$

It can be shown (Seber, 1984) that H^* contains the mean $\boldsymbol{\mu}$ and has the directions of the first q eigenvectors $\mathbf{b}_1, \ldots, \mathbf{b}_q$, and so H^* is the set of translated linear combinations of $\mathbf{b}_1, \ldots, \mathbf{b}_q$:

$$H^* = \left\{ \boldsymbol{\mu} + \sum_{k=1}^{q} \alpha_k \mathbf{b}_k : \alpha_1, \ldots, \alpha_q \in R \right\}. \tag{6.75}$$

Then

$$z_j = (\mathbf{x} - \boldsymbol{\mu})'\mathbf{b}_j \quad (j = 1, \ldots, q) \tag{6.76}$$

are the coordinates of the centered $\mathbf{x}$ in the coordinate system of the $\mathbf{b}_j$, and

$$d_H = \sum_{j=p+1}^{p} z_j^2 = \|\mathbf{x}-\widehat{\mathbf{x}}_H\|^2, \quad d_C = \sum_{j=1}^{q} z_j^2 \tag{6.77}$$

are the squared distances from $\mathbf{x}$ to H and from $\widehat{\mathbf{x}}_H$ to $\boldsymbol{\mu}$, respectively.

Note that the results of PCA are not invariant under general affine transformations, in particular under changes in the units of the variables. Doing PCA implies that we consider the Euclidean norm to be a sensible measure of distance, and this may require a previous rescaling of the variables. PCA is, however, invariant under *orthogonal* transformations, i.e., transformations that do not change Euclidean distances.

Given a data set $X = \{\mathbf{x}_1, \ldots, \mathbf{x}_n\}$, the sample principal components are computed by replacing $\boldsymbol{\mu}$ and $\boldsymbol{\Sigma}$ by the sample mean and covariance matrix. For each observation $\mathbf{x}_i$ we compute the scores $z_{ij} = (\mathbf{x}_i - \overline{\mathbf{x}})'\mathbf{b}_j$ and the distances

$$d_{\widehat{H},i} = \sum_{j=q+1}^{p} z_{ij}^2 = \left\|\mathbf{x}_i - \widehat{\mathbf{x}}_{\widehat{H},i}\right\|^2, \quad d_{C.i} = \sum_{j=1}^{q} z_{ij}^2, \tag{6.78}$$

where $\widehat{H}$ is the estimated hyperplane. A simple data analytic tool similar to the plot of residuals versus fitted values in regression is to plot $d_{\widehat{H},i}$ vs. $d_{C,i}$.

As can be expected, outliers may have a distorting effect on the results. For instance, in Example 6.1, the first principal component of the correlation matrix of the data explains 75% of the variability, while after deleting the atypical point it explains 90%. The simplest way to deal with this problem is to replace $\overline{\mathbf{x}}$ and $\mathbf{Var}(X)$ with robust estimates $\widehat{\boldsymbol{\mu}}$ and $\widehat{\boldsymbol{\Sigma}}$ of multivariate location and dispersion. Campbell (1980) uses M-estimates. Croux and Haesbroeck (2000) discuss several properties of this approach. Note that the results depend only on the shape of $\widehat{\boldsymbol{\Sigma}}$ (Section 6.3.2).

However, better results can be obtained by taking advantage of the particular features of PCA. For affine equivariant estimation the "natural" metric is that given by squared Mahalanobis distances $d_i = (\mathbf{x}_i - \widehat{\boldsymbol{\mu}})'\widehat{\boldsymbol{\Sigma}}^{-1}(\mathbf{x}_i - \widehat{\boldsymbol{\mu}})$, which depends on the data through $\widehat{\boldsymbol{\Sigma}}^{-1}$, while for PCA we have a fixed metric given by Euclidean distances. This implies that the concept of outliers changes. In the first case an outlier which should be downweighted is a point with a large squared Mahalanobis distance to the center of the data, while in the second it is a point with a large Euclidian distance to the hyperplane ("large" as compared to the majority of points). For instance, consider two independent variables with zero means and standard deviations 10 and 1, and $q = 1$. The first principal component corresponds to the first coordinate axis. Two data values, one at (100,1) and one at (10,10), have identical large squared Mahalanobis distances of 101, but their Euclidean distances to the first axis are 1 and 10 respectively. The second one would be harmful to the estimation of the principal components, but the first one is a "good" point for that purpose.

Boente (1983, 1987) studied M-estimates for PCA. An alternative approach to robust PCA is to replace the variance in (6.72) by a robust scale. This approach was first proposed by Li and Chen (1985), who found serious computational problems.

Croux and Ruiz-Gazen (1996) proposed an approximation based on a finite number of directions. The next sections describe two new preferred approaches to robust PCA. One is based on robust fitting of a hyperplane H by minimization of a robust scale, while the other is a simple and fast "spherical" principal components method that works well for large data sets.

6.10.1 Robust PCA based on a robust scale

The proposed approach is based on replacing the expectation in (6.74) by a robust M-scale (Maronna, 2005). For given q-dimensional hyperplane H call $\widehat{\sigma}(H)$ an M-scale estimate of the $d_{H,i}$ in (6.78); that is, $\widehat{\sigma}(H)$ satisfies the M-scale equation

$$\frac{1}{n}\sum_{i=1}^{n}\rho\left(\frac{\left\|\mathbf{x}_i-\widehat{\mathbf{x}}_{H,i}\right\|^2}{\widehat{\sigma}(H)}\right)=\delta. \tag{6.79}$$

Then we search for $\widehat{H}$ having the form of the right-hand-side of (6.75) such that $\widehat{\sigma}(\widehat{H})$ is minimum. For a given H let

$$\widehat{\boldsymbol{\mu}}=\frac{1}{\sum_{i=1}^{n}w_i}\sum_{i=1}^{n}w_i\mathbf{x}_i,\quad \widehat{\mathbf{V}}=\sum_{i=1}^{n}w_i\left(\mathbf{x}_i-\widehat{\boldsymbol{\mu}}\right)\left(\mathbf{x}_i-\widehat{\boldsymbol{\mu}}\right)', \tag{6.80}$$

with

$$w_i=W\left(\frac{\left\|\mathbf{x}_i-\widehat{\mathbf{x}}_{H,i}\right\|^2}{\widehat{\sigma}}\right) \tag{6.81}$$

where $W=\rho'$. It can be shown by differentiating (6.79) with respect to $\boldsymbol{\mu}$ and $\mathbf{b}_j$ that the optimal $\widehat{H}$ has the form (6.75) where $\boldsymbol{\mu}=\widehat{\boldsymbol{\mu}}$ and $\mathbf{b}_1,\ldots,\mathbf{b}_q$ are the eigenvectors of $\widehat{\mathbf{V}}$ corresponding to its q largest eigenvalues. That is, the hyperplane is defined by a weighted mean of the data and the principal directions of a weighted covariance matrix, where points distant from $\widehat{H}$ receive small weights.

This result suggests an iterative procedure, in the spirit of the iterative reweighting approach of Sections 6.7.1 and 6.7.2. Starting with some initial $\widehat{H}_0$, compute the weights with (6.81), then compute $\widehat{\boldsymbol{\mu}}$ and $\widehat{\mathbf{V}}$ with (6.80) and the corresponding principal components, which yield a new $\widehat{H}$. It follows from the results by Boente (1983) that if W is nondecreasing, then σ decreases at each step of this procedure and the method converges to a local minimum.

There remains the problem of starting values. Simulations in Maronna (2005) show that rather than subsampling, it is better to directly choose the initial directions $\mathbf{b}_j$ at random on the unit sphere. The whole procedure is based on the strategy in Section 5.7.3. For each random start, a small number of iterations are performed, the best candidates are kept and are used as starting points for full iteration, and the solution with minimum $\widehat{\sigma}$ is chosen. The resulting procedure is fast for large p, and simulations show it to be very competitive with other proposals, in both robustness and efficiency. See Maronna (2005) for further details.

A similar estimate can be based on an L-scale. For $h < n$ compute the L-scale estimate

$$\widehat{\sigma}(H) = \sum_{i=1}^{h} d_{H,(i)},$$

where the $d_{H,(i)}$'s are the ordered values of $\left\|\mathbf{x}_i - \widehat{\mathbf{x}}_{H,i}\right\|^2$. Then the hyperplane $\widehat{H}$ minimizing $\widehat{\sigma}(H)$ corresponds to the principal components of (6.80) with $w_i = \mathrm{I}\left(d_{\widehat{H},i} \leq d_{\widehat{H},(h)}\right)$. This amounts to "trimmed" principal components, in which the $\mathbf{x}_i$'s with the h smallest values of $d_{\widehat{H},i}$ are trimmed. The analogous iterative procedure converges to a local minimum. The results obtained in the simulations in Maronna (2005) for the L-scale are not as good as those corresponding to the M-scale.

6.10.2 Spherical principal components

In this section we describe a simple but effective approach proposed by Locantore, Marron, Simpson, Tripoli, Zhang and Cohen (1999).

Let $\mathbf{x}$ have an elliptical distribution (6.7), in which case if $\mathbf{Var}(\mathbf{x})$ exists it is a constant multiple of $\boldsymbol{\Sigma}$. Let $\mathbf{y} = (\mathbf{x} - \boldsymbol{\mu})/\left\|\mathbf{x} - \boldsymbol{\mu}\right\|$, i.e., $\mathbf{y}$ is the normalization of $\mathbf{x}$ to the surface of the unit sphere centered at $\boldsymbol{\mu}$. Boente and Fraiman (1999) showed that the eigenvectors $\mathbf{t}_1, \ldots, \mathbf{t}_p$ (but not the eigenvalues!) of the covariance matrix of $\mathbf{y}$ (i.e., its principal axes) coincide with those of $\boldsymbol{\Sigma}$. They showed furthermore that if σ (.) is any dispersion statistic then the values $\sigma\left(\mathbf{x}'\mathbf{t}_j\right)^2$ are proportional to the eigenvalues of $\boldsymbol{\Sigma}$. Proofs are given in Section 6.12.11.

This result is the basis for a simple robust approach to PCA, called *spherical principal components* (SPC). Let $\widehat{\boldsymbol{\mu}}$ be a robust multivariate location estimate, and compute

$$\mathbf{y}_i = \begin{cases} (\mathbf{x}_i - \widehat{\boldsymbol{\mu}}) / \left|\left|\mathbf{x}_i - \widehat{\boldsymbol{\mu}}\right|\right| & \text{if} \quad \mathbf{x}_i \neq \widehat{\boldsymbol{\mu}} \\ \mathbf{0} & \text{otherwise.} \end{cases}$$

Let $\widehat{\mathbf{V}}$ be the sample covariance matrix of the $\mathbf{y}_i$'s with corresponding eigenvectors $\mathbf{b}_j$ ($j = 1, \ldots, p$). Now compute $\widehat{\lambda}_j = \widehat{\sigma}(\mathbf{x}'\mathbf{b}_j)^2$ where $\widehat{\sigma}$ is a robust dispersion estimate (such as the MAD). Call $\widehat{\lambda}_{(j)}$ the sorted λ's, $\widehat{\lambda}_{(1)} \geq \ldots \geq \widehat{\lambda}_{(p)}$, and $\mathbf{b}_{(j)}$ the corresponding eigenvectors. Then the first q principal directions are given by the $\mathbf{b}_{(j)}$'s, $j = 1, \ldots, q$, and the respective "proportion of unexplained variance" is given by (6.73), where λ_j is replaced by $\lambda_{(j)}$.

In order for the resulting robust PCA to be invariant under orthogonal transformations of the data, it is not necessary that $\widehat{\boldsymbol{\mu}}$ be affine equivariant, but only orthogonal equivariant, i.e., such that $\widehat{\boldsymbol{\mu}}\,(\mathbf{T}X) = \mathbf{T}\widehat{\boldsymbol{\mu}}\,(X)$ for all *orthogonal* $\mathbf{T}$. The simplest choice for $\widehat{\boldsymbol{\mu}}$ is the "space median":

$$\widehat{\boldsymbol{\mu}} = \arg\min_{\boldsymbol{\mu}} \sum_{i=1}^{n} \left|\left|\mathbf{x}_i - \boldsymbol{\mu}\right|\right|.$$

Table 6.3 Bus data: proportion of unexplained variability for q components

q	Classical	Robust
1	0.188	0.549
2	0.083	0.271
3	0.044	0.182
4	0.026	0.135
5	0.018	0.100
6	0.012	0.069

Note that this is an M-estimate since it corresponds to (6.9) with $\rho(t) = \sqrt{t}$ and $\Sigma = \mathrm{I}$. Thus the estimate can be easily computed iteratively through the first equation in (6.52) with $W_1(t) = 1/\sqrt{t}$, starting with the coordinate-wise medians. It follows from Section 6.12.4 that this estimate has BP = 0.5.

This procedure is deterministic and very fast, and it can be computed with collinear data without any special adjustments. Despite its simplicity, simulations by Maronna (2005) show that this SPC method performs very well.

Example 6.3 *The data set* **bus** *(Hettich and Bay, 1999) corresponds to a study in automatic vehicle recognition (Siebert, 1987). Each of the 218 rows corresponds to a view of a bus silhouette, and contains 18 attributes of the image. The SDs are in general much larger than the respective MADNs. The latter vary between 0 (for variable 9) and 34. Hence it was decided to exclude variable 9 and divide the remaining variables by their MADNs.*

Table 6.3 (code **bus***) shows the proportions of unexplained variability (6.73) as a function of the number q of components, for the classical PCA and for SPC.*

It would seem that since the classical method has smaller unexplained variability than the robust method, classical PCA gives a better representation. However, this is not the case. Table 6.4 gives the quantiles of the distances $d_{H,i}$ in (6.78) for $q = 3$, and Figure 6.11 compares the logs of the respective ordered values (the log scale was used because of the extremely large outliers).

It is seen in the figure that the hyperplane from the robust fit has in general smaller distances to the data points, except for some clearly outlying ones. On the other hand, in Table 6.3 the classical estimate seems to perform better than the robust one. The

Table 6.4 Bus data: quantiles of distances to hyperplane

	0.1	0.2	0.3	0.4	0.5	0.6	0.7	0.8	0.9	Max
Classical	1.9	2.3	2.8	3.2	3.7	4.4	5.5	6.4	8.1	23
Robust	1.2	1.6	1.8	2.2	2.5	3.1	3.8	5.2	9.0	1039

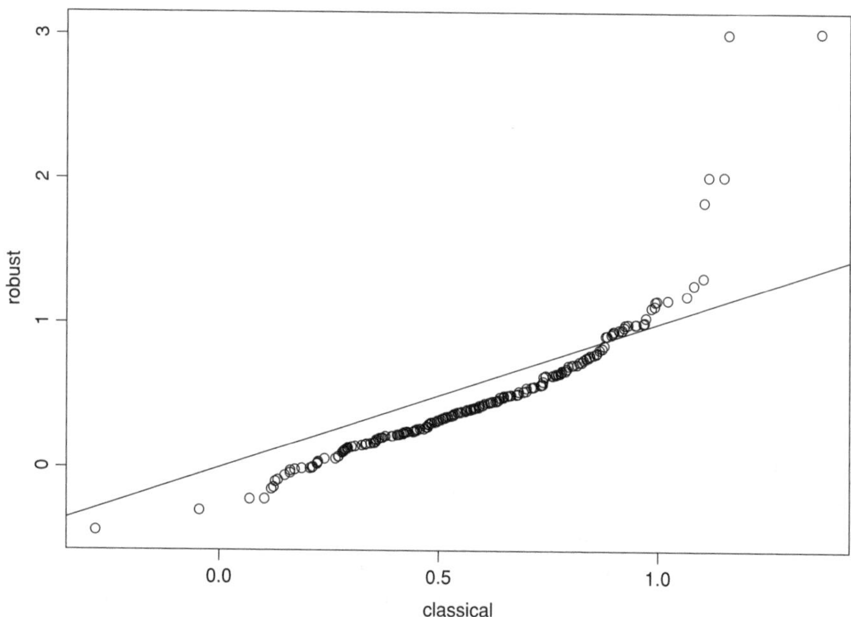

Figure 6.11 Bus image data: Q–Q plot of logs of distances to hyperplane ($q = 3$) from classical and robust estimates. The line is the identity diagonal

reason is that the two estimates use different *measures of variability. The classical procedure uses variances which are influenced by the outliers, and so large outliers in the direction of the first principal axes will inflate the corresponding variances and hence increase their proportion of explained variability. On the other hand, the robust SPC uses a robust dispersion measure which is free of this drawback, and gives a more accurate measure of the unexplained variability for the bulk of the data.*

6.11 *Other estimates of location and dispersion

6.11.1 Projection estimates

Note that, if $\widehat{\mathbf{\Sigma}}$ is the sample covariance matrix of $\mathbf{x}$ and $\widehat{\sigma}(.)$ denotes the sample SD, then

$$\widehat{\sigma}(\mathbf{a}'\mathbf{x})^2 = \mathbf{a}'\widehat{\mathbf{\Sigma}}\mathbf{a} \quad \forall\, \mathbf{a} \in R^p. \tag{6.82}$$

It would be desirable to have a robust $\widehat{\mathbf{\Sigma}}$ fulfilling (6.82) when $\widehat{\sigma}$ is a robust dispersion like the MAD. It can be shown that the SD is the only dispersion measure satisfying (6.82), and hence this goal is unattainable. To overcome this difficulty, dispersion P-estimates (Maronna, Stahel and Yohai, 1992) were proposed as "best" approximations to (6.82), in analogy to the approach in Section 5.14.2. Specifically, a dispersion P-estimate is a matrix $\widehat{\boldsymbol{\Sigma}}$ that satisfies

$$\sup_{\mathbf{a}\neq\mathbf{0}} \left| \log \left(\frac{\widehat{\sigma}(\mathbf{a}'\mathbf{x})^2}{\mathbf{a}'\widehat{\boldsymbol{\Sigma}}\mathbf{a}} \right) \right| = \min . \tag{6.83}$$

A similar idea for location was proposed by Tyler (1994). If $\widehat{\mu}(.)$ denotes the sample mean, then the sample mean of $\mathbf{x}$ may be characterized as a vector $\boldsymbol{\nu}$ satisfying

$$\widehat{\mu}(\mathbf{a}'(\mathbf{x}-\boldsymbol{\nu})) = \mathbf{0} \quad \forall\, \mathbf{a} \in R^p . \tag{6.84}$$

Let $\widehat{\mu}$ now be a robust univariate location statistic. It would be desirable to find $\boldsymbol{\nu}$ satisfying (6.84); this unfeasible goal is replaced by defining a location P-estimate as a vector $\boldsymbol{\nu}$ such that

$$\max_{\|\mathbf{a}\|=1} \frac{\left|\widehat{\mu}(\mathbf{a}'(\mathbf{x}-\boldsymbol{\nu}))\right|}{\widehat{\sigma}(\mathbf{a}'\mathbf{x})} = \min, \tag{6.85}$$

where $\widehat{\sigma}$ is a robust dispersion (the condition $\|\mathbf{a}\| = 1$ is equivalent to $\mathbf{a} \neq \mathbf{0}$). Note that $\boldsymbol{\nu}$ is the point minimizing the outlyingness measure (6.43). The estimate with $\widehat{\mu}$ and $\widehat{\sigma}$ equal to the median and MAD respectively is called "MP-estimate" by Adrover and Yohai (2002).

It is easy to verify that both location and dispersion P-estimates are equivariant. It can be shown that their maximum asymptotic biases at the normal model do not depend on p. The maximum bias corresponding to the location MP-estimate is 0.32 (Adrover and Yohai, 2002), which is clearly smaller than the values in Table 6.2.

6.11.2 Constrained M-estimates

Kent and Tyler (1996) define robust efficient estimates, called *constrained M-estimates* (CM-estimates, for short), as in Section 5.14.3

$$(\widehat{\boldsymbol{\mu}}, \widehat{\boldsymbol{\Sigma}}) = \arg\min_{\mu,\Sigma} \left\{ \frac{1}{n} \sum_{i=1}^{n} \rho(d_i) + \frac{1}{2} \log |\boldsymbol{\Sigma}| \right\},$$

with the constraint

$$\frac{1}{n} \sum_{i=1}^{n} \rho(d_i) \leq \varepsilon,$$

where $d_i = (\mathbf{x}_i - \mu)' \boldsymbol{\Sigma}^{-1} (\mathbf{x}_i - \mu)$, ρ is a bounded ρ-function and $\boldsymbol{\Sigma}$ ranges over the symmetric positive definite $p \times p$ matrices.

They show the FBP for data in general position to be

$$\varepsilon^* = \frac{1}{n} \min\left([n\varepsilon], [n(1-\varepsilon) - p]\right),$$

and hence the bound (6.27) is attained when $[n\varepsilon] = (n-p)/2$.

These estimates satisfy M-estimating equations (6.11)–(6.12). By an adequate choice of ρ, they can be tuned to attain a desired efficiency.

6.11.3 Multivariate MM- and τ-estimates

Lopuhaã (1991, 1992) defined multivariate estimates that attain both a given asymptotic efficiency and a given BP, extending to the multivariate case the approaches of τ- and MM-estimates described for regression in Sections 5.5 and 5.14.1, respectively. Details are given in the respective articles.

6.11.4 Multivariate depth

Another approach for location is based on extending the notion of order statistics to multivariate data, and then defining $\boldsymbol{\mu}$ as a "multivariate median" or, more generally, a multivariate L-estimator. Among the large amount of literature on the subject, we cite the work of Tukey (1975b), Liu (1990), Zuo and Serfling (2000) and Bai and He (1999). The maximum BP of this type of estimate is 1/3, which is much lower than the maximum BP for equivariant estimates given by (6.27); see Donoho and Gasko (1992) and Chen and Tyler (2002).

6.12 Appendix: proofs and complements

6.12.1 Why affine equivariance?

Let $\mathbf{x}$ have an elliptical density $f(\mathbf{x}, \boldsymbol{\mu}, \boldsymbol{\Sigma})$ of the form (6.7). Here $\boldsymbol{\mu}$ and $\boldsymbol{\Sigma}$ are the distribution parameters. Then if $\mathbf{A}$ is nonsingular, the usual formula for the density of transformed variables yields that $\mathbf{y} = \mathbf{A}\mathbf{x} + \mathbf{b}$ has density

$$f(\mathbf{A}^{-1}(\mathbf{y} - \mathbf{b}), \boldsymbol{\mu}, \boldsymbol{\Sigma}) = f(\mathbf{y}, \mathbf{A}\boldsymbol{\mu} + \mathbf{b}, \mathbf{A}\boldsymbol{\Sigma}\mathbf{A}'), \tag{6.86}$$

and hence the location and dispersion parameters of $\mathbf{y}$ are $\mathbf{A}\boldsymbol{\mu} + \mathbf{b}$ and $\mathbf{A}\boldsymbol{\Sigma}\mathbf{A}'$ respectively.

Denote by $\left(\widehat{\boldsymbol{\mu}}(X), \widehat{\boldsymbol{\Sigma}}(X)\right)$ the values of the estimates corresponding to a sample $X = \{\mathbf{x}_1, \ldots, \mathbf{x}_n\}$. Then it is desirable that the estimates $\left(\widehat{\boldsymbol{\mu}}(Y), \widehat{\boldsymbol{\Sigma}}(Y)\right)$ corresponding to $Y = \{\mathbf{y}_1, \ldots, \mathbf{y}_n\}$ with $\mathbf{y}_i = \mathbf{A}\mathbf{x}_i + \mathbf{b}$ transform in the same manner as the parameters do in (6.86): that is:

$$\widehat{\boldsymbol{\mu}}(Y) = \mathbf{A}\widehat{\boldsymbol{\mu}}(X) + \mathbf{b}, \quad \widehat{\boldsymbol{\Sigma}}(Y) = \mathbf{A}\widehat{\boldsymbol{\Sigma}}(X)\mathbf{A}', \tag{6.87}$$

which corresponds to (6.3).

Affine equivariance is natural in those situations where it is desirable that the result remains essentially unchanged under *any* nonsingular linear transformations, like linear discriminant analysis, canonical correlations and factor analysis. This does not happen in PCA, since it is based on a fixed metric which is invariant only under orthogonal transformations.

6.12.2 Consistency of equivariant estimates

We shall show that affine equivariant estimates are consistent for elliptical distributions, in the sense that if $\mathbf{x} \sim f(\mathbf{x}, \boldsymbol{\mu}, \boldsymbol{\Sigma})$ then the asymptotic values $\widehat{\boldsymbol{\mu}}_\infty$ and $\widehat{\boldsymbol{\Sigma}}_\infty$ satisfy

$$\widehat{\boldsymbol{\mu}}_\infty = \boldsymbol{\mu}, \ \widehat{\boldsymbol{\Sigma}}_\infty = c\boldsymbol{\Sigma}, \tag{6.88}$$

where c is a constant.

Denote again for simplicity $\left(\widehat{\boldsymbol{\mu}}_\infty(\mathbf{x}), \widehat{\boldsymbol{\Sigma}}_\infty(\mathbf{x})\right)$ as the asymptotic values of the estimates corresponding to the distribution of $\mathbf{x}$. Note that the asymptotic values share the affine equivariance of the estimates, i.e., (6.87) holds also for $\widehat{\boldsymbol{\mu}}$ and $\widehat{\boldsymbol{\Sigma}}$ replaced with $\widehat{\boldsymbol{\mu}}_\infty$ and $\widehat{\boldsymbol{\Sigma}}_\infty$.

We first prove (6.88) for the case $\boldsymbol{\mu} = \mathbf{0}$, $\boldsymbol{\Sigma} = \mathrm{I}$. Then the distribution is spherical, and so $\mathcal{D}(\mathbf{Tx}) = \mathcal{D}(\mathbf{x})$ for any orthogonal matrix $\mathbf{T}$. In particular for $\mathbf{T} = -\mathbf{I}$ we have

$$\widehat{\boldsymbol{\mu}}_\infty(\mathbf{Tx}) = \widehat{\boldsymbol{\mu}}_\infty(-\mathbf{x}) = -\widehat{\boldsymbol{\mu}}_\infty(\mathbf{x}) = \widehat{\boldsymbol{\mu}}_\infty(\mathbf{x}),$$

which implies $\widehat{\boldsymbol{\mu}}_\infty = \mathbf{0}$. At the same time we have

$$\widehat{\boldsymbol{\Sigma}}_\infty(\mathbf{Tx}) = \mathbf{T}\widehat{\boldsymbol{\Sigma}}_\infty(\mathbf{x})\mathbf{T}' = \widehat{\boldsymbol{\Sigma}}_\infty(\mathbf{x}) \tag{6.89}$$

for all orthogonal $\mathbf{T}$. Write $\widehat{\boldsymbol{\Sigma}}_\infty = \mathbf{U}\Lambda\mathbf{U}'$, where $\mathbf{U}$ is orthogonal and $\Lambda = \operatorname{diag}\left(\lambda_1, \ldots, \lambda_p\right)$. Putting $\mathbf{T} = \mathbf{U}^{-1}$ in (6.89) yields $\Lambda = \widehat{\boldsymbol{\Sigma}}_\infty(\mathbf{x})$, so that $\widehat{\boldsymbol{\Sigma}}_\infty(\mathbf{x})$ is diagonal. Now let $\mathbf{T}$ be the transformation that interchanges the first two coordinate axes. Then $\mathbf{T}\widehat{\boldsymbol{\Sigma}}_\infty(\mathbf{x})\mathbf{T}' = \widehat{\boldsymbol{\Sigma}}_\infty(\mathbf{x})$ implies that $\lambda_1 = \lambda_2$, and the same procedure shows that $\lambda_1 = \ldots = \lambda_p$. Thus $\widehat{\boldsymbol{\Sigma}}_\infty(\mathbf{x})$ is diagonal with all diagonal elements equal; that is, $\widehat{\boldsymbol{\Sigma}}_\infty(\mathbf{x}) = c\mathbf{I}$.

To complete the proof of (6.88), put $\mathbf{y} = \boldsymbol{\mu} + \mathbf{Ax}$ where $\mathbf{x}$ is as before and $\mathbf{AA}' = \boldsymbol{\Sigma}$, so that $\mathbf{y}$ has distribution (6.7). Then the equivariance implies that

$$\widehat{\boldsymbol{\mu}}_\infty(\mathbf{y}) = \boldsymbol{\mu} + \mathbf{A}\widehat{\boldsymbol{\mu}}_\infty(\mathbf{x}) = \boldsymbol{\mu}, \ \widehat{\boldsymbol{\Sigma}}_\infty(\mathbf{y}) = \mathbf{A}\widehat{\boldsymbol{\Sigma}}_\infty(\mathbf{x})\mathbf{A}' = c\boldsymbol{\Sigma}.$$

The same approach can be used to show that the asymptotic covariance matrix of $\widehat{\boldsymbol{\mu}}$ verifies (6.6), noting that if $\widehat{\boldsymbol{\mu}}$ has asymptotic covariance matrix $\mathbf{V}$, then $\mathbf{A}\widehat{\boldsymbol{\mu}}$ has asymptotic covariance matrix $\mathbf{AVA}'$ (see Section 6.12.7 for further details).

6.12.3 The estimating equations of the MLE

We shall prove (6.11)–(6.12). As a generic notation, if $g(\mathbf{T})$ is a function of the $p \times q$ matrix $\mathbf{T} = [t_{ij}]$, then $\partial g/\partial\mathbf{T}$ will denote the $p \times q$ matrix with elements $\partial g/\partial t_{ij}$; a vector argument corresponds to $q = 1$. It is well known (see Seber, 1984) that

$$\frac{\partial\,|\mathbf{A}|}{\partial\mathbf{A}} = |\mathbf{A}|\,\mathbf{A}^{-1} \tag{6.90}$$

and the reader can easily verify that

$$\frac{\partial\mathbf{b}'\mathbf{A}\mathbf{b}}{\partial\mathbf{b}} = (\mathbf{A} + \mathbf{A}')\mathbf{b} \ \text{ and } \ \frac{\partial\mathbf{b}'\mathbf{A}\mathbf{b}}{\partial\mathbf{A}} = \mathbf{bb}'. \tag{6.91}$$

Put $\mathbf{V} = \boldsymbol{\Sigma}^{-1}$. Then (6.9) becomes

$$\text{ave}_i(\rho(d_i)) - \log|\mathbf{V}| = \min, \tag{6.92}$$

with $d_i = (\mathbf{x}_i - \boldsymbol{\mu})' \mathbf{V} (\mathbf{x}_i - \boldsymbol{\mu})$. It follows from (6.91) that

$$\frac{\partial d_i}{\partial \boldsymbol{\mu}} = 2\mathbf{V}(\mathbf{x}_i - \boldsymbol{\mu}) \text{ and } \frac{\partial d_i}{\partial \mathbf{V}} = (\mathbf{x}_i - \boldsymbol{\mu})(\mathbf{x}_i - \boldsymbol{\mu})'. \tag{6.93}$$

Differentiating (6.92) yields

$$2\mathbf{V}\text{ave}_i\{W(d_i)(\mathbf{x}_i - \boldsymbol{\mu})\} = \mathbf{0} \text{ and ave}_i \left\{W(d_i)(\mathbf{x}_i - \boldsymbol{\mu})(\mathbf{x}_i - \boldsymbol{\mu})'\right\} - \mathbf{V}^{-1} = \mathbf{0},$$

which are equivalent to (6.11)–(6.12).

6.12.4 Asymptotic BP of monotone M-estimates

Location with Σ known

It will be shown that the BP of the location estimate given by (6.17) with $\boldsymbol{\Sigma}$ known is $\varepsilon^* = 0.5$. It may be supposed without loss of generality that $\boldsymbol{\Sigma} = \mathrm{I}$ (Problem 6.4) so that $d(\mathbf{x}, \boldsymbol{\mu}, \boldsymbol{\Sigma}) = \|\mathbf{x} - \boldsymbol{\mu}\|^2$. Let $v(d) = \sqrt{d} W_1(d)$. It is assumed that for all d

$$v(d) \le K = \lim_{s\to\infty} v(s) < \infty. \tag{6.94}$$

For a given ε and a contaminating sequence G_m, call $\boldsymbol{\mu}_m$ the solution of (6.17) corresponding to the mixture $(1-\varepsilon)F + \varepsilon G_m$. Then the scalar product of (6.17) with $\boldsymbol{\mu}_m$ yields

$$(1-\varepsilon)\mathrm{E}_F v(\|\mathbf{x} - \boldsymbol{\mu}_m\|^2)\frac{(\mathbf{x} - \boldsymbol{\mu}_m)'\boldsymbol{\mu}_m}{\|\mathbf{x} - \boldsymbol{\mu}_m\|\|\boldsymbol{\mu}_m\|} + \varepsilon \mathrm{E}_{G_m} v(\|\mathbf{x} - \boldsymbol{\mu}_m\|^2)\frac{(\mathbf{x} - \boldsymbol{\mu}_m)'\boldsymbol{\mu}_m}{\|\mathbf{x} - \boldsymbol{\mu}_m\|\|\boldsymbol{\mu}_m\|} = 0. \tag{6.95}$$

Assume that $\left\|\boldsymbol{\mu}_m\right\| \to \infty$. Then since for each $\mathbf{x}$ we have

$$\lim_{m\to\infty} \|\mathbf{x} - \boldsymbol{\mu}_m\|^2 = \infty, \quad \lim_{m\to\infty} \frac{(\mathbf{x} - \boldsymbol{\mu}_m)'\boldsymbol{\mu}_m}{\|\mathbf{x} - \boldsymbol{\mu}_m\|\|\boldsymbol{\mu}_m\|} = -1,$$

(6.95) yields

$$0 \le (1-\varepsilon)K(-1) + \varepsilon K$$

which implies $\varepsilon \ge 0.5$.

The assumption (6.94) holds in particular if v is monotone. The case with v not monotone has the complications already described for univariate location in Section 3.2.3.

Dispersion

To prove (6.25), we deal only with $\mathbf{\Sigma}$, and hence assume $\boldsymbol{\mu}$ is known and equal to $\mathbf{0}$. Thus $\widehat{\mathbf{\Sigma}}_\infty$ is defined by an equation of the form

$$\mathrm{E}W\left(\mathbf{x}'\widehat{\mathbf{\Sigma}}_\infty^{-1}\mathbf{x}\right)\mathbf{x}\mathbf{x}' = \widehat{\mathbf{\Sigma}}_\infty. \tag{6.96}$$

Let $\alpha = \mathrm{P}(\mathbf{x} = \mathbf{0})$. It will be shown first that in order for (6.96) to have a solution, it is necessary that

$$K(1-\alpha) \geq p. \tag{6.97}$$

Let $\mathbf{A}$ be any matrix such that $\widehat{\mathbf{\Sigma}}_\infty = \mathbf{A}\mathbf{A}'$. Multiplying (6.96) by $\mathbf{A}^{-1}$ on the left and by $\mathbf{A}^{-1\prime}$ on the right yields

$$\mathrm{E}W\left(\mathbf{y}'\mathbf{y}\right)\mathbf{y}\mathbf{y}' = \mathbf{I}, \tag{6.98}$$

where $\mathbf{y} = \mathbf{A}^{-1}\mathbf{x}$. Taking the trace in (6.98) yields

$$p = \mathrm{E}W\left(\|\mathbf{y}\|^2\right)\|\mathbf{y}\|^2 \mathrm{I}\left(\mathbf{y} \neq \mathbf{0}\right) \leq \mathrm{P}\left(\mathbf{y} \neq \mathbf{0}\right) \sup_d \left(dW(d)\right) = K(1-\alpha), \tag{6.99}$$

which proves (6.97).

Now let F attribute zero mass to the origin, and consider a proportion ε of contamination with distribution G. Then (6.96) becomes

$$(1-\varepsilon)\mathrm{E}_F W\left(\mathbf{x}'\widehat{\mathbf{\Sigma}}_\infty^{-1}\mathbf{x}\right)\mathbf{x}\mathbf{x}' + \varepsilon \mathrm{E}_G W\left(\mathbf{x}'\widehat{\mathbf{\Sigma}}_\infty^{-1}\mathbf{x}\right)\mathbf{x}\mathbf{x}' = \widehat{\mathbf{\Sigma}}_\infty. \tag{6.100}$$

Assume $\varepsilon < \varepsilon^*$. Take G concentrated at $\mathbf{x}_0$:

$$(1-\varepsilon)\mathrm{E}_F W\left(\mathbf{x}'\widehat{\mathbf{\Sigma}}_\infty^{-1}\mathbf{x}\right)\mathbf{x}\mathbf{x}' + \varepsilon W\left(\mathbf{x}_0'\widehat{\mathbf{\Sigma}}_\infty^{-1}\mathbf{x}_0\right)\mathbf{x}_0\mathbf{x}_0' = \widehat{\mathbf{\Sigma}}_\infty. \tag{6.101}$$

Put $\mathbf{x}_0 = \mathbf{0}$ first. Then the distribution $(1-\varepsilon)F + \varepsilon G$ attributes mass ε to $\mathbf{0}$, and hence in order for a solution to exist, we must have (6.97), i.e., $K(1-\varepsilon) \geq p$, and hence $\varepsilon^* \leq 1 - p/K$.

Let $\mathbf{x}_0$ now be arbitrary and let $\mathbf{A}$ again be as above; then

$$(1-\varepsilon)\mathrm{E}_F W(\mathbf{y}'\mathbf{y})\mathbf{y}\mathbf{y}' + \varepsilon W(\mathbf{y}_0'\mathbf{y}_0)\mathbf{y}_0\mathbf{y}_0' = \mathbf{I}, \tag{6.102}$$

where $\mathbf{y} = \mathbf{A}^{-1}\mathbf{x}$ and $\mathbf{y}_0 = \mathbf{A}^{-1}\mathbf{x}_0$. Let $\mathbf{a} = \mathbf{y}_0/\|\mathbf{y}_0\|$. Then multiplying in (6.102) by $\mathbf{a}'$ on the left and by $\mathbf{a}$ on the right yields

$$(1-\varepsilon)\mathrm{E}_F W\left(\|\mathbf{y}\|^2\right)\left(\mathbf{y}'\mathbf{a}\right)^2 + \varepsilon W\left(\|\mathbf{y}_0\|^2\right)\|\mathbf{y}_0\|^2 = 1 \geq \varepsilon W\left(\|\mathbf{y}_0\|^2\right)\|\mathbf{y}_0\|^2. \tag{6.103}$$

Call λ_p and λ_1 the smallest and largest eigenvalues of $\widehat{\mathbf{\Sigma}}_\infty$. Let $\mathbf{x}_0$ now tend to infinity. Since $\varepsilon < \varepsilon^*$, the eigenvalues of $\widehat{\mathbf{\Sigma}}_\infty$ are bounded away from zero and infinity. Since

$$\|\mathbf{y}_0\|^2 = \mathbf{x}_0'\widehat{\mathbf{\Sigma}}_\infty^{-1}\mathbf{x}_0 \geq \frac{\|\mathbf{x}_0\|^2}{\lambda_p}$$

it follows that $\mathbf{y}_0$ tends to infinity. Hence the right-hand member of (6.103) tends to εK, and this implies $\varepsilon \leq 1/K$.

Now let $\varepsilon > \varepsilon^*$. Then either $\lambda_p \to 0$ or $\lambda_1 \to \infty$. Call $\mathbf{a}_1$ and $\mathbf{a}_p$ the unit eigenvectors corresponding to λ_1 and λ_p. Multiplying (6.100) by $\mathbf{a}_1'$ on the left and by $\mathbf{a}_1$ on the right yields

$$(1-\varepsilon)\mathrm{E}_F W\left(\mathbf{x}'\widehat{\mathbf{\Sigma}}_\infty^{-1}\mathbf{x}\right)\left(\mathbf{x}'\mathbf{a}_1\right)^2 + \varepsilon \mathrm{E}_G W\left(\mathbf{x}'\widehat{\mathbf{\Sigma}}_\infty^{-1}\mathbf{x}\right)(\mathbf{x}'\mathbf{a}_1)^2 = \lambda_1.$$

Suppose that $\lambda_1 \to \infty$. Divide the above expression by λ_1; recall that the first expectation is bounded and that

$$\frac{(\mathbf{x}'\mathbf{a}_1)^2}{\lambda_1} \le \mathbf{x}'\widehat{\mathbf{\Sigma}}_\infty^{-1}\mathbf{x}.$$

Then in the limit we have

$$1 = \varepsilon \mathrm{E}_G W\left(\mathbf{x}'\widehat{\mathbf{\Sigma}}_\infty^{-1}\mathbf{x}\right)\frac{(\mathbf{x}'\mathbf{a}_1)^2}{\lambda_1} \le \varepsilon K,$$

and hence $\varepsilon \ge 1/K$.

On the other hand, taking the trace in (6.103) and proceeding as in the proof of (6.99) yields

$$p = (1-\varepsilon)\mathrm{E}_F W\left(\|\mathbf{y}\|^2\right)\|\mathbf{y}\|^2 + \varepsilon \mathrm{E}_G W\left(\|\mathbf{y}\|^2\right)\|\mathbf{y}\|^2 \ge (1-\varepsilon)\mathrm{E}_F W\left(\|\mathbf{y}\|^2\right)\|\mathbf{y}\|^2.$$

Note that

$$\|\mathbf{y}\|^2 \ge \frac{\left(\mathbf{x}'\mathbf{a}_p\right)^2}{\lambda_p}.$$

Hence $\lambda_p \to 0$ implies $\|\mathbf{y}\|^2 \to \infty$ and thus the right-hand side of the equation above tends to $(1-\varepsilon)K$, which implies $\varepsilon \ge 1 - p/K$.

6.12.5 The estimating equations for S-estimates

We are going to prove (6.31)–(6.32). Put for simplicity

$$\mathbf{V} = \mathbf{\Sigma}^{-1} \text{ and } d_i = d(\mathbf{x}_i, \boldsymbol{\mu}, \mathbf{\Sigma}) = (\mathbf{x}_i - \boldsymbol{\mu})'\mathbf{V}(\mathbf{x}_i - \boldsymbol{\mu})$$

and call $\sigma(\boldsymbol{\mu}, \mathbf{V})$ the solution of

$$\mathrm{ave}_i\left\{\rho\left(\frac{d_i}{\sigma}\right)\right\} = \delta. \tag{6.104}$$

Then (6.28) amounts to minimizing $\sigma(\boldsymbol{\mu}, \mathbf{V})$ with $|\mathbf{V}| = 1$. Solving this problem by the method of Lagrange's multipliers becomes

$$g(\boldsymbol{\mu}, \mathbf{V}, \lambda) = \sigma(\boldsymbol{\mu}, \mathbf{V}) + \lambda\left(|\mathbf{V}| - 1\right) = \min.$$

Differentiating g with respect to λ, $\boldsymbol{\mu}$ and $\mathbf{V}$, and recalling (6.90), we have $|\mathbf{V}| = 1$ and

$$\frac{\partial \sigma}{\partial \boldsymbol{\mu}} = \mathbf{0}, \quad \frac{\partial \sigma}{\partial \mathbf{V}} + \lambda \mathbf{V}^{-1} = \mathbf{0}. \tag{6.105}$$

Differentiating (6.104) and recalling the first equations in (6.105) and in (6.93) yields

$$\mathrm{ave}_i\left\{W\left(\frac{d_i}{\sigma}\right)\left(\sigma\frac{\partial d_i}{\partial\boldsymbol{\mu}}-d_i\frac{\partial\sigma}{\partial\boldsymbol{\mu}}\right)\right\}=2\sigma\,\mathrm{ave}_i\left\{W\left(\frac{d_i}{\sigma}\right)(\mathbf{x}_i-\boldsymbol{\mu})\right\}=\mathbf{0},$$

which implies (6.31). Proceeding similarly with respect to $\mathbf{V}$ we have

$$\mathrm{ave}_i\left\{W\left(\frac{d_i}{\sigma}\right)\left(\sigma\frac{\partial d_i}{\partial\mathbf{V}}+d_i\lambda\mathbf{V}^{-1}\right)\right\}=\sigma\,\mathrm{ave}_i\left\{W\left(\frac{d_i}{\sigma}\right)(\mathbf{x}_i-\boldsymbol{\mu})(\mathbf{x}_i-\boldsymbol{\mu})'\right\}+b\mathbf{V}^{-1}=\mathbf{0},$$

with

$$b=\lambda\;\mathrm{ave}_i\left\{W\left(\frac{d_i}{\sigma}\right)d_i\right\};$$

and this implies (6.32) with $c=-b/\sigma$.

6.12.6 Behavior of S-estimates for high p

It will be shown that an S-estimate with continuous ρ becomes increasingly similar to the classical estimate when $p\to\infty$. For simplicity, this property is proved here only for normal data. However, it can be proved under more general conditions which include finite fourth moments.

Because of the equivariance of S-estimates, we may assume that the true parameters are $\boldsymbol{\Sigma}=\mathbf{I}$ and $\boldsymbol{\mu}=\mathbf{0}$. Then, since the estimate is consistent, its asymptotic values are $\widehat{\boldsymbol{\mu}}_\infty=\mathbf{0},\ \ \widehat{\boldsymbol{\Sigma}}_\infty=\mathbf{I}$.

For each p let $d^{(p)}=d(\mathbf{x},\widehat{\boldsymbol{\mu}}_\infty,\widehat{\boldsymbol{\Sigma}}_\infty)$. Then $d^{(p)}=||\mathbf{x}||^2\sim\chi^2_p$ and hence

$$\mathrm{E}\left(\frac{d^{(p)}}{p}\right)=1,\ \ \mathrm{SD}\left(\frac{d^{(p)}}{p}\right)=\sqrt{\frac{2}{p}},\tag{6.106}$$

which implies that the distribution of $d^{(p)}/p$ is increasingly concentrated around 1, and $d^{(p)}/p\to 1$ in probability when $p\to\infty$.

Since ρ is continuous, there exists $a>0$ such that $\rho(a)=0.5$. Call σ_p the scale corresponding to $d^{(p)}$

$$0.5=\mathrm{E}\rho\left(\frac{d^{(p)}}{\sigma_p}\right).\tag{6.107}$$

We shall show that

$$\frac{d^{(p)}}{\sigma_p}\to_p a.\tag{6.108}$$

Since $d^{(p)}/p\to_p 1$, we have for any $\varepsilon>0$

$$\mathrm{E}\rho\left(\frac{d^{(p)}}{p/(a(1+\varepsilon))}\right)\to\rho\left(a(1+\varepsilon)\right)>0.5$$

and

$$\mathrm{E}\rho\left(\frac{d^{(p)}}{p/(a(1-\varepsilon))}\right) \rightarrow \rho\left(a(1-\varepsilon)\right) < 0.5.$$

Then (6.107) implies that for large enough p

$$\frac{p}{a(1+\varepsilon)} \leq \sigma_p \leq \frac{p}{a(1-\varepsilon)}$$

and hence $\lim_{p\rightarrow\infty}\left(a\sigma_p/p\right) = 1$, which implies

$$\frac{d^{(p)}}{\sigma_p} = a\frac{d^{(p)}/p}{a\sigma_p/p} \rightarrow_p a$$

as stated. This implies that for large n and p the weights of the observations $W\left(d_i/\widehat{\sigma}\right)$ are

$$W\left(\frac{d_i}{\widehat{\sigma}}\right) = W\left(\frac{d(\mathbf{x}_i, \widehat{\boldsymbol{\mu}}, \widehat{\boldsymbol{\Sigma}})}{\widehat{\sigma}}\right) \simeq W\left(\frac{d(\mathbf{x}_i, \widehat{\boldsymbol{\mu}}_\infty, \widehat{\boldsymbol{\Sigma}}_\infty)}{\sigma_p}\right) \approx W(a);$$

that is, they are practically constant. Hence $\widehat{\boldsymbol{\mu}}$ and $\widehat{\boldsymbol{\Sigma}}$, which are weighted means and covariances, will be very similar to $\mathrm{E}\mathbf{x}$ and $\mathbf{Var}(\mathbf{x})$, and hence very efficient for normal data.

6.12.7 Calculating the asymptotic covariance matrix of location M-estimates

Recall that the covariance matrix of the classical location estimate is a constant multiple of the covariance matrix of the observations, since $\mathbf{Var}(\overline{\mathbf{x}}) = n^{-1}\mathbf{Var}(\mathbf{x})$. We shall show a similar result for M-estimates at elliptically distributed data. Let the estimates be defined by (6.14)–(6.15). As explained at the end of Section 6.12.2, it can be shown that if $\mathbf{x}_i$ has an elliptical distribution (6.7), the asymptotic covariance matrix of $\widehat{\boldsymbol{\mu}}$ has the form (6.6): $\mathbf{V} = v\boldsymbol{\Sigma}$, where v is a constant that we shall now calculate.

It can be shown that in the elliptical case the asymptotic distribution of $\widehat{\boldsymbol{\mu}}$ is the same as if $\boldsymbol{\Sigma}$ were assumed known. In view of the equivariance of the estimate, we may consider only the case $\boldsymbol{\mu} = \mathbf{0}$, $\boldsymbol{\Sigma} = \mathbf{I}$. Then it follows from (6.88) that $\widehat{\boldsymbol{\Sigma}}_\infty = c\mathbf{I}$, and taking the trace in (6.15) we have that c is the solution of (6.21).

It will be shown that

$$v = \frac{pa}{b^2},$$

where (writing $z = ||\mathbf{x}||^2$)

$$a = \mathrm{E}W_1\left(\frac{z}{c}\right)^2 z, \quad b = 2\mathrm{E}W_1'\left(\frac{z}{c}\right)\frac{z}{c} + p\mathrm{E}W_1\left(\frac{z}{c}\right). \tag{6.109}$$

We may write (6.14) as

$$\sum_{i=1}^{n} \Psi(\mathbf{x}_i, \boldsymbol{\mu}) = \mathbf{0},$$

with

$$\Psi(\mathbf{x}, \boldsymbol{\mu}) = W_1\left(\frac{\|\mathbf{x}-\boldsymbol{\mu}\|^2}{c}\right)(\mathbf{x}-\boldsymbol{\mu}).$$

It follows from (3.48) that $\mathbf{V} = \mathbf{B}^{-1}\mathbf{A}\mathbf{B}'^{-1}$, where

$$\mathbf{A} = \mathrm{E}\Psi(\mathbf{x}, \mathbf{0})\Psi(\mathbf{x}, \mathbf{0})', \ \mathbf{B} = \mathrm{E}\dot{\Psi}(\mathbf{x}_i, \mathbf{0}),$$

where $\dot{\Psi}$ is the derivative of Ψ with respect to $\boldsymbol{\mu}$, i.e., the matrix $\dot{\Psi}$ with elements $\dot{\Psi}_{jk} = \partial \Psi_j / \partial \mu_k$.

We have

$$\mathbf{A} = \mathrm{E}W_1\left(\frac{\|\mathbf{x}\|^2}{c}\right)^2 \mathbf{x}\mathbf{x}'.$$

Since $\mathcal{D}(\mathbf{x})$ is spherical, $\mathbf{A}$ is a multiple of the identity: $\mathbf{A} = t\mathbf{I}$. Taking the trace and recalling that $\mathrm{tr}(\mathbf{x}\mathbf{x}') = \mathbf{x}'\mathbf{x}$, we have

$$\mathrm{tr}(\mathbf{A}) = \mathrm{E}W_1\left(\frac{\|\mathbf{x}\|^2}{c}\right)^2 \|\mathbf{x}\|^2 = a = tp,$$

and hence $\mathbf{A} = (a/p)\,\mathbf{I}$.

To calculate $\mathbf{B}$, recall that

$$\frac{\partial\, \|\mathbf{a}\|^2}{\partial \mathbf{a}} = 2\mathbf{a}, \quad \frac{\partial \mathbf{a}}{\partial \mathbf{a}} = \mathbf{I},$$

and hence

$$\dot{\Psi}(\mathbf{x}_i, \boldsymbol{\mu}) = -\left\{2W_1'\left(\frac{\|\mathbf{x}-\boldsymbol{\mu}\|^2}{c}\right)\frac{(\mathbf{x}-\boldsymbol{\mu})}{c}(\mathbf{x}-\boldsymbol{\mu})' + W_1\left(\frac{\|\mathbf{x}-\boldsymbol{\mu}\|^2}{c}\right)\mathbf{I}\right\}.$$

Then the same reasoning yields $\mathbf{B} = -(b/p)\mathbf{I}$, which implies

$$\mathbf{V} = \frac{pa}{b^2}\mathbf{I},$$

as stated.

To compute c for normal data, note that it depends only on the distribution of $\|\mathbf{x}\|^2$, which is χ^2_p.

This approach can also be used to calculate the efficiency of location S-estimates.

6.12.8 The exact fit property

Let the data set X contain $q \geq n - m^*$ points on the hyperplane $H = \left\{\mathbf{x} : \beta'\mathbf{x} = \gamma\right\}$. It will be shown that $\widehat{\mu}(X) \in H$ and $\widehat{\Sigma}(X)\beta = \mathbf{0}$.

Without loss of generality we can take $\|\beta\| = 1$ and $\gamma = 0$. In fact, the equation defining H does not change if we divide both sides by $\|\beta\|$; and since

$$\widehat{\mu}(X + \mathbf{a}) = \widehat{\mu}(X) + \mathbf{a},\ \widehat{\Sigma}(X + \mathbf{a}) = \widehat{\Sigma}(X),$$

we may replace $\mathbf{x}$ by $\mathbf{x} + \mathbf{a}$ where $\beta'\mathbf{a} = 0$.

Now $H = \left\{\mathbf{x} : \beta'\mathbf{x} = 0\right\}$ is a subspace. Call $\mathbf{P}$ the matrix corresponding to the orthogonal projection on the subspace orthogonal to H, i.e., $\mathbf{P} = \beta\beta'$. Define for $t \in R$

$$\mathbf{y}_i = \mathbf{x}_i + t\beta\beta'\mathbf{x}_i = (\mathbf{I}+t\mathbf{P})\mathbf{x}_i.$$

Then $Y = \{\mathbf{y}_1, \ldots, \mathbf{y}_n\}$ has at least q elements in common with $\mathbf{x}$, since $\mathbf{Pz} = \mathbf{0}$ for $\mathbf{z} \in H$. Hence by the definition of BP, $\widehat{\mu}(Y)$ remains bounded for all t. Since

$$\widehat{\mu}(Y) = \widehat{\mu}(X) + t\beta\beta'\widehat{\mu}(X)$$

the left-hand side is a bounded function of t, while the right-hand side tends to infinity with t unless $\beta'\widehat{\mu}(X) = 0$, i.e., $\widehat{\mu}(X) \in H$.

In the same way

$$\widehat{\Sigma}(Y) = (\mathbf{I}+t\mathbf{P})\widehat{\Sigma}(X)(\mathbf{I}+t\mathbf{P}) = \widehat{\Sigma}(X)+t^2\left(\beta'\widehat{\Sigma}(X)\beta\right)\beta\beta' + t\left(\mathbf{P}\widehat{\Sigma}(X)+\widehat{\Sigma}(X)\mathbf{P}\right)$$

is a bounded function of t, which implies that $\widehat{\Sigma}(X)\beta = \mathbf{0}$.

6.12.9 Elliptical distributions

A random vector $\mathbf{r} \in R^p$ is said to have a spherical distribution if its density f depends only on $\|\mathbf{r}\|$; that is, it has the form

$$f(\mathbf{r}) = h(\|\mathbf{r}\|) \tag{6.110}$$

for some nonnegative function h. It follows that for any orthogonal matrix $\mathbf{T}$

$$\mathcal{D}(\mathbf{Tr}) = \mathcal{D}(\mathbf{r}), \tag{6.111}$$

Actually, (6.111) may be taken as the general definition of a spherical distribution, without requiring the existence of a density; but we prefer the present definition for reasons of simplicity.

The random vector $\mathbf{x}$ will be said to have an elliptical distribution if

$$\mathbf{x} = \mu + \mathbf{Ar} \tag{6.112}$$

where $\mu \in R^p$, $\mathbf{A} \in R^{p\times p}$ is nonsingular and $\mathbf{r}$ has a spherical distribution. Let

$$\Sigma = \mathbf{AA}'.$$

We shall call μ and Σ the location vector and the dispersion matrix of $\mathbf{x}$, respectively. We now state the most relevant properties of elliptical distributions. If $\mathbf{x}$ is given by (6.112), then:

1. The distribution of $\mathbf{Bx}+\mathbf{c}$ is also elliptical, with location vector $\mathbf{B}\mu+\mathbf{c}$ and dispersion matrix $\mathbf{B}\Sigma\mathbf{B}'$.
2. If the mean and variances of $\mathbf{x}$ exist, then

$$\mathrm{E}\mathbf{x}=\mu,\quad \mathbf{Var}\,(\mathbf{x})=c\Sigma,$$

where c is a constant.
3. The density of $\mathbf{x}$ is

$$|\Sigma|^{-p/2}h((\mathbf{x}-\mu)'\Sigma^{-1}(\mathbf{x}-\mu)).$$

4. The distributions of linear combinations of $\mathbf{x}$ belong to the same location-scale family; more precisely, for any $\mathbf{a}\in R^p$

$$\mathcal{D}(\mathbf{a}'\mathbf{x})=\mathcal{D}(\mathbf{a}'\mu+\sqrt{\mathbf{a}'\Sigma\mathbf{a}}r_1), \tag{6.113}$$

where r_1 is the first coordinate of $\mathbf{r}$.
The proofs of (1) and (3) are immediate. The proof of (2) follows from the fact that, if the mean and variances of $\mathbf{r}$ exist, then

$$\mathrm{E}\mathbf{r}=\mathbf{0},\quad \mathbf{Var}\,(\mathbf{r})=c\mathbf{I}$$

for some constant c.

Proof of (4): It will be shown that the distribution of a linear combination of $\mathbf{r}$ does not depend on its direction; more precisely, for all $\mathbf{a}\in R^p$

$$\mathcal{D}(\mathbf{a}'\mathbf{r})=\mathcal{D}(\|\mathbf{a}\|\,r_1). \tag{6.114}$$

In fact, let $\mathbf{T}$ be an orthogonal matrix with columns $\mathbf{t}_1,\ldots,\mathbf{t}_p$ such that $\mathbf{t}_1=\mathbf{a}/\|\mathbf{a}\|$. Then $\mathbf{Ta}=(\|\mathbf{a}\|,0,0,\ldots,0)'$.

Then by (6.111)

$$\mathcal{D}(\mathbf{a}'\mathbf{r})=\mathcal{D}(\mathbf{a}'\mathbf{T}'\mathbf{r})=\mathcal{D}((\mathbf{Ta})'\,\mathbf{r})=\mathcal{D}(\|\mathbf{a}\|\,r_1)$$

as stated; and (6.113) follows from (6.114) and (6.112).

6.12.10 Consistency of Gnanadesikan–Kettenring correlations

Let the random vector $\mathbf{x}=(x_1,\ldots,x_p)$ have an elliptical distribution: that is, $\mathbf{x}=\mathbf{Az}$ where $\mathbf{z}=(z_1,\ldots,z_p)$ has a spherical distribution. This implies that for any $\mathbf{u}\in R^p$,

$$\mathcal{D}(\mathbf{u}'\mathbf{x})=\mathcal{D}(\|\mathbf{b}\|\,z_1)\ \text{ with }\ \mathbf{b}=\mathbf{A}'\mathbf{u},$$

and hence

$$\sigma(\mathbf{u}'\mathbf{x})=\sigma_0\,\|\mathbf{b}\|\ \text{ with }\ \sigma_0=\sigma(z_1). \tag{6.115}$$

Let $U_j = \mathbf{u}_j'\mathbf{x}$ $(j = 1, 2)$ be two linear combinations of $\mathbf{x}$. It will be shown that their robust correlation (6.63) coincides with the ordinary one.

Assume that $\mathbf{z}$ has finite second moments. We may assume that $\mathbf{Var}(\mathbf{z}) = \mathbf{I}$. Then

$$\mathrm{Cov}(U_1, U_2) = \mathbf{b}_1'\mathbf{b}_2, \quad \mathrm{Var}(U_j) = \left\|\mathbf{b}_j\right\|^2,$$

where $\mathbf{b}_j = \mathbf{A}'\mathbf{u}_j$; and hence the ordinary correlation is

$$\mathrm{Corr}(U_1, U_2) = \frac{\mathbf{b}_1'\mathbf{b}_2}{\|\mathbf{b}_1\| \, \|\mathbf{b}_2\|}.$$

Put $\sigma_j = \sigma(U_j)$ for brevity. It follows from (6.115) that

$$\sigma_j = \left\|\mathbf{b}_j\right\| \sigma_0, \quad \sigma\left(\frac{U_1}{\sigma_1} \pm \frac{U_2}{\sigma_2}\right) = \left\|\frac{\mathbf{b}_1}{\|\mathbf{b}_1\|} \pm \frac{\mathbf{b}_2}{\|\mathbf{b}_2\|}\right\|,$$

and hence (6.63) yields

$$\mathrm{RCorr}(U_1, U_2) = \frac{\mathbf{b}_1'\mathbf{b}_2}{\|\mathbf{b}_1\| \, \|\mathbf{b}_2\|} = \mathrm{Corr}(U_1, U_2).$$

6.12.11 Sperical principal components

We may assume without loss of generality that $\boldsymbol{\mu} = \mathbf{0}$. The covariance matrix of $\mathbf{x}/\left\|\mathbf{x}\right\|$ is

$$\mathbf{U} = \mathrm{E}\frac{\mathbf{x}\mathbf{x}'}{\|\mathbf{x}\|^2}. \tag{6.116}$$

It will be shown that $\mathbf{U}$ and $\boldsymbol{\Sigma}$ have the same eigenvectors.

It will be first assumed that $\boldsymbol{\Sigma}$ is diagonal: $\boldsymbol{\Sigma} = \mathrm{diag}\{\lambda_1, \ldots, \lambda_p\}$ where $\lambda_1, \ldots, \lambda_p$ are its eigenvalues. Then the eigenvectors of $\boldsymbol{\Sigma}$ are the vectors of the canonical basis $\mathbf{b}_1, \ldots, \mathbf{b}_p$ with $b_{jk} = \delta_{jk}$. It will be shown that the $\mathbf{b}_j$s' are also the eigenvectors of $\mathbf{U}$, i.e.,

$$\mathbf{U}\mathbf{b}_j = \alpha_j\mathbf{b}_j \tag{6.117}$$

for some α_j. For a given j put $\mathbf{u} = \mathbf{U}\mathbf{b}_j$; then we must show that $k \neq j$ implies $u_k = 0$. In fact, for $k \neq j$,

$$u_k = \mathrm{E}\frac{x_j x_k}{\|\mathbf{x}\|^2},$$

where x_j $(j = 1, \ldots, p)$ are the coordinates of $\mathbf{x}$. The symmetry of the distribution implies that $\mathcal{D}(x_j, x_k) = \mathcal{D}(x_j, -x_k)$, which implies

$$u_k = \mathrm{E}\frac{x_j(-x_k)}{\|\mathbf{x}\|^2} = -u_k$$

and hence $u_k = 0$. This proves (6.117). It follows from (6.117) that $\mathbf{U}$ is diagonal.

Now let $\mathbf{\Sigma}$ have arbitrary eigenvectors $\mathbf{t}_1, \ldots, \mathbf{t}_p$. Call λ_j $(j = 1, \ldots, p)$ its eigenvalues, and let $\mathbf{T}$ be the orthogonal matrix with columns $\mathbf{t}_1, \ldots, \mathbf{t}_p$, so that

$$\mathbf{\Sigma} = \mathbf{T\Lambda T}',$$

where $\mathbf{\Lambda} = \text{diag}\{\lambda_1, \ldots, \lambda_p\}$. We must show that the eigenvectors of $\mathbf{U}$ in (6.116) are the $\mathbf{t}_j$'s.

Let $\mathbf{z} = \mathbf{T}'\mathbf{x}$. Then $\mathbf{z}$ has an elliptical distribution with location vector $\mathbf{0}$ and dispersion matrix Λ. The orthogonality of $\mathbf{T}$ implies that $\|\mathbf{z}\| = \|\mathbf{x}\|$. Let

$$\mathbf{V} = \mathrm{E}\frac{\mathbf{zz}'}{\|\mathbf{z}\|^2} = \mathbf{T}'\mathbf{UT}. \tag{6.118}$$

It follows from (6.117) that the $\mathbf{b}_j$'s are the eigenvectors of $\mathbf{V}$, and hence that $\mathbf{V}$ is diagonal. Then (6.118) implies that $\mathbf{U} = \mathbf{TVT}'$, which implies that the eigenvectors of $\mathbf{U}$ are the columns of $\mathbf{T}$, which are the eigenvectors of $\mathbf{\Sigma}$. This completes the proof of the equality of the eigenvectors of $\mathbf{U}$ and $\mathbf{\Sigma}$.

Now let σ (.) be a dispersion statistic. We shall show that the values of $\sigma\left(\mathbf{x}'\mathbf{t}_j\right)^2$ are proportional to the eigenvalues of $\mathbf{\Sigma}$. In fact, it follows from (6.113) and $\mathbf{t}_j'\mathbf{\Sigma}\boldsymbol{t}_j = \lambda_j$ that for all j,

$$\mathcal{D}(\mathbf{t}_j'\mathbf{x}) = \mathcal{D}(\mathbf{t}_j'\boldsymbol{\mu} + \sqrt{\lambda_j}r_1),$$

and hence

$$\sigma\left(\mathbf{t}_j'\mathbf{x}\right) = \sigma\left(\sqrt{\lambda_j}r_1\right) = \sqrt{\lambda_j}d,$$

with $d = \sigma\left(r_1\right)$.

6.13 Problems

6.1. Show that if $\mathbf{x}$ has distribution (6.7), then $\mathrm{E}\mathbf{x} = \boldsymbol{\mu}$ and $\text{var}\,(\mathbf{x}) = c\mathbf{\Sigma}$.

6.2. Prove that M-estimates (6.14)–(6.15) are affine equivariant.

6.3. Show that the asymptotic value of an equivariant $\widehat{\mathbf{\Sigma}}$ at a spherical distribution is a scalar multiple of $\mathbf{I}$.

6.4. Show that the result of Section 6.12.4 is valid for any $\mathbf{\Sigma}$.

6.5. Prove that if $\rho(t)$ is a bounded nondecreasing function, then $t\rho'(t)$ cannot be nondecreasing.

6.6. Let $\widehat{\mu}$ and $\widehat{\mathbf{\Sigma}}$ be S-estimates of location and dispersion based on the scale $\widehat{\sigma}$ and let $\widehat{\sigma}_0 = \widehat{\sigma}(\mathbf{d}(X, \widehat{\mu}, \widehat{\mathbf{\Sigma}}))$. Given a constant σ_0 define $\widehat{\mu}^*$ and $\widehat{\mathbf{\Sigma}}^*$ as the values μ and $\mathbf{\Sigma}$ that minimize $|\mathbf{\Sigma}|$ subject to $\widehat{\sigma}(\mathbf{d}(\mathbf{x}, \mu, \mathbf{\Sigma})) = \sigma_0$. Prove that $\widehat{\mu}^* = \widehat{\mu}$ and $\widehat{\mathbf{\Sigma}}^* = (\widehat{\sigma}_0/\sigma_0)\widehat{\mathbf{\Sigma}}$.

6.7. Show that $\overline{\mathbf{x}}$ and $\mathbf{Var}\,(X)$ are the values of $\widehat{\boldsymbol{\mu}}$ and $\widehat{\mathbf{\Sigma}}$ minimizing $|\mathbf{\Sigma}|$ subject to $(1/n)\sum_{i=1}^n d(\mathbf{x}_i, \boldsymbol{\mu}, \mathbf{\Sigma}) = p$.

6.8. Let $\widehat{\boldsymbol{\mu}}$ and $\widehat{\mathbf{\Sigma}}$ be the MCD estimates of location and dispersion, which minimize the scale $\widehat{\sigma}(d_1, \ldots, d_n) = \sum_{i=1}^h d_{(i)}$. For each subsample $A = \{\mathbf{x}_{i_1}, \ldots, \mathbf{x}_{i_h}\}$ of size h call $\overline{\mathbf{x}}_A$ and $\mathbf{C}_A$ the sample mean and covariance matrix corresponding to A. Let A^* be a subsample of size h that minimizes $|\mathbf{C}_A|$. Show that A^* is the set of observations corresponding to the h smallest values $\mathbf{d}(\mathbf{x}_i, \widehat{\boldsymbol{\mu}}, \widehat{\boldsymbol{\Sigma}})$, and that $\widehat{\boldsymbol{\mu}} = \overline{\mathbf{x}}_{A^*}$ and $\widehat{\boldsymbol{\Sigma}} = \left|\mathbf{C}_{A^*}\right|^{-1/p} \mathbf{C}_{A^*}$.

6.9. Let $\widehat{\boldsymbol{\mu}}$ and $\widehat{\boldsymbol{\Sigma}}$ be the MVE estimates of location and dispersion. Let $\widehat{\boldsymbol{\mu}}^*$, $\widehat{\boldsymbol{\Sigma}}^*$ be the values of $\boldsymbol{\mu}$ and $\boldsymbol{\Sigma}$ minimizing $|\boldsymbol{\Sigma}|$ under the constraint that the ellipsoid

$$\{\mathbf{x} \in R^p : (\mathbf{x} - \boldsymbol{\mu})'\boldsymbol{\Sigma}^{-1}(\mathbf{x} - \boldsymbol{\mu}) \leq 1\}$$

of volume $|\boldsymbol{\Sigma}|$ contains at least $n/2$ sample points. Show that $\widehat{\boldsymbol{\mu}}^* = \widehat{\boldsymbol{\mu}}$ and $\widehat{\boldsymbol{\Sigma}}^* = \lambda\widehat{\boldsymbol{\Sigma}}$ where $\lambda = \mathrm{Med}\{\mathbf{d}(X, \widehat{\boldsymbol{\mu}}, \widehat{\boldsymbol{\Sigma}})\}$.

6.10. Prove (6.30).

6.11. Let (x, y) be bivariate normal with zero means, unit variances and correlation ρ, and let ψ be a monotone ψ-function. Show that $\mathrm{E}(\psi(x)\psi(y))$ is an increasing function of ρ [hint: $y = \rho x + \sqrt{1-\rho^2}z$ with $z \sim (0, 1)$ independent of x].

6.12. Prove (6.19).

6.13. The data set **glass** from Hettich and Bay (1999) contains measurements of the presence of seven chemical constituents in 76 pieces of glass from nonfloat windows. Compute the classical estimate and the bisquare S-estimate of location and dispersion and the respective Mahalanobis distances. For both, do the Q–Q plots of distances and the plots of distances vs. index numbers, and compare the results.

6.14. The first principal component is often used to represent multispectral images. The data set **image** (Frery, 2005) contains the values corresponding to three frequency bands for each of 1573 pixels of a radar image. Compute the classical and spherical principal components and compare the directions of the respective eigenvectors and the fits given by the first component.

7

Generalized Linear Models

In Chapter 4 we considered regression models where the response variable y depends linearly on several explanatory variables $x_1, \ldots, x_p$. In this case y was a quantitative variable, i.e., it could take on any real value, and the regressors—which could be quantitative or qualitative—affected only its mean.

In this chapter we shall consider more general situations in which the regressors affect the distribution function of y; but to retain parsimony, it is assumed that this distribution depends on them only through a linear combination $\sum_j \beta_j x_j$ where the β_j's are unknown.

The first situation that we shall treat is that when y is a 0–1 variable.

7.1 Logistic regression

Let y be a 0–1 variable representing the death or survival of a patient after heart surgery. Here $y = 1$ and $y = 0$ represent death and survival, respectively. We want to predict this outcome by means of different regressors such as $x_1 =$ age, $x_2 =$ diastolic pressure, etc.

We observe $(\mathbf{x}, y)$ where $\mathbf{x} = (x_1, \ldots, x_p)'$ is the vector of explanatory variables. Assume first that $\mathbf{x}$ is fixed (i.e., nonrandom). To model the dependency of y on $\mathbf{x}$, we assume that $\mathrm{P}(y = 1)$ depends on $\boldsymbol{\beta}'\mathbf{x}$ for some unknown $\boldsymbol{\beta} \in R^p$. Since $\mathrm{P}(y = 1) \in [0, 1]$ and $\boldsymbol{\beta}'\mathbf{x}$ may take on any real value, we make the further assumption that

$$\mathrm{P}(y = 1) = F(\boldsymbol{\beta}'\mathbf{x}), \tag{7.1}$$

where the *link function* F is any continuous distribution function. If instead $\mathbf{x}$ is random, it will be assumed that the probabilities are conditional, i.e.,

$$\mathrm{P}(y = 1|\mathbf{x}) = F(\boldsymbol{\beta}'\mathbf{x}). \tag{7.2}$$

Robust Statistics – Theory and Methods Ricardo A. Maronna, R. Douglas Martin and Víctor J. Yohai

In the frequent case of a model with an intercept, the first coordinate of each $\mathbf{x}_i$ is one, and the prediction may be written as

$$\boldsymbol{\beta}'\mathbf{x}_i = \beta_0 + \underline{\mathbf{x}}_i\boldsymbol{\beta}_1, \tag{7.3}$$

with $\underline{\mathbf{x}}_i$ and $\boldsymbol{\beta}_1$ as in (4.6).

The most popular link functions are those corresponding to the logistic distribution

$$F(y) = \frac{e^y}{1+e^y} \tag{7.4}$$

("logistic model") and to the standard normal distribution $F(y) = \Phi(y)$ ("probit model"). For the logistic model we have

$$\log \frac{\mathrm{P}(y=1)}{1-\mathrm{P}(y=1)} = \boldsymbol{\beta}'\mathbf{x}.$$

The left-hand side is called the *log odds ratio*, and is seen to be a linear function of $\mathbf{x}$.

Now let $(\mathbf{x}_1, y_1), \ldots, (\mathbf{x}_n, y_n)$ be a sample from model (7.1), where $\mathbf{x}_1, \ldots, \mathbf{x}_n$ are fixed. From now on we shall write for simplicity

$$p_i(\boldsymbol{\beta}) = F(\boldsymbol{\beta}'\mathbf{x}_i).$$

Then $y_1, \ldots, y_n$ are response random variables which take on values 1 and 0 with probabilities $p_i(\boldsymbol{\beta})$ and $1 - p_i(\boldsymbol{\beta})$ respectively, and hence their frequency function is

$$p(y_{i,}\boldsymbol{\beta}) = p_i^{y_i}(\boldsymbol{\beta})(1-p_i(\boldsymbol{\beta}))^{1-y_i}.$$

Hence the log-likelihood function of the sample $L(\boldsymbol{\beta})$ is given by

$$\log L(\boldsymbol{\beta}) = \sum_{i=1}^n \left[y_i \log p_i(\boldsymbol{\beta}) + (1-y_i)\log(1-p_i(\boldsymbol{\beta}))\right]. \tag{7.5}$$

Differentiating (7.5) yields the estimating equations for the maximum likelihood estimate (MLE):

$$\sum_{i=1}^n \frac{y_i - p_i(\boldsymbol{\beta})}{p_i(\boldsymbol{\beta})\,(1-p_i(\boldsymbol{\beta}))} F'(\boldsymbol{\beta}'\mathbf{x}_i)\mathbf{x}_i = \mathbf{0}. \tag{7.6}$$

In the case of random $\mathbf{x}_i$'s, (7.2) yields

$$\log L(\boldsymbol{\beta}) = \sum_{i=1}^n \left[y_i \log p_i(\boldsymbol{\beta}) + (1-y_i)\log(1-p_i(\boldsymbol{\beta}))\right] + \sum_{i=1}^n \log g(\mathbf{x}_i), \tag{7.7}$$

where g is the density of the $\mathbf{x}_i$'s. Differentiating this log likelihood again yields (7.6).

For predicting the values y_i from the corresponding regressor vector $\mathbf{x}_i$, the ideal situation would be that of "perfect separation", i.e., when there exist $\boldsymbol{\gamma} \in R^p$ and $\alpha \in R$ such that

$$\begin{aligned} \boldsymbol{\gamma}'\mathbf{x}_i > \alpha \quad &\text{if} \quad y_i = 1 \\ \boldsymbol{\gamma}'\mathbf{x}_i < \alpha \quad &\text{if} \quad y_i = 0, \end{aligned} \tag{7.8}$$

and therefore $\gamma'\mathbf{x} = \alpha$ is a "separating hyperplane". It is intuitively clear that if one such hyperplane exists, there must be infinite ones. However, this has the consequence that the MLE becomes undetermined. More precisely, let $\beta(k) = k\gamma$. Then

$$\lim_{k\to+\infty} p_i(\beta(k)) = \lim_{u\to+\infty} F(u) = 1 \text{ if } y = 1$$

and

$$\lim_{k\to+\infty} p_i(\beta(k)) = \lim_{u\to-\infty} F(u) = 1 \text{ if } y = 0.$$

Therefore

$$\lim_{k\to\infty} \sum_{i=1}^{n} \left[y_i \log p_i(\beta(k)) + (1 - y_i) \log((1 - p_i(\beta(k))) \right] = 0.$$

Since for all finite β

$$\sum_{i=1}^{n} y_i \log p_i(\beta(k)) + (1 - y_i) \log(1 - p_i(\beta(k))) < 0,$$

then, according to (7.5)–(7.7), the MLE does not exist for either fixed or random $\mathbf{x}_i$'s.

Albert and Anderson (1984) showed that the MLE is unique and finite if and only if no $\gamma \in R^p$ and $\alpha \in R$ exist such that

$$\begin{array}{lll} \gamma'\mathbf{x}_i \geq \alpha & \text{if} & y_i = 1 \\ \gamma'\mathbf{x}_i \leq \alpha & \text{if} & y_i = 0. \end{array}$$

For $\gamma \in R^p$ call $K(\gamma)$ the number of points in the sample which do not satisfy (7.8), and define

$$k_0 = \min_{\gamma \in R^p} K(\gamma), \quad \gamma_0 = \arg\min_{\gamma \in R^p} K(\gamma). \tag{7.9}$$

Then replacing the k_0 points which do not satisfy (7.8) for $\gamma = \gamma_0$ (called "overlapping points") by other k_0 points lying on the correct side of the hyperplane $\gamma_0'\mathbf{x} = 0$, the MLE goes to infinity. Then we can say that the breakdown point of the MLE in this case is k_0/n. Observe that the points which replace the k_0 misclassified points are not "atypical". They follow the pattern of the majority: those with $\gamma_0'\mathbf{x}_i > 0$ have $y_i = 1$ and those with $\gamma_0'\mathbf{x}_i < 0$ have $y_i = 0$. The fact that the points that produce breakdown to infinity are not outliers was observed for the first time by Croux, Flandre and Haesbroeck (2002). They also showed that the effect produced by outliers on the MLE is quite different; it will be described later in this section.

It is easy to show that the function (7.4) verifies $F'(y) = F(y)(1 - F(y))$. Hence in the logistic case, (7.6) simplifies to

$$\sum_{i=1}^{n} (y_i - p_i(\beta))\mathbf{x}_i = \mathbf{0}. \tag{7.10}$$

We shall henceforth treat only the logistic case, which is probably the most commonly used one, and which, as we shall now see, is easier to robustify.

According to (3.47), the influence function of the MLE for the logistic model is

$$\mathrm{IF}(y, \mathbf{x}, \boldsymbol{\beta}) = \mathbf{M}^{-1}(y - F(\boldsymbol{\beta}'\mathbf{x}))\mathbf{x},$$

where $\mathbf{M} = \mathrm{E}(F'(\boldsymbol{\beta}'\mathbf{x})\mathbf{x}\mathbf{x}')$. Since the factor $(y - F(\boldsymbol{\beta}'\mathbf{x}))$ is bounded, the only outliers that make this influence large are those such that $||\mathbf{x}_i|| \to \infty$, $y_i = 1$ and $\boldsymbol{\beta}'\mathbf{x}_i$ is bounded away from ∞, or those such that $||\mathbf{x}_i|| \to \infty$, $y_i = 0$ and $\boldsymbol{\beta}'\mathbf{x}_i$ is bounded away from $-\infty$. Croux et al. (2002) showed that if the model has an intercept (see (7.3)), then unlike the case of ordinary linear regression, this kind of outliers make the MLE of $\boldsymbol{\beta}_1$ tend to zero and not to infinity. More precisely, they show that by conveniently choosing not more than $2(p-1)$ outliers, the MLE $\widehat{\boldsymbol{\beta}}_1$ of $\boldsymbol{\beta}_1$ can be made as close to zero as desired. This is a situation where, although the estimate remains bounded, we may say that it breaks down since its values are determined by the outliers rather than by the bulk of the data, and in this sense the breakdown point to zero of the MLE is $\leq 2(p-1)/n$.

To exemplify this lack of robustness we consider a sample of size 100 from the model

$$\log \frac{\mathrm{P}(y=1)}{1-\mathrm{P}(y=1)} = \beta_0 + \beta_1 x,$$

where $\beta_0 = -2$, $\beta_1 = 3$ and x is uniform in the interval $[0, 1]$. Figure 7.1 (code **logregsim1**) shows the sample, and we find as expected that for low values of x, a

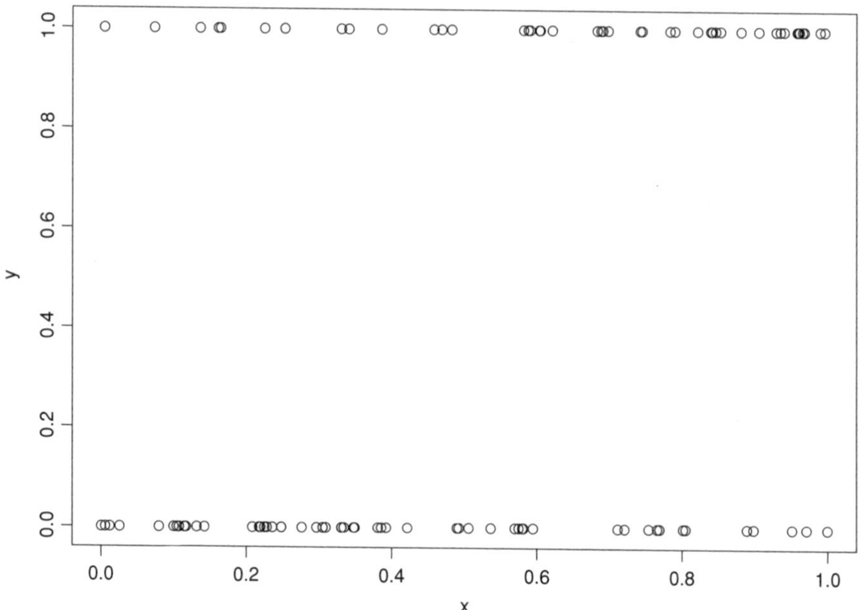

Figure 7.1 Simulated data: plot of y vs. x

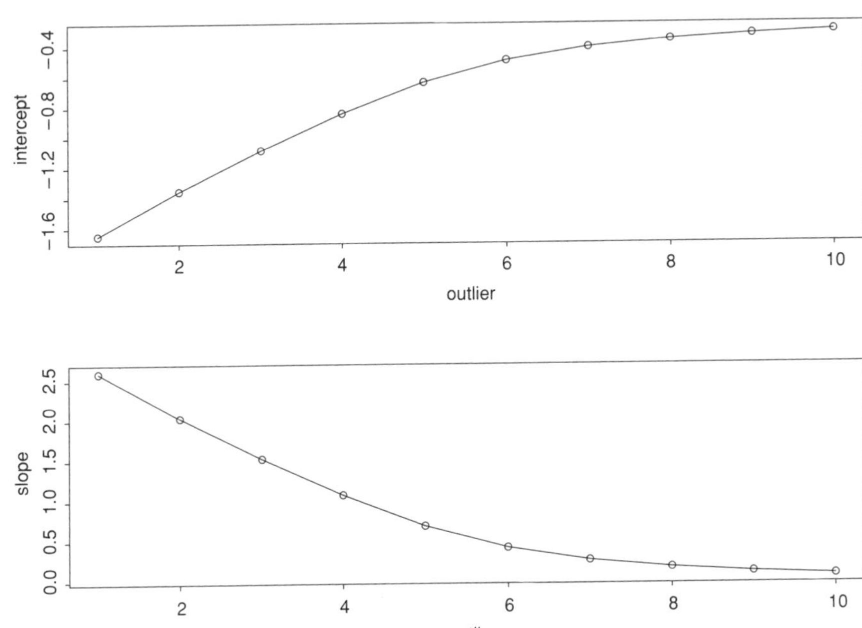

Figure 7.2 Simulated data: effect of one outlier

majority of the y's are zero and the opposite occurs for large values of x. The MLE is $\beta_0 = -1.72$, $\beta_1 = 2.76$.

Now we add to this sample one outlier of the form $x_0 = i$ and $y_0 = 0$ for $i = 1, \ldots, 10$, and we plot in Figure 7.2 the values of β_0 and β_1. Observe that the value of β_1 tends to zero and β_0 converges to $\log(\alpha/(1-\alpha))$, where $\alpha = 45/101 \approx 0.45$ is the frequency of ones in the contaminated sample.

7.2 Robust estimates for the logistic model

7.2.1 Weighted MLEs

Carroll and Pederson (1993) proposed a simple way to turn the MLE into an estimate with bounded influence, by downweighting high-leverage observations. A measure of the leverage of observation $\mathbf{x}$ similar to (5.46) is defined as

$$h_n(\mathbf{x}) = ((\mathbf{x} - \widehat{\mu}_n)'\widehat{\Sigma}_n^{-1}(\mathbf{x} - \widehat{\mu}_n))^{1/2},$$

where $\widehat{\mu}_n$ and $\widehat{\Sigma}_n$ are respectively a robust location vector and a robust dispersion matrix estimate of $\mathbf{x}_1, \ldots, \mathbf{x}_n$. Note that if $\widehat{\mu}_n$ and $\widehat{\Sigma}_n$ are affine equivariant, this measure is invariant under affine transformations.

Then robust estimates can be obtained by minimizing

$$\sum_{i=1}^{n} w_i \left[y_i \log p_i(\beta) + (1 - y_i) \log(1 - p_i(\beta)) \right],$$

with

$$w_i = W(h_n(\mathbf{x}_i)), \tag{7.11}$$

where W is a nonincreasing function such that $W(u)u$ is bounded. Carroll and Pederson (1993) proposed choosing W in the following family which depends on a parameter $c > 0$:

$$W(u) = \left(1 - \frac{u^2}{c^2}\right)^3 \mathrm{I}(|u| \le c).$$

In S-PLUS the estimates are implemented using these weights or the Huber family of weights defined in (2.32). This estimate will be called *weighted maximum likelihood estimate* (WMLE). According to (3.47), its influence function is

$$\mathrm{IF}(y, \mathbf{x}, \beta) = \mathbf{M}^{-1}(y - F(\beta'\mathbf{x}))\mathbf{x}W(h(\mathbf{x})),$$

with

$$h(\mathbf{x}) = ((\mathbf{x} - \boldsymbol{\mu})'\Sigma^{-1}(\mathbf{x} - \boldsymbol{\mu}))^{1/2},$$

where $\boldsymbol{\mu}$ and Σ are the limit values of $\widehat{\boldsymbol{\mu}}_n$ and $\widehat{\Sigma}_n$, and

$$\mathbf{M} = \mathrm{E}(W(h(\mathbf{x}))F'(\beta\mathbf{x})\mathbf{x}\mathbf{x}').$$

These estimates are asymptotically normal and their asymptotic covariance matrix can be found using (3.49).

7.2.2 Redescending M-estimates

The MLE for the logistic model can also be defined as minimizing the total *deviance*

$$D(\beta) = \sum_{i=1}^{n} d^2(p_i(\beta), y_i),$$

where $d(\mathbf{u}, y)$ is given by

$$d(u, y) = \left\{-2\left[y \log u + (1 - y)\log(1 - u)\right]\right\}^{1/2} \mathrm{sgn}(y - u) \tag{7.12}$$

and is a signed measure of the discrepancy between a Bernoulli variable y and its expected value u. Observe that

$$d(u, y) = \begin{cases} 0 & \text{if} \quad u = y \\ -\infty & \text{if} \quad u = 1,\ y = 0 \\ \infty & \text{if} \quad u = 0,\ y = 1. \end{cases}$$

In the logistic model, the values $d(p_i(\boldsymbol{\beta}), y_i)$ are called *deviance residuals*, and they measure the discrepancies between the probabilities fitted using the regression coefficients $\boldsymbol{\beta}$ and the observed values. In Section 7.3 we define the deviance residuals for a larger family of models.

Pregibon (1981) proposed robust M-estimates for the logistic model based on minimizing

$$M(\beta) = \sum_{i=1}^{n} \rho(d^2(p_i(\boldsymbol{\beta}), y_i)),$$

where $\rho(u)$ is a function which increases more slowly than the identity function. Bianco and Yohai (1996) observed that for random $\mathbf{x}_i$ these estimates are not Fisher-consistent, i.e., the respective score function does not satisfy (3.31), and found that this difficulty may be overcome by using a correction term. They proposed to estimate $\boldsymbol{\beta}$ by minimizing

$$M(\boldsymbol{\beta}) = \sum_{i=1}^{n} \left[\rho(d^2(p_i(\boldsymbol{\beta}), y_i)) + q(p_i(\boldsymbol{\beta}))\right], \tag{7.13}$$

where $\rho(u)$ is nondecreasing and bounded and

$$q(u) = \upsilon(u) + \upsilon(1-u),$$

with

$$\upsilon(u) = 2\int_0^u \psi(-2\log t)dt$$

and $\psi = \rho'$.

Croux and Haesbroeck (2003) described sufficient conditions on ρ to guarantee a finite minimum of $M(\boldsymbol{\beta})$ for all samples with overlapping observations ($k_0 > 0$). They proposed to choose ψ in the family

$$\psi_c^{CH}(u) = \exp\left(-\sqrt{\max(u, c)}\right). \tag{7.14}$$

Differentiating (7.13) with respect to $\boldsymbol{\beta}$ and using the facts that

$$q'(u) = 2\psi(-2\log u) - 2\psi(-2\log(1-u))$$

and that in the logistic model $F' = F(1-F)$, we get

$$2\sum_{i=1}^{n} \psi(d_i^2(\boldsymbol{\beta}))(y_i - p_i(\boldsymbol{\beta}))\mathbf{x}_i$$

$$-2\sum_{i=1}^{n} p_i(\boldsymbol{\beta})(1-p_i(\boldsymbol{\beta}))[\psi(-2\log p_i(\boldsymbol{\beta})) - \psi(-2\log(1-p_i(\boldsymbol{\beta})))]\mathbf{x}_i = \mathbf{0},$$

where $d_i(\boldsymbol{\beta}) = d(p_i(\boldsymbol{\beta}), y_i)$ are the deviance residuals given in (7.12). This equation can also be written as

$$\sum_{i=1}^{n}[\psi(d_i^2(\boldsymbol{\beta}))(y_i - p_i(\boldsymbol{\beta})) - \mathrm{E}_{\beta}(\psi(d_i^2(\boldsymbol{\beta}))(y_i - p_i(\boldsymbol{\beta}))|\mathbf{x}_i)]\mathbf{x}_i = \mathbf{0}, \tag{7.15}$$

where E_{β} denotes the expectation when $\mathrm{P}(y_i = 1|\mathbf{x}_i) = p_i(\boldsymbol{\beta})$. Putting

$$\Psi(y_i, \mathbf{x}_i, \boldsymbol{\beta}) = [\psi(d_i^2(\boldsymbol{\beta}))(y_i - p_i(\boldsymbol{\beta})) - \mathrm{E}_{\beta}(\psi(d_i^2(\boldsymbol{\beta}))(y_i - p_i(\boldsymbol{\beta}))|\mathbf{x}_i)]\mathbf{x}_i, \tag{7.16}$$

equation (7.15) can also be written as

$$\sum_{i=1}^{n}\Psi(y_i, \mathbf{x}_i, \boldsymbol{\beta}) = \mathbf{0}.$$

From (7.16) it is clear that $\mathrm{E}_{\beta}(\Psi(y_i, \mathbf{x}_i, \boldsymbol{\beta})) = \mathbf{0}$, and therefore these estimates are Fisher-consistent. Their influence function can again be obtained from (3.47). Bianco and Yohai (1996) proved that under general conditions these estimates are asymptotically normal. The asymptotic covariance matrix can be obtained from (3.49).

In Figure 7.3 we repeat the same graph as in Figure 7.2 using both estimates: the MLE and a redescending M-estimate with $\psi_{0.5}^{CH}$ (code **logregsim2**). We observe that the changes in both the slope and intercept of the M-estimate are very small compared to those of the MLE.

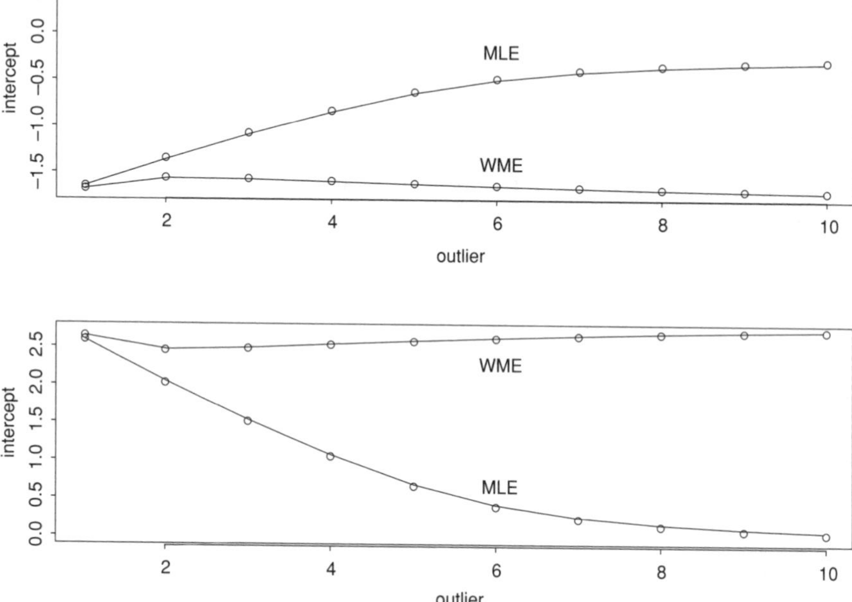

Figure 7.3 Effect of an outlier on the M-estimate of slope and intercept

Since the function $\Psi(y_i, \mathbf{x}_i, \beta)$ is not bounded, the M-estimate does not have bounded influence. To obtain bounded influence estimates, Croux and Haesbroeck (2003) proposed to downweight high-leverage observations. They define a Mallows-type GM-estimate by minimizing

$$M(\beta) = \sum_{i=1}^{n} w_i \left[\rho(d^2(p_i(\beta), y_i)) + q(p_i(\beta))\right], \tag{7.17}$$

where the weights are given by (7.11).

Example 7.1 *The data set* ***leukemia*** *has been considered by Cook and Weisberg (1982, Chapter 5, p. 193) and Johnson (1985) to illustrate the identification of influential observations. The data consist of 33 leukemia patients. The response variable is one when the patient survives more than 52 weeks. Two covariates are considered: white blood cell count (WBC) and presence or absence of certain morphological characteristics in the white cells (AG). The model also includes an intercept.*

Cook and Weisberg detected an observation (#15) corresponding to a patient with WBC = 100.000 who survived for a long period of time. This observation was very influential on the MLE. They also noticed that after removing this observation a much better overall fit was obtained, and that the fitted survival probabilities of those observations corresponding to patients with small values of WBC increased significantly.

In Table 7.1 (code **leukemia**) we give the estimated slopes and their asymptotic standard deviations corresponding to

- the MLE with the complete sample (MLE)
- the MLE after removing the influential observation (MLE_{-15})
- the weighted MLE (WMLE)
- the M-estimate (M) corresponding to the Croux and Haesbroeck family ψ_c^{CH} with $c = 0.5$

Table 7.1 Estimates for leukemia data and their standard errors

Estimate	Intercept	WBC ($\times 10^4$)	AG
MLE	−1.31 (0.81)	−0.32 (0.18)	2.26 (0.95)
MLE_{-15}	0.21 (1.08)	−2.35 (1.35)	2.56 (1.23)
WMLE	0.17 (1.08)	−2.25 (1.32)	2.52 (1.22)
M	0.16 (1.66)	−1.77 (2.33)	1.93 (1.16)
WM	0.20 (1.19)	−2.21 (0.98)	2.40 (1.30)
CUBIF	−0.68 (0.91)	−0.91 (0.50)	2.25 (1.03)

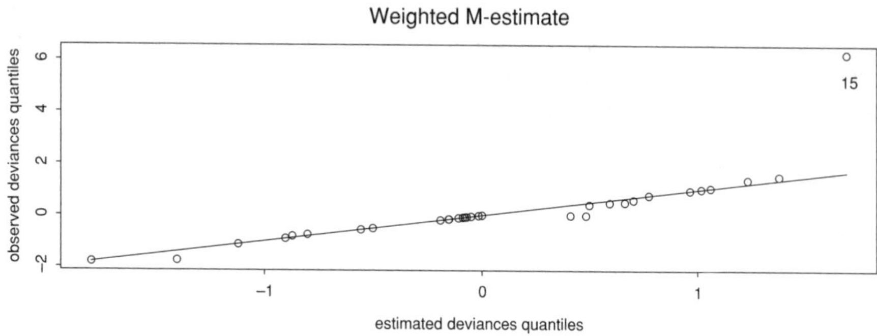

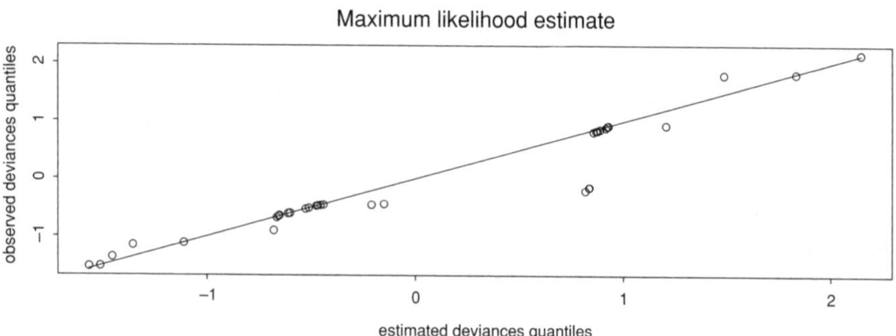

Figure 7.4 Leukemia data: Q–Q plots of deviances

- the weighted M-estimate (WM)
- the optimal conditionally unbiased bounded influence estimate (CUBIF) that will be described for a more general family of models in Section 7.3.

We can observe that coefficients fitted with MLE_{-15} are very similar to those of the WMLE, M- and WM-estimates. The CUBIF gives results intermediate between MLE and MLE_{-15}.

Figure 7.4 shows the Q–Q plots of the observed deviances $d(p_i(\widehat{\boldsymbol{\beta}}), y_i)$ when $\widehat{\boldsymbol{\beta}}$ is the MLE and the WM-estimate. Also plotted in the same figure is the identity line. In ordinary linear regression, we can theoretically calculate the distribution of the residuals assuming a distribution of the errors (say, normal); this is the basis for the Q–Q plot. For logistic regression, García Ben and Yohai (2004) proposed to calculate the theoretical distribution (which depends on that of $\mathbf{x}$) of $d(F(\widehat{\boldsymbol{\beta}}'\mathbf{x}), y)$, assuming $\widehat{\boldsymbol{\beta}}$ to be the true parameter vector and approximating the distribution of $\mathbf{x}$ by empirical one. These Q–Q plots display the empirical quantiles of $d(F(\widehat{\boldsymbol{\beta}}'\mathbf{x}_i), y_i)$ against the theoretical ones. It is seen that the WM-estimate gives a better fit than the MLE, and that its Q–Q plot pinpoints observation 15 as an outlier.

Table 7.2 Estimates for skin data

Estimate	Intercept	Log VOL	Log RATE
MLE	−9.53 (3.21)	3.88 (1.42)	2.65 (0.91)
WMLE	−9.36 (3.18)	3.83 (1.41)	2.59 (0.90)
M	−14.21 (10.88)	5.82 (4.52)	3.72 (2.70)
WM	−14.21 (10.88)	5.82 (4.53)	3.72 (2.70)
CUBIF	−9.47 (3.22)	3.85 (1.42)	2.63 (0.91)

Example 7.2 *The data set* **skin** *was introduced by Finney (1947) and later studied by Pregibon (1982) and Croux and Haesbroeck (2003). The response is the presence or absence of vasoconstriction of the skin of the digits after air inspiration, and the explanatory variables are the logarithms of the volume of air inspired (log VOL) and of the inspiration rate (log RATE).*

Table 7.2 gives the estimated coefficients and standard errors for the MLE, WMLE, M-, WM- and CUBIF estimates (code **skin**). Since there are no outliers in the regressors, the weighted versions give similar results to the unweighted ones. This also explains in part why the CUBIF estimate is very similar to the MLE.

Figure 7.5 shows the Q–Q plots of deviances of the MLE and the M-estimate. The identity line is also plotted for reference. The latter lets us more neatly detect the observations 4 and 18 as outliers. On deleting these observations, the remaining data set has only one overlapping observation. This is the reason why the M- and WM-estimates that downweight these observations have large standard errors.

7.3 Generalized linear models

The logistic model is included in a more general class called *generalized linear models* (GLMs). If $\mathbf{x}$ is fixed, the distribution of y is given by a density $f(y, \lambda)$ depending on a parameter λ, and there is a known one-to-one function l, called *the link function*, and an unknown vector $\boldsymbol{\beta}$, such that $l(\lambda) = \boldsymbol{\beta}'\mathbf{x}$. If $\mathbf{x}$ is random, the *conditional* distribution of y given $\mathbf{x}$ is given by $f(y, \lambda)$ with $l(\lambda) = \boldsymbol{\beta}'\mathbf{x}$. In the previous section we had $\lambda \in (0, 1)$ and

$$f(y, \lambda) = \lambda^y (1 - \lambda)^{1-y}, \tag{7.18}$$

and $l = F^{-1}$.

A convenient framework is the *exponential family* of distributions:

$$f(y, \lambda) = \exp[m(\lambda)y - G(m(\lambda)) - t(y)], \tag{7.19}$$

where m, G and t are given functions. When $l = m$, the link function is called *canonical*.

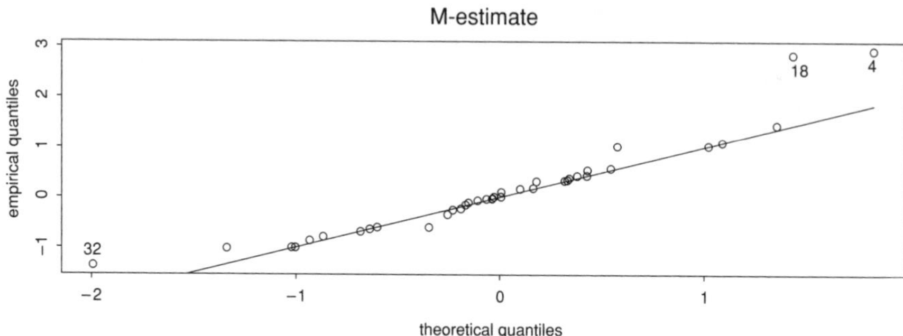

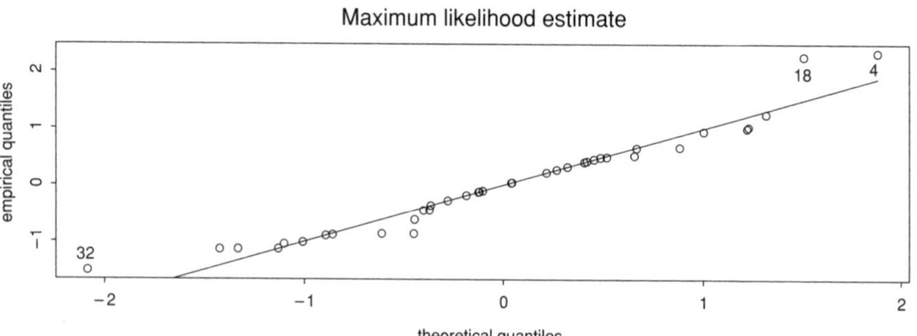

Figure 7.5 Skin data: Q–Q plots of deviances of the maximum likelihood and the weighted M-estimates

It is easy to show that if y has distribution (7.19), then

$$\mathrm{E}_\lambda(y) = g(m(\lambda)),$$

where $g = G'$.

This family contains the Bernoulli distribution (7.18) which corresponds to

$$m(\lambda) = \log \frac{\lambda}{1-\lambda},\ G(u) = \log(1+e^u) \text{ and } t(y) = 0.$$

In this case the canonical link corresponds to the logistic model and $\mathrm{E}_\lambda(y) = \lambda$.

Another example is the Poisson family with

$$f(y, \lambda) = \frac{\lambda^y e^{-\lambda}}{y!}.$$

This family corresponds to (7.19) with

$$m(\lambda) = \log \lambda,\ G(u) = e^u \text{ and } t(y) = \log y!.$$

This yields $\mathrm{E}y = g(m(\lambda)) = \lambda$. The canonical link in this case is $l(\lambda) = \log \lambda$.

Define $\widehat{\lambda}(y)$ as the value of λ that maximizes $f(y,\lambda)$ or equivalently that maximizes

$$\log f(y,\lambda) = m(\lambda)y - G(m(\lambda)) - t(y).$$

Differentiating we obtain that $\lambda = \widehat{\lambda}(y)$ should satisfy

$$m'(\lambda)y - g(m(\lambda))m'(\lambda) = 0$$

and therefore $g(m(\widehat{\lambda}(y))) = y$. Define the deviance residual function by

$$\begin{aligned} d(y,\lambda) &= \left\{2\log\left[f(y,\lambda)/f(y,\widehat{\lambda}(y))\right]\right\}^{1/2}\operatorname{sgn}(y - g(m(\lambda)) \\ &= \left\{2\left[m(\lambda)y - G(m(\lambda)) - m(\widehat{\lambda}(y))y + G(m(\widehat{\lambda}(y)))\right]\right\}^{1/2}\operatorname{sgn}(y - g(m(\lambda))). \end{aligned}$$

It is easy to check that when y is a Bernoulli variable, this definition coincides with (7.12).

Consider now a sample $(\mathbf{x}_1, y_1), \ldots, (\mathbf{x}_n, y_n)$ from a generalized linear model with the canonical link function and fixed $\mathbf{x}_i$. Then the log likelihood is

$$\begin{aligned} \log L(\beta) &= \sum_{i=1}^{n} \log f(y_i, m^{-1}(\beta'\mathbf{x}_i)) \\ &= \sum_{i=1}^{n}(\beta'\mathbf{x}_i)y_i - \sum_{i=1}^{n} G(\beta'\mathbf{x}_i) - \sum_{i=1}^{n} t(y_i). \end{aligned} \tag{7.20}$$

The MLE maximizes $\log L(\beta)$ or equivalently

$$\begin{aligned} &\sum_{i=1}^{n} 2\left(\log f(y_i, m^{-1}(\beta'\mathbf{x}_i)) - \log f(y_i, \widehat{\lambda}(y_i))\right) \\ &\quad = \sum_{i=1}^{n} d^2(y_i, m^{-1}(\beta'\mathbf{x}_i)). \end{aligned}$$

Differentiating (7.20) we get the equations for the MLE:

$$\sum_{i=1}^{n}(y_i - g(\beta'\mathbf{x}_i))\mathbf{x}_i = \mathbf{0}. \tag{7.21}$$

For example, for the Poisson family this equation is

$$\sum_{i=1}^{n}(y_i - e^{\beta'\mathbf{x}_i})\mathbf{x}_i = \mathbf{0}. \tag{7.22}$$

7.3.1 Conditionally unbiased bounded influence estimates

To robustify the MLE, Künsch, Stefanski and Carroll (1989) considered M-estimates of the form

$$\sum_{i=1}^{n} \Psi(y_i, \mathbf{x}_i, \boldsymbol{\beta}) = \mathbf{0},$$

where $\Psi : R^{1+p+p} \to R^p$ such that

$$\mathrm{E}(\Psi(y, \mathbf{x}, \boldsymbol{\beta})|\mathbf{x}_i) = \mathbf{0}. \tag{7.23}$$

These estimates are called conditionally Fisher-consistent. Clearly these estimates are also Fisher-consistent, i.e., $\mathrm{E}(\Psi(y_i, \mathbf{x}_i, \boldsymbol{\beta})) = 0$. Künsch et al. (1989) found the estimate in this class which solves an optimization problem similar to Hampel's one studied in Section 3.5.4. This estimate minimizes a measure of efficiency—based on the asymptotic covariance matrix under the model—subject to a bound on a measure of infinitesimal sensitivity similar to the gross-error sensitivity. Since these measures are quite complicated and may be controversial, we do not give more details about their definition.

The optimal score function Ψ has the following form:

$$\Psi(y, \mathbf{x}, \boldsymbol{\beta}, b, \mathbf{B}) = W(\boldsymbol{\beta}, y, \mathbf{x}, b, \mathbf{B}) \left\{ y - g(\boldsymbol{\beta}'\mathbf{x}) - c\left(\boldsymbol{\beta}'\mathbf{x}, \frac{b}{h(\mathbf{x}, \mathbf{B})}\right) \right\} \mathbf{x},$$

where b is the bound on the measure of infinitesimal sensitivity, $\mathbf{B}$ is a dispersion matrix that will be defined below, and $h(\mathbf{x}, \mathbf{B}) = (\mathbf{x}'\mathbf{B}^{-1}\mathbf{x})^{1/2}$ is a leverage measure. Observe the similarity with (7.21). The function W downweights atypical observations and makes Ψ bounded, and therefore the corresponding M-estimate has bounded influence. The function $c(\boldsymbol{\beta}'\mathbf{x}, b/h(\mathbf{x}, \mathbf{B}))$ is a bias correction term chosen so that (7.23) holds. Call $r(y, \mathbf{x}, \boldsymbol{\beta})$ the corrected residual

$$r(y, \mathbf{x}, \boldsymbol{\beta}, b, \mathbf{B}) = y_i - g(\boldsymbol{\beta}'\mathbf{x}) - c\left(\boldsymbol{\beta}'\mathbf{x}, \frac{b}{h(\mathbf{x}, \mathbf{B})}\right). \tag{7.24}$$

Then the weights are of the form

$$W(\boldsymbol{\beta}, y, \mathbf{x}, b, \mathbf{B}) = W_b(r(y, \mathbf{x}, \boldsymbol{\beta})h(\mathbf{x}, \mathbf{B})),$$

where W_b is the Huber weight function (2.32) given by

$$W_b(x) = \min\left\{1, \frac{b}{|x|}\right\}.$$

Then, as in the Schweppe-type GM-estimates of Section 5.11, W downweights observations with a high product of corrected residuals and leverage.

Finally the matrix $\mathbf{B}$ should satisfy

$$\mathrm{E}(\Psi(y, \mathbf{x}, \boldsymbol{\beta}, b, \mathbf{B})\Psi'(y, \mathbf{x}, \boldsymbol{\beta}, b, \mathbf{B})) = \mathbf{B}.$$

Table 7.3 Estimates for epilepsy data

Estimate	Intercept	Age10	Base4	Trt	Base4*Trt
MLE	1.84 (0.13)	0.24 (0.04)	0.09 (0.002)	−0.13 (0.04)	0.004 (0.002)
MLE_{-49}	1.60 (0.15)	0.29 (0.04)	0.10 (0.004)	−0.24 (0.05)	0.018 (0.004)
CUBIF	1.63 (0.27)	0.13 (0.08)	0.15 (0.022)	−0.22 (0.12)	0.015 (0.022)

Details of how to implement these estimates, in particular of how to estimate **B**, and a more precise description of their optimal properties can be found in Künsch et al. (1989). We shall call these estimates *optimal conditionally unbiased bounded influence* ("optimal CUBIF") estimates

These estimates are implemented in S-PLUS for the logistic and Poisson models.

Example 7.3 *Breslow (1996) used a Poisson GLM to study the effect of drugs in epilepsy patients using a sample of size 59 (data set* **epilepsy***). The response variable is the number of attacks during four weeks (sumY) in a given time interval and the explanatory variables are: patient age divided by 10 (Age10), the number of attacks in the four weeks prior to the study (Base4), a dummy variable that takes values 1 or 0 if the patient received the drug or a placebo respectively (Trt) and Base4*Trt, to take account of the interaction between these two variables. We fit a Poisson GLM with log link using the MLE and the optimal CUBIF. Since, as we shall see below, observation 49 appears as a neat outlier, we also give the MLE without this observation (MLE_{-49}).*

The coefficient estimates and their standard errors are shown in Table 7.3 (code **epilepsy**). Figure 7.6 shows the Q–Q plots of the deviances corresponding to the optimal CUBIF estimate and to the MLE. The identity line is also plotted. This plot shows that the CUBIF estimate gives a much better fit and that observation 49 is a clear outlier.

7.3.2 Other estimates for GLMs

The redescending M-estimates of Section 7.2.2 can be extended to other GLMs. Bianco, García Ben and Yohai (2005) considered M-estimates for the case that the distribution of y is gamma and the link function is the logarithm. They showed that in this case no correction term is needed for Fisher-consistency.

Cantoni and Ronchetti (2001) robustified the quasi-likelihood approach to estimate GLMs. The quasi-likelihood estimates proposed by Wedderburn (1974) are defined as solutions of the equation

$$\sum_{i=1}^{n} \frac{y_i - \mu(\boldsymbol{\beta}'\mathbf{x}_i)}{V(\boldsymbol{\beta}'\mathbf{x}_i)} \mu'(\boldsymbol{\beta}'\mathbf{x}_i)\mathbf{x}_i = 0,$$

where

$$\mu(\boldsymbol{\lambda}) = \mathrm{E}_{\lambda}(y), \ \ \mathbf{V}(\boldsymbol{\lambda}) = \mathrm{Var}_{\lambda}(y).$$

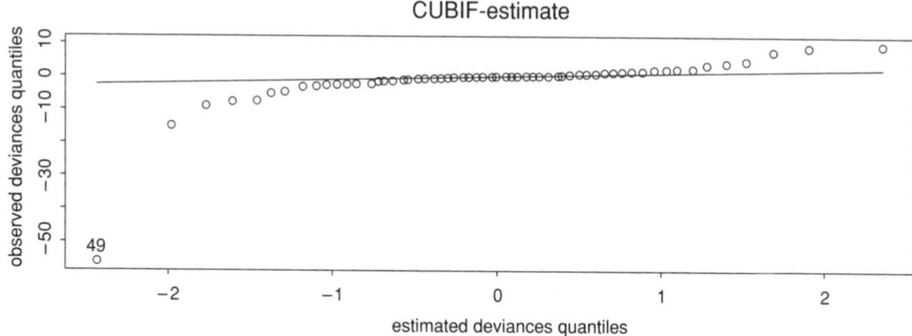

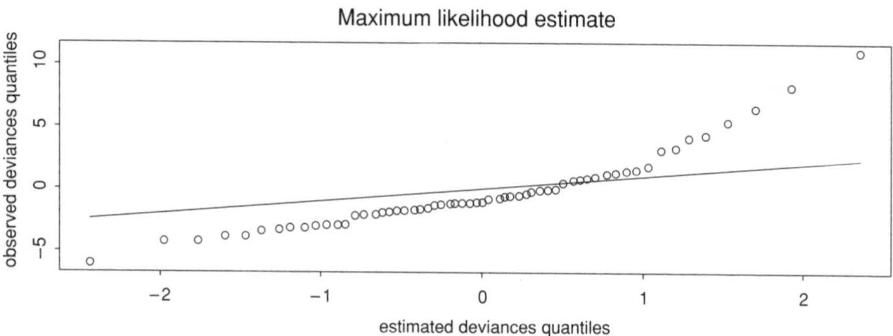

Figure 7.6 Epilepsy data: Q–Q plots of deviances

The robustification proposed by Cantoni and Ronchetti is performed by bounding and centering the quasi-likelihood score function

$$\psi(y, \boldsymbol{\beta}) = \frac{y - \mu(\boldsymbol{\beta}'\mathbf{x})}{V(\boldsymbol{\beta}'\mathbf{x})}\mu'(\boldsymbol{\beta}'\mathbf{x})\mathbf{x},$$

similarly to what was done with the maximum likelihood score function in Section 7.3.1. The purpose of centering is to obtain conditional Fisher-consistent estimates and that of bounding is to bound the IF.

7.4 Problems

7.1. The data set **neuralgia** (Piegorsch, 1992) contains the values of four predictors for 18 patients, the outcome being whether the patient experienced pain relief after a treatment. The data are described in Chapter 12. Compare the fits given by the different logistic regression estimates discussed in this chapter.

7.2. The data set **aptitude** (Miles and Shevlin, 2000) contains the values of two predictors for 27 subjects, the outcome being whether the subject passed an aptitude test. The data are described in Chapter 12. Compare the fits given by the different logistic regression estimates discussed in this chapter.

7.3. Prove (7.10).

7.4. Consider the univariate logistic model without intercept for the sample $(x_1, y_1)\ldots(x_n, y_n)$ with $x_i \in R$, i.e., $\boldsymbol{\beta}'\mathbf{x} = \beta x$. Let

$$p(x, \beta) = e^{\beta x} / \left(1 + e^{\beta x}\right) = \mathrm{P}(y = 1).$$

(a) Show that

$$A_n(\beta) = \sum_{i=1}^{n} (y_i - p(x_i, \beta))x_i$$

is decreasing in β.

(b) Call $\widehat{\beta}_n$ the ML estimate. Assume $\widehat{\beta}_n > 0$. Add one outlier $(K, 0)$ where $K > 0$; call $\widehat{\beta}_{n+1}(K)$ the MLE computed with the enlarged sample. Show that $\lim_{K\to\infty} \widehat{\beta}_{n+1}(K) = 0$. State a similar result when $\widehat{\beta}_n < 0$.

7.5. Let $Z_n = \{(\mathbf{x}_1, y_1), \ldots, (\mathbf{x}_n, y_n)\}$ be a sample for the logistic model, where the first coordinate of each $\mathbf{x}_i$ is one if the model contains an intercept. Consider a new sample $Z_n^* = \{(-\mathbf{x}_1, 1 - y_1), \ldots, (-\mathbf{x}_n, 1 - y_n)\}$.

(a) Explain why it is desirable that an estimate $\widehat{\boldsymbol{\beta}}$ satisfies the equivariance property $\widehat{\boldsymbol{\beta}}(Z_n^*) = \widehat{\boldsymbol{\beta}}(Z_n)$.

(b) Show that M-, WM- and CUBIF estimates satisfy this property.

7.6. For the model in Problem 7.4 define an estimate by the equation

$$\sum_{i=1}^{n} (y_i - p(x_i, \beta))\mathrm{sgn}(x_i) = 0.$$

Since deleting all $x_i = 0$ yields the same estimate, it will henceforth be assumed that $x_i \neq 0$ for all i.

(a) Show that this estimate is Fisher-consistent.

(b) Show that the estimate is a weighted ML-estimate.

(c) Given the sample $Z_n = \{(x_i, y_i), i = 1, \ldots, n\}$, define the sample $Z_n^* = \{(x_i^*, y_i^*), i = 1, \ldots, n\}$, where $(x_i^*, y_i^*) = (x_i, y_i)$ if $x_i > 0$ and $(x_i^*, y_i^*) = (-x_i, 1 - y_i)$ if $x_i < 0$. Show that $\widehat{\beta}_n(Z_n) = \widehat{\beta}_n(Z_n^*)$.

(d) Show that $\widehat{\beta}_n(Z_n^*)$ fulfills the equation $\sum_{i=1}^n y_i^* = \sum_{i=1}^n p(x_i^*, \beta)$. Hence $\widehat{\beta}_n(Z_n^*)$ is the value of β that matches the empirical frequency of ones with the theoretical one.

(e) Prove that $\sum_i p(x_i^*, 0) = n/2$.

(f) Show that if n is even, then the minimum number of observations that it is necessary to change in order to make $\widehat{\beta}_n = 0$ is $|n/2 - \sum_{i=1}^n y_i^*|$.

(g) Discuss the analog property for odd n.

(h) Show that the influence function of this estimate is

$$\mathrm{IF}(y, x, \beta) = \frac{(y - p(x, \beta))\mathrm{sgn}(x)}{A}$$

where $A = \mathrm{E}(p(x, \beta)(1 - p(x, \beta))|x|)$, and hence $\mathrm{GES} = 1/A$.

7.7. Consider the CUBIF estimate defined in Section 7.3.

(a) Show that the correction term $c(a,b)$ defined above (7.24) is a solution of the equation

$$\mathrm{E}_{m^{-1}(a)}(\psi_b^H(y - g(a) - c(a,b))) = 0.$$

(b) In the case of the logistic model for the Bernoulli family put $g(a) = e^a/(1+e^a)$. Then prove that $c(a,b) = c^*(g(a), b)$, where

$$c^*(p,b) = \begin{cases} (1-p)(p-b)/p & \text{if } p > \max\left(\frac{1}{2}, b\right) \\ p(b-1+p)/(1-p) & \text{if } p < \min\left(\frac{1}{2}, b\right) \\ 0 & \text{elsewhere.} \end{cases}$$

(c) Show that the limit when $b \to 0$ of the CUBIF estimate for the model in Problem 7.4 satisfies the equation

$$\sum_{i=1}^{n} \frac{(y - p(x_i, \beta))}{\max(p(x_i,\beta), 1 - p(x_i,\beta))} \operatorname{sgn}(x_i) = 0.$$

Compare this estimate with the one of Problem 7.6.

(d) Show that the influence function of this estimate is

$$\mathrm{IF}(y, x, \beta) = \frac{1}{A} \frac{(y - p(x,\beta))\operatorname{sgn}(x_i)}{\max(p(x,\beta), 1 - p(x,\beta))}$$

with $A = \mathrm{E}(\min(p(x,\beta), (1 - p(x,\beta)))|x|)$; and that the gross-error sensitivity is $\mathrm{GES}(\beta) = 1/A$.

(e) Show that this GES is smaller than the GES of the estimate given in Problem 7.6. Explain why this may happen.

8

Time Series

Throughout this chapter we shall focus on time series in *discrete time* whose time index t is integer valued, i.e., $t = 0, \pm 1, \pm 2, \ldots$. We shall typically label the observed values of time series as x_t or y_t, etc.

We shall assume that our time series either is stationary in some sense or may be reduced to stationarity by a combination of elementary differencing operations and regression trend removal. Two types of stationarity are in common use, *second-order* stationarity and *strict* stationarity. The sequence is said to be second-order (or *wide-sense*) stationary if the first- and second-order moments $\mathrm{E}y_t$ and $\mathrm{E}(y_{t_1} y_{t_2})$ exist and are finite, with $\mathrm{E}y_t = \mu$ a constant independent of t, and the covariance of y_{t+l} and y_t depends only on the lag l:

$$\mathrm{Cov}(y_{t+l}, y_t) = C(l) \text{ for all } t, \tag{8.1}$$

where C is called the *covariance function.*

The time series is said to be strictly stationary if for every integer $k \geq 1$ and every subset of times $t_1, t_2, \ldots, t_k$, the joint distribution of $y_{t_1}, y_{t_2}, \ldots, y_{t_k}$ is invariant with respect to shifts in time, i.e., for every positive integer k and every integer l we have

$$\mathcal{D}(y_{t_1+l}, y_{t_2+l}, \ldots, y_{t_k+l}) = \mathcal{D}(y_{t_1}, y_{t_2}, \ldots, y_{t_k}),$$

where $\mathcal{D}$ denotes the joint distribution. A strictly stationary time series with finite second moments is obviously second-order stationary, and we shall assume unless otherwise stated that our time series is at least second-order stationary.

8.1 Time series outliers and their impact

Outliers in time series are more complex than in the situations dealt with in the previous chapters, where there is no temporal dependence in the data. This is because in the

Robust Statistics – Theory and Methods Ricardo A. Maronna, R. Douglas Martin and Víctor J. Yohai

time series setting we encounter several different types of outliers, as well as other important behaviors that are characterized by their temporal structure. Specifically, in fitting time series models we may have to deal with one or more of the following:

- Isolated outliers
- Patchy outliers
- Level shifts in mean value.

While level shifts have a different character than outliers, they are a frequently occurring phenomenon that must be dealt with in the context of robust model fitting, and so we include them in our discussion of robust methods for time series. The following figures display time series which exhibit each of these types of behavior. Figure 8.1 shows a time series of stock returns for a company with stock ticker NHC that contains an isolated outlier. Here we define stock returns r_t as the relative change in price $r_t = (p_t - p_{t-1})/p_{t-1}$.

Figure 8.2 shows a time series of stock prices (for a company with stock ticker WYE) which has a patch outlier of length 4 with roughly constant size. Patch outliers can have different shapes or "configurations". For example, the stock returns for the company with ticker GHI in Figure 8.3 have a "doublet"patch outlier. The doublet outlier in the GHI returns arises because of the differencing operation in the two returns computations that involve the isolated outlier in the GHI price series.

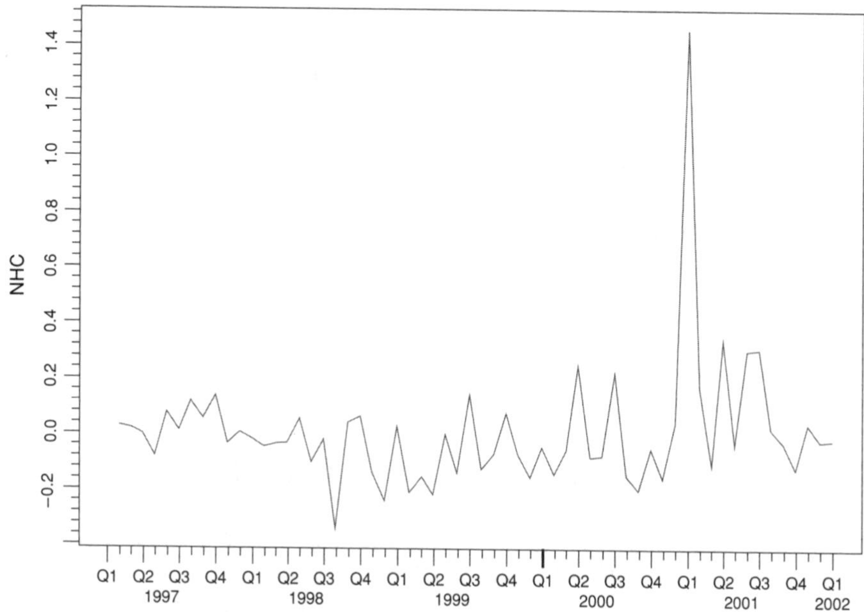

Figure 8.1 Stock returns (NHC) with isolated outlier

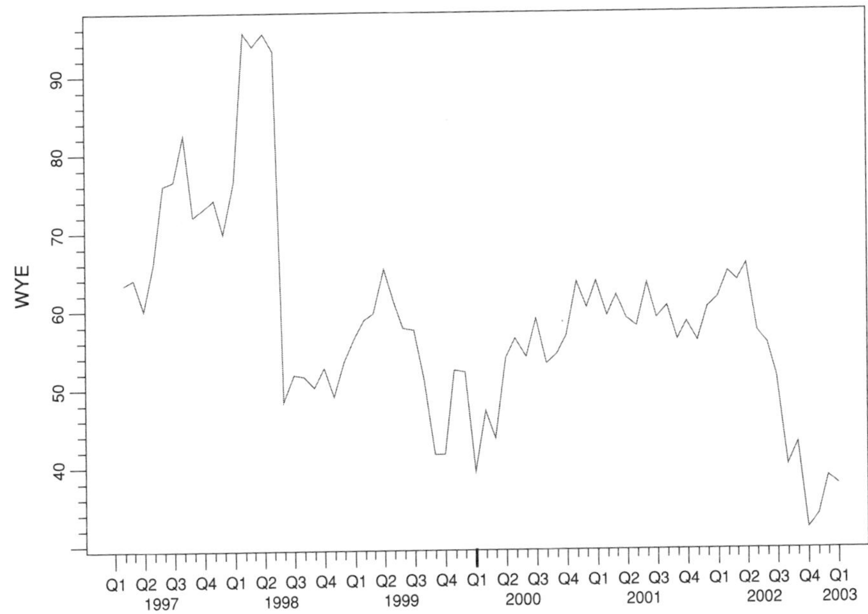

Figure 8.2 Stock prices (WYE) with patch outliers

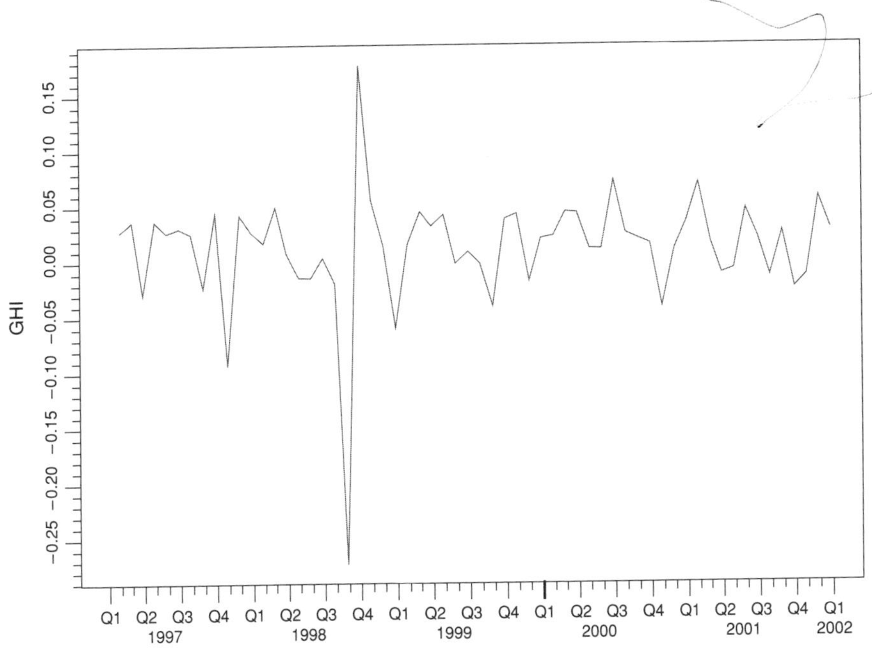

Figure 8.3 Stock returns (GHI) with doublet outlier

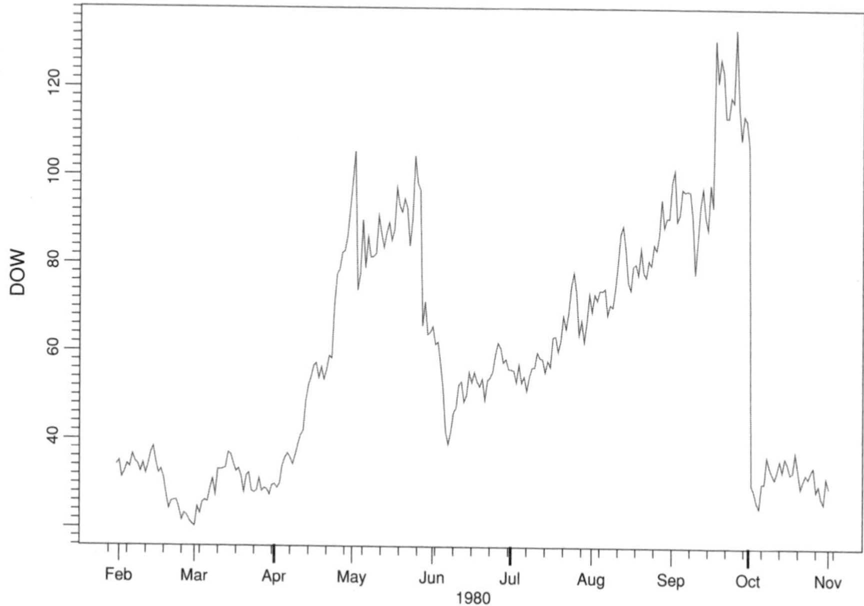

Figure 8.4 Stock prices with level shift

Figure 8.4 shows a price time series for Dow Jones (ticker DOW) that has a large level shift at the begining of October. Note that this level shift will produce an isolated outlier in the Dow Jones returns series.

Finally, Figure 8.5 shows a time series of tobacco sales in the UK (West and Harrison, 1997) that contains both an isolated outlier and two or three level shifts. The series also appears to contain trend segments at the beginning and end of the series. It is important to note that since isolated outliers, patch outliers and level shifts can all occur in a single time series, it will not suffice to discuss robustness toward outliers without taking into consideration handling of patch outliers and level shifts. Note also that when one first encounters an outlier, i.e., as the most recent observation in a time series, then lacking side information we do not know whether it is an isolated outlier, or a level shift or a short patch outlier. Consequently it will take some amount of future data beyond the time of occurrence of the outlier in order to resolve this uncertainty.

8.1.1 Simple examples of outliers influence

Time series outliers can have an arbitrarily adverse influence on parameter estimates for time series models, and the nature of this influence depends on the type of outlier. We focus on the lag-k autocorrelation

$$\rho(k) = \frac{\mathrm{Cov}(y_{t+k}, y_t)}{\mathrm{Var}(y_t)} = \frac{C(k)}{C(0)}. \tag{8.2}$$

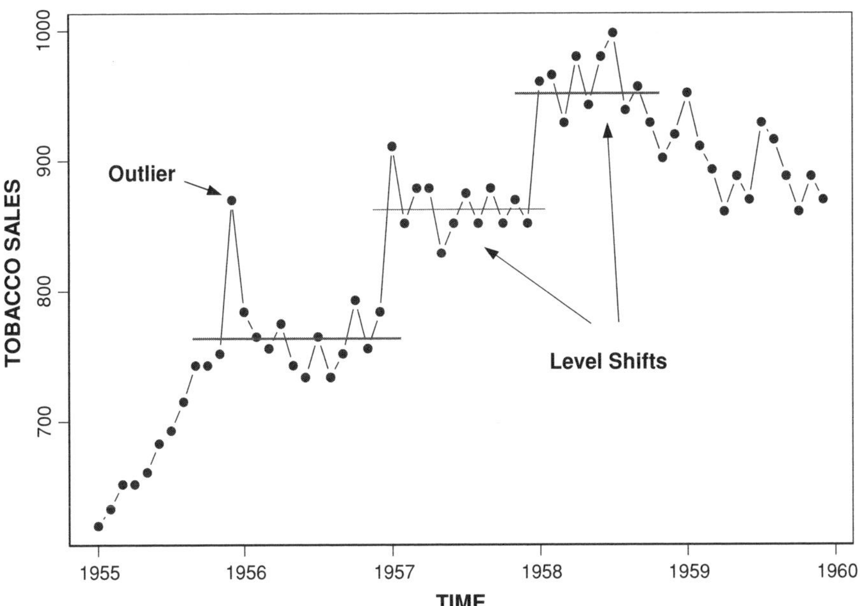

Figure 8.5 Tobacco and related sales in the UK

Here we take a simple first look at the impact of time series outliers of different types by focusing on the special case of the estimation of $\rho(1)$. Let $y_1, y_2, \ldots, y_T$ be the observed values of the series. We initially assume for simplicity that $\mu = \mathrm{E}y = 0$. In that case a natural estimate $\widehat{\rho}(1)$ of $\rho(1)$ is given by the lag-1 sample autocorrelation coefficient

$$\widehat{\rho}(1) = \frac{\sum_{t=1}^{T-1} y_t y_{t+1}}{\sum_{t=1}^{T} y_t^2}. \tag{8.3}$$

It may be shown that $|\widehat{\rho}(1)| \leq 1$, which is certainly a reasonable property for such an estimate (see Problem 8.1).

Now suppose that for some t_0, the true value y_{t_0} is replaced by an arbitrary value A, where $2 \leqslant t_0 \leqslant T-1$. In this case the estimate becomes

$$\widehat{\rho}(1) = \frac{\sum_{t=1}^{T-1} y_t y_{t+1} \mathrm{I}(t \notin \{t_0 - 1, t_0\})}{\sum_{t=1}^{T} y_t^2 \mathrm{I}(t \neq t_0) + A^2} + \frac{y_{t_0-1} A + A\, y_{t_0+1}}{\sum_{t=1}^{T} y_t^2 \mathrm{I}(t \neq t_0) + A^2}.$$

Since A appears quadratically in the denominator and only linearly in the numerator of the above estimate, $\widehat{\rho}(1)$ goes to zero as $A \to \infty$ with all other values of y_t for $t \neq t_0$ held fixed. So whatever the original value of $\widehat{\rho}(1)$, the alteration of an original value y_{t_0} to an outlying value $y_{t_0} = A$ results in a "bias"of $\widehat{\rho}(1)$ toward zero, the more so the larger the magnitude of the outlier.

Now consider the case of a patch outlier of constant value A and patch length k, where the values y_i for $i = t_0, \ldots, t_0 + k - 1$ are replaced by A. In this case the above estimate has the form

$$\widehat{\rho}(1) = \frac{\sum_{t=1}^{T-1} y_t y_{t+1} \mathrm{I}\{t \notin [t_0 - 1, t_0 + k - 1]\}}{\sum_{t=1}^{T} y_t^2 \mathrm{I}\{t \notin [t_0, t_0 + k - 1]\} + k A^2}$$
$$+ \frac{y_{t_0 - 1} A + (k-1)A^2 + A\, y_{t_0+k}}{\sum_{t=1}^{T} y_t^2 \mathrm{I}\{t \notin [t_0, t_0 + k - 1]\} + k A^2},$$

and therefore

$$\lim_{A\to\infty} \widehat{\rho}(1) = \frac{k-1}{k}.$$

Hence, the limiting value of $\widehat{\rho}(1)$ with the patch outlier can either increase or decrease relative to the original value, depending on the value of k and the value of $\widehat{\rho}(1)$ without the patch outlier. For example, if $k = 10$ with $\widehat{\rho}(1) = 0.5$ without the patch outlier, then $\widehat{\rho}(1)$ increases to the value 0.9 as $A \to \infty$.

In some applications one may find that outliers come in pairs of opposite signs. For example, when computing first differences of a time series that has isolated outliers we get a "doublet" outlier as shown in Figure 8.3. We leave it to the reader to show that for a "doublet"outlier with adjacent values having equal magnitude but opposite signs, i.e., values $\pm A$, the limiting value of $\widehat{\rho}(1)$ as $A \to \infty$ is $\widehat{\rho}(1) = -0.5$ (Problem 8.2).

Of course one can seldom make the assumption that the time series has zero mean, so one usually defines the lag-1 sample autocorrelation coefficient using the definition

$$\widehat{\rho}(1) = \frac{\sum_{t=1}^{T-1} (y_t - \bar{y})(y_{t+1} - \bar{y})}{\sum_{t=1}^{T} (y_t - \bar{y})^2}. \tag{8.4}$$

Determining the influence of outliers in this more realistic case is often algebraically quite messy but doable. For example, in the case of an isolated outlier of size A, it may be shown that the limiting value of $\widehat{\rho}(1)$ as $A \to \infty$ is approximately $-1/T$ for large T (Problem 8.3). However, it is usually easier to resort to some type of influence function calculation as described in Section 8.11.1.

8.1.2 Probability models for time series outliers

In this section we describe several probability models for time series outliers, including *additive* outliers (AOs), *replacement* outliers (ROs) and *innovations* outliers (IOs). Let x_t be a wide-sense stationary "core" process of interest, and let v_t be a stationary outlier process which is non-zero a fraction ε of the time, i.e., $\mathrm{P}(v_t = 0) = 1 - \varepsilon$. In practice the fraction ε is often positive but small.

Under an AO model, instead of x_t one actually observes

$$y_t = x_t + v_t \tag{8.5}$$

where the processes x_t and v_t are assumed to be independent of one another. A special case of the AO model was originally introduced by Fox (1972), who called them Type I outliers. Fox attributed such outliers to a "gross-error of observation or a recording error that affects a single observation". The AO model will generate mostly isolated outliers if v_t is an independent and identically distributed (i.i.d.) process, with standard deviation (or scale) much larger than that of x_t. For example, suppose, that x_t is a zero-mean normally distributed process with $\mathrm{Var}(x_t) = \sigma_x^2$, and v_t has a normal mixture distribution with degenerate central component

$$v_t \sim (1-\varepsilon)\delta_0 + \varepsilon \mathrm{N}(\mu_v, \sigma_v^2). \tag{8.6}$$

Here δ_0 is a point mass distribution located at zero, and we assume that the normal component $\mathrm{N}(\mu_v, \sigma_v^2)$ has variance $\sigma_v^2 \gg \sigma_x^2$. In this case y_t will contain an outlier at any fixed time t with probability ε and the probability of getting two outliers in a row is the much smaller ε^2. It will be assumed that $\mu_v = 0$ unless otherwise stated.

Additive patch outliers can be obtained by specifying that at any given t, $v_t = 0$ with probability $1-\varepsilon$; and with probability ε, v_t is the first observation of a patch outlier having a particular structure that persists for k time periods. We leave it for the reader (Problem 8.4) to construct a probability model to generate patch outliers.

RO models have the form

$$y_t = (1-z_t)x_t + z_t w_t \tag{8.7}$$

where z_t is a zero–one process with $\mathrm{P}(z_t = 0) = 1-\varepsilon$, and w_t is a "replacement"process that is not necessarily independent of x_t. Actually, RO models contain AO models as a special case in which $w_t = x_t + v_t$, and z_t is a Bernoulli process, i.e., z_t and z_u are independent for $t \neq u$. Outliers that are mostly isolated are obtained for example when z_t is a Bernoulli process, and x_t and w_t are zero-mean normal processes with $\mathrm{Var}(w_t) = \sigma_w^2 \gg \sigma_x^2$. For the reader familiar with Markov chains, we can say that patch outliers may be obtained by letting z_t be a Markov process that remains in the "one"state for more than one time period (of fixed or random duration), and w_t has an appropriately specified probability model.

IOs are a highly specialized form of outlier that occur in linear processes such as AR, ARMA and ARIMA models, which will be discussed in subsequent sections. IO models were first introduced by Fox (1972), who termed them Type II outliers, and noted that an IO "will affect not only the current observation but also subsequent observations". For the sake of simplicity we illustrate IOs here in the special case of a first-order autoregression model, which is adequate to reveal the character of these type of outliers. A stationary first-order AR model is given by

$$x_t = \phi x_{t-1} + u_t \tag{8.8}$$

where the *innovations* process u_t is i.i.d. with zero mean and finite variance, and $|\phi| < 1$. An IO is an outlier in the u_t process. IOs are obtained for example when the

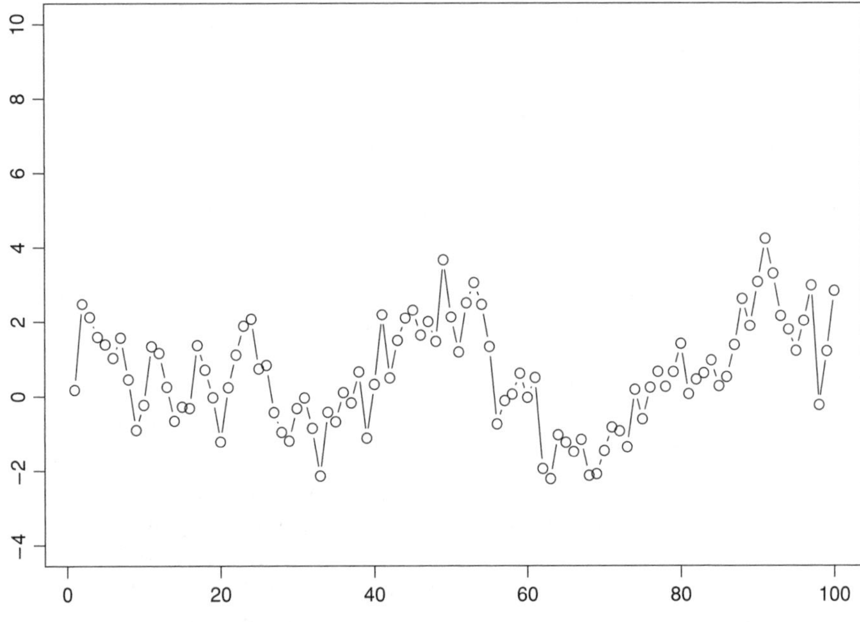

Figure 8.6 Gaussian AR(1) series with $\phi = 0.5$

innovations process has a zero-mean normal mixture distribution

$$(1-\varepsilon)\mathrm{N}(0, \sigma_0^2) + \varepsilon\mathrm{N}(0, \sigma_1^2) \tag{8.9}$$

where $\sigma_1^2 \gg \sigma_0^2$. More generally, we can say that the process has IOs when the distribution of u_t is heavy tailed (e.g., a Student t-distribution).

Example 8.1 *AR(1) with an IO and AO.*

To illustrate the impact of AO's versus IO's, we display in Figure 8.6 a Gaussian first-order AR series of length 100 with parameter $\phi = 0.5$. The same series with AO's is shown in Figure 8.7, and the same series with a single IO in Figure 8.8. The AOs in Figure 8.7 are indicated by crosses over the circles. The IO in Figure 8.8 was created by replacing (only) the normal innovation u_{20} with an atypical innovation having value $u_{20} = 10$. The persistent effect of the IO at $t = 20$ on subsequent observations is quite clear. The effect of this outlier decays roughly like ϕ^{t-20} for times $t > 20$.

One may think of an IO as an "impulse" input to a dynamic system driven by a background of uncorrelated or i.i.d. white noise. Consequently the output of the system behaves like a system *impulse response*, a concept widely used in linear systems theory, at nearby times subsequent to the time of occurrence of the outlier. It will be seen in Section 8.4.3 that IOs are "good" outliers in the sense that they can

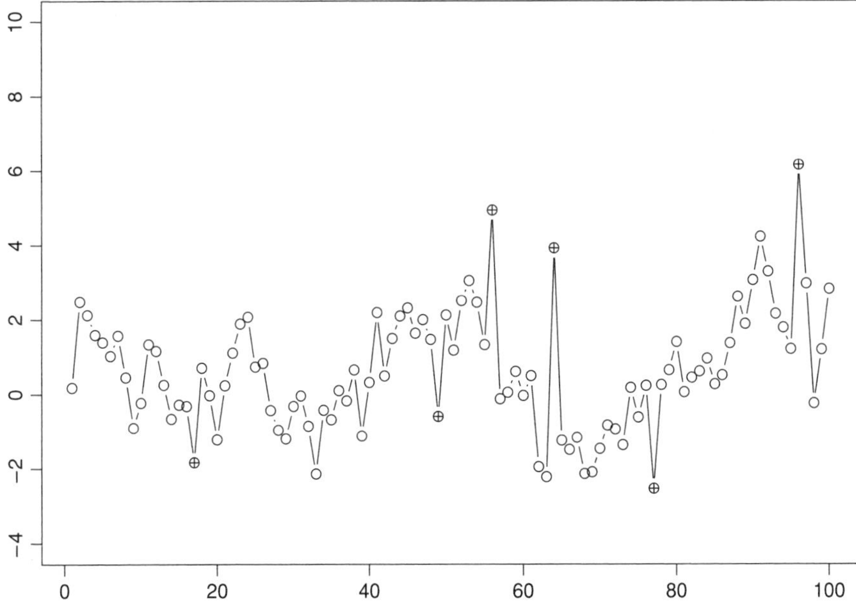

Figure 8.7 Gaussian AR(1) series with AOs

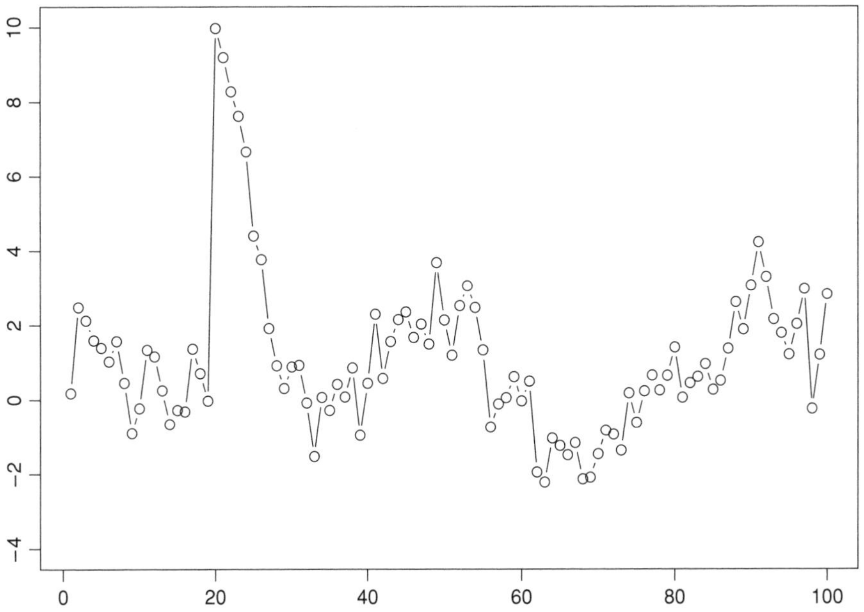

Figure 8.8 AR(1) series with an IO

improve the precision of the estimation of the parameters in AR and ARMA models, e.g., the parameter ϕ in the AR(1) model.

8.1.3 Bias impact of AOs

In this section we provide a simple illustrative example of the bias impact of AOs on the estimation of a first-order zero-mean AR model.

The reader may easily check that the lag-1 autocorrelation coefficient $\rho(1)$ for the AR(1) model (8.8) is equal to ϕ (see Problem 8.6). Furthermore, a natural least-squares (LS) estimate $\widehat{\phi}$ of ϕ in the case of perfect observations $y_t \equiv x_t$ is obtained by solving the minimization problem

$$\min_{\phi} \sum_{t=2}^{T} (y_t - \phi y_{t-1})^2. \tag{8.10}$$

Differentiation with respect to ϕ gives the estimating equation

$$\sum_{t=2}^{T} y_{t-1}(y_t - \widehat{\phi} y_{t-1}) = 0$$

and solving for $\widehat{\phi}$ gives the LS estimate

$$\widehat{\phi} = \frac{\sum_{t=2}^{T} y_t y_{t-1}}{\sum_{t=1}^{T-1} y_t^2}. \tag{8.11}$$

A slightly different estimate is

$$\phi^* = \frac{\sum_{t=2}^{T} y_t y_{t-1}}{\sum_{t=1}^{T} y_t^2}, \tag{8.12}$$

which coincides with (8.3). The main difference between these two estimates is that $\left|\phi^*\right|$ is bounded by one, while $\left|\widehat{\phi}\right|$ can take on values larger than one. Since the true autocorrelation coefficient ϕ has $|\phi| \leq 1$, and actually $|\phi| < 1$ except in case of perfect linear dependence, the estimate ϕ^* is usually preferred.

Let $\rho_y(1)$ be the lag-1 autocorrelation coefficient for the AO observations $y_t = x_t + v_t$. It may be shown that when $T \to \infty$, ϕ^* converges to $\rho_y(1)$ under mild regularity conditions, and the same is true of $\widehat{\phi}$ (Brockwell and Davis, 1991). If we assume that v_t is independent of x_t, and that v_t has lag-1 autocorrelation coefficient $\rho_v(1)$, then

$$\begin{aligned}\rho_y(1) = \frac{\mathrm{Cov}(y_t, y_{t-1})}{\mathrm{Var}(y_t)} &= \frac{\mathrm{Cov}(x_t, x_{t-1}) + \mathrm{Cov}(v_t, v_{t-1})}{\sigma_x^2 + \sigma_v^2} \\ &= \frac{\sigma_x^2 \phi + \sigma_v^2 \rho_v(1)}{\sigma_x^2 + \sigma_v^2}.\end{aligned}$$

The large-sample bias of ϕ^* is

$$\text{Bias}(\phi^*) = \rho_y(1) - \phi = \frac{\sigma_v^2}{\sigma_x^2 + \sigma_v^2}(\rho_v(1) - \phi)$$

$$= \frac{R}{1+R}(\rho_v(1) - \phi) \tag{8.13}$$

where $R = \sigma_v^2/\sigma_x^2$ is the "noise-to-signal"ratio. We see that the bias is zero when R is zero, i.e., when the AOs have zero variance. The bias is bounded in magnitude by $|\rho_v(1) - \phi|$ and approaches $|\rho_v(1) - \phi|$ as R approaches infinity. When the AOs have lag-1 autocorrelation $\rho_v(1) = 0$ and R is very large, the bias is approximately $-\phi$ and correspondingly the estimate ϕ^* has a value close to zero. As an intermediate example, suppose that $\rho_v(1) = 0$, $\phi = 0.5$, $\sigma_x^2 = 1$ and that v_t has distribution (8.6) with $\varepsilon = 0.1$, $\mu_v = 0$ and $\sigma_v^2 = 0.9$. Then the bias is negative and equal to -0.24. On the other hand if the values of $\rho_v(1)$ and ϕ are interchanged, with the other parameter values remaining fixed, then the bias is positive and has value $+0.24$.

8.2 Classical estimates for AR models

In this section we describe the properties of classical estimates of the parameters of an autoregression model. In particular we describe the form of these estimates, state the form of their limiting multivariate normal distribution, and describe their efficiency and robustness in the absence of outliers.

An autoregression model of order p, called an AR(p) model for short, generates a time series according to the stochastic difference equation

$$y_t = \gamma + \phi_1 y_{t-1} + \phi_2 y_{t-2} + \ldots + \phi_p y_{t-p} + u_t \tag{8.14}$$

where the innovations u_t are an i.i.d. sequence of random variables with mean 0 and finite variance σ_u^2. The innovations are assumed to be independent of past values of the y_t's. It is known that the time series y_t is stationary if all the roots of the characteristic polynomial

$$\phi(z) = 1 - \phi_1 z - \phi_2 z^2 - \ldots - \phi_p z^p \tag{8.15}$$

lie outside the unit circle in the complex plane. When y_t is stationary it has a mean value $\mu = \text{E}(y_t)$ that is determined by taking the mean value of both sides of (8.14), giving

$$\mu = \gamma + \phi_1\mu + \phi_2\mu + \ldots + \phi_p\mu + 0,$$

which implies

$$\mu\left(1 - \sum_{i=1}^{p}\phi_i\right) = \gamma, \tag{8.16}$$

and hence

$$\mu = \frac{\gamma}{1 - \sum_{i=1}^{p} \phi_i}. \tag{8.17}$$

In view of (8.16), the AR(p) model may also be written in the form

$$y_t - \mu = \phi_1(y_{t-1} - \mu) + \phi_2(y_{t-2} - \mu) + \ldots + \phi_p(y_{t-p} - \mu) + u_t. \tag{8.18}$$

There are several asymptotically equivalent forms of LS estimates of the AR(p) model parameters that are asymptotically efficient when the distribution of u_t is normal. Given a sample of observations $y_1, y_2, \ldots, y_T$, it seems natural at first glance to compute LS estimates of the parameters by choosing $\gamma, \phi_1, \phi_2, \ldots, \phi_p$ to minimize the sum of squares:

$$\sum_{t=p+1}^{T} \widehat{u}_t^2(\phi, \gamma) \tag{8.19}$$

where $\widehat{u}_t$ are the prediction residuals defined by

$$\widehat{u}_t = \widehat{u}_t(\phi, \gamma) = y_t - \gamma - \phi_1 y_{t-1} - \phi_2 y_{t-2} - \ldots - \phi_p y_{t-p}. \tag{8.20}$$

This is equivalent to applying ordinary LS to the linear regression model

$$\mathbf{y} = \mathbf{G}\beta + \mathbf{u} \tag{8.21}$$

where

$$\begin{aligned} \mathbf{y}' &= (y_{p+1}, y_{p+2}, \ldots, y_T) \\ \mathbf{u}' &= (u_{p+1}, u_{p+2}, \ldots, u_T) \\ \beta' &= (\phi', \gamma) = (\phi_1, \phi_2, \ldots, \phi_p, \gamma) \end{aligned} \tag{8.22}$$

and

$$\mathbf{G} = \begin{bmatrix} y_p & y_{p-1} & \cdots & y_1 & 1 \\ y_{p+1} & y_p & & y_2 & 1 \\ \vdots & \vdots & \vdots & \vdots & \vdots \\ y_{T-1} & y_{T-2} & \cdots & y_{T-p} & 1 \end{bmatrix}. \tag{8.23}$$

This form of LS estimate may also be written as

$$\widehat{\beta} = (\widehat{\phi}_1, \widehat{\phi}_2, \ldots, \widehat{\phi}_p, \widehat{\gamma})' = \left(\mathbf{G}'\mathbf{G}\right)^{-1} \mathbf{G}'\mathbf{y} \tag{8.24}$$

and the mean value estimate can be computed as

$$\widehat{\mu} = \frac{\widehat{\gamma}}{1 - \sum_{i=1}^{p} \widehat{\phi}_i}. \tag{8.25}$$

An alternative approach is to estimate μ by the sample mean $\overline{y}$ and compute the LS estimate of $\widehat{\phi}$ as

$$\widehat{\phi} = \left(\mathbf{G}^{*\prime}\mathbf{G}^{*}\right)^{-1}\mathbf{G}^{*\prime}\mathbf{y}^{*}, \tag{8.26}$$

where $\mathbf{y}^{*}$ is now the vector of centered observations $y_t - \overline{y}$, and $\mathbf{G}^{*}$ is defined as in (8.23) but replacing the y_t's by the y_t^{*} and omitting the last column of ones.

Unfortunately the above forms of the LS estimate do not ensure that the estimates $\widehat{\phi} = (\widehat{\phi}_1, \widehat{\phi}_2, \ldots, \widehat{\phi}_p)'$ correspond to a stationary autoregression, i.e., it can happen that one or more of the roots of the estimated characteristic polynomial

$$\widehat{\phi}(z) = 1 - \widehat{\phi}_1 z - \widehat{\phi}_2 z^2 - \ldots - \widehat{\phi}_p z^p$$

lie inside the unit circle. A common way around this is to use the so-called Yule–Walker equations to estimate ϕ. Let $C(l)$ be the autocovariance (8.1) at lag l. The Yule–Walker equations relating the autocovariances and the parameters of an AR(p) process are obtained from (8.18) as

$$C(k) = \sum_{i=1}^{p} \phi_i C(k-i) \quad (k \geq 1). \tag{8.27}$$

For $1 \leq k \leq p$, (8.27) may be expressed in matrix equation form as

$$\mathbf{C}\phi = \mathbf{g} \tag{8.28}$$

where $\mathbf{g}' = (C(1), C(2), \ldots, C(p))$ and the $p \times p$ matrix $\mathbf{C}$ has elements $C_{ij} = C(i-j)$. It is left for the reader (Problem 8.5) to verify the above equations for an AR(p) model.

The Yule–Walker equations can also be written in terms of the autocorrelations as

$$\rho(k) = \sum_{i=1}^{p} \phi_i \rho(k-i) \quad (k \geq 1). \tag{8.29}$$

The Yule–Walker estimate $\widehat{\phi}_{YW}$ of ϕ is obtained by replacing the unknown lag-l covariances in $\mathbf{C}$ and $\mathbf{g}$ of (8.28) by the covariance estimates

$$\widehat{C}(l) = \frac{1}{T}\sum_{t=1}^{T-|l|} (y_{t+l} - \overline{y})(y_t - \overline{y}), \tag{8.30}$$

and solving for $\widehat{\phi}_{YW}$. It can be shown that the above lag-l covariance estimates are biased and that unbiased estimates can be obtained under normality by replacing the denominator T by $T - |l| - 1$. However, the covariance matrix estimate $\widehat{\mathbf{C}}$ based on the above biased lag-l estimates is preferred since it is known to be positive definite (with probability 1), and furthermore the resulting Yule–Walker parameter estimate $\widehat{\phi}_{YW}$ corresponds to a stationary autoregression (see, e.g., Brockwell and Davis, 1991).

The Durbin–Levinson algorithm, to be described in Section 8.2.1, is a convenient recursive procedure to compute the sequence of Yule–Walker estimates for AR(1), AR(2), etc.

8.2.1 The Durbin–Levinson algorithm

We shall describe the Durbin–Levinson algorithm, which is a recursive method to derive the best memory-$(m-1)$ linear predictor from the best memory-$(m-1)$ predictor.
Let y_t be a second-order stationary process. It will be assumed that $\mathrm{E}y_t = 0$. Otherwise, we apply the procedure to the centered process $y_t - \mathrm{E}y_t$ rather than y_t.

Let

$$\widehat{y}_{t,m} = \phi_{m,1} y_{t-1} + \ldots + \phi_{m,m} y_{t-m} \tag{8.31}$$

be the minimum mean-square error (MMSE) memory-m linear predictor of y_t based on $y_{t-1}, \ldots, y_{t-m}$. The "diagonal" coefficients $\phi_{m,m}$ are the so-called *partial autocorrelations*, which are very useful for the identification of AR models, as will be seen in Section 8.7.1.

The $\phi_{m,m}$ satisfy

$$|\phi_{m,m}| < 1, \tag{8.32}$$

except when the process is deterministic.

The MMSE *forward* prediction residuals are defined as

$$\widehat{u}_{t,m} = y_t - \phi_{m,1} y_{t-1} - \ldots - \phi_{m,m} y_{t-m}. \tag{8.33}$$

The memory-m backward linear predictor of y_t, i.e., the MMSE predictor of y_t as a linear function of $y_{t+1}, \ldots, y_{t+m}$, can be shown to be

$$\widehat{y}^{*}_{t,m} = \phi_{m,1} y_{t+1} + \ldots + \phi_{m,m} y_{t+m},$$

and the *backward* MMSE prediction residuals are

$$\overset{*}{u}_{t,m} = y_t - \phi_{m,1} y_{t+1} - \ldots - \phi_{m,m} y_{t+m}. \tag{8.34}$$

Note that $\widehat{u}_{t,m-1}$ and $\overset{*}{u}_{t-m,m-1}$ are both orthogonal to the linear space spanned by $y_{t-1}, \ldots, y_{t-m+1}$, with respect to the expectation inner product, i.e.,

$$\mathrm{E}\widehat{u}_{t,m-1}\, y_{t-k} = \mathrm{E}\overset{*}{u}_{t-m,m-1} y_{t-k} = 0, \; k = 1, \ldots, m-1.$$

We shall first derive the form of the memory-m predictor assuming that the true memory-$(m-1)$ predictor and the true values of the autocorrelations $\rho(k)$, $k = 1, \ldots, m$; are known.

Let $\zeta^{*}\overset{*}{u}_{t-m,m-1}$ be the MMSE linear predictor of $\widehat{u}_{t,m-1}$ based on $\overset{*}{u}_{t-m,m-1}$. Then

$$\mathrm{E}\big(\widehat{u}_{t,m-1} - \zeta^{*}\overset{*}{u}_{t-m,m-1}\big)^2 = \min_{\zeta} \mathrm{E}\big(\widehat{u}_{t,m-1} - \zeta \overset{*}{u}_{t-m,m-1}\big)^2. \tag{8.35}$$

It can be proved that the MMSE memory-m predictor is given by

$$\begin{aligned}\widehat{y}_{t,m} &= \widehat{y}_{t,m-1} + \zeta^* u^*_{t-m,m-1} \\ &= \left(\phi_{m-1,1} - \zeta^* \phi_{m-1,m-1}\right) y_{t-1} + \ldots \\ &\quad + \left(\phi_{m-1,i} - \zeta^* \phi_{m-1,1}\right) y_{t-m+1} + \zeta^* y_{t-m}. \end{aligned} \tag{8.36}$$

To show (8.36) it suffices to prove that

$$\mathrm{E}\left[(y_t - \widehat{y}_{t,m} - \zeta^* u^*_{t-m,m}) y_{t-i}\right] = 0, \; i = 1, \ldots, m, \tag{8.37}$$

which we leave as Problem 8.8. Then from (8.36) we have

$$\phi_{m,i} = \begin{cases} \zeta^* & \text{if} \quad i = m \\ \phi_{m-1,i} - \zeta^* \phi_{m-1,m-i} & \text{if} \quad 1 \le i \le m-1. \end{cases} \tag{8.38}$$

According to (8.38), if we already know $\phi_{m-1,i}$, $1 \le i \le m-1$, then to compute all the $\phi_{m,i}$'s, we only need $\zeta^* = \phi_{m,m}$.

It is easy to show that

$$\begin{aligned}\phi_{m,m} &= \mathrm{Corr}\left(\widehat{u}_{t,m-1}, u^*_{t-m,m-1}\right) \\ &= \frac{\rho(m) - \sum_{i=1}^{m-1} \rho(m-i)\phi_{m-1,i}}{1 - \sum_{i=1}^{m-1} \rho(i)\phi_{m-1,i}}. \end{aligned} \tag{8.39}$$

The first equality above justifies the term "partial autocorrelation": it is the correlation between y_t and y_{t-m} after the linear contribution of y_i $(i = t-1, \ldots, t-m+1)$ has been subtracted out.

If y_t is a stationary AR(p) process with parameters $\phi_1, \ldots, \phi_p$, it may be shown (Problem 8.11) that

$$\phi_{p,i} = \phi_i, \quad 1 \le i \le p, \tag{8.40}$$

and

$$\phi_{m,m} = 0, \quad m > p. \tag{8.41}$$

In the case that we have only a sample from a process, the unknown autocorrelations $\rho(k)$ can be estimated by their empirical counterparts, and the Durbin–Levinson algorithm can be used to estimate the predictor coefficients $\phi_{m,j}$ in a recursive way. In particular, if the process is assumed to be AR(p), then the AR coefficient estimates are obtained by substituting estimates in (8.40).

It is easy to show that $\phi_{1,1} = \rho(1)$ or equivalently

$$\phi_{1,1} = \arg\min_{\zeta} \mathrm{E}(y_t - \zeta y_{t-1})^2. \tag{8.42}$$

We shall now describe the classical Durbin–Levinson algorithm in such a way as to clarify the basis for its robust version to be given in Section 8.6.4.

The first step is to set $\widehat{\phi}_{1,1} = \widehat{\rho}(1)$, which is equivalent to

$$\widehat{\phi}_{1,1} = \arg\min_{\zeta} \sum_{t=2}^{T} (y_t - \zeta y_{t-1})^2 . \tag{8.43}$$

Assuming that estimates $\widehat{\phi}_{m-1,i}$ of $\phi_{m-1,i}$, for $1 \le i \le m-1$, have already been computed, $\widehat{\phi}_{m,m}$ can be computed from (8.39), where the ρ's and ϕ's are replaced by their estimates. Alternatively, $\widehat{\phi}_{m,m}$ is obtained as

$$\widehat{\phi}_{m,m} = \arg\min_{\zeta} \sum_{t=m+1}^{T} \widehat{u}_{t,m}^2(\zeta), \tag{8.44}$$

with

$$\begin{aligned}\widehat{u}_{t,m}(\zeta) &= y_t - \widehat{y}_{t,m-1} - \zeta u^*_{t-m,m-1} \\ &= y_t - \left(\widehat{\phi}_{m-1,1} - \zeta\widehat{\phi}_{m-1,m-1}\right) y_{t-1} - \dots \\ &\quad - \left(\widehat{\phi}_{m-1,m-1} - \zeta\widehat{\phi}_{m-1,1}\right) y_{t-m+1} - \zeta y_{t-m},\end{aligned} \tag{8.45}$$

and where the backward residuals $u^*_t - m, m-1$ are computed here by

$$u^*_{t-m,m-1} = y_{t-m} - \widehat{\phi}_{m-1,1} y_{t-m+1} - \dots - \widehat{\phi}_{m-1,m-1} y_{t-1}.$$

The remaining $\widehat{\phi}_{m,i}$'s are computed using the recursion (8.38). It is easy to verify that this sample Durbin–Levinson method is essentially equivalent to obtaining the AR(m) estimate $\widehat{\phi}_m$ by solving the Yule–Walker equations for $m = 1, 2, \dots, p$.

8.2.2 Asymptotic distribution of classical estimates

The LS and Yule–Walker estimates have the same asymptotic distribution, which will be studied in this section. Call $\widehat{\lambda} = (\widehat{\phi}_1, \widehat{\phi}_2, \dots, \widehat{\phi}_p, \widehat{\mu})$ the LS or Yule–Walker estimate of $\boldsymbol{\lambda} = (\phi_1, \phi_2, \dots, \phi_p, \mu)$ based on a sample of size T. Here $\widehat{\mu}$ can be either the sample mean estimate or the estimate of (8.25) based on the LS estimate $\widehat{\beta}$ defined in (8.19). It is known that $\widehat{\lambda}$ converges in distribution to a $(p+1)$-dimensional multivariate normal distribution

$$\sqrt{T}(\widehat{\boldsymbol{\lambda}} - \boldsymbol{\lambda}) \to_d \mathrm{N}_{p+1}(\mathbf{0}, \mathbf{V}_{LS}) \tag{8.46}$$

where the asymptotic covariance matrix $\mathbf{V}_{LS}$ is given by

$$\mathbf{V}_{LS} = \begin{bmatrix} \mathbf{V}_{LS,\phi} & \mathbf{0}' \\ \mathbf{0} & V_{LS,\mu} \end{bmatrix} \tag{8.47}$$

with

$$V_{LS,\mu} = \frac{\sigma_u^2}{\left(1 - \sum_{i=1}^{p} \phi_i\right)^2} \tag{8.48}$$

and

$$\mathbf{V}_{LS,\phi} = \mathbf{V}_{LS}(\phi) = \sigma_u^2 \mathbf{C}^{-1}, \tag{8.49}$$

where $\mathbf{C}$ is the $p \times p$ covariance matrix of $(y_{t-1}, \ldots, y_{t-p})$, which does not depend on t (due to the stationarity of y_t), and $\sigma_u^2 \mathbf{C}^{-1}$ depends only on the AR parameters ϕ. See for example Anderson (1994) or Brockwell and Davis (1991). In Section 8.15 we give a heuristic derivation of this result and the expression for $D = \mathbf{C}/\sigma_u{}^2$.

Remark: Note that if we apply formula (5.6) for the asymptotic covariance matrix of the LS estimate under a linear model with random predictors to the regression model (8.21), then the result coincides with (8.49).

The block-diagonal structure of $\mathbf{V}$ shows that $\widehat{\mu}$ and $\widehat{\phi} = \left(\widehat{\phi}_1, \widehat{\phi}_2, \ldots, \widehat{\phi}_p\right)'$ are asymptotically uncorrelated. The standard estimate of the innovations variance σ_u^2 is

$$\widehat{\sigma}_u^2 = \frac{1}{T-p} \sum_{t=p+1}^{T} \left(y_t - \widehat{\gamma} - \widehat{\phi}_1 y_{t-1} - \widehat{\phi}_2 y_{t-2} - \ldots - \widehat{\phi}_p y_{t-p}\right)^2 \tag{8.50}$$

or alternatively

$$\widehat{\sigma}_u^2 = \frac{1}{T-p} \sum_{t=p+1}^{T} \left(\widetilde{y}_t - \widehat{\phi}_1 \widetilde{y}_{t-1} - \widehat{\phi}_2 \widetilde{y}_{t-2} - \ldots - \widehat{\phi}_p \widetilde{y}_{t-p}\right)^2 \tag{8.51}$$

where $\widetilde{y}_{t-i} = y_{t-i} - \widehat{\mu}$, $i = 0, 1, \ldots, p$. It is known that $\widehat{\sigma}_u^2$ is asymptotically uncorrelated with $\widehat{\boldsymbol{\lambda}}$ and has asymptotic variance

$$\mathrm{AsVar}\left(\widehat{\sigma}_u^2\right) = \mathrm{E}(u^4) - \sigma_u^4. \tag{8.52}$$

In the case of normally distributed u_t this expression reduces to $\mathrm{AsVar}\left(\widehat{\sigma}_u^2\right) = 2\sigma_u^4$.

What is particularly striking about the asymptotic covariance matrix $\mathbf{V}_{LS}$ is that the $p \times p$ submatrix $\mathbf{V}_{LS,\phi}$ is a constant that depends only on ϕ, and not at all on the distribution F_u of the innovations u_t (assuming finite variance innovations!). This distribution-free character of the estimate led Whittle (1962) to use the term *robust* to describe the LS estimates of AR parameters. With hindsight, this was a rather misleading use of this term because the constant character of $\mathbf{V}_{LS,\phi}$ holds only under *perfectly* observed autoregressions, i.e., with no AOs or ROs. Furthermore it turns out that the LS estimate will not be efficiency robust with respect to heavy-tailed deviations of the IOs from normality, as we discuss in Section 8.4. It should also be noted that the variance V_μ is not constant with respect to changes in the variance of the innovations, and $\mathrm{AsVar}\left(\widehat{\sigma}_u^2\right)$ depends upon the fourth moment as well as the variance of the innovations.

8.3 Classical estimates for ARMA models

A time series y_t is called an autoregressive moving-average model of orders p and q, or ARMA(p, q) for short, if it obeys the stochastic difference equation

$$(y_t - \mu) - \phi_1(y_{t-1} - \mu) - \ldots - \phi_p(y_{t-p} - \mu) = -\theta_1 u_{t-1} - \ldots - \theta_q u_{t-q} + u_t, \tag{8.53}$$

where the i.i.d. innovations u_t have mean 0 and finite variance σ_u^2. This equation may be written in more compact form as

$$\phi(B)(y_t - \mu) = \theta(B)u_t \tag{8.54}$$

where B is the back-shift operator, i.e., $By_t = y_{t-1}$, and $\phi(B)$ and $\theta(B)$ are polynomial back-shift operators given by

$$\phi(B) = 1 - \phi_1 B - \phi_2 B^2 - \ldots - \phi_p B^p \tag{8.55}$$

and

$$\theta(B) = 1 - \theta_1 B - \theta_2 B^2 - \ldots - \theta_q B^q. \tag{8.56}$$

The process is called *invertible* if y_t can be expressed as an infinite linear combination of the y_s's for $s < t$ plus the innovations:

$$y_t = u_t + \sum_{i=1}^{\infty} \eta_i y_{t-i} + \gamma.$$

It will henceforth be assumed that the ARMA process is stationary and invertible. The first assumption requires that all roots of the polynomial $\phi(B)$ lie outside the unit circle and the second requires the same of the roots of $\theta(B)$.

Let $\boldsymbol{\lambda} = (\boldsymbol{\phi}, \boldsymbol{\theta}, \mu) = (\phi_1, \phi_2, \ldots, \phi_p, \theta_1, \theta_2, \ldots, \theta_q, \mu)$ and consider the sum of squared residuals

$$\sum_{t=p+1}^{T} \widehat{u}_t^2(\boldsymbol{\lambda}) \tag{8.57}$$

where the residuals $\widehat{u}_t(\boldsymbol{\lambda})$ may be computed recursively as

$$\begin{aligned}\widehat{u}_t(\boldsymbol{\lambda}) = {} & (y_t - \mu) - \phi_1(y_{t-1} - \mu) - \ldots - \phi_p(y_{t-p} - \mu) \\ & + \theta_1 \widehat{u}_{t-1}(\boldsymbol{\lambda}) + \ldots + \theta_q \widehat{u}_{t-q}(\boldsymbol{\lambda})\end{aligned} \tag{8.58}$$

with the initial conditions

$$\widehat{u}_p(\boldsymbol{\lambda}) = \widehat{u}_{p-1}(\boldsymbol{\lambda}) = \ldots = \widehat{u}_{p-q+1}(\boldsymbol{\lambda}) = 0. \tag{8.59}$$

Minimizing the sum of squared residuals (8.57) with respect to $\boldsymbol{\lambda}$ produces a LS estimate $\widehat{\boldsymbol{\lambda}}_{LS} = (\widehat{\boldsymbol{\phi}}, \widehat{\boldsymbol{\theta}}, \widehat{\mu})$. When the innovations u_t have a normal distribution with mean 0 and finite variance σ_u^2, this LS estimate is a *conditional* maximum likelihood estimate, conditioned on $y_1, y_2, \ldots, y_p$ and on

$$u_{p-q+1} = u_{p-q+2} = \ldots = u_{p-1} = u_p = 0.$$

See for example Harvey and Philips (1979), where it is also shown how to compute the exact Gaussian maximum likelihood estimate of ARMA model parameters.

It is known that under the above assumptions for the ARMA(p, q) process, the LS estimate, as well as the conditional and exact maximum likelihood estimates, converge asymptotically to a multivariate normal distribution:

$$\sqrt{T}(\widehat{\boldsymbol{\lambda}}_{LS} - \boldsymbol{\lambda}) \to_d \mathrm{N}_{p+q+1}(\mathbf{0}, \mathbf{V}_{LS}) \tag{8.60}$$

where

$$\mathbf{V}_{LS} = \mathbf{V}_{LS}(\boldsymbol{\phi}, \boldsymbol{\theta}, \sigma_u^2) = \begin{bmatrix} \mathbf{D}^{-1}(\boldsymbol{\phi}, \boldsymbol{\theta}) & \mathbf{0}' \\ \mathbf{0} & V_{LS,\mu} \end{bmatrix}, \tag{8.61}$$

with $V_{LS,\mu}$ the asymptotic variance of the location estimate $\widehat{\mu}$ and $\mathbf{D}(\boldsymbol{\phi}, \boldsymbol{\theta})$ the $(p+q) \times (p+q)$ asymptotic covariance matrix of $(\widehat{\boldsymbol{\phi}}, \widehat{\boldsymbol{\theta}})$. Expressions for the elements of $\mathbf{D}(\boldsymbol{\phi}, \boldsymbol{\theta})$ are given in Section 8.15. As the notation indicates, $\mathbf{D}(\boldsymbol{\phi}, \boldsymbol{\theta})$ depends only on $\boldsymbol{\phi}$ and $\boldsymbol{\theta}$ and so the LS estimate $(\widehat{\boldsymbol{\phi}}, \widehat{\boldsymbol{\theta}})$ has the same distribution-free property as in the AR case, described at the end of Section 8.2.2. The expression for $V_{LS,\mu}$ is

$$V_{LS,\mu} = \frac{\sigma_u^2}{\xi^2} \tag{8.62}$$

with

$$\xi = -\frac{1 - \phi_1 - \ldots - \phi_p}{1 - \theta_1 - \ldots - \theta_q}, \tag{8.63}$$

which depends upon the variance of the innovations σ_u^2 as well as on $\boldsymbol{\phi}$ and $\boldsymbol{\theta}$.

The conditional MLE of the innovations variance σ_u^2 for an ARMA(p, q) model is given by

$$\widehat{\sigma}_u^2 = \frac{1}{T - p} \sum_{t=p+1}^{T} \widehat{u}_t^2(\widehat{\boldsymbol{\lambda}}). \tag{8.64}$$

The estimate $\widehat{\sigma}_u^2$ is asymptotically uncorrelated with $\widehat{\boldsymbol{\lambda}}_{LS}$ and has the same asymptotic distribution as in the AR case, namely $\mathrm{AsVar}\left(\widehat{\sigma}_u^2\right) = \mathrm{E}(u_t^4) - \sigma_u^4$.

Note that the asymptotic distribution of $\widehat{\boldsymbol{\lambda}}$ does not depend on the distribution of the innovations, and hence the precision of the estimates does not depend on their variance, as long as it is finite.

A natural estimate of the variance of the estimate $\widehat{\mu}$ is obtained by plugging parameter estimates into (8.62).

8.4 M-estimates of ARMA models

8.4.1 M-estimates and their asymptotic distribution

An M-estimate $\widehat{\boldsymbol{\lambda}}_M$ of the parameter vector $\boldsymbol{\lambda}$ for an ARMA(p, q) model is obtained by minimizing

$$\sum_{t=p+1}^{T} \rho\left(\frac{\widehat{u}_t(\boldsymbol{\lambda})}{\widehat{\sigma}}\right) \tag{8.65}$$

where ρ is a ρ-function already used for regression in (5.7). The residuals $\widehat{u}_t(\boldsymbol{\lambda})$ are defined as in the case of the LS estimate, and $\widehat{\sigma}$ is a robust scale estimate that is obtained either simultaneously with $\boldsymbol{\lambda}$ (e.g., as an M-scale of the $\widehat{u}_t$'s as in Section 2.6.2) or previously as with MM-estimates in Section 5.5. We assume that when $T \to \infty$, $\widehat{\sigma}$ converges in probability to a value σ which is a scale parameter of the innovations. It is also assumed that σ is standardized so that if the innovations are normal, σ coincides with the standard deviation σ_u of u_t, as explained for the location case at the end of Section 2.5.

Let $\psi = \rho'$ and assume that $\widehat{\sigma}$ has a limit in probability σ and

$$\mathrm{E}\psi\left(\frac{u_t}{\sigma}\right) = 0, \tag{8.66}$$

Note that this condition is analogous to (4.41) used in regression. Under the assumptions concerning the ARMA process made in Section 8.3 and under reasonable regularity conditions, which include that $\sigma_u^2 = \mathrm{Var}(u_t) < \infty$, the M-estimate $\widehat{\boldsymbol{\lambda}}_M$ has an asymptotic normal distribution given by

$$\sqrt{T}(\widehat{\boldsymbol{\lambda}}_M - \boldsymbol{\lambda}) \to_d \mathrm{N}_{p+q+1}(\mathbf{0}, \mathbf{V}_M), \tag{8.67}$$

with

$$\mathbf{V}_M = \mathbf{V}_M(\boldsymbol{\phi}, \boldsymbol{\theta}, \sigma^2) = a\mathbf{V}_{LS} \tag{8.68}$$

where a depends on the distribution F of the u_t's:

$$a = a(\psi, F) = \frac{\sigma^2 \mathrm{E}\psi\,(u_t/\sigma)^2}{\sigma_u^2\,(\mathrm{E}\psi'\,(u_t/\sigma))^2}. \tag{8.69}$$

A heuristic proof is given in Section 8.15. In the normal case, $\sigma = \sigma_u$ and a coincides with the reciprocal of the efficiency of a location or regression M-estimate (see (4.45)).

In the case that $\rho(t) = -\log f(t)$ where f is the density of the innovations, the M-estimate $\widehat{\boldsymbol{\lambda}}_M$ is a conditional MLE, and in this case the M-estimate is asymptotically efficient.

8.4.2 The behavior of M-estimates in AR processes with AOs

We already know from the discussion of Sections 8.1.1 and 8.1.3 that LS estimates of AR models are not robust in the presence of AOs or ROs. Such outliers cause both bias and inflated variability of LS estimates.

The LS estimate (8.19) proceeds as an ordinary regression, where y_t is regressed on the "predictors" $y_{t-1}, \ldots, y_{t-p}$. Similarly, any robust regression estimate based on the minimization of a function of the residuals can be applied to the AR model, in particular the S-, M- and MM-estimates defined in Chapter 5. In order to obtain some degree of robustness it is necessary, just as in the treatment of ordinary regression in that chapter, that ρ be bounded, and in addition an adequate algorithm must be used to help insure a good local minimum.

This approach has the advantage that the existing software for regression can be readily used. It has, however, the drawback that if the observations y_t are actually an AR(p) process contaminated with an AO or RO, the robustness of the estimates decreases with increasing p. The reason for this is that in the estimation of AR(p) parameters, the observation y_t is used in computing the $p+1$ residuals $\widehat{u}_t(\gamma, \phi), \widehat{u}_{t+1}(\gamma, \phi), \ldots, \widehat{u}_{t+p}(\gamma, \phi)$. Each time that an outlier appears in the series it may spoil $p+1$ residuals. In an informal way we can say that the breakdown point of an M-estimate is not larger than $0.5/(p+1)$. Correspondingly the bias due to an AO can be quite high and one expects only a limited degree of robustness.

Example 8.2 *Simulated AR(3) data with AOs.*

To demonstrate the effect of contamination on these estimates, we generated $T = 200$ observations x_t from a stationary normal AR(3) model with $\sigma_u = 1$, $\gamma = 0$ and $\phi = (8/6, -5/6, 1/6)'$. We then modified k evenly spaced observations by adding four to each, for $k = 10$ and 20. Table 8.1 shows the results for LS and for the MM regression estimate with bisquare function and efficiency 0.85 (code **AR3**).

It is seen that the LS estimate is much affected by 10 outliers. The MM-estimate is similar to the LS estimate when there are no outliers. It is less biased and so better

Table 8.1 LS and MM-estimates of the parameters of AR(3) simulated process

Estimate	#(outliers)	ϕ_1	ϕ_2	ϕ_3	γ	σ_u
LS	0	1.35	−0.83	0.11	0.11	0.99
	10	0.78	−0.10	−0.25	0.30	1.66
	20	0.69	0.40	0.04	0.69	2.63
MM	0	1.35	−0.79	0.10	0.11	0.99
	10	1.10	−0.36	−0.13	0.19	1.18
	20	0.84	−0.10	−0.12	0.31	1.58
	True values	1.333	−0.833	0.166	0.00	1.00

than LS when there are outliers, but is affected by them when there are 20 outliers. The reason is that in this case the proportion of outliers is 20/200 = 0.1, which is near the heuristic BP value of $0.125 = 0.5/(p+1)$, which was discussed in Section 8.4.2.

8.4.3 The behavior of LS and M-estimates for ARMA processes with infinite innovations variance

The asymptotic behavior of the LS and M-estimates for ARMA processes has been discussed in the previous sections under the assumption that the innovations u_t have finite variance. When this is not true, it may be surprising to know that under certain conditions the LS estimate not only is still consistent, but also converges to the true value at a faster rate than it would under finite innovations variance, with the consistency rate depending on the rate at which $\mathrm{P}(|u_t| > k)$ tends to zero as $k \to \infty$.

For the case of an M-estimate with bounded ψ, and assuming that a good robust scale estimate $\widehat{\sigma}$ is used, a heavy-tailed f can lead to ultra-precise estimation of the ARMA parameters $(\boldsymbol{\phi}, \boldsymbol{\theta})$ (but not of μ), in the sense that $\sqrt{T}(\widehat{\boldsymbol{\phi}}-\boldsymbol{\phi}) \to_p \mathbf{0}$ and $\sqrt{T}(\widehat{\boldsymbol{\theta}}-\boldsymbol{\theta}) \to_p \mathbf{0}$. This fact can be understood by noting that if u_t has a heavy-tailed distribution like the Cauchy distribution, then the expectations in (8.69) and σ are finite, while σ_u is infinite.

To fix ideas, consider fitting an AR(1) model. Estimating ϕ is equivalent to fitting a straight line to the lag-1 scatterplot of y_t vs. y_{t-1}. Each IO appears twice in the scatterplot: as y_{t-1} and as y_t. In the first case it is a "good" leverage point, and in the second it is an outlier. Both LS and M-estimates take advantage of the leverage point. But the LS estimate is affected by the outlier, while the M-estimate is not.

The main LS results were derived by Kanter and Steiger (1974), Yohai and Maronna (1977), Knight (1987, 1989) and Hannan and Kanter (1977) for AR processes, and by Mikosch, Gadrich, Kluppelberg and Adler (1995), Davis (1996) and Rachev and Mittnik (2000) in the ARMA case.

Results for monotone M-estimates were obtained by Davis, Knight and Liu (1992) and Davis (1996).

The challenges of establishing results in time series with infinite variance innovations has been of such considerable interest to academics that it has resulted in many papers, particularly in the econometrics and finance literature. See for example applications to unit root tests (Samarakoon and Knight, 2005), and references therein, and applications to GARCH models (Rachev and Mittnik, 2000). One of the most interesting of the latter is the application to option pricing by Menn and Rachev (2005).

To help understand the intuitive reasons for the behavior of the LS estimate under heavy-tailed innovations, we present a simple example with IOs. Figure 8.9 shows a simulated Gaussian AR(1) process with $\phi = 0.8$ and four IOs, and Figure 8.10 displays the respective lag-1 scatterplot. When appearing as y_t, the four IOs stand out as clear outliers. When appearing as y_{t-1}, each IO and the subsequent values appear as "good" leverage points that increase the precision of $\widehat{\phi}$.

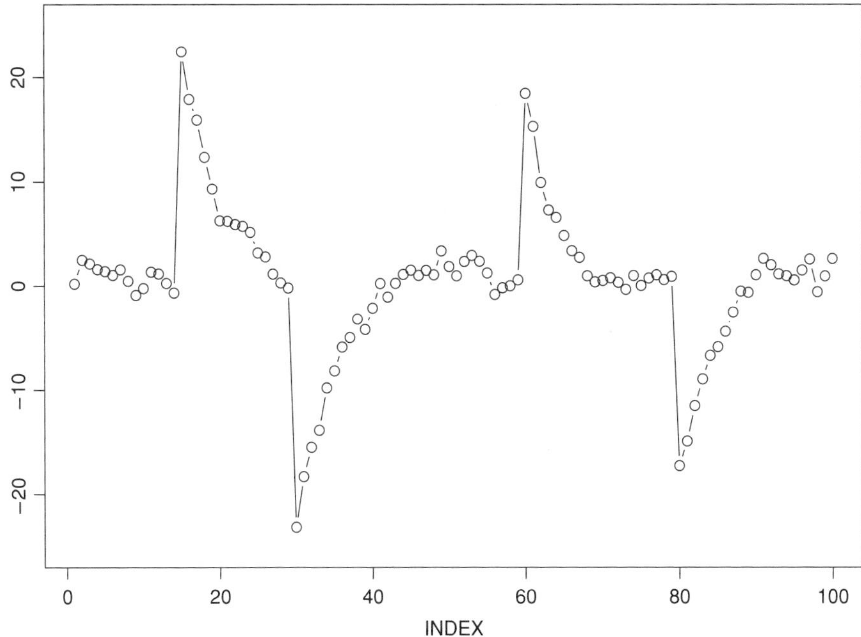

Figure 8.9 AR(1) time series with four IOs

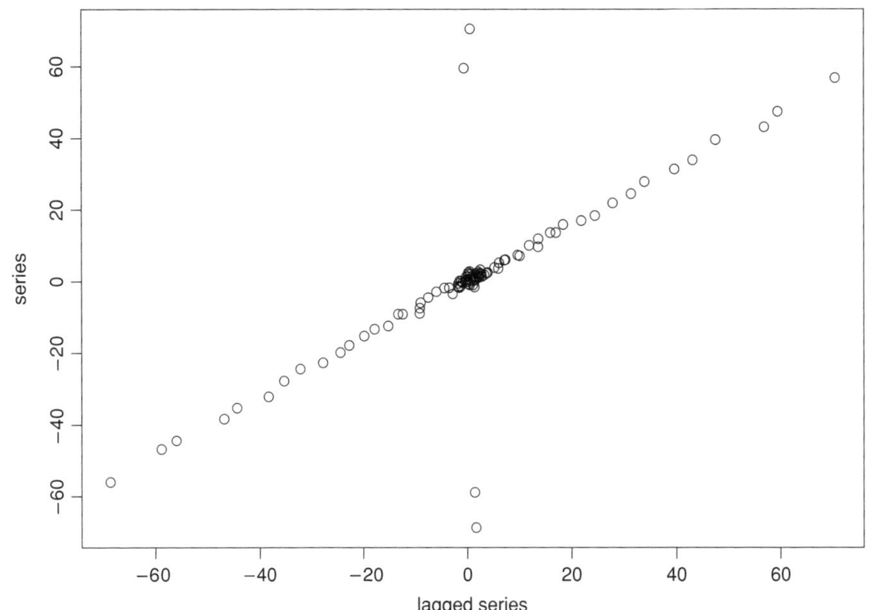

Figure 8.10 Lag-1 scatterplot of AR(1) series with IO (**Code 8.7**)

The LS estimate is 0.790 with estimated SD of 0.062, which is near the approximate finite sample value $\mathrm{SD}(\widehat{\phi}) \approx \sqrt{(1-\phi^2)/T} = 0.060$. The MM-estimate is 0.798 with estimated SD of 0.016. This exemplifies the distribution-free character of the LS estimate and the higher precision of the MM-estimate under IOs.

8.5 Generalized M-estimates

One approach to curb the effect of "bad leverage points" due to outliers is to modify M-estimates in a way similar to ordinary regression. Note first that the estimating equation of an M-estimate, obtained by differentiating the objective function with respect to $(\widehat{\gamma}, \widehat{\phi})$, is

$$\sum_{t=p+1}^{T} \mathbf{z}_{t-1}\psi\left(\frac{\widehat{u}_t}{\widehat{\sigma}_u}\right) = 0 \tag{8.70}$$

where $\psi = \rho'$ is bounded and $\mathbf{z}_t = (1, y_t, y_{t-1}, \ldots, y_{t-p+1})'$.

One attempt to improve the robustness of the estimates is to modify the above equation (8.70) by bounding the influence of outliers in $\mathbf{z}_{t-1}$ as well as in the residuals $\widehat{u}_t(\widehat{\gamma},\widehat{\phi})$. This results in the class of *generalized M-estimates* (GM-estimates), similar to those defined for regression in Section 5.11. A GM-estimate $(\widehat{\gamma},\widehat{\phi})$ is obtained by solving

$$\sum_{t=p+1}^{T} \eta\left(d_T(\mathbf{y}_{t-1}), \frac{\widehat{u}_t(\widehat{\gamma},\widehat{\phi})}{\widehat{\sigma}_u}\right)\mathbf{z}_{t-1} = 0 \tag{8.71}$$

where the function $\eta(.,.)$ is bounded and continuous in both arguments (e.g., of Mallows or Schweppe type defined in Section 5.11) and $\widehat{\sigma}$ is obtained by a simultaneous M-equation of the form

$$\frac{1}{n}\sum_{i=1}^{n} \rho\left(\frac{\widehat{u}_t(\widehat{\gamma},\widehat{\phi})}{\widehat{\sigma}_u}\right) = \delta. \tag{8.72}$$

Here

$$d_T(\mathbf{y}_{t-1}) = \frac{1}{p}\mathbf{y}_{t-1}'\widehat{\mathbf{C}}^{-1}\mathbf{y}_{t-1} \tag{8.73}$$

with $\widehat{\mathbf{C}}$ an estimate of the $p \times p$ covariance matrix $\mathbf{C}$ of $\mathbf{y}_{t-1} = (y_{t-1}, y_{t-2}, \ldots, y_{t-p})'$.

In the remark above (8.50) it was pointed out that the asymptotic distribution of LS estimates for AR models coincides with that of LS estimates for the regression model (8.21). The same can be said of GM-estimates.

GM-estimates for AR models were introduced by Denby and Martin (1979) and Martin (1980, 1981), who called them (BIFAR) bounded influence autoregressive estimates. Bustos (1982) showed that GM-estimates for AR(p) models are asymptotically normal, with covariance matrix given by the analog of the regression case (5.49). Künsch (1984) derived Hampel-optimal GM-estimates.

There are two main possibilities for $\widehat{\mathbf{C}}$. The first is to use the representation $\mathbf{C} = \sigma_u^2 \mathbf{D}(\phi)$ given by the matrix $\mathbf{D}$ in Section 8.15, where ϕ is the parameter vector of the p-th order autoregression, and put $\widehat{\mathbf{C}} = \sigma_u^2 \mathbf{D}(\widehat{\phi})$ in (8.73). Then $\widehat{\phi}$ appears twice in (8.71): in d_T and in $\widehat{u}_t$. This is a natural approach when fitting a single autoregression of given order p.

The second possibility is convenient in the commonly occurring situation where one fits a sequence of autoregressions of increasing order, with a view toward determining a "best" order p_{opt}.

Let $\phi_{k,1}, \ldots, \phi_{k,k}$ be the coefficients of the "best-fitting" autoregression of order k, given in Section 8.2.1. The autocorrelations $\rho(1), \ldots, \rho(p-1)$ depend only on $\phi_{p-1,1}, \ldots, \phi_{p-1,p-1}$ and can be obtained from the Yule–Walker equations by solving a linear system. Therefore the correlation matrix $\mathbf{R}$ of $\mathbf{y}_{t-1}$ also depends only on $\phi_{p-1,1}, \ldots, \phi_{p-1,p-1}$. We also have that

$$\mathbf{C} = \gamma(0)\mathbf{R}.$$

Then we can estimate $\phi_{p,1}, \ldots, \phi_{p,p}$ recursively as follows. Suppose that we have already computed estimates $\widehat{\phi}_{p-1,1}, \ldots, \widehat{\phi}_{p-1,p-1}$. Then, we estimate $\phi_{p,1}, \ldots, \phi_{p,p}$ by solving (8.71) and (8.72) with $\widehat{\mathbf{C}} = \widehat{\gamma}(0)\widehat{\mathbf{R}}$, where $\widehat{\gamma}(0)$ is a robust estimate of the variance of the y_t's (e.g., the square of the MADN) and $\widehat{\mathbf{R}}$ is computed from the Yule–Walker equations using $\widehat{\phi}_{p-1,1}, \ldots, \widehat{\phi}_{p-1,p-1}$.

Table 8.2 shows the results of applying a Mallows-type GM-estimate to the data of Example 8.2 (code **AR3**). It is seen that the performance of the GM-estimate is not better than that of the MM-estimate shown in Table 8.1.

Table 8.2 GM-estimates of the parameters of AR(3) simulated process

#(outliers)	ϕ_1	ϕ_2	ϕ_3	γ	σ_u
0	1.31	−0.79	0.10	0.11	0.97
10	1.15	−0.52	−0.03	0.17	1.06
20	0.74	−0.16	−0.09	0.27	1.46
True values	1.333	−0.833	0.166	0.00	1.00

8.6 Robust AR estimation using robust filters

In this section we assume that the observations process y_t has the AO form $y_t = x_t + v_t$ with x_t an AR(p) process as given in (8.14) with parameters $\boldsymbol{\lambda} = (\phi_1, \phi_2, \ldots, \phi_p, \gamma)'$ and v_t independent of x_t. An attractive approach is to define robust estimates by minimizing a robust scale of the prediction residuals, as with regression S-estimates in Section 5.6.1. It turns out that this approach is not sufficiently robust. A more robust method is obtained by minimizing a robust scale of prediction residuals obtained with a robust filter that curbs the effect of outliers. We begin by explaining why the simple approach of minimizing a robust scale of the prediction residuals is not adequate. Most of the remainder of the section is devoted to describing the robust filtering method, the scale minimization approach using prediction residuals from a robust filter, and the computational details for the whole procedure. The section concludes with an application example and an extension of the method to integrated AR(p) models.

8.6.1 Naive minimum robust scale AR estimates

In this section we deal with the robust estimation of the AR parameters by minimizing a robust scale estimate $\widehat{\sigma}$ of prediction residuals. Let y_t, $1 \leq t \leq T$, be observations corresponding to an AO model $y_t = x_t + v_t$ where x_t is an AR(p) process. For any $\boldsymbol{\lambda} = (\phi_1, \phi_2, \ldots, \phi_p, \mu)' \in R^{p+1}$, define the residual vector as

$$\widehat{\mathbf{u}}(\boldsymbol{\lambda}) = (\widehat{u}_{p+1}(\boldsymbol{\lambda}), \ldots, \widehat{u}_T(\boldsymbol{\lambda}))',$$

where

$$\widehat{u}_t(\boldsymbol{\lambda}) = (y_t - \mu) - \phi_1(y_{t-1} - \mu) - \ldots - \phi_p(y_{t-p} - \mu). \tag{8.74}$$

Given a scale estimate $\widehat{\sigma}$, an estimate of $\boldsymbol{\lambda}$ can be defined by

$$\widehat{\boldsymbol{\lambda}} = \arg\min_{\boldsymbol{\lambda}} \widehat{\sigma}(\widehat{\mathbf{u}}(\boldsymbol{\lambda})). \tag{8.75}$$

If $\widehat{\sigma}$ is a high-BP M-estimate of scale we would have the AR analog of regression S-estimates. Boente, Fraiman and Yohai (1987) generalized the notion of qualitative robustness (Section 3.7) to time series, and proved that S-estimates for autoregression are qualitatively robust and have the same efficiency as regression S-estimates. As happens in the regression case (see (5.28)), estimates based on the minimization of an M-scale are M-estimates, where the scale is the minimum scale, and therefore all the asymptotic theory of M-estimates applies under suitable regularity conditions.

If $\widehat{\sigma}$ is a τ-estimate of scale (Section 5.14.1), it can be shown that, as in the regression case, the resulting AR estimates have a higher normal efficiency than that corresponding to an M-scale.

For the reasons given in Section 8.4.2, any estimate based on the prediction residuals has a BP not larger than $0.5/(p+1)$ for AR(p) models. Since invertible MA and ARMA models have infinite AR representations, the BP of estimates based on the prediction residuals will be zero for such models.

The next subsection shows how to obtain an improved S-estimate through the use of robust filtering.

8.6.2 The robust filter algorithm

Let y_t be an AO process (8.5), where x_t is a stationary AR(p) process with mean 0 and $\{v_t\}$ are i.i.d. independent of $\{x_t\}$ with distribution (8.9). To avoid the propagation of outliers to many residuals, as described above, we shall replace the prediction residuals $\widehat{u}_t(\boldsymbol{\lambda})$ in (8.74) by the *robust prediction residuals*

$$\widetilde{u}_t(\lambda) = (y_t - \mu) - \phi_1(\widehat{x}_{t-1|t-1} - \mu) - \ldots - \phi_p(\widehat{x}_{t-p|t-1} - \mu) \tag{8.76}$$

obtained by replacing the AO observations $y_{t-i}, i = 1, \ldots, p$, in (8.74) by the *robust filtered values* $\widehat{x}_{t-i|t-1} = \widehat{x}_{t-i|t-1}(\boldsymbol{\lambda}), i = 1, \ldots, p$, which are approximations to the values $\mathrm{E}(x_{t-i}|y_1, \ldots, y_t)$.

These approximated conditional expectations were derived by Masreliez (1975) and are obtained by means of a *robust filter*. To describe this filter we need the so-called *state-space representation* of the x_t's (see, e.g., Brockwell and Davis, 1991), which for an AR(p) model is

$$\mathbf{x}_t = \boldsymbol{\mu} + \Phi(\mathbf{x}_{t-1} - \boldsymbol{\mu}) + \mathbf{d}u_t \tag{8.77}$$

where $\mathbf{x}_t = (x_t, x_{t-1}, \ldots, x_{t-p+1})'$ is called the *state vector*, $\mathbf{d}$ is defined by

$$\mathbf{d} = (1, 0, \ldots, 0)', \quad \boldsymbol{\mu} = (\mu, \ldots, \mu)', \tag{8.78}$$

and Φ is the *state-transition matrix* given by

$$\Phi = \begin{bmatrix} \phi_1 \ldots \phi_{p-1} & \phi_p \\ \mathbf{I}_{p-1} & \mathbf{0}_{p-1} \end{bmatrix}. \tag{8.79}$$

Here $\mathbf{I}_k$ is the $k \times k$ identity matrix and $\mathbf{0}_k$ the zero vector in R^k.

The following recursions compute robust filtered vectors $\widehat{\mathbf{x}}_{t|t}$ which are approximations of $\mathrm{E}(\mathbf{x}_t|\, y_1, y_2, \ldots, y_t)$ and robust one-step-ahead predictions $\widehat{\mathbf{x}}_{t|t-1}$ which are approximations of $\mathrm{E}(\mathbf{x}_t|y_1, y_2, \ldots, y_{t-1})$. At each time $t-1$ the robust prediction vectors $\widehat{\mathbf{x}}_{t|t-1}$ are computed from the robustly filtered vectors $\widehat{\mathbf{x}}_{t-1|t-1}$ as

$$\widehat{\mathbf{x}}_{t|t-1} = \boldsymbol{\mu} + \Phi(\widehat{\mathbf{x}}_{t-1|t-1} - \boldsymbol{\mu}). \tag{8.80}$$

Then the prediction vector $\widehat{x}_{t|t-1}(\boldsymbol{\lambda})$ and the AO observation y_t are used to compute the residual $\widetilde{u}_t(\boldsymbol{\lambda})$ and $\widehat{\mathbf{x}}_{t|t}$ using the recursions

$$\widetilde{u}_t(\boldsymbol{\lambda}) = (y_t - \mu) - \boldsymbol{\phi}'(\widehat{\mathbf{x}}_{t-1|t-1} - \boldsymbol{\mu}) \tag{8.81}$$

and

$$\widehat{\mathbf{x}}_{t|t} = \widehat{\mathbf{x}}_{t|t-1} + \frac{1}{s_t}\mathbf{m}_t \psi\left(\frac{\widetilde{u}_t(\boldsymbol{\lambda})}{s_t}\right), \tag{8.82}$$

where s_t is an estimate of the scale of the prediction residual $\widetilde{u}_t$ and $\mathbf{m}_t$ is a vector. Recursions for s_t and $\mathbf{m}_t$ are provided in Section 8.16. Here ψ is a bounded ψ-function that for some constants $a < b$ satisfies

$$\psi(u) = \begin{cases} u \text{ if } & |u| \leq a \\ 0 \text{ if } & |u| > b. \end{cases} \tag{8.83}$$

It turns out that the first element of $\mathbf{m}_t$ is s_t^2, and hence the first coordinate of the vector recursion (8.82) gives the scalar version of the filter. Hence if

$\widehat{\mathbf{x}}_{t|t} = (\widehat{x}_{t|t}, \ldots, \widehat{x}_{t-p+1|t})'$ and $\widehat{\mathbf{x}}_{t|t-1} = (\widehat{x}_{t|t-1}, \ldots, \widehat{x}_{t-p+1|t-1})'$, we have

$$\widehat{x}_{t|t} = \widehat{x}_{t|t-1} + s_t \psi \left(\frac{\widetilde{u}_t(\boldsymbol{\lambda})}{s_t} \right). \tag{8.84}$$

It follows that

$$\widehat{x}_{t|t} = \widehat{x}_{t|t-1} \quad \text{if} \quad |\widetilde{u}_t| > b s_t \tag{8.85}$$

and

$$\widehat{x}_{t|t} = y_t \quad \text{if} \quad |\widetilde{u}_t| \le a s_t. \tag{8.86}$$

Equation (8.85) shows that the robust filter rejects observations with scaled absolute robust prediction residuals $|\widetilde{u}_t/s_t| \ge b$, and replaces them with predicted values based on previously filtered data. Equation (8.86) shows that observations with $|\widetilde{u}_t/s_t| \le a$ remain unaltered. Observations for which $|\widetilde{u}_t/s_t| \in (a, b)$ are modified depending on how close the values are to a or b. Consequently the action of the robust filter is to "clean" the data of outliers by replacing them with predictions (one-sided interpolates) while leaving most of the remaining data unaltered. As such the robust filter might well be called an "outlier-cleaner".

The above robust filter recursions have the same general form as the class of approximate conditional mean robust filters introduced by Masreliez (1975). See also Masreliez and Martin (1977), Kleiner, Martin and Thomson (1979), Martin and Thomson (1982), Martin, Samarov and Vandaele (1983), Martin and Yohai (1985), Brandt and Künsch (1988) and Meinhold and Singpurwalla (1989). In order that the filter $\widehat{x}_{t|t}$ be robust in a well-defined sense, it is sufficient that the functions ψ and $\psi(u)/u$ be bounded and continuous (Martin and Su, 1985).

The robust filtering algorithm, which we have just described for the case of a true AR(p) process x_t, can also be used for data cleaning and prediction based on cleaned data for a memory-l predictor, $1 \le l < p$. Such use of the robust filter algorithm is central to the robustified Durbin–Levinson algorithm that we describe shortly.

Remark 1: Note that the filter as described modifies all observations which are far enough from their predicted values, including IOs. But this may damage the output of the filter, since altering one IO spoils the prediction of the ensuing values. The following modification of the above procedure deals with this problem. When a sufficiently large number of consecutive observations have been corrected, i.e., $\widehat{x}_{t|t} \ne y_t$ for $t = t_0, \ldots, t_0 + h$, the procedure goes back to t_0 and redefines $\widehat{x}_{t_0|t_0} = y_t$ and then goes on with the recursions.

Remark 2: Note that the robust filter algorithm replaces large outliers with predicted values based on the past, and as such produces "one-sided" interpolated values. One can improve the quality of the outlier treatment by using a two-sided interpolation at outlier positions by means of a robust smoother algorithm. One such algorithm is described by Martin (1979), who derives the robust smoother as an approximate conditional mean smoother analog to Masreliez's approximate conditional mean filter.

8.6.3 Minimum robust scale estimates based on robust filtering

If $\widehat{\sigma}$ is a robust scale estimate, an estimate based on robust filtering may be defined as

$$\widehat{\boldsymbol{\lambda}} = \arg\min_{\boldsymbol{\lambda}} \widehat{\sigma}(\widetilde{\mathbf{u}}(\boldsymbol{\lambda})) \tag{8.87}$$

where $\widetilde{u}(\boldsymbol{\lambda}) = (\widetilde{u}_{p+1}(\boldsymbol{\lambda}), \ldots, \widetilde{u}_T(\boldsymbol{\lambda}))'$ is the vector of robust prediction residuals $\widetilde{u}_t$ given by (8.76). The use of these residuals in place of the raw prediction residuals (8.74) prevents the smearing effect of isolated outliers, and therefore will result in an estimate that is more robust than M-estimates or estimates based on a scale of the raw residuals $\widehat{u}_t$.

One problem with this approach is that the objective function $\widehat{\sigma}(\widetilde{u}(\boldsymbol{\lambda}))$ in (8.87) typically has multiple local minima, making it difficult to find a global minimum. Fortunately there is a computational approach based on a different parameterization in which the optimization is performed one parameter at a time. This procedure amounts to a robustified Durbin–Levinson algorithm to be described in Section 8.6.4.

8.6.4 A robust Durbin–Levinson algorithm

There are two reasons why the Durbin–Levinson procedure is not robust:

- The quadratic loss function in (8.35) is unbounded.
- The residuals $\widehat{u}_{t,m}(\phi)$ defined in (8.45) are subject to an outliers "smearing" effect: if y_t is an isolated outlier, it spoils the $m+1$ residuals $\widehat{u}_{t,m}(\phi)$, $\widehat{u}_{t+1,m}(\phi), \ldots, \widehat{u}_{t+m,m}(\phi)$.

We now describe a modification of the standard sample-based Durbin–Levinson method that eliminates the preceding two sources of nonrobustness. The observations y_t are assumed to have been previously robustly centered by the subtraction of the median or another robust location estimate.

A robust version of (8.31) will be obtained in a recursive way analogous to the classical Durbin–Levinson algorithm, as follows.

Let $\widetilde{\phi}_{m-1,1}, \ldots, \widetilde{\phi}_{m-1,m-1}$ be robust estimates of the coefficients $\phi_{m-1,1}, \ldots, \phi_{m-1,m-1}$ of the memory-$(m-1)$ linear predictor. If we knew that $\phi_{m,m} = \zeta$, then according to (8.38), we could estimate the memory-m predictor coefficients as

$$\widetilde{\phi}_{m,i}(\zeta) = \widetilde{\phi}_{m-1,i} - \zeta\widetilde{\phi}_{m-1,m-i},\ i = 1, \ldots, m-1. \tag{8.88}$$

Therefore it would only remain to estimate ζ.

The robust memory-m prediction residuals $\widetilde{u}_{t,m}(\zeta)$ may be written in the form

$$\begin{aligned}\widetilde{u}_{t,m}(\zeta) = y_t - \widetilde{\phi}_{m,1}(\zeta)\,\widehat{x}^{(m)}_{t-1|t-1}(\zeta) - \ldots - \widetilde{\phi}_{m,m-1}(\zeta)\,\widehat{x}^{(m)}_{t-m+1|t-1}(\zeta) \\ - \zeta\,\widehat{x}^{(m)}_{t-m|t-1}(\zeta)\end{aligned} \tag{8.89}$$

where $\widehat{x}^{(m)}_{t-i|t-1}(\zeta)$, $i = 1, \ldots, m$, are the components of the robust state vector estimate $\widehat{\mathbf{x}}^{(m)}_{t-1|t-1}$ obtained using the robust filter (8.82) corresponding to an order-m autoregression with parameters

$$(\widetilde{\phi}_{m,1}(\zeta), \widetilde{\phi}_{m,2}(\zeta), \ldots, \widetilde{\phi}_{m,m-1}(\zeta), \zeta). \tag{8.90}$$

Observe that $\widetilde{u}_{t,m}(\zeta)$ is defined as $\widehat{u}_{t,m}(\zeta)$ in (8.45), except for the replacement of $y_{t-1}, \ldots, y_{t-m}$ by the robustly filtered values $\widehat{x}^{(m)}_{t-1|t-1}(\zeta), \ldots, \widehat{x}^{(m)}_{t-m|t-1}(\zeta)$. Now an outlier y_t may spoil only a single residual $\widetilde{u}_{t,m}(\zeta)$, rather than $p+1$ residuals as in the case of the usual AR(p) residuals in (8.74).

The standard Durbin–Levinson algorithm computes $\widehat{\phi}_{m,m}$ by minimizing the sum of squares (8.44), which in the present context is equivalent to minimizing the sample standard deviation of the $\widetilde{u}_{t,m}(\zeta)$'s defined by (8.89). Since the $\widetilde{u}_{t,m}(\zeta)$'s may have outliers in the y_t term and the sample standard deviation is not robust, we replace it with a highly robust scale estimate $\widehat{\sigma} = \widehat{\sigma}(\widetilde{u}_{m+1,m}(\zeta), \ldots, \widetilde{u}_{T,m}(\zeta))$. We have thus eliminated the two sources of non-robustness of the standard Durbin–Levinson algorithm. Finally, the robust partial autocorrelation coefficient estimates $\widehat{\phi}_{m,m}$, $m = 1, 2, \ldots, p$, are obtained sequentially by solving

$$\widehat{\phi}_{m,m} = \arg\min_{\zeta} \widehat{\sigma}(\widetilde{u}_{m+1,m}(\zeta), \ldots, \widetilde{u}_{T,m}(\zeta)), \tag{8.91}$$

where for each m the values $\widehat{\phi}_{m,i}$, $i = 1, \ldots, m-1$, are obtained from (8.38). This minimization can be performed by a grid search on $(-1, 1)$.

The first step of the procedure is to compute a robust estimate $\widetilde{\phi}_{1,1}$ of $\phi_{1,1}$ by means of a robust version of (8.43), namely (8.91) with $m = 1$, where

$$\widetilde{u}_{t,1}(\zeta) = y_t - \zeta \widehat{x}_{t-1|t}(\zeta).$$

8.6.5 Choice of scale for the robust Durbin–Levinson procedure

One possibility for the choice of a robust scale in (8.91) is to use an M-scale with a BP of 0.5, in which case the resulting estimator is an S-estimate of autoregression using robustly filtered values. However, it was pointed out earlier in Section 5.6.1 that Hössjer (1992) proved that an S-estimate of regression with a BP of 0.5 cannot have a large-sample efficiency greater than 0.33 when the errors have a normal distribution. This fact provided the motivation for using τ-estimates of regression as defined in equations (5.58), (5.59) and (5.60) of Section 5.14.1. These estimates can attain a high efficiency, e.g., 95%, when the errors have a normal distribution, while at the same time having a high BP of 0.5. The relative performance of a τ-estimate versus an S-estimate with regard to BP and normal efficiency is expected to carry over to a large extent to the present case of robust AR model fitting using robustly filtered observations. Thus we recommend that the robust scale estimate $\widehat{\sigma}$ in (8.91) be a τ-scale defined as in

Table 8.3 Fτ-estimates of the parameters of AR(3) simulated process

#(outliers)	ϕ_1	ϕ_2	ϕ_3	γ	σ_u
0	1.40	−0.90	0.19	0.32	0.98
10	1.27	−0.76	0.07	0.15	1.14
20	1.40	−0.90	0.19	0.18	1.14
True values	1.333	−0.833	0.166	0.00	1.00

(5.58), (5.59) and (5.60), but with residuals given by (8.89) and (8.88). The examples we show for robust fitting of AR, ARMA, ARIMA and REGARIMA models in the remainder of this chapter are all computed with an algorithm that uses a τ-scale applied to robustly filtered residuals. We shall call such estimates *filtered* τ*- (or F*τ*-) estimates*. These estimates were studied by Bianco, García Ben, Martínez and Yohai (1996).

Table 8.3 shows the results of applying an Fτ-estimate to the data of Example 8.2. It is seen that the impact of outliers is slight, and comparison with Tables 8.1 and 8.2 shows the performance of the Fτ-estimate to be superior to that of the MM- and GM-estimates.

8.6.6 Robust identification of AR order

The classical approach based on Akaike's information criterion, AIC (Akaike, 1973, 1974b), when applied to the choice of the order of AR models, leads to the minimization of

$$\text{AIC}_p = \log\left(\frac{1}{T-p}\sum_{t=p+1}^{T}\widehat{u}_t^2(\widehat{\boldsymbol{\lambda}}_{p,\text{LS}})\right) + \frac{2p}{T-p},$$

where $\widehat{\boldsymbol{\lambda}}_{p,\text{LS}}$ is the LS estimate corresponding to an AR(p) model, and $\widehat{u}_t$ are the respective residuals. The robust implementation of this criterion in the S-PLUS program **arima.rob,** described in Chapter 11, is based on the minimization of

$$\text{RAIC}_p = \log\left(\tau^2\left(\widetilde{u}_{p+1}\left(\widehat{\boldsymbol{\lambda}}_{p,\text{rob}}\right),\ldots,\widetilde{u}_T\left(\widehat{\boldsymbol{\lambda}}_{p,\text{rob}}\right)\right)\right) + \frac{2p}{T-p},$$

where $\widetilde{u}_i\left(\widehat{\boldsymbol{\lambda}}_{p,\text{rob}}\right)$ are the filtered residuals corresponding to the Fτ-estimate and τ is the respective scale.

As with the RFPE criterion in (5.53), we consider that it would be better to multiply the penalty term $2p/(T-p)$ by a factor depending on the choice of the scale and on the distribution of the residuals. This area requires further research.

8.7 Robust model identification

Time series autocorrelations are often computed for exploratory purposes without reference to a parametric model. In addition autocorrelations are often computed along with partial autocorrelations for use in identification of ARMA and ARIMA models; see for example Brockwell and Davis (1991). We know already from Sections 8.1.1 and 8.1.3 that additive outliers can have considerable influence and cause bias and inflated variability in the case of a lag-one correlation estimate, and Section 8.2.1 indicates that additive outliers can have a similar adverse impact on partial autocorrelation estimates. Thus one needs robust estimates of autocorrelations and partial autocorrelations in the presence of AOs or ROs.

8.7.1 Robust autocorrelation estimates

Suppose we want to estimate an unknown lag-k autocorrelation (8.2) of the time series x_t based on AO observations $y_t = x_t + v_t$. One may think of several ways of doing this, the simplest being to use robust pairwise correlation estimates $\widehat{\rho}(k)$, as in Section 6.9.1, based on the two-dimensional sets of observations (y_t, y_{t-k}), $t = k+1, \ldots, T$. As we saw in that section, this approach has the drawback that the resulting correlation matrix need not be positive semidefinite.

While other methods for robust estimation of autocorrelations and partial autocorrelations have been discussed in the literature (see e.g., Ma and Genton, 2000), our recommendation is to use one of the following two approaches based on robust fitting of a "long" AR model of order p^* using the robustly filtered τ-scale estimate:

(a) compute classical autocorrelations and partial autocorrelations based on the robustly filtered values $\widehat{x}_{t|t}$ for the AR(p^*) model, or
(b) compute the robust partial autocorrelation estimates $\widehat{\phi}_{k,k}$, $k = 1, \ldots, p^*$, from the sequence of robust AR(k) fits, and use the robustly estimated AR(p^*) coefficients $\widehat{\phi}_1, \ldots, \widehat{\phi}_{p^*}$ to compute for a given K the robust autocorrelation estimates $\widehat{\rho}(k)$ ($k = 1, \ldots, K$) by solving the Yule–Walker equations (8.29)

$$\rho(k) = \sum_{i=1}^{p^*} \phi_i \rho(k-i) \quad (k \geq 1) \tag{8.92}$$

for the values of the unknown $\rho(k)$, where the unknown ϕ_i's are replaced by the estimates $\widehat{\phi}_i$. Note that the first $p^* - 1$ equations of the above set suffice to determine $\rho(k)$, $k = 1, \ldots, p^* - 1$, and that $\rho(k)$ for $k \geq p^*$ are obtained recursively from (8.92).

Method (a) is attractive in that, aside from the robust filtering that depends upon a parametric model approximation, it is nonparametric in character. On the other hand there may be some small biases introduced by the local linear dependency of the robust filter predictions at outlier positions. Method (b) is quite natural, but it is a parametric method and so relies on the goodness of the long AR approximation.

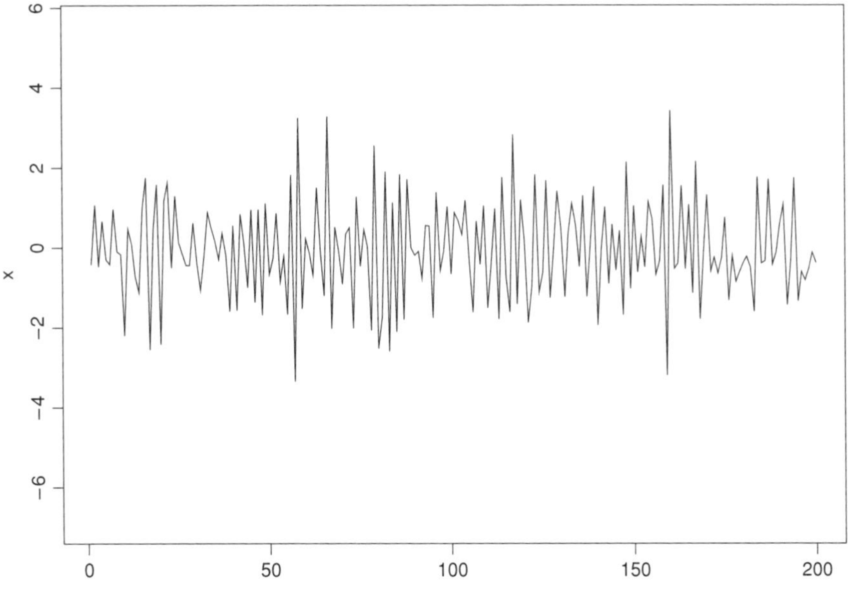

Figure 8.11 Gaussian AR(2) series

Example 8.3 *Simulated AR(2) AO model.*

Consider an AO model $y_t = x_t + v_t$ where x_t is a zero-mean Gaussian AR(2) model with parameters $\phi = (\phi_1, \phi_2)' = (-0.65, -0.3)'$ and innovations variance $\sigma_u^2 = 1$; and $v_t = z_t w_t$, where z_t and w_t are independent, $\mathrm{P}(z_t = \pm 1) = 0.5$, and w_t has the mixture distribution (8.6) with $\varepsilon = 0.1$, $\sigma_v = 1$ and $\mu_v = 4$. The following results are computed with code **AR2plots**. Figure 8.11 shows a series of x_t of length 200, and Figure 8.12 shows the same series with the AOs. There are 24 AOs, as compared to an expected value of 20.

Table 8.4 displays the LS estimate (Yule–Walker version) applied to the AO data, which is quite far from the true values, but quite consistent with the dominant white-noise character of the AOs that result in a process with little correlation at any lag. The Fτ-estimate is seen to be much closer to the true values. The $F\tau$-estimate applied to the outlier-free series x_t and the LS estimate based on the robustly filtered data $\widehat{x}_{t|t}$ are also close to the true values.

Use of the Yule–Walker equations (8.92) as implemented in code **AR2-YW** shows that the lag-k autocorrelations $\rho(k)$ have values -0.50, 0.025, 0.13, -0.09 for $k = 1, 2, 3, 4$ respectively. Figure 8.13 shows the classical autocorrelation estimate based on the lag-k sample autocorrelations, for the outlier-free series x_t. The results reveal significant autocorrelations at lags 1, 3 and marginally 4, with the lag 1 and 3 values being reasonably close to the true values of -0.5 and 0.13. The classical autocorrelation estimate based on the AO series y_t in Figure 8.14 does not detect any

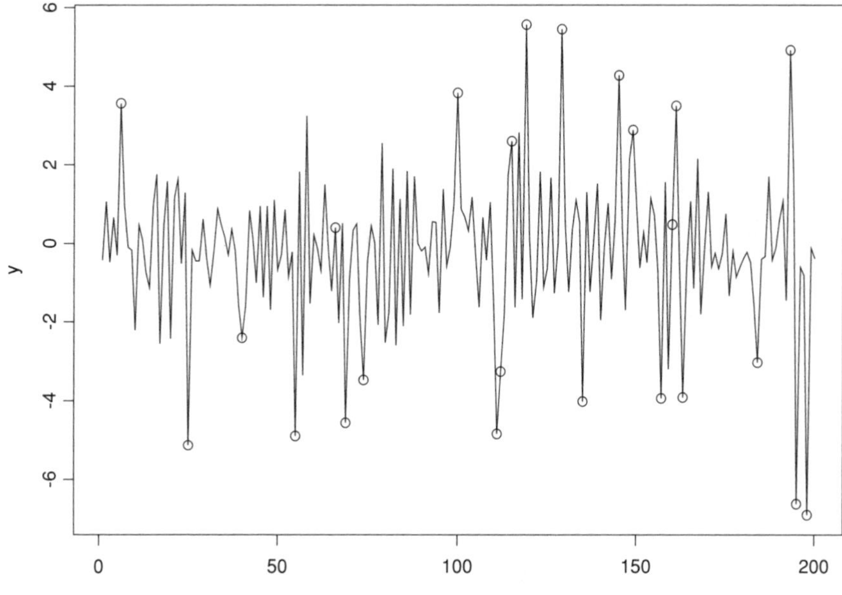

Figure 8.12 Gaussian AR(2) series with AOs

of the significant autocorrelations indicated for x_t, while yielding smaller autocorrelation values at most other lags along with a spurious nonzero autocorrelation at lag 8. The zero autocorrelation indications at lags 1, 3 and 4 and smaller autocorrelations at most other lags are a consequence of the fact that the AOs are a white-noise process, whose true autocorrelations are zero at all lags, and that dominates the x_t series values.

Figure 8.15 shows the autocorrelation function estimate based on the robustly filtered data $\widehat{x}_{t|t}$ which is seen to be similar to the estimate based on the outlier-free series x_t, shown in Figure 8.13.

Alternatively, we can use the Yule–Walker equations to compute the autocorrelation estimates based on the robust parameter estimates $\widehat{\phi}_{y,\mathrm{rob}} = (-0.69, -0.39)'$

Table 8.4 Estimates for AR(2) with AOs

Estimate	Data	ϕ_1	ϕ_2
LS	y_t	−0.12	−0.03
Fτ	y_t	−0.69	−0.39
Fτ	x_t	−0.74	−0.29
LS	$\widehat{x}_{t\|t}$	−0.76	−0.38
True		−0.65	−0.30

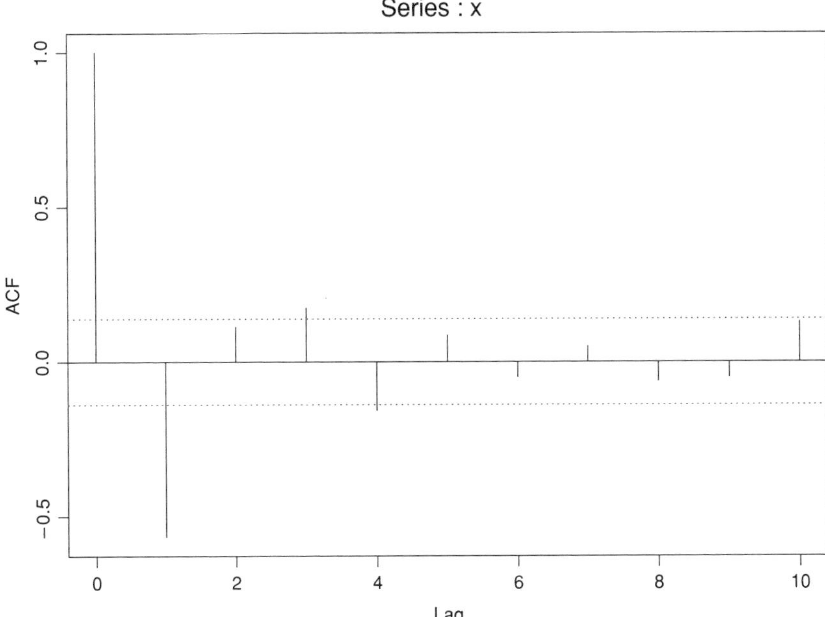

Figure 8.13 Correlation estimates for x_t

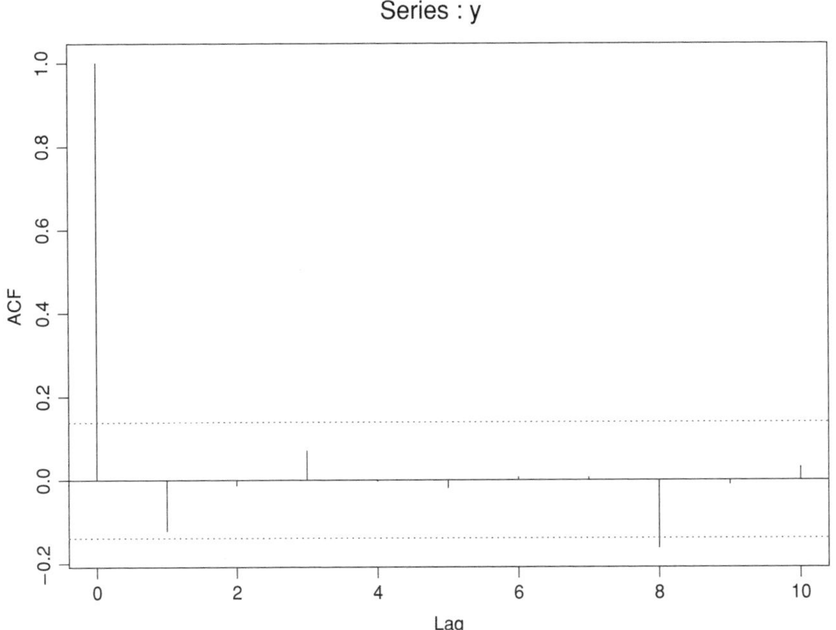

Figure 8.14 Correlation estimates for y_t

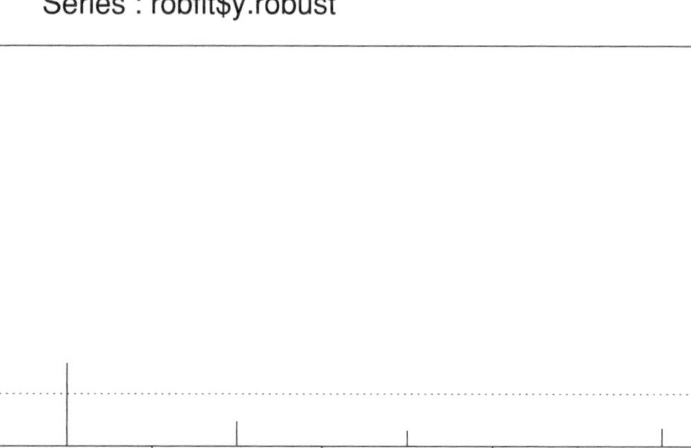

Figure 8.15 Correlation estimates for $\hat{x}_{t|t}$

obtained with the Fτ-estimates. Use of code **AR2-YW** with these parameter values gives autocorrelation estimates $-0.50, -0.05, 0.23, -0.14$ for $k = 1, 2, 3, 4$ respectively, which are quite consistent with the estimates of Figure 8.15.

Example 8.4 *Robust autocorrelation estimates for MA(1) series.*

Now consider an AO model $y_t = x_t + v_t$ where x_t is a Gaussian MA(1) process $x_t = u_t - \theta u_{t-1}$. It is easy to show that the lag-k autocorrelations $\rho(k)$ of x_t are zero except for $k = 1$ where $\rho(1) = -\theta/(1+\theta^2)$, and that $\rho(1) = \rho(1, \theta)$ is bounded in magnitude by $1/2$ for $-1 < \theta < 1$. Code **MA1** generates the figures and other results for this example. Figure 8.16 shows a series of length 200 of an invertible Gaussian MA(1) process x_t with $\theta = -0.9$ and $\sigma_u^2 = 1$, for which $\rho(1) = 0.497$, and Figure 8.17 shows the series y_t with AOs as in Example 8.3, except that now $\mu_v = 6$ instead of 4.

Figure 8.18 shows the classical autocorrelations computed from the outlier-free x_t series and Figure 8.19 shows the classical autocorrelations computed from the y_t series. The horizontal dashed lines give the approximate 95% confidence interval for an autocorrelation coefficient under the null hypothesis that the autocorrelation coefficient is zero, i.e., the lines are located at $\pm 1.96T^{-1/2}$. The former figure gives a fairly accurate estimate of the true lag-1 autocorrelation value of 0.497 along with

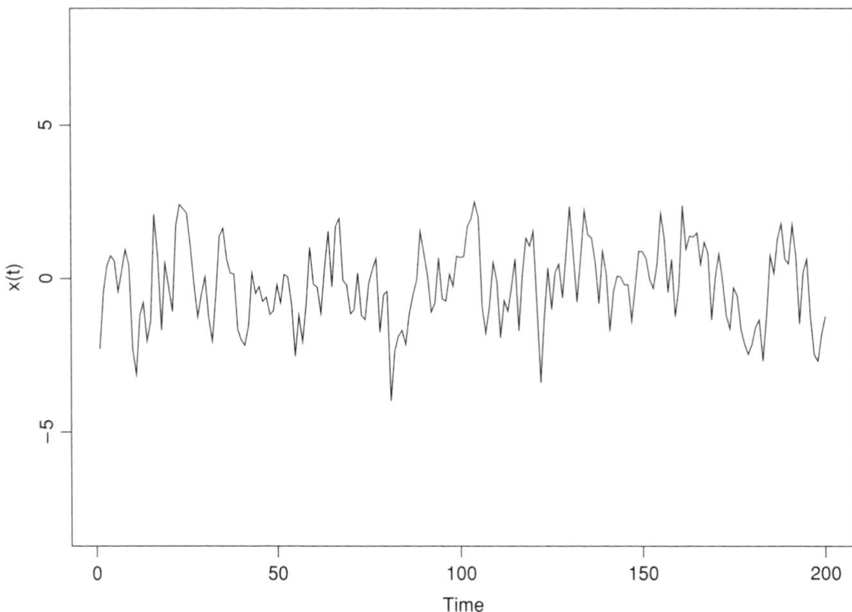

Figure 8.16 Gaussian MA(1) series with $\theta = 0.9$

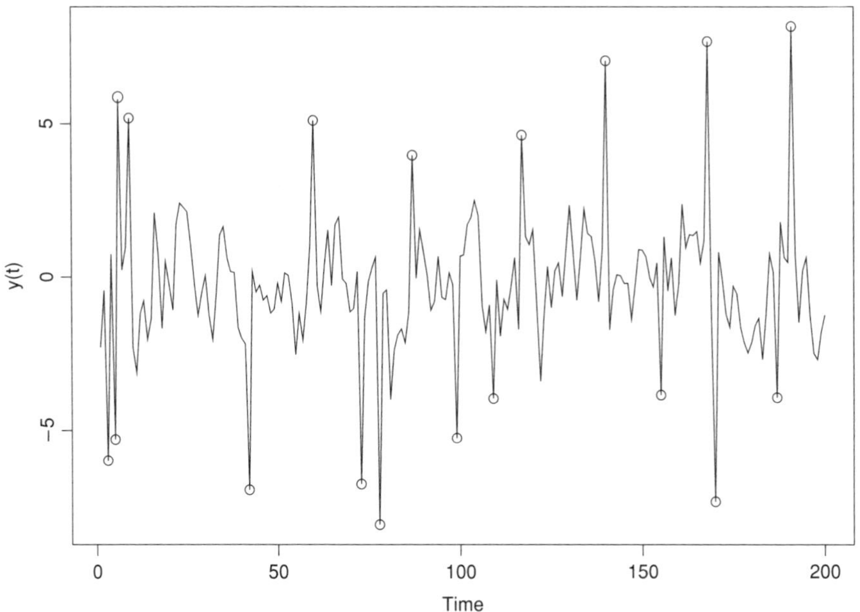

Figure 8.17 MA(1) series with AOs

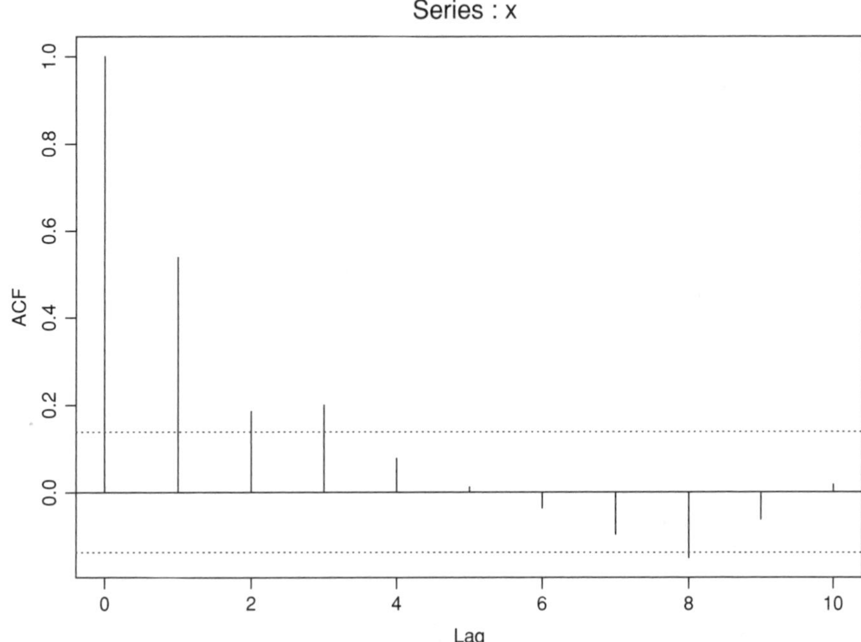

Figure 8.18 Correlations based on the Gaussian x_t MA(1) series

marginal indication of positive autocorrelations at lags 2 and 3, while the latter figure fails to estimate the significant autocorrelation at lag 1 and has no other significant autocorrelations. Figure 8.20 shows the robust autocorrelations obtained as classical autocorrelations for the robustly filtered data using an AR model of the MA(1) process. The order of the autoregression was obtained with the robust selection method described in Section 8.6.6. The method estimated an AR(5) model with parameters

$$(\widehat{\phi}_1, \ldots, \widehat{\phi}_5) = (0.69, -0.53, 0.56, -0.29, 0.10)$$

and filtered 17 out of 18 outliers. The resulting robust autocorrelation estimates are almost identical to those obtained in Figure 8.18 based on the Gaussian MA(1) series x_t.

8.7.2 Robust partial autocorrelation estimates

We may consider two approaches to obtain robust partial autocorrelations:

(a) in a nonparametric way, by applying the usual Durbin–Levinson algorithm in Section 8.2.1 to the robustly filtered data $\widehat{x}_{t|t}$ based on a long robust AR fit of order p^* as discussed previously for autocorrelations, or

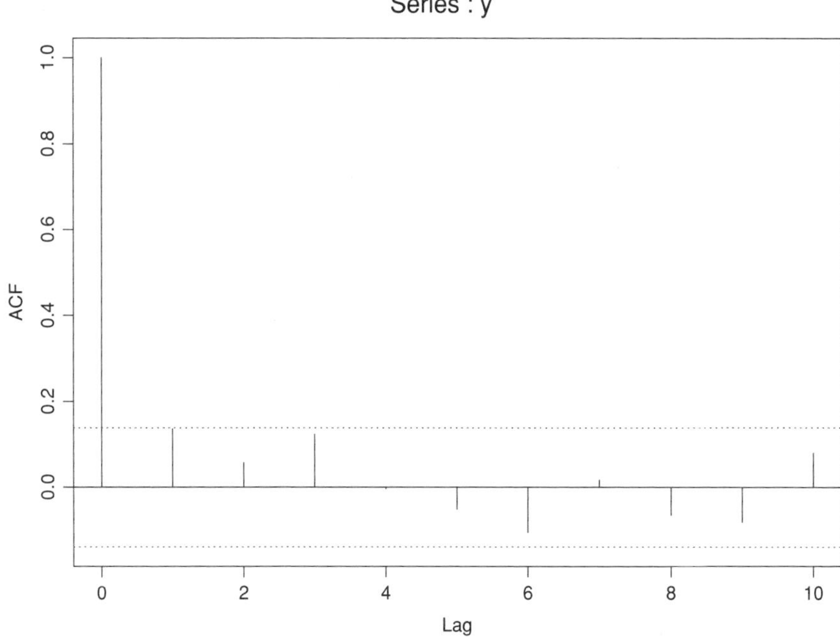

Figure 8.19 Correlations based on y_t with AOs

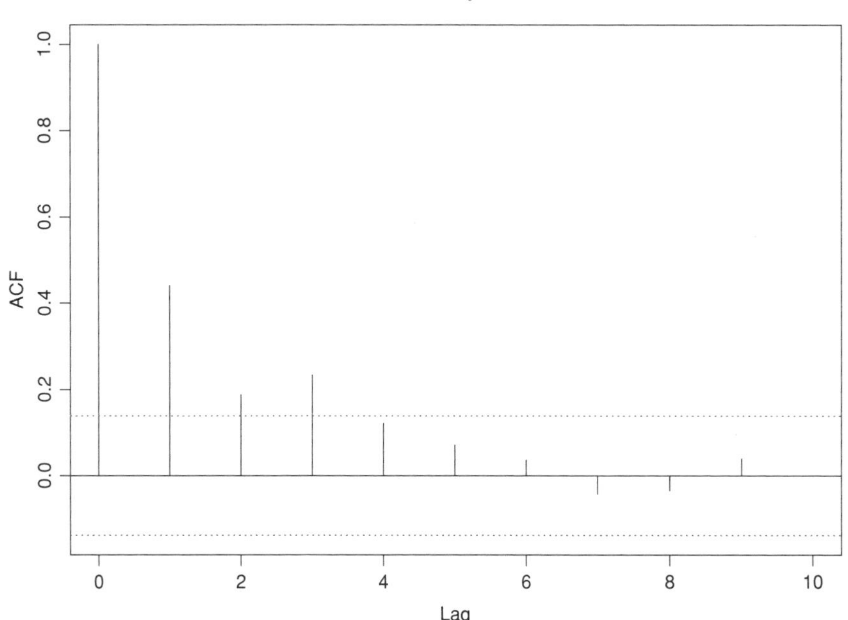

Figure 8.20 Robust correlations from filtered y_t

(b) in a parametric way by using the estimates $\widehat{\phi}_{k,k}$ from a sequence of robust AR(k) $F\tau$-estimates $\widehat{\boldsymbol{\phi}}_k = (\widehat{\phi}_{k,1}, \ldots, \widehat{\phi}_{k,k})$ of orders $k = 1, \ldots, p^*$.

We illustrate both approaches for the same two examples we used above to illustrate robust autocorrelation estimation.

Example 8.5 *Robust partial autocorrelation estimates for an AR(2) series.*

Consider again the AR(2) process of Example 8.3. It follows from (8.39) that for an AR(2) process the partial autocorrelation coefficients $\phi_{k,k}$ are given by

$$\phi_{1,1} = \rho(1), \quad \phi_{2,2} = \frac{\rho(2) - \rho(1)^2}{1 - \rho(1)^2},$$

and $\phi_{k,k} = 0$ for $k \geq 3$. The true autocorrelations are $\rho(1) = -0.500$ and $\rho(2) = 0.025$, giving partial autocorrelations $\phi_{1,1} = -0.5$ and $\phi_{2,2} = -0.3$. Code **AR2-PACF-a** uses the robustly filtered series $\widehat{x}_{t|t}$ from the $F\tau$-estimate, with robust automatic order selection (which yielded $p^* = 2$), to compute robust partial autocorrelation estimates

$$(\widehat{\phi}_{1,1}, \ldots, \widehat{\phi}_{4,4}) = (-0.55, \ -0.38, \ 0.07, \ 0.06).$$

It is known that the large-sample standard deviations of the classical estimated partial autocorrelations are $1/\sqrt{T}$. Assuming this result to hold for the classical autocovariances computed from the filtered data, we would have SD $\left(\widehat{\phi}_{k,k}\right) \approx 0.07$. Hence the estimated values fall within about one SD of the true values.

On the other hand, code **AR2-PACF-b** yields for approach (b):

$$(\widehat{\phi}_{1,1}, \ldots, \widehat{\phi}_{4,4}) = (-0.46, -0.39, 0.022, 0.07).$$

Example 8.6 *Robust partial autocorrelation estimates for an MA(1) series.*

Consider again Example 8.4 of an AO model $y_t = x_t + v_t$, where x_t is an MA(1) process $x_t = u_t - \theta u_{t-1}$ with $\theta = -0.9$ and $\sigma_u^2 = 1$. It may be shown that the partial autocorrelation coefficients for this process are

$$\phi_{k,k} = \frac{-\theta^k(1 - \theta^2)}{1 - \theta^{2(k+1)}}. \tag{8.93}$$

See for example Brockwell and Davis (1991, p.100). For $\theta = -0.9$ this gives

$$(\phi_{1,1}, \ldots, \phi_{5,5}) = (0.50, -0.33, 0.24, -0.19, 0.16)$$

rounded to two decimal places.

For this example we use the same simulated series of length 200 as in Example 8.4. Code **MA1-PACF-a** applies the classical Durbin–Levinson algorithm to the robustly filtered series $\widehat{x}_t$, resulting in the partial autocorrelation coefficient estimates

$$(\widehat{\phi}_{1,1}, \ldots, \widehat{\phi}_{5,5}) = (0.54, -0.15, 0.24, -0.20, 0.11),$$

which are close to the true values except for $k = 2$. The same code applies the standard Durbin–Levinson algorithm to the robustly filtered data, based on a robust autoregression order estimate of 5, resulting in the estimates

$$(0.44, -0.01, 0.19, -0.07, 0.03),$$

which is not very satisfactory except for $k = 1$ and $k = 3$ (at least the alternating signs are preserved). Code **MA1-PACF-b** uses $F\tau$-estimates to fit robust AR models of orders 1 through 5, resulting in the parametric robust estimates

$$(\widehat{\phi}_{1,1}, \ldots, \widehat{\phi}_{5,5}) = (0.56, -0.12, 0.27, -0.24, 0.10).$$

These appear to be more accurate estimates of the true partial autocorrelations, suggesting that the robust filtering operation may introduce some bias in the estimates and that perhaps the robust parametric estimates are to be more trusted. This is a topic in need of further study.

8.8 Robust ARMA model estimation using robust filters

In this section we assume that we observe $y_t = x_t + v_t$, where x_t follows an ARMA model given by (8.53). The parameters to be estimated are given by the vector $\boldsymbol{\lambda} = (\boldsymbol{\phi}, \boldsymbol{\theta}, \mu) = (\phi_1, \phi_2, \ldots, \phi_p, \theta_1, \theta_2, \ldots, \theta_q, \mu)$.

8.8.1 τ-estimates of ARMA models

In order to motivate the use of Fτ-estimates for fitting ARMA models, we first describe naive τ-estimates that do not involve the use of robust filters. Assume first there is no contamination, i.e., $y_t = x_t$. For $t \geq 2$ call $\widehat{y}_{t|t-1}(\boldsymbol{\lambda})$ the optimal linear predictor of y_t based on $y_1, \ldots, y_{t-1}$ when the true parameter is $\boldsymbol{\lambda}$, as described in Section 8.2.1. For $t = 1$ put $\widehat{y}_{t|t-1}(\boldsymbol{\lambda}) = \mu = \mathrm{E}(y_t)$. Then if u_t are normal we also have

$$\widehat{y}_{t|t-1}(\boldsymbol{\lambda}) = \mathrm{E}(y_t | y_1, \ldots, y_{t-1}), \quad t > 1. \tag{8.94}$$

Define the prediction errors as

$$\widehat{u}_t(\boldsymbol{\lambda}) = y_t - \widehat{y}_{t|t-1}(\boldsymbol{\lambda}). \tag{8.95}$$

Note that these errors are not the same as $\widehat{u}_t(\boldsymbol{\lambda})$ defined by (8.58).

The variance of $\widehat{u}_t(\boldsymbol{\lambda})$, $\sigma_t^2(\boldsymbol{\lambda}) = \mathrm{E}(y_t - \widehat{y}_{t|t-1}(\boldsymbol{\lambda}))^2$, has the form

$$\sigma_t^2(\boldsymbol{\lambda}) = a_t^2(\boldsymbol{\lambda})\sigma_u^2, \tag{8.96}$$

with $\lim_{t\to\infty} a_t^2(\boldsymbol{\lambda}) = 1$ (see Brockwell and Davis, 1991). In the AR case we have $a = 1$ for $t \geq p + 1$.

Suppose that the innovations u_t have a $\mathrm{N}(0, \sigma_u^2)$ distribution. Let $L(y_1, \ldots, y_T, \boldsymbol{\lambda}, \sigma_u)$ be the likelihood and define

$$Q(\boldsymbol{\lambda}) = -2 \max_{\sigma_u} \log L(y_1, \ldots, y_T, \boldsymbol{\lambda}, \sigma_u). \tag{8.97}$$

Except for a constant, we have (see Brockwell and Davis, 1991)

$$Q(\boldsymbol{\lambda}) = \sum_{t=1}^{T} \log a_t^2(\boldsymbol{\lambda}) + T \log \left(\frac{1}{T} \sum_{t=1}^{T} \frac{\widehat{u}_t^2(\boldsymbol{\lambda})}{a_t^2(\boldsymbol{\lambda})} \right). \tag{8.98}$$

Then the MLE of $\boldsymbol{\lambda}$ is given by

$$\widehat{\boldsymbol{\lambda}} = \arg\min_{\boldsymbol{\lambda}} Q(\boldsymbol{\lambda}). \tag{8.99}$$

Observe that

$$\frac{1}{T} \sum_{t=1}^{T} \frac{\widehat{u}_t^2(\boldsymbol{\lambda})}{a_t^2(\boldsymbol{\lambda})}$$

is the square of an estimate of σ_u based on the values $\widehat{u}_t(\boldsymbol{\lambda})/a_t(\boldsymbol{\lambda})$, $t = 1, \ldots, T$. Then it seems natural to define a τ-estimate $\widehat{\boldsymbol{\lambda}}$ of $\boldsymbol{\lambda}$ by minimizing

$$Q^*(\boldsymbol{\lambda}) = \sum_{t=1}^{T} \log a_t^2(\boldsymbol{\lambda}) + T \log \left(\tau^2 \left(\frac{\widehat{u}_1(\boldsymbol{\lambda})}{a_1(\boldsymbol{\lambda})}, \ldots, \frac{\widehat{u}_T(\boldsymbol{\lambda})}{a_T(\boldsymbol{\lambda})} \right) \right), \tag{8.100}$$

where for any $\mathbf{u} = (u_1, \ldots, u_T)'$ a τ-scale estimate is defined by

$$\tau^2(\mathbf{u}) = s^2(\mathbf{u}) \sum_{t=1}^{T} \rho_2 \left(\frac{u_t}{s(\mathbf{u})} \right), \tag{8.101}$$

with $s(\mathbf{u})$ an M-scale estimate based on a bounded ρ-function ρ_1. See Section 5.14.1 for further details in the context of regression τ-estimates.

While regression τ-estimates can simultaneously have a high BP value of 0.5 and a high efficiency at the normal distribution, the τ-estimate $\widehat{\boldsymbol{\lambda}}$ of $\boldsymbol{\lambda}$ has a BP of at most $0.5/(p+1))$ in the AR(p) case, and is zero in the MA and ARMA cases, for the reasons given at the end of Section 8.6.1.

8.8.2 Robust filters for ARMA models

One way to achieve a positive (and hopefully reasonably high) BP for ARMA models with AO is to extend the AR robust filter method of Section 8.6.2 to ARMA models based on a state-space representation of these models.

The extension consists of modifying the state-space representation (8.77)–(8.79) as follows. Let x_t be an ARMA(p, q) process and $k = \max(p, q+1)$. Then in Section 8.17 we show that it is possible to define a k-dimensional state-space vector $\boldsymbol{\alpha}_t = (\alpha_{1,t}, \ldots, \alpha_{k,t})'$ with $\alpha_{1,t} = x_t - \mu$, so that the following representation holds:

$$\boldsymbol{\alpha}_t = \Phi \boldsymbol{\alpha}_{t-1} + \mathbf{d} u_t, \tag{8.102}$$

where

$$\mathbf{d} = (1, -\theta_1, \ldots, -\theta_{k-1})', \tag{8.103}$$

with $\theta_i = 0$ for $i > q$ in case $p > q$. The state-transition matrix Φ is now given by

$$\Phi = \begin{bmatrix} \boldsymbol{\phi}_{k-1} & \mathbf{I}_{k-1} \\ \phi_k & \mathbf{0}_{k-1} \end{bmatrix} \tag{8.104}$$

and where $\boldsymbol{\phi}_{k-1} = (\phi_1, \ldots, \phi_{k-1})$ and $\phi_i = 0$ for $i > p$.

Suppose now that the observations y_t follow the AO process (8.5). The Masreliez approximate robust filter can be derived in a way similar to the AR(p) case in Section 8.6.2. The filter yields approximations

$$\widehat{\boldsymbol{\alpha}}_{t|t} = (\widehat{\alpha}_{t,1|t}, \ldots, \widehat{\alpha}_{t,k|t}) \quad \text{and} \quad \widehat{\boldsymbol{\alpha}}_{t|t-1} = (\widehat{\alpha}_{t,1|t-1}, \ldots, \widehat{\alpha}_{t,k|t-1})$$

of $\mathrm{E}(\boldsymbol{\alpha}_t | y_1, \ldots, y_t)$ and $\mathrm{E}(\boldsymbol{\alpha}_t | y_1, \ldots, y_{t-1})$ respectively. Observe that

$$\widehat{x}_{t|t} = \widehat{x}_{t|t}(\boldsymbol{\lambda}) = \widehat{\alpha}_{t,1|t}(\boldsymbol{\lambda}) + \mu$$

and

$$\widehat{x}_{t|t-1} = \widehat{x}_{t|t-1}(\boldsymbol{\lambda}) = \widehat{\alpha}_{t,1|t-1}(\boldsymbol{\lambda}) + \mu$$

approximate $\mathrm{E}(x_t | y_1, \ldots, y_t)$ and $\mathrm{E}(x_t | y_1, \ldots, y_{t-1})$, respectively.

The recursions to obtain $\widehat{\boldsymbol{\alpha}}_{t|t}$ and $\widehat{\boldsymbol{\alpha}}_{t|t-1}$ are as follows:

$$\begin{aligned} \widehat{\boldsymbol{\alpha}}_{t|t-1} &= \Phi \widehat{\boldsymbol{\alpha}}_{t-1|t-1}, \\ \widetilde{u}_t(\lambda) &= y_t - \widehat{x}_{t|t-1} = y_t - \widehat{\alpha}_{t,1|t-1}(\boldsymbol{\lambda}) - \mu, \end{aligned} \tag{8.105}$$

and

$$\widehat{\boldsymbol{\alpha}}_{t|t} = \widehat{\boldsymbol{\alpha}}_{t|t-1} + \frac{1}{s_t} \mathbf{m}_t \psi \left(\frac{\widetilde{u}_t(\boldsymbol{\lambda})}{s_t} \right). \tag{8.106}$$

Taking the first component in the above equation, adding μ to each side, and using the fact that the first component of $\mathbf{m}_t$ is s_t^2 yields

$$\widehat{x}_{t|t} = \widehat{x}_{t|t-1} + s_t \psi \left(\frac{\widetilde{u}_t(\boldsymbol{\lambda})}{s_t} \right),$$

and therefore (8.85) and (8.86) hold.

Further details on the recursions of s_t and $\mathbf{m}_t$ are provided in Section 8.16. The recursions for this filter are the same as (8.80), (8.82) and the associated filter covariance recursions in Section 8.16, with $\widehat{\mathbf{x}}_{t|t}$ replaced with $\widehat{\boldsymbol{\alpha}}_{t|t}$ and $\widehat{\mathbf{x}}_{t|t-1}$ replaced with $\widehat{\boldsymbol{\alpha}}_{t|t-1}$. Further details are provided in Section 8.17.

As we shall see in Section 8.17, in order to implement the filter, a value for the ARMA innovations variance σ_u^2 is needed as well as a value for $\boldsymbol{\lambda}$. We deal with this issue as in the case of an AR model by replacing this unknown variance with an estimate $\widehat{\sigma}_u^2$ in a manner described subsequently.

IO's are dealt with as described in Remark1 at the end of Section 8.6.2.

8.8.3 Robustly filtered τ-estimates

A τ-estimate $\widehat{\boldsymbol{\lambda}}$ based on the robustly filtered observations y_t can be obtained now by replacing the raw unfiltered residuals (8.95) in $Q^*(\boldsymbol{\lambda})$ of (8.100) with the new robustly filtered residuals (8.105) and then minimizing $Q^*(\boldsymbol{\lambda})$. Then by defining

$$Q^*(\boldsymbol{\lambda}) = \sum_{t=1}^{T} \log a_t^2(\boldsymbol{\lambda}) + T \log\left(\tau^2\left(\frac{\widetilde{u}_1(\boldsymbol{\lambda})}{a_1(\boldsymbol{\lambda})}, \ldots, \frac{\widetilde{u}_T(\boldsymbol{\lambda})}{a_T(\boldsymbol{\lambda})}\right)\right),$$

the filtered τ-estimate (Fτ-estimate) is defined by

$$\widehat{\boldsymbol{\lambda}} = \arg\min_{\boldsymbol{\lambda}} \ Q^*(\boldsymbol{\lambda}). \tag{8.107}$$

Since the above $Q^*(\boldsymbol{\lambda})$ may have several local minima, a good robust initial estimate is required. Such an estimate is obtained by the following steps:

1. Fit an AR(p^*) model using the robust filtered τ-estimate of Section 8.6.3, where p^* is selected by the robust order selection criterion RAIC described in Section 8.6.6. The value of p^* will almost always be larger than p, and sometimes considerably larger. This fit gives the needed estimate $\widehat{\sigma}_u^2$, as well as robust parameter estimates $(\widehat{\phi}_1^o, \ldots, \widehat{\phi}_{p^*}^o)$ and robustly filtered values $\widehat{x}_{t|t}$.
2. Compute estimates of the first p autocorrelations of x_t and of η_i, $1 \leq i \leq q$, where
$$\eta_i = \frac{\mathrm{Cov}(x_t, u_{t-i})}{\sigma_u^2}, \tag{8.108}$$
using the estimates $(\widehat{\phi}_1^o, \ldots, \widehat{\phi}_{p^*}^o)$ and $\widehat{\sigma}_u^2$.
3. Finally compute the initial parameter estimates of the ARMA(p, q) model by matching the first p autocorrelations and the q values η_i with those obtained in step 2.

Example 8.7 *A simulated MA(1) series with AO.*

As an example, we generated an MA(1) series of 200 observations with 10 equally spaced AOs as follows (code **MA1-AO**):

$$y_t = \begin{cases} x_t + 4 & \text{if } t = 20i,\ i = 1, \ldots, 10 \\ x_t & \text{otherwise} \end{cases}$$

where $x_t = 0.8u_{t-1} + u_t$ and the u_t's are i.i.d. N(0,1) variables.

The model parameters were estimated using the Fτ- and the LS estimates, and the results are shown in Table 8.5. We observe that the robust estimate is very close to the true value, while the LS estimate is very much influenced by the outliers. Figure 8.21 shows the observed series y_t and the filtered series $\widehat{x}_{t|t}$. It is seen that the latter is almost coincident with y_t except for the 10 outliers, which are replaced by the predicted values.

Table 8.5 Estimates of the parameters of MA(1) simulated process

	θ	μ	σ_u^2
Fτ	−0.82	−0.05	1.16
LS	−0.45	0.41	1.96
True values	−0.80	0.00	1.00

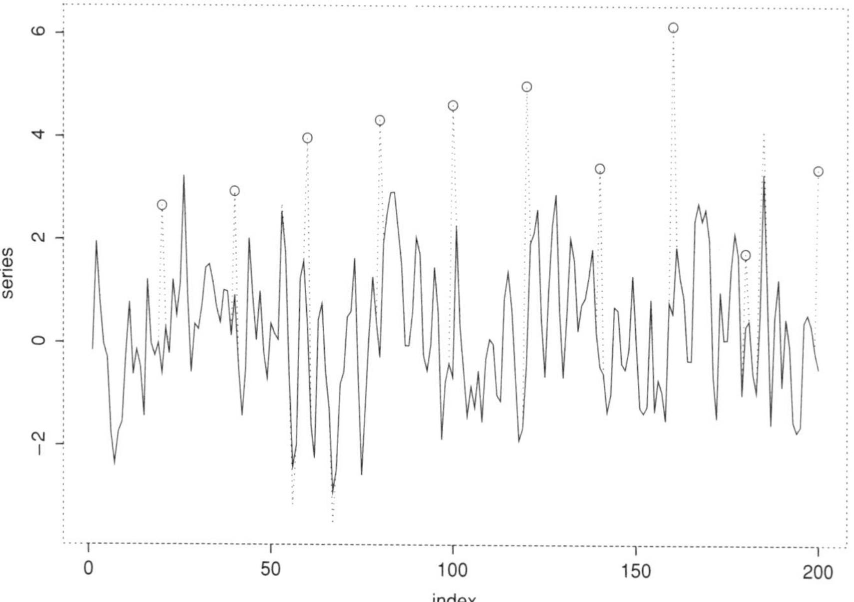

Figure 8.21 Simulated MA(1) series with 10 AOs: observed (- - -) and filtered (—) data

8.9 ARIMA and SARIMA models

We define an autoregression integrated moving-average process y_t of orders p, d, q (ARIMA(p, d, q) for short) as one such that its order-d differences are a stationary ARMA(p, q) process, and hence satisfies

$$\phi(B)(1-B)^d y_t = \gamma + \theta(B)u_t, \tag{8.109}$$

where ϕ and θ are polynomials of order p and q, and u_t are the innovations.

A seasonal ARIMA process y_t of regular orders p, d, q, seasonal period s, and seasonal orders P, D, Q (SARIMA$(p, d, q) \times (P, D, Q)_s$ for short) fulfills the equation

$$\phi(B)\Phi(B^s)(1-B)^d \left(1-B^s\right)^D y_t = \gamma + \theta(B)\Theta(B^s)u_t, \tag{8.110}$$

where ϕ and θ are as above, and Φ and Θ are polynomials of order P and Q respectively. It is assumed that the roots of ϕ, θ, Φ and Θ lie outside the unit circle and then the differenced series $(1-B)^d(1-B^s)^D y_t$ is a stationary and invertible ARMA process.

In the sequel we shall restrict ourselves to the case $P=0$ and $Q \leq 1$. Then $\Theta(B) = 1 - \Theta_s B$ and (8.110) reduces to

$$(1-B)^d\left(1-B^s\right)^D\phi(B)y_t = \gamma + \theta(B)(1-\Theta_s B^s)u_t. \tag{8.111}$$

The reason for this limitation is that, although the Fτ-estimates already defined for ARMA models can be extended to arbitrary SARIMA models, there is a computational difficulty in finding a suitable robust initial estimate for the iterative optimization process. At present this problem has been solved only for $P=0$ and $Q \leq 1$.

Assume now that we have observations $y_t = x_t + v_t$ where x_t fulfills an ARIMA model and v_t is an outlier process. A naive way to estimate the parameters is to difference y_t, thereby reducing the model to an ARMA(p, q) model, and then apply the Fτ-estimate already described. The problem with this approach is that the differencing operations will result in increasing the number of outliers. For example, with an ARIMA$(p, 1, q)$ model, the single regular difference operation will convert isolated outliers into two consecutive outliers of opposite sign (a so called"doublet"). However, one need not difference the data and may instead use the robust filter on the observations y_t as in the previous section, but based on the appropriate state-space model for the process (8.111).

The state-space representation is of the same form as (8.102), except that it uses a state-transition matrix Φ^* based on the coefficients of the polynomial operator of order $p^* = p + d + sD$

$$\phi^*(B) = (1-B)^d\left(1-B^s\right)^D\phi(B). \tag{8.112}$$

For example, in the case of an ARIMA$(1, 1, q)$ model with AR polynomial operator $\phi(B) = 1 - \phi_1 B$, we have

$$\phi^*(B) = 1 - \phi_1^* B - \phi_2^* B^2$$

with coefficients $\phi_1^* = 1 + \phi_1$ and $\phi_2^* = -\phi_1$. And for model (8.111) with $p = D = 1, d = q = Q = 0$ and seasonal period $s = 12$, we have

$$\phi^*(B) = 1 - \phi_1^* B - \phi_{12}^* B^{12} - \phi_{13}^* B^{13}$$

with

$$\phi_1^* = \phi_1, \quad \phi_{12}^* = 1, \quad \phi_{13}^* = -\phi_1.$$

Therefore, for each value of $\boldsymbol{\lambda} = (\boldsymbol{\phi}, \boldsymbol{\theta}, \gamma, \Theta_s)$ (where Θ_s is the seasonal MA parameter when $Q = 1$) the filtered residuals corresponding to the operators ϕ^* and θ^* are computed, yielding the residuals $\widetilde{u}_t(\boldsymbol{\lambda})$. Then the F$\tau$-estimate is defined by the $\boldsymbol{\lambda}$ minimizing $Q^*(\boldsymbol{\lambda})$, with Q^* defined as in (8.100) but with $\phi^*(B)$ instead of $\phi(B)$.

More details can be found in Bianco et al. (1996).

Example 8.8 *Residential telephone extensions (RESEX) series.*

This example deals with a monthly series of inward movement of residential telephone extensions in a fixed geographic area from January 1966 to May 1973 (RESEX). The series was analyzed by Brubacher (1974), who identified a SARIMA(2,0,0) $\times$ $(0,1,0)_{12}$ model, and by Martin et al. (1983).

Table 8.6 displays the LS, GM- and Fτ-estimates of the parameters (code **RESEX**). We observe that they are quite different, and the estimation of the SD of the innovation corresponding to the LS estimate is much larger than the ones obtained with the GM- and the filtered τ-estimates.

Table 8.6 Estimates of the parameters of RESEX series

Estimates	ϕ_1	ϕ_2	γ	σ_u
Fτ	0.27	0.49	0.41	1.12
GM	0.41	0.33	0.39	1.75
LS	0.48	−0.17	1.86	6.45

Figure 8.22 shows the observed data y_t and the filtered values $\widehat{x}_{t|t}$, which are seen to be almost coincident with y_t except at outlier locations.

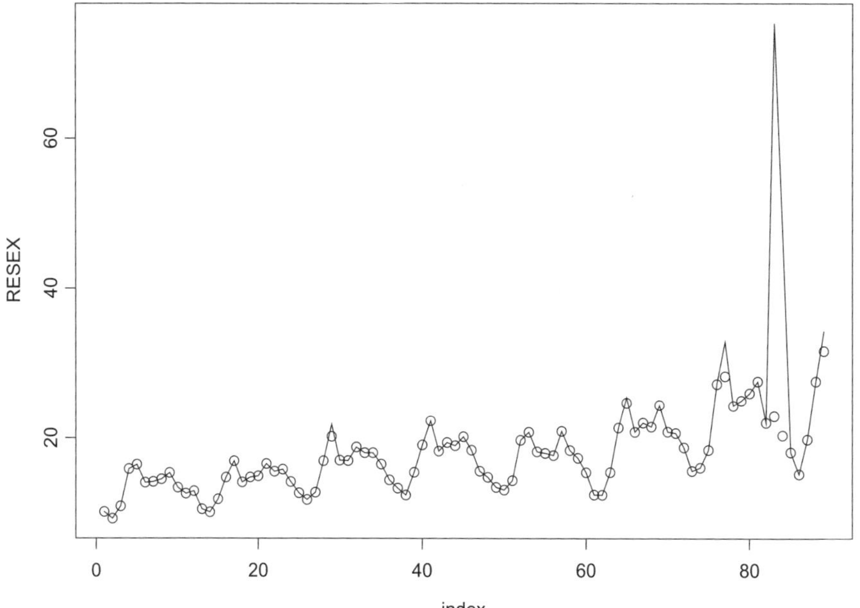

Figure 8.22 RESEX series: observed (solid line) and filtered (circles) values

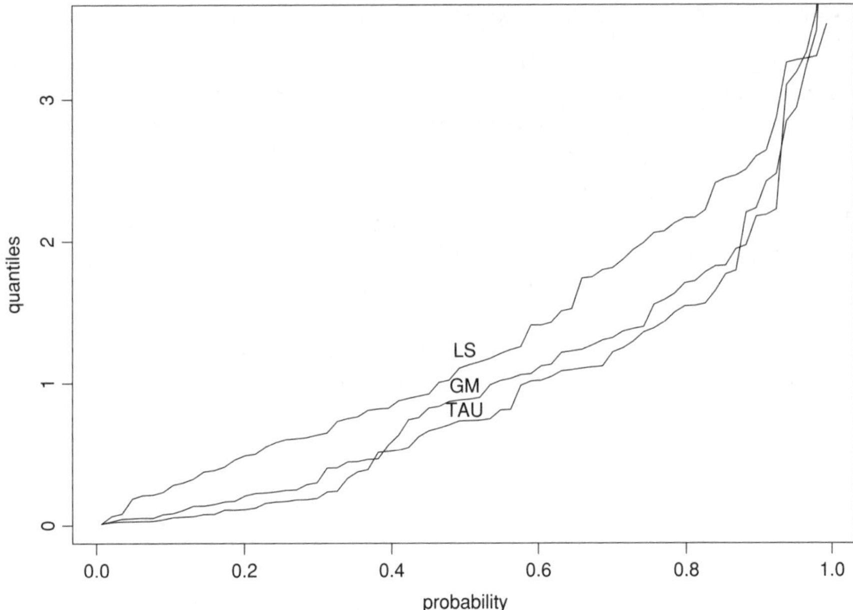

Figure 8.23 Quantiles of absolute residuals of estimates for RESEX series

In Figure 8.23 we show the quantiles of the absolute values of the residuals of the three estimates. The three largest residuals are huge and hence were not included to improve graph visibility. It is seen that the Fτ-estimate yields the smallest quantiles, and hence gives the best fit to the data.

8.10 Detecting time series outliers and level shifts

In many situations it is important to identify the type of perturbations that the series undergo. In this section we describe classical and robust diagnostic methods to detect outliers and level shifts in ARIMA models. As for the diagnostic procedures described in Chapter 4 for regression, the classical procedures are based on residuals obtained using nonrobust estimates. In general, these procedures succeed only when the proportion of outliers is very low and the outliers are not very large. Otherwise, due to masking effects, the outliers may not be detected.

Let y_t, $1 \leq t \leq T$, be an observed time series. We consider perturbed models of the form

$$y_t = x_t + \omega \, \xi_t^{(t_0)}, \tag{8.113}$$

where the unobservable series x_t is an ARIMA process satisfying

$$\phi(B)(1 - B)^d x_t = \theta(B) u_t, \tag{8.114}$$

and the term $\omega\xi_t^{(t_0)}$ represents the effect on period t of the perturbation occurring at time t_0.

The value of ω in (8.113) measures the size of the or level shift at time t_0 and the form of $\xi_t^{(t_0)}$ depends on the type of the outlier. Let $o_t^{(t_0)}$ be an indicator variable for time t_0 ($o_t^{(t_0)} = 1$ for $t = t_0$ and 0 otherwise). Then an AO at time t_0 can be modeled by

$$\xi_t^{(t_0)} = o_t^{(t_0)} \tag{8.115}$$

and a level shift at time t_0 by

$$\xi_t^{(t_0)} = \begin{cases} 0 & \text{if } t < t_0 \\ 1 & \text{if } t \geq t_0. \end{cases}$$

To model an IO at time t_0, the observed series y_t is given by

$$\phi(B)(1 - B)^d \; y_t = \theta(B)\big(u_t + \omega \, o_t^{(t_0)}\big).$$

Then, for an IO we get

$$\xi_t^{(t_0)} = \phi(B)^{-1}(1 - B)^{-d}\theta(B) \, o_t^{(t_0)}. \tag{8.116}$$

We know that robust estimates are not very much influenced by a small fraction of atypical observations in the cases of IO or AO. The case of level shifts is different. A level shift at period t_0 modifies all the observations y_t with $t \geq t_0$. However, if the model includes a difference, differencing (8.113) we get

$$(1 - B)y_t = (1 - B)x_t + \omega(1 - B)\xi_t^{(t_0)},$$

and since $(1 - B)\xi_t^{(t_0)} = o_t^{(t_0)}$, the differenced series has an AO at time t_0. Then a robust estimate applied to the differenced series is not going to be very much influenced by the level shift. Therefore, the only case in which a robust procedure may be influenced by a level shift is when the model does not contain any difference.

8.10.1 Classical detection of time series outliers and level shifts

In this subsection, we shall describe the basic ideas of Chang, Tiao and Chen (1988) for outlier detection in ARIMA models. Similar approaches were considered by Tsay (1988) and Chen and Liu (1993). Procedures based on deletion diagnostics were proposed by Peña (1987, 1990), Abraham and Chuang (1989), Bruce and Martin (1989) and Ledolter (1991).

For the sake of simplicity, we start by assuming that the parameters of the ARIMA model, $\boldsymbol{\lambda}$ and σ_u^2, are known.

Let $\pi(B)$ be the filter defined by

$$\pi(B) = \theta(B)^{-1}\phi(B)(1 - B)^d \; = 1 - \pi_1 B - \pi_2 B^2 - \ldots - \pi_k B^k - \ldots. \tag{8.117}$$

Then, from (8.114), $\pi(B)x_t = u_t$. Since $\pi(B)$ is a linear operator, we can apply it to both sides of (8.113), obtaining

$$\pi(B) \; y_t = u_t + \omega \, \pi(B) \; \xi_t^{(t_0)}, \tag{8.118}$$

which is a simple linear regression model with independent errors and regression coefficient ω.

Therefore, the LS estimate of ω is given by

$$\widehat{\omega} = \frac{\sum_{t=t_0}^{T} (\pi(B)\, y_t) \left(\pi(B)\, \xi_t^{(t_0)}\right)}{\sum_{t=t_0}^{T} \left(\pi(B)\, \xi_t^{(t_0)}\right)^2}, \tag{8.119}$$

with variance

$$\mathrm{Var}(\widehat{\omega}) = \frac{\sigma_u^2}{\sum_{t=t_0}^{T} (\pi(B)\, \xi_t^{(t_0)})^2}, \tag{8.120}$$

where σ_u^2 is the variance of u_t.

In practice, since the parameters of the ARIMA model are unknown, (8.119) and (8.120) are computed using LS or ML estimates of the ARIMA parameter. Let $\widehat{\pi}$ be defined as in (8.117) but using the estimates instead of the true parameters. Then (8.119) and (8.120) are replaced by

$$\widehat{\omega} = \frac{\sum_{t=t_0}^{T} \widehat{u}_t \left(\widehat{\pi}(B)\, \xi_t^{(t_0)}\right)}{\sum_{t=t_0}^{T} (\widehat{\pi}(B)\, \xi_t^{(t_0)})^2}, \tag{8.121}$$

and

$$\widehat{\mathrm{Var}}(\widehat{\omega}) = \frac{\widehat{\sigma}_u^2}{\sum_{t=t_0}^{T} (\widehat{\pi}(B)\, \xi_t^{(t_0)})^2}, \tag{8.122}$$

where

$$\widehat{u}_t = \widehat{\pi}(B) y_t$$

and

$$\widehat{\sigma}_u^2 = \frac{1}{T - t_0} \sum_{t=t_0}^{T} \left(\widehat{u}_t - \widehat{\omega}\left(\widehat{\pi}(B)\, \xi_t^{(t_0)}\right)\right)^2.$$

In the case of IO, the estimator of the outlier size given by (8.119) reduces to the innovation residual at t_0, i.e., $\hat{\omega} = \widehat{u}_{t_0}$.

A test to detect the presence of an outlier at a given t_0 can be based on the t-like statistic

$$U = \frac{|\widehat{\omega}|}{(\widehat{\mathrm{Var}}(\widehat{\omega}))^{1/2}}. \tag{8.123}$$

Since, in general, neither t_0 nor the type of outlier are known, in order to decide if there is an outlier at any position, the statistic

$$U_0 = \max_{t_0} \max\{U_{t_0,\mathrm{AO}},\ U_{t_0,\mathrm{LvS}},\ U_{t_0,\mathrm{IO}}\}$$

is used, where $U_{t_0,\mathrm{AO}}$, $U_{t_0,\mathrm{LvS}}$ and $U_{t_0,\mathrm{IO}}$ are the statistics defined by (8.123) corresponding to an AO, level shift (LvS) and IO at time t_0, respectively. If $U_0 > M$, where M is a conveniently chosen constant, one declares that there is an outlier or level shift. The time t_0 when the outlier or level shift occurs and and whether the additive effect is an AO, IO or LvS is determined by where the double maximum is attained.

Since the values

$$\widehat{u}_t = \sum_{i=0}^{\infty} \widehat{\pi}_i y_{t-i}$$

can only be computed from a series extending into the infinite past, in practice, with data observed for $t = 1, \ldots, T$, they are approximated by

$$\widehat{u}_t = \sum_{i=0}^{t-1} \widehat{\pi}_i y_{t-i}.$$

As we mentioned above, this type of procedure may fail due to the presence of a large fraction of outliers and/or level shifts. This failure may be due to two facts. On one hand, the outliers or level shift may have a large influence on the MLE, and therefore the residuals may not reveal the outlying observations. This drawback may be overcome by using robust estimates of the ARMA coefficients. On the other hand, if y_{t_0} is an outlier or level shift, as we noted before, not only is $\widehat{u}_{t_0}$ affected, but the effect of the outlier or level shift is propagated to the subsequent innovation residuals. Since the statistic U_0 is designed to detect the presence of an outlier or level shift at time t_0, it is desirable that U_0 be influenced by only an outlier at t_0. Outliers or level shift at previous locations, however, may have a misleading influence on U_0. In the next subsection we show how to overcome this problem by replacing the innovation residuals $\widehat{u}_t$ by the filtered residuals studied in Section 8.8.3.

8.10.2 Robust detection of outliers and level shifts for ARIMA models

In this section we describe an iterative procedure introduced by Bianco (2001) for the detection of AO, level shifts and IO in an ARIMA model. The algorithm is similar to the one described in the previous subsection. The main difference is that the new method uses innovation residuals based on the filtered τ-estimates of the ARIMA parameters instead of a Gaussion MLE, and uses a robust filter instead of the filter π to obtain an equation analogous to (8.118).

A detailed description of the procedure follows:

1. Estimate the parameters $\boldsymbol{\lambda}$ and σ_u robustly using an Fτ-estimator. These estimates will be denoted by $\widehat{\boldsymbol{\lambda}}$ and $\widehat{\sigma}_u$ respectively.
2. Apply the robust filter described in Section 8.8.3 to y_t using the estimates computed in step 1. This step yields the filtered residuals $\widetilde{u}_t$ and the scales s_t.
3. In order to make the procedure less costly in terms of computing time, a preliminary set of outlier locations is determined in the following way: declare that time t_0 is

a candidate for an outlier or level shift location if

$$|\widetilde{u}_{t_0}| > M_1 s_{t_0}, \tag{8.124}$$

where M_1 is a conveniently chosen constant, and denote by C the set of t_0's where (8.124) holds.

4. For each $t_0 \in C$, let $\widehat{\pi}^*$ be a robust filter similar to the one applied in step 2, but such that for $t_0 \leq t \leq t_0 + h$ the function ψ is replaced by the identity for a conveniently chosen value of h. Call $\widetilde{u}_t^* = \widehat{\pi}^*(B)y_t$ the residuals obtained with this filter. Since these residuals now have different variances, we estimate ω by weighted LS, with weights proportional to $1/s_t^2$. Then (8.121), (8.122) and (8.123) are now replaced by

$$\tilde{\omega} = \frac{\sum_{t=t_0}^{T} \widetilde{u}_t^* \; \widehat{\pi}^*(B) \; \xi_t^{(t_0)} / s_t^2}{\sum_{t=t_0}^{T} \left(\widehat{\pi}^*(B) \; \xi_t^{(t_0)}\right)^2 / s_t^2}, \tag{8.125}$$

$$\widehat{\text{Var}}(\tilde{\omega}) = \frac{1}{\sum_{t=t_0}^{T} (\widehat{\pi}^*(B) \; \xi^{(t_0)})^2 / s_t^2} \tag{8.126}$$

and

$$U^* = \frac{|\tilde{\omega}|}{\left(\widehat{\text{Var}}(\tilde{\omega})\right)^{1/2}}. \tag{8.127}$$

The purpose of replacing $\widehat{\pi}$ by $\widehat{\pi}^*$ is to eliminate the effects of outliers at positions different from t_0. For this reason the effect of those outliers before t_0 and after $t_0 + h$ is reduced by means of the robust filter. Since we want to detect a possible outlier at time t_0, and the effect of an outlier propagates to the subsequent observations, we do not downweight the effect of possible outliers between t_0 and $t_0 + h$.

5. Compute

$$U_0^* = \max_{t_0 \in C} \max\{U_{t_0,\text{AO}}^*, \; U_{t_0,\text{LvS}}^*, \; U_{t_0,\text{IO}}^*\},$$

where $U_{t_0,\text{AO}}^*$, $U_{t_0,\text{LvS}}^*$ and $U_{t_0,\text{IO}}^*$ are the statistics defined by (8.127) corresponding to an AO, level shift and IO at time t_0, respectively. If $U_0^* \leq M_2$, where M_2 is a conveniently chosen constant, no new outliers are detected and the iterative procedure is stopped. Instead, if $U_0^* > M_2$, a new AO, level shift or IO is detected, depending on where the maximum is attained.

6. Clean the series of the detected AO, level shifts or IO by replacing y_t with $y_t - \tilde{\omega}\xi_t^{(t_0)}$, where $\xi_t^{(t_0)}$ corresponds to the perturbation at t_0 pointed out by the test. Then, the procedure is iterated going back to step 2 until no new perturbations are found.

The constant M_1 should be chosen rather small (e.g., $M_1 = 2$) to increase the power of the procedure for the detection of outliers. Based on simulations we recommend using $M_2 = 3$.

Table 8.7 Outliers detected with the robust procedure in simulated MA(1) series

Index	Type	Size	U_0^*
20	AO	2.58	4.15
40	AO	3.05	4.48
60	AO	3.75	5.39
80	AO	3.97	5.70
100	AO	3.05	4.84
120	AO	4.37	6.58
140	AO	3.97	5.64
160	AO	4.66	6.62
180	AO	2.99	4.44
200	AO	3.85	3.62

As we have already mentioned, this procedure will be reliable to detect level shifts only if the ARIMA model includes at least an ordinary difference ($d > 0$).

Example 8.9 *Continuation of Example 8.7.*

On applying the robust procedure just described to the data, all the outliers were detected. Table 8.7 shows the outliers found with this procedure as well as their corresponding type, size and value of the test statistic.

The classical procedure of Section 8.10.1 detects only two outliers: observations 120 and 160. The LS estimates of the parameters after removing the effect of these two outliers are $\widehat{\theta} = -0.48$ and $\widehat{\mu} = 0.36$, which are also far from the true values.

Example 8.10 *Continuation of Example 8.8.*

Table 8.8 shows the outliers and level shifts found on applying the robust procedure to the RESEX data.

Table 8.8 Detected outliers in the RESEX series

Index	Date	Type	Size	U_0^*
29	5/68	AO	2.52	3.33
47	11/69	AO	-1.80	3.16
65	5/71	LvS	1.95	3.43
77	5/72	AO	4.78	5.64
83	11/72	AO	52.27	55.79
84	12/72	AO	27.63	27.16
89	5/73	AO	4.95	3.12

We observe two very large outliers corresponding to the last two months of 1972. The explanation is that November 1972 was a "bargain" month, i.e., free installation of resident extensions, with a spillover effect since not all November orders could be fulfilled that month.

8.10.3 REGARIMA models: estimation and outlier detection

A REGARIMA model is a regression model where the errors are an ARIMA time series. Suppose that we have T observations $(\mathbf{x}_1, y_1), \ldots, (\mathbf{x}_T, y_T)$ with $\mathbf{x}_i \in R^k$, $y_i \in R$ satisfying

$$y_t = \boldsymbol{\beta}'\mathbf{x}_t + e_t$$

where $e_1, \ldots, e_T$ follow an ARIMA(p, d, q) model

$$\phi(B)(1 - B)^d e_t = \theta(B)u_t.$$

As in the preceding cases, we consider the situation when the actual observations are described by a REGARIMA model plus AO, IO and level shifts. That is, instead of observing y_t we observe

$$y_t^* = y_t + \omega\,\xi_t^{(t_0)}, \tag{8.128}$$

where $\xi_t^{(t_0)}$ is as in the ARIMA model.

All the procedures for ARIMA models described in the preceding sections can be extended to REGARIMA models.

Define for each value of $\boldsymbol{\beta}$,

$$\widehat{e}_t(\boldsymbol{\beta}) = y_t^* - \boldsymbol{\beta}'\mathbf{x}_t,\ t = 1, \ldots, T$$

and put

$$\widehat{w}_t(\boldsymbol{\beta}) = (1 - B)^d \widehat{e}_t(\boldsymbol{\beta}),\quad t = d + 1, \ldots, T.$$

When $\boldsymbol{\beta}$ is the true parameter, $\widehat{w}_t(\boldsymbol{\beta})$ follows an ARMA(p, q) model with an AO, IO or level shift. Then it is natural to define for any $\boldsymbol{\beta}$ and $\boldsymbol{\lambda} = (\boldsymbol{\phi}, \boldsymbol{\theta})$ the residuals $\widehat{u}_t(\boldsymbol{\beta}, \boldsymbol{\lambda})$ as in (8.58), but replacing y_t by $\widehat{w}_t(\boldsymbol{\beta})$, i.e.,

$$\widehat{u}_t(\boldsymbol{\beta}, \boldsymbol{\lambda}) = \widehat{w}_t(\boldsymbol{\beta}) - \phi_1\widehat{w}_{t-1}(\boldsymbol{\beta}) - \ldots - \phi_p\widehat{w}_{t-p}(\boldsymbol{\beta}) + \theta_1\widehat{u}_{t-1}(\boldsymbol{\beta}, \boldsymbol{\lambda}) + \ldots$$
$$+ \theta_q\widehat{u}_{t-q}(\boldsymbol{\beta}, \boldsymbol{\lambda})\quad (t = p + d + 1, \ldots, T).$$

Then the LS estimate is defined as $(\boldsymbol{\beta}, \boldsymbol{\lambda})$ minimizing

$$\sum_{t=p+d+1}^{T} \widehat{u}_t^2(\boldsymbol{\beta}, \boldsymbol{\lambda}),$$

and an M-estimate is defined as $(\boldsymbol{\beta}, \boldsymbol{\lambda})$ minimizing

$$\sum_{t=p+d+1}^{T} \rho\left(\frac{\widehat{u}_t(\boldsymbol{\beta}, \boldsymbol{\lambda})}{\widehat{\sigma}}\right),$$

where $\widehat{\sigma}$ is the scale estimate of the innovations u_t. As in the case of regression with independent errors, the LS estimate is very sensitive to outliers, and M-estimates with a bounded ρ are robust when the u_t's are heavy tailed, but not for other types of outliers like AOs.

Let $\widetilde{u}_t(\boldsymbol{\beta}, \boldsymbol{\lambda})$ be the filtered residuals corresponding to the series $\widehat{e}_t(\boldsymbol{\beta})$ using the ARIMA(p, d, q) model with parameter $\boldsymbol{\lambda}$. Then we can define Fτ-estimates as in Section 8.8.3, i.e.,

$$(\hat{\boldsymbol{\beta}}, \widehat{\boldsymbol{\lambda}}) = \arg\min_{\boldsymbol{\lambda}} \; Q^*(\hat{\boldsymbol{\beta}}, \boldsymbol{\lambda}),$$

where

$$Q^*(\hat{\boldsymbol{\beta}}, \boldsymbol{\lambda}) = \sum_{t=1}^{T} \log a_t^2(\boldsymbol{\lambda}) + T \log\left(\tau^2\left(\frac{\widetilde{u}_1(\hat{\boldsymbol{\beta}}, \boldsymbol{\lambda})}{a_1(\boldsymbol{\lambda})}, \ldots, \frac{\widetilde{u}_T(\hat{\boldsymbol{\beta}}, \boldsymbol{\lambda})}{a_T(\boldsymbol{\lambda})}\right)\right).$$

The robust procedure for detecting the outliers and level shifts of Section 8.10.2 can also easily be extended to REGARIMA models. For details on the Fτ-estimates and outliers and level shift detection procedures for REGARIMA models, see Bianco et al. (2001).

8.11 Robustness measures for time series

8.11.1 Influence function

In all situations considered so far we have a finite-dimensional vector $\boldsymbol{\lambda}$ of unknown parameters (e.g., $\boldsymbol{\lambda} = (\phi_1, \ldots, \phi_p, \theta_1, \ldots, \theta_q, \mu)'$ for ARMA models) and an estimate $\widehat{\boldsymbol{\lambda}}_T = \widehat{\boldsymbol{\lambda}}_T(y_1, \ldots, y_T)$. When y_t is a strictly stationary process, it holds under very general conditions that $\widehat{\boldsymbol{\lambda}}_T$ converges in probability to a vector $\widehat{\boldsymbol{\lambda}}_\infty$ which depends on the joint (infinite-dimensional) distribution F of $\{y_t : t = 1, 2, \ldots\}$.

Künsch (1984) extends Hampel's definition (3.3) of the influence function to time series in the case that $\widehat{\boldsymbol{\lambda}}_T$ is defined by M-estimating equations which depend on a fixed number k of observations:

$$\sum_{t=k}^{n} \Psi(\mathbf{y}_t, \widehat{\boldsymbol{\lambda}}_T) = 0, \tag{8.129}$$

where $\mathbf{y}_t = (y_t, \ldots, y_{t-k+1})'$. Strict stationarity implies that the distribution F_k of $\mathbf{y}_t$ does not depend on t. Then for a general class of stationary processes, $\widehat{\boldsymbol{\lambda}}_\infty$ exists and depends only on F_k, and is the solution of the equation

$$\mathrm{E}_{F_k} \Psi(\mathbf{y}_t, \boldsymbol{\lambda}) = \mathbf{0}. \tag{8.130}$$

For this type of time series, the Hampel influence function could be defined as

$$\mathrm{IF_H}(\mathbf{y}; \widehat{\boldsymbol{\lambda}}, F_k) = \lim_{\varepsilon \downarrow 0} \frac{\widehat{\boldsymbol{\lambda}}_\infty \left[(1-\varepsilon) F_k + \varepsilon \delta_{\mathbf{y}}\right] - \widehat{\boldsymbol{\lambda}}_\infty(F_k)}{\varepsilon}, \tag{8.131}$$

where $\mathbf{y} = (y_k, \ldots, y_1)'$, and the subscript "H" stands for the Hampel definition. Then proceeding as in Section 5.11 it can be shown that for estimates of the form (8.129) the analog of (3.47) holds. Then by analogy with (3.28) the gross-error sensitivity is defined as $\sup_{\mathbf{y}} \left\| \mathrm{IF_H}(\mathbf{y}; \widehat{\boldsymbol{\lambda}}, F_k) \right\|$, where $\|.\|$ is a convenient norm.

If y_t is an AR(p) process, it is natural to generalize LS through M-estimates of the form (8.130) with $k = p + 1$. Künsch (1984) found the Hampel-optimal estimate for this situation, which turns out to be a GM-estimate of Schweppe form (5.45).

However, this definition has several drawbacks:

- This form of contamination is not a realistic one. The intuitive idea of a contamination rate $\varepsilon = 0.05$ is that about 5% of the observations are altered. But in the definition (8.131) ε is the proportion of outliers in each k-dimensional marginal. In general, given ε and $\mathbf{y}$, there exists no process such that all its k-dimensional marginals are $(1-\varepsilon) F_k + \varepsilon \delta_{\mathbf{y}}$.
- The definition cannot be applied to processes such as ARMA model in which the natural estimating equations do not depend on finite-dimensional distributions.

An alternative approach was taken by Martin and Yohai (1986) who introduced a new definition of *influence functional* for time series, which we now briefly discuss. We assume that observations y_t are generated by the general RO model

$$y_t^\varepsilon = \left(1 - z_t^\varepsilon\right) x_t + z_t^\varepsilon w_t \tag{8.132}$$

where x_t is a stationary process (typically normally distributed) with joint distribution F_x, w_t is an outlier-generating process and z_t^ε is a 0–1 process with $\mathrm{P}\left(z_t^\varepsilon = 1\right) = \varepsilon$. This model encompasses the AO model through the choice $w_t = x_t + v_t$ with v_t independent of x_t, and provides a pure replacement model when w_t is independent of x_t. The model can generate both isolated and patch outliers of various lengths through appropriate choices of the process z_t^ε. Assume that $\widehat{\boldsymbol{\lambda}}_\infty \left(F_y^\varepsilon\right)$ is well defined for the distribution F_y^ε of y_t^ε. Then the *time series influence function* $\mathrm{IF}\left(\{F_{x,w,z}^\varepsilon\}; \widehat{\boldsymbol{\lambda}}\right)$ is the directional derivative at F_x:

$$\mathrm{IF}(\{F_{x,z,w}^\varepsilon\}; \widehat{\boldsymbol{\lambda}}) = \lim_{\varepsilon \downarrow 0} \frac{1}{\varepsilon} \left(\widehat{\boldsymbol{\lambda}}_\infty \left(F_y^\varepsilon\right) - \widehat{\boldsymbol{\lambda}}_\infty(F_x)\right), \tag{8.133}$$

where $F_{x,z,w}^\varepsilon$ is the joint distribution of the processes x_t, z_t^ε and w_t.

The first argument of IF is a distribution, and so in general the time series IF is a functional on a distribution space, which is to be contrasted with $\mathrm{IF_H}$ which is a function on a finite-dimensional space. However, in practice we often choose special forms of the outlier-generating process w_t such as constant amplitude outliers, e.g., for AOs we may let $w_t \equiv x_t + v$ and for pure ROs we let $w_t = v$ where v is a constant.

Although the time series IF is similar in spirit to IF_H, it coincides with the latter only in the very restricted case that $\widehat{\boldsymbol{\lambda}}_T$ is permutation invariant and (8.132) is restricted to an i.i.d. pure RO model (see Corollary 4.1 in Martin and Yohai, 1986).

While IF is generally different from IF_H, there is a close relationship between both. That is, if $\widehat{\boldsymbol{\lambda}}$ is defined by (8.129), then under regularity conditions:

$$\text{IF}(\{F_{x,z,w}\}; \widehat{\boldsymbol{\lambda}}) = \lim_{\varepsilon \downarrow 0} \frac{\text{E}\left[\text{IF}_\text{H}(\mathbf{y}_k; \widehat{\boldsymbol{\lambda}}, F_{x,k})\right]}{\varepsilon} \tag{8.134}$$

where $F_{x,k}$ is the k-dimensional marginal of F_x and the distribution of $\mathbf{y}_k$ is the k-dimensional marginal of F_y^{ε}.

The above result is proved in Theorem 4.1 of Martin and Yohai (1986), where a number of other results concerning the time series IF are presented. In particular:

- conditions are established which aid in the computation of time series IFs
- IFs are computed for LS and robust estimates of AR(1) and MA(1) models, and the results reveal the differing behaviors of the estimators for both isolated and patchy outliers
- it is shown that for MA models, bounded ψ-functions do not yield bounded IFs, whereas redescending ψ-functions do yield bounded IFs
- optimality properties are established for the class of RA estimates described in 8.12.1.

8.11.2 Maximum bias

In Chapter 3 we defined the maximum asymptotic bias of an estimator $\widehat{\theta}$ at a distribution F in an ε-contamination neighborhood of a parametric model. This definition made sense for i.i.d. observations, but cannot be extended in a straightforward manner to time series. A basic difficulty is that the simple mixture model $(1-\varepsilon)\,F_\theta + \varepsilon G$ that suffices for independent observations is not adequate for time series for the reasons given in the previous section.

As a simple case consider estimation of ϕ in the AR(1) model $x_t = \phi x_{t-1} + u_t$, where u_t has $\text{N}(0, \sigma_u^2)$ distribution. The asymptotic value of the LS estimate and of the M- and GM-estimates depends only on the joint distribution $F_{2,y}$ of y_1 and y_2. Specification of $F_{2,y}$ is more involved than the two-term mixture distribution $(1-\varepsilon)\,F_\theta + \varepsilon G$ used in the definition of bias given in Section 3.3. For example, suppose we have the AO model given by (8.5), where v_t is an i.i.d. series independent of x_t with contaminated normal distribution (8.6). Denote by $\text{N}_2(\mu_1, \mu_2, \sigma_1^2, \sigma_2^2, \gamma)$ the bivariate normal distribution with means μ_1 and μ_2, variances σ_1^2 and σ_2^2 and covariance γ and call $\sigma_x^2 = \text{Var}\,(x_t) = \sigma_u^2/(1-\phi^2)$. Then the joint distribution $F_{2,y}$ is a normal mixture distribution with four components:

$$\begin{aligned}
&(1-\varepsilon)^2 \text{N}_2\left(0, 0, \sigma_x^2, \sigma_x^2, \phi\sigma_x^2\right) + \varepsilon(1-\varepsilon)\text{N}_2\left(0, 0, \sigma_x^2 + \sigma_v^2, \sigma_x^2, \phi\sigma_x^2\right) \\
&+ \varepsilon(1-\varepsilon)\text{N}_2\left(0, 0, \sigma_x^2, \sigma_x^2 + \sigma_v^2, \phi\sigma_x^2\right) \\
&+ \varepsilon^2 \text{N}_2\left(0, 0, \sigma_x^2 + \sigma_v^2, \sigma_x^2 + \sigma_v^2, \phi\sigma_x^2\right).
\end{aligned} \tag{8.135}$$

The four terms correspond to the cases of no outliers, an outlier in y_1, an outlier in y_2, and outliers in both y_1 and y_2, respectively.

This distribution is even more complicated when modeling patch outliers in v_t, and things get much more challenging for estimators that depend on joint distributions of order greater than two such as AR(p), MA(q) and ARMA(p, q) models, where one must consider either p-dimensional joint distributions or joint distributions of all orders.

Martin and Jong (1977) took the above joint distribution modeling approach in computing maximum bias curves for a particular class of GM-estimates of an AR(1) parameter under both isolated and patch AO models. But it seems difficult to extend such calculations to higher-order models, and typically one has to resort to simulation methods to estimate maximum bias and BP (see Section 8.11.4 for an example of simulation computation of maximum bias curves).

A simple example of bias computation was given in Section 8.1.3. The asymptotic value of the LS estimate is the correlation between y_1 and y_2, and as such can be computed from the mixture expression (8.135), as the reader may verify (Problem 8.12).

Note that the maximum bias in (8.13) is $|\rho_v(1) - \phi|$, which depends upon the value of ϕ, and this feature holds in general for ARMA models.

8.11.3 Breakdown Point

Extending the notion of BP given in Section 3.2 to the time series setting presents some difficulties.

The first is how "contamination" is defined. One could simply consider the finite BP for observations $y_1, \ldots, y_T$ as defined in Section 3.2.5, and then define the asymptotic BP by letting $T \to \infty$. The drawback of this approach is that it is intractable except in very simple cases. We are thus led to consider contamination by a process such as AO or RO, with the consequence that the results will depend on the type of contaminating process considered.

The second is how "breakdown" is defined. This difficulty is due to the fact that in time series models the parameter space is generally bounded, and moreover the effect of outliers is more complicated than with location, regression or scale.

This feature can be seen more easily in the AR(1) case. It was seen in Section 8.1.3 that the effect on the LS estimate of contaminating a process x_t with an AO process v_t is that the estimate may take on any value between the lag-one autocorrelations of x_t and v_t. If v_t is arbitrary, then the asymptotic value of the estimate may be arbitrarily close to the boundary $\{-1, 1\}$ of the parameter space, and thus there would be breakdown according to the definitions of Section 3.2.

However, in some situations it is considered more reasonable to take only isolated (i.e., i.i.d.) AOs into account. In this case the worst effect of the contamination is to shrink the estimate toward zero, and this could be considered as "breakdown" if the true parameter is not null. One could define breakdown as the estimate approaching ± 1 or 0, but it would be unsatisfactory to tailor the definition in an ad-hoc manner to each estimate and type of contamination.

A completely general definition taking these problems into account was given by Genton and Lucas (2003). The intuitive idea is that breakdown occurs for some contamination rate ε_0 if further increasing the contamination rate does not further enlarge the range of values taken on by the estimate over the contamination neighborhood. In particular, for the case of AR(1) with independent AOs, it follows from the definition that breakdown occurs if the estimate can be taken to zero.

The details of the definition are very elaborate and are therefore omitted here.

8.11.4 Maximum bias curves for the AR(1) model

Here we present maximum bias curves from Martin and Yohai (1991) for three estimates of ϕ for a centered AR(1) model with RO:

$$y_t = x_t(1 - z_t) + w_t z_t, \quad x_t = \phi x_{t-1} + u_t$$

where z_t are i.i.d. with

$$\mathrm{P}(z_t = 1) = \gamma, \quad \mathrm{P}(w_t = c) = \mathrm{P}(w_t = -c) = 0.5.$$

The three considered estimates are

- the estimate obtained by modeling the outliers found using the procedure described in Section 8.10.1 (Chang et al., 1988),
- the median of slopes estimate $\mathrm{Med}(y_t/y_{t-1})$, which as mentioned in Section 5.14.2 has bias-optimality properties, and
- a filtered M-scale robust estimate, which is the same as the $F\tau$-estimate except that an M-scale was used by Martin and Yohai instead of the τ-scale which is the one recommended in this book.

The curves were computed by a Monte Carlo procedure. Let $\widehat{\phi}_T(\varepsilon, c)$ be the value of any of the three estimates for sample size T. For sufficiently large T, the value of $\widehat{\phi}_T(\varepsilon, c)$ will be negligibly different from its asymptotic value $\widehat{\phi}_\infty(\varepsilon, c)$ and $T = 2000$ was used for the purpose of this approximation. Then the maximum asymptotic bias was approximated as

$$B(\varepsilon) = \sup_c \left|\widehat{\phi}_T(\varepsilon, c) - \phi\right| \tag{8.136}$$

by search on a grid of c values from 0 to 6 with a step size of 0.02. We plot the signed value of $B(\varepsilon)$ in Figure 8.24 for the case $\phi = 0.9$. The results clearly show the superiority of the robust filtered M-scale estimate, which has relatively small bias over the entire range of ε from 0 to 0.4, with estimator breakdown (not shown) occurring about $\varepsilon = 0.45$. Similar results would be expected for the $F\tau$-estimate. The estimate obtained using the classical outlier detection procedure of Chang et al. (1988) has quite poor global performance, i.e., while its maximum bias behaves similarly to that of the robust filtered M-scale estimate for small ε, the estimator breaks down in the presence of white-noise contamination, with a bias of essentially -0.9 for ε a little less than 0.1. The GM-estimator has a maximum bias behavior in between that of the other two estimates, with rapidly increasing maximum bias as ε increases

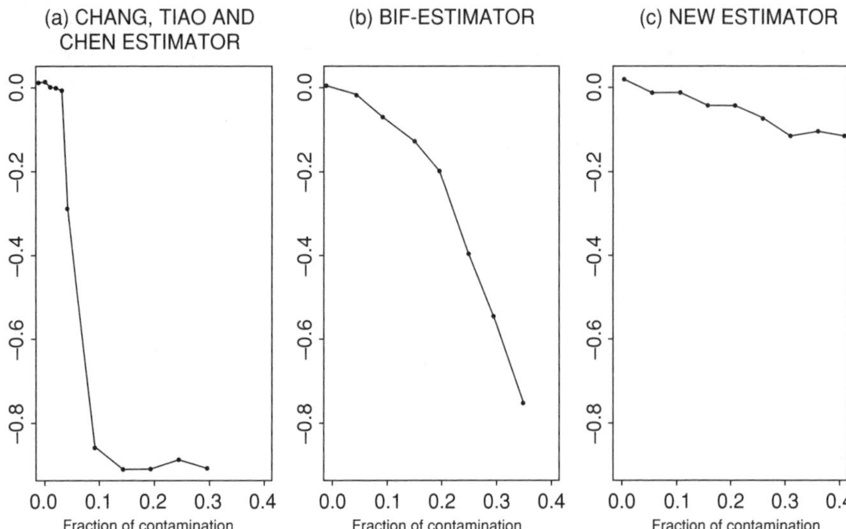

Figure 8.24 Maximum bias curves ("BIF" indicates the GM-estimate). Reproduced from IMA Volumes in Mathematics and its Applications, Vol. 33, Stahel and Weisberg, eds., Directions in Robust Statistics and Diagnostics, Part I, Springer-Verlag, 1991, page 244, "Bias robust estimation of autoregression parameters", Martin, R. D. and Yohai, V. J., Figure 3. With kind permission of Springer Science and Business Media.

beyond roughly 0.1, but is not quite broken down at $\varepsilon = 0.35$. However, one can conjecture from the maximum bias curve that breakdown to zero occurs by $\varepsilon = 0.35$. We note that other types of bounded influence GM-estimates that use redescending functions can apparently achieve better maximum bias behavior than this particular GM-estimate (see Martin and Jong, 1977).

8.12 Other approaches for ARMA models

8.12.1 Estimates based on robust autocovariances

The class of robust estimates based on robust autocovariances (RA estimates) was proposed by Bustos and Yohai (1986). We are going to motivate these estimates by a convenient robustification of the estimating LS equations.

Let $\boldsymbol{\lambda} = (\boldsymbol{\phi}, \boldsymbol{\theta}, \gamma)$. As a particular case of the results to be proved in Section 8.15, the equations for the LS estimate can be reexpressed as

$$\sum_{t=p+i+1}^{T} \widehat{u}_t(\boldsymbol{\lambda}) \sum_{j=0}^{t-p-i-1} \pi_j(\phi)\widehat{u}_{t-j-i}(\lambda) = 0,\ i = 1, \ldots, p,$$

$$\sum_{t=p+i+1}^{T} \widehat{u}_t(\boldsymbol{\lambda}) \sum_{j=0}^{t-p-i-1} \zeta_j(\theta)\widehat{u}_{t-j-i}(\lambda) = 0,\ i = 1, \ldots, q$$

and

$$\sum_{t=p+1}^{T} \widehat{u}_t(\boldsymbol{\lambda}) = 0, \tag{8.137}$$

where $\widehat{u}_t(\boldsymbol{\lambda})$ is defined in (8.58), and π_j and ζ_j are the coefficients of the inverses of the AR and MA polynomials, i.e.,

$$\phi^{-1}(B) = \sum_{j=0}^{\infty} \pi_j(\boldsymbol{\phi}) B^j$$

and

$$\theta^{-1}(B) = \sum_{j=0}^{\infty} \zeta_j(\boldsymbol{\theta}) B^j.$$

This system of equations can be written as

$$\sum_{j=0}^{T-p-i-1} \pi_j(\boldsymbol{\phi}) M_{i+j}(\boldsymbol{\lambda}) = 0, \ i = 1, \ldots, p, \tag{8.138}$$

$$\sum_{j=0}^{T-p-i-1} \zeta_j(\boldsymbol{\theta}) M_{i+j}(\boldsymbol{\lambda}) = 0, \ i = 1, \ldots, q \tag{8.139}$$

and (8.137), where

$$M_j(\boldsymbol{\lambda}) = \sum_{t=p+j+1}^{T} \widehat{u}_t(\boldsymbol{\lambda}) \widehat{u}_{t-j}(\boldsymbol{\lambda}) = 0. \tag{8.140}$$

The RA estimates are obtained by replacing the term $M_j(\boldsymbol{\lambda})$ in (8.138) and (8.139) with

$$M_j^*(\boldsymbol{\lambda}) = \sum_{t=p+j+1}^{T} \eta(\widehat{u}_t(\boldsymbol{\lambda}), \widehat{u}_{t-j}(\boldsymbol{\lambda})), \tag{8.141}$$

and (8.137) with

$$\sum_{t=p+1}^{T} \psi(\widehat{u}_t(\boldsymbol{\lambda})) = 0,$$

where ψ is a bounded ψ-function.

The name of this family of estimates comes from the fact that $M_j/(T-j-p)$ is an estimate of the autocovariance of the residuals $\widehat{u}_t(\boldsymbol{\lambda})$, and $M_j^*/(T-j-p)$ is a robust version thereof.

Two types of η-functions are considered by Bustos and Yohai (1986): Mallows-type functions of the form $\eta(u, v) = \psi^*(u)\psi^*(v)$ and Schweppe-type functions of the form $\eta(u, v) = \psi^*(uv)$, where ψ^* is a bounded ψ-function. The functions ψ and ψ^* can be taken for example in the Huber or bisquare families. These estimates have good robustness properties for AR(p) models with small p. However, the fact that they use regular residuals makes them vulnerable to outliers when p is large or $q > 0$. They are consistent and asymptotically normal. A heuristic proof is given in Section 8.15. The asymptotic covariance matrix is of the form $b(\psi, F)V_{LS}$, where $b(\psi, F)$ is a scalar term (Bustos and Yohai, 1986).

8.12.2 Estimates based on memory-m prediction residuals

Suppose that we want to fit an ARMA(p, q) model using the series y_t, $1 \leq t \leq T$. It is possible to define M-estimates, GM-estimates and estimates based on the minimization of a residual scale using residuals based on a memory-m predictor, where $m \geq p + q$. This last condition is needed to ensure that the estimates are well defined.

Consider the memory-m best linear predictor when the true parameter is $\boldsymbol{\lambda} = (\boldsymbol{\phi}, \boldsymbol{\theta}, \boldsymbol{\mu})$:

$$\widehat{y}_{t,m}(\boldsymbol{\lambda}) = \mu + \varphi_{m,1}(\boldsymbol{\phi}, \boldsymbol{\theta})(y_{t-1} - \mu) + \ldots + \varphi_{m,m}(\boldsymbol{\phi}, \boldsymbol{\theta})(y_{t-m} - \mu),$$

where $\varphi_{m,i}$ are the coefficients of the predictor defined in (8.31) (here we call them $\varphi_{m,i}$ rather than $\phi_{m,i}$ to avoid confusion with the parameter vector $\boldsymbol{\phi}$). Masarotto (1987) proposed to estimate the parameters using memory-m residuals defined by

$$\widehat{u}_{t,m}(\boldsymbol{\lambda}) = y_t - \widehat{y}_{t,m}(\boldsymbol{\lambda}),\ t = m + 1, \ldots, T.$$

Masarotto proposed this approach for GM-estimates, but actually it can be used with any robust estimate, such as MM-estimates, or estimates based on the minimization of a robust residual scale

$$\widehat{\sigma}(\widehat{u}_{p+1,m}(\boldsymbol{\lambda}), \ldots, \widehat{u}_{T,m}(\boldsymbol{\lambda})),$$

where $\widehat{\sigma}$ is an M- or τ-scale. Since one outlier spoils $m + 1$ memory-m residuals, the robustness of these procedures depends on how large is m.

For AR(p) models, the memory-p residuals are the regular residuals $\widehat{u}_t(\boldsymbol{\lambda})$ given in (8.20), and therefore no new estimates are defined here.

One shortcoming of the estimates based on memory-m residuals is that the convergence to the true values holds only under the assumption that the process y_t is Gaussian.

8.13 High-efficiency robust location estimates

In Section 8.2 we wrote the AR(p) model in the two equivalent forms (8.14) and (8.18). We have been somewhat cavalier about which of these two forms to use in fitting the model, implicitly thinking that the location parameter μ is a nuisance parameter that we don't very much care about. In that event one is tempted to use a simple robust location estimate $\widehat{\mu}$ for the centering, e.g., one may use for $\widehat{\mu}$ an ordinary location M-estimate as described in Section 2.2. However, one may be interested in the location parameter for its own sake, and in addition one may ask whether there are any disadvantages in using an ordinary location M-estimate for the centering approach to fitting an AR model.

Use of the relationship (8.17) leads naturally to the location estimate

$$\widehat{\mu} = \frac{\widehat{\gamma}}{1 - \sum_{i=0}^{p} \widehat{\phi}_i}. \tag{8.142}$$

It is easy to check that the same form of location estimate is obtained for an ARMA(p, q) model in intercept form. In the context of M-estimates or GM-estimates of AR and ARMA models we call the estimate (8.142) a *proper* location M- (or GM-)estimate.

It turns out that use of an ordinary location M-estimate has two problems when applied to ARMA models. The first is that selection of the tuning constant to achieve a desired high efficiency when the innovations are normally distributed depends upon the model parameters, which are not known in advance. This problem is most severe for ARMA(p, q) models with $q > 0$. The second problem is that the efficiency of the ordinary M-estimate can be exceedingly low relative to the proper M-estimate. Details are provided by Lee and Martin (1986), who show that

- for an AR(1) model, the efficiency of the ordinary M-estimate relative to the proper M-estimate is between 10% and 20% for $\phi = \pm 0.9$ and approximately 60% for $\phi = \pm 0.5$, and
- for an MA(1) model the relative efficiency is above approximately 80% for positive θ but is around 50% for $\theta = -0.5$ and is arbitrarily low as θ approaches -1. The latter was shown by Grenander (1981) to be a point of super-efficiency.

The conclusion is that one should not use the ordinary location M-estimate for AR and ARMA processes when one is interested in location for its own sake. Furthermore the severe loss of efficiency of the ordinary location M-estimate that is obtained for some parameter values gives pause to its use for centering purposes, even when one is not interested in location for its own sake. It seems from the evidence at hand that it is prudent to fit the intercept form of AR and ARMA models, and when the location estimate is needed it can be computed from expression (8.142).

8.14 Robust spectral density estimation

8.14.1 Definition of the spectral density

Any second-order stationary process y_t defined for integer t has a spectral representation

$$y_t = \int_{-1/2}^{1/2} \exp(i2\pi t f) dZ(f) \tag{8.143}$$

where $Z(f)$ is a complex orthogonal increments process on $(-1/2, 1/2]$, i.e., for any $f_1 < f_2 \leq f_3 < f_4$

$$\mathrm{E}\left\{(Z(f_2) - Z(f_1))\overline{(Z(f_4) - Z(f_3))}\right\} = 0,$$

where $\overline{z}$ denotes the conjugate of the complex number z. See for example Brockwell and Davis (1991). This result says that any stationary time series can be interpreted as the limit of a sum of sinusoids $A_i \cos(2\pi f_i t + \Phi_i)$ with random amplitudes A_i and random phases Φ_i. The process $Z(f)$ defines an increasing function $G(f) = \mathrm{E}\,|Z(f)|^2$ with $G(-1/2) = 0$ and $G(1/2) = \sigma^2 = \mathrm{Var}(y_t)$. The function $G(f)$ is called the *spectral distribution* function, and when its derivative $S(f) = G'(f)$ exists it is called the *spectral density* function of y_t. Other commonly used terms for $S(f)$ are *power spectral density*, *spectrum* and *power spectrum*. We assume for purposes of this discussion that $S(f)$ exists, which implies that y_t has been "de-meaned", i.e., y_t has been centered by subtracting its mean. The more general case of a discrete time process on time intervals of length Δ is easily handled with slight modifications to the above (see for example Bloomfield, 1976; Percival and Walden, 1993).

Using the orthogonal increments property of $Z(f)$ one immediately finds that the lag-k covariances of y_t are given by

$$C(k) = \int_{-1/2}^{1/2} \exp(i2\pi kf)S(f)df. \tag{8.144}$$

Thus the $C(k)$ are the Fourier coefficients of $S(f)$ and so we have the Fourier series representation

$$S(f) = \sum_{k=-\infty}^{\infty} C(k)\exp(-i2\pi fk). \tag{8.145}$$

8.14.2 AR spectral density

It is easy to show that for a zero-mean AR(p) process with parameters $\phi_1, \ldots, \phi_p$ and innovations variance σ_u^2 the spectral density is given by

$$S_{AR,p}(f) = \frac{\sigma_u^2}{|H(f)|^2} \tag{8.146}$$

where

$$H(f) = 1 - \sum_{k=1}^{p} \phi_k \exp(i2\pi fk). \tag{8.147}$$

The importance of this result is that any continuous and nonzero spectral density $S(f)$ can be approximated arbitrarily closely and uniformly in f by an AR(p) spectral density $S_{AR,p}(f)$ for sufficiently large p (Grenander and Rosenblatt, 1957).

8.14.3 Classic spectral density estimation methods

The classic, most frequently used method for estimating a spectral density is a non-parametric method based on smoothing the periodogram. The steps are as follows. Let y_t, $t = 1, \ldots, T$, be the observed data, let d_t, $t = 1, \ldots, T$, be a *data taper* that goes smoothly to zero at both ends, and form the modified data $\widetilde{y}_t = d_t y_t$. Then use the fast Fourier transform (FFT) (Bloomfield, 1976) to compute the discrete Fourier transform

$$X(f_k) = \sum_{t=1}^{T} \widetilde{y}_t \exp(-i2\pi f_k t) \tag{8.148}$$

where $f_k = k/T$ for $k = 0, 1, \ldots, [T/2]$, and use the result to form the *periodogram*:

$$\widehat{S}(f_k) = \frac{1}{T} |X(f_k)|^2 . \tag{8.149}$$

It is known that the periodogram is an approximately unbiased estimate of $S(f)$ for large T, but it is not a consistent estimate. For this reason $\widehat{S}(f_k)$ is smoothed in the frequency domain to obtain an improved estimate of reduced variability, namely

$$\overline{S}(f_k) = \sum_{m=-M}^{M} w_m \widehat{S}(f_m), \tag{8.150}$$

where the smoothing weights w_m are symmetric with

$$w_m = w_{-m} \text{ and } \sum_{m=-M}^{M} w_m = 1.$$

The purpose of the data taper is to reduce the so-called *leakage* effect of implicit truncation of the data with a rectangular window, and originally data tapers such as a cosine window or Parzen window were used. For details on this and other aspects of spectral density estimation see Bloomfield (1976). A much preferred method is to use a prolate spheroidal taper, whose application in spectral analysis was pioneered by Thomson (1977). See also Percival and Walden (1993).

Given the result in Section 8.14.2 one can also use a parametric AR($\widehat{p}$) approximation approach to estimating the spectral density based on parameter estimates $\widehat{\phi}_1, \ldots, \widehat{\phi}_{\widehat{p}}$ and $\widehat{\sigma}_u^2$; here $\widehat{p}$ is an estimate of the order p, obtained through a selection criterion such as AIC, BIC or FPE which are discussed in Brockwell and Davis (1991). In this case we compute

$$\widehat{S}_{AR,\widehat{p}}(f) = \frac{\widehat{\sigma}_u^2}{\left|1 - \sum_{k=1}^{\widehat{p}} \widehat{\phi}_k \exp(i2\pi f k)\right|^2} \tag{8.151}$$

on a grid of frequency values $f = f_k$.

8.14.4 Prewhitening

Prewhitening is a filtering technique introduced by Blackman and Tukey (1958), in order to transform a time series into one whose spectrum is nearly flat. One then estimates the spectral density of the prewhitened series, with a greatly reduced impact of leakage bias, and then transforms the prewhitened spectral density back, using the frequency domain equivalent of inverse filtering in order to obtain an estimate of the spectrum for the original series. Tukey (1967) says: "If low frequencies are 10^3, 10^4, or 10^5 times as active as high ones, a not infrequent phenomenon in physical situations, even a fairly good window is too leaky for comfort. The cure is not to go in for fancier windows, but rather to preprocess the data toward a flatter spectrum, to analyze this *prewhitened* series, and then to adjust its estimated spectrum for the easily computable effects of preprocessing." The classic (nonrobust) way to accomplish the overall estimation method is to use the following modified form of the AR spectrum estimate (8.151):

$$\overline{S}_{AR,\widehat{p}}(f) = \frac{\overline{S}_{\widehat{u},\widehat{p}}(f)}{\left|1 - \sum_{k=1}^{\widehat{p}} \widehat{\phi}_k \exp(i2\pi fk)\right|^2} \tag{8.152}$$

where $\overline{S}_{\widehat{u},\widehat{p}}(f)$ is a smoothed periodogram estimate as described above, but applied to the fitted AR *residuals* $\widehat{u}_t = y_t - \widehat{\phi}_1 y_{t-1} - \ldots - \widehat{\phi}_{\widehat{p}} y_{t-\widehat{p}}$. The estimate $\overline{S}_{AR,\widehat{p}}(f)$ provides substantial improvement on the simpler estimate $\widehat{S}_{AR,\widehat{p}}(f)$ in (8.151) by replacing the numerator estimate $\widehat{\sigma}_u^2$ that is fixed independent of frequency with the frequency-varying estimate $\overline{S}_{r,\widehat{p}}(f)$. The order estimate $\widehat{p}$ may be obtained with an AIC or BIC order selection method (the latter is known to be preferable). Experience indicates that use of moderately small fixed orders p_o in the range from two to six will often suffice for effective prewhitening, suggesting that automatic order selection will often result in values of $\widehat{p}$ in a similar range.

8.14.5 Influence of outliers on spectral density estimates

Suppose the AO model $y_t = x_t + v_t$ contains a single additive outlier v_{t_0} of size A. Then the periodogram $\widehat{S}_y(f_k)$ based on the observations y_t will have the form

$$\widehat{S}_y(f_k) = \widehat{S}_x(f_k) + \frac{A^2}{T} + 2\frac{A}{T}\,\mathrm{Re}\left[X(f_k)\exp(i2\pi f_k)\right], \tag{8.153}$$

where $\widehat{S}_x(f_k)$ is the periodogram based on the outlier-free series x_t and Re denotes the real part. Thus the outlier causes the estimate to be raised by the constant amount A^2/T at all frequencies plus the amount of the oscillatory term

$$\frac{2A}{T}\,\mathrm{Re}\left[X(f_k)\exp(i2\pi f_k)\right]$$

that varies with frequency. If the spectrum amplitude varies over a wide range with frequency then the effect of the outlier can be to obscure small but important peaks (corresponding to small-amplitude oscillations in the x_t series) in low-amplitude regions of the spectrum. It can be shown that a pair of outliers can generate an oscillation whose frequency is determined by the time separation of the outliers, and whose impact can also obscure features in the low-amplitude region of the spectrum (Problem 8.13).

To get an idea of the impact of AOs more generally we focus on the mean and variance of the smoothed periodogram estimates $\overline{S}(f_k)$ under the assumption that x_t and v_t are independent, and that the conditions of consistency and asymptotic normality of $\overline{S}(f_k)$ hold. Then for moderately large sample sizes the mean and variance of $\overline{S}(f_k)$ are given approximately by

$$\mathrm{E}\overline{S}(f_k) = S_y(f_k) = S_x(f_k) + S_v(f_k) \tag{8.154}$$

and

$$\mathrm{Var}(\overline{S}(f_k)) = S_y(f_k)^2 = S_x(f_k)^2 + S_v(f_k)^2 + 2S_x(f_k)S_v(f_k). \tag{8.155}$$

Thus AOs cause both bias and inflated variability of the smoothed periodogram estimate. If v_t is i.i.d. with variance σ_v^2 the bias is just σ_v^2 and the variance is inflated by the amount $\sigma_v^2 + 2S_x(f_k)\sigma_v^2$.

Striking examples of the influence that outliers can have on spectral density estimates were given by Kleiner et al. (1979) and Martin and Thomson (1982). The most dramatic and compelling of these examples is the one in the former paper, where the data consist of 1000 measurements of diameter distortions along a section of an advanced wave-guide designed to carry over 200000 simultaneous telephone conversations. In this case the data are a "space" series but it can be treated in the same manner as a time series as far as spectrum analysis is concerned. Two relatively minor outliers due to a malfunctioning of the recording instrument, and not noticeable in simple plots of the data, obscure important features of a spectrum having a very wide dynamic range (in this case the ratio of the prediction variance to the process variance of an AR(7) fit is approximately 10^{-6}!). Figure 8.25 (from Kleiner et al., 1979) shows the diameter distortion measurements as a function of distance along the wave-guide, and points out that the two outliers are noticeable only in a considerably amplified local section of the data. Figure 8.26 shows the differenced series (a "poor man's prewhitening") which clearly reveals the location of the two outliers as doublets; Figure 8.27 shows the classic periodogram-based estimate (dashed line) with the oscillatory artifact caused by the outliers, along with a robust estimate (solid line) that we describe next. Note in the latter figure that the classic estimate has an eight-decade dynamic range while the robust estimate has a substantially increased dynamic range of close to eleven decades, and reveals features that have known physical interpretations that are totally obscured in the classical estimate (see Kleiner et al. (1979) for details).

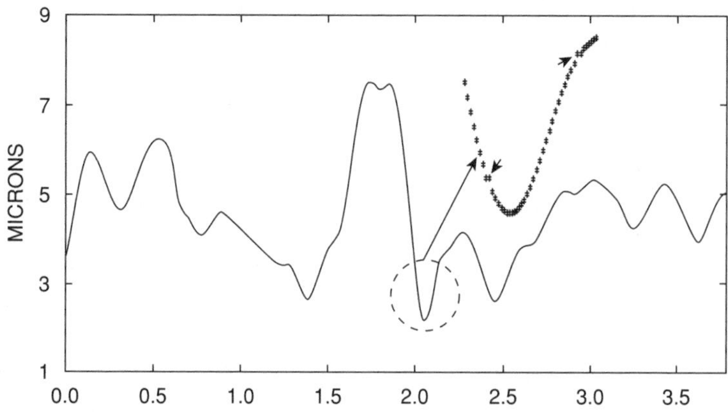

Figure 8.25 Wave-guide data: diameter distortion measurements vs. distance. Reproduced from Jour. Royal Statistical Society, B, 41, No. 3, 1979, pp 313–351, Blackwell Publishing, "Robust estimation of power spectra", Kleiner, B., Martin, R. D., and Thomson, D. J., Figures 4A, 4B, 4C. With kind permission of Blackwell Publishing.

8.14.6 Robust spectral density estimation

Our recommendation is to compute robust spectral density estimates by robustifying the prewhitened spectral density (8.152) as follows. The AR parameter estimates $\widehat{\phi}_1, \widehat{\phi}_2, \ldots, \widehat{\phi}_{\widehat{p}}$ and $\widehat{\sigma}_u^2$ are computed using the $F\tau$-estimate, and $\widehat{p}$ is computed using the robust order selection method of Section 8.6.6. Then to compute a robust smoothed spectral density estimate $\overline{S}_{\widetilde{u}^*,\widehat{p}}(f)$, the nonrobust residual estimates

$$\widehat{u}_t = y_t - \widehat{\phi}_1 y_{t-1} - \ldots - \widehat{\phi}_{\widehat{p}} y_{t-\widehat{p}}$$

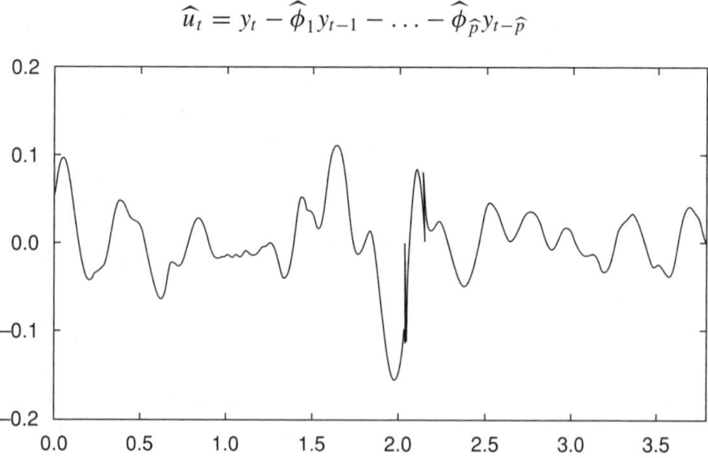

Figure 8.26 Wave-guide data: differenced series. Reproduced from Jour. Royal Statistical Society, B, 41, No. 3, 1979, pp 313–351, Blackwell Publishing, "Robust estimation of power spectra", Kleiner, B., Martin, R. D., and Thomson, D. J., Figures 4A, 4B, 4C. With kind permission of Blackwell Publishing.

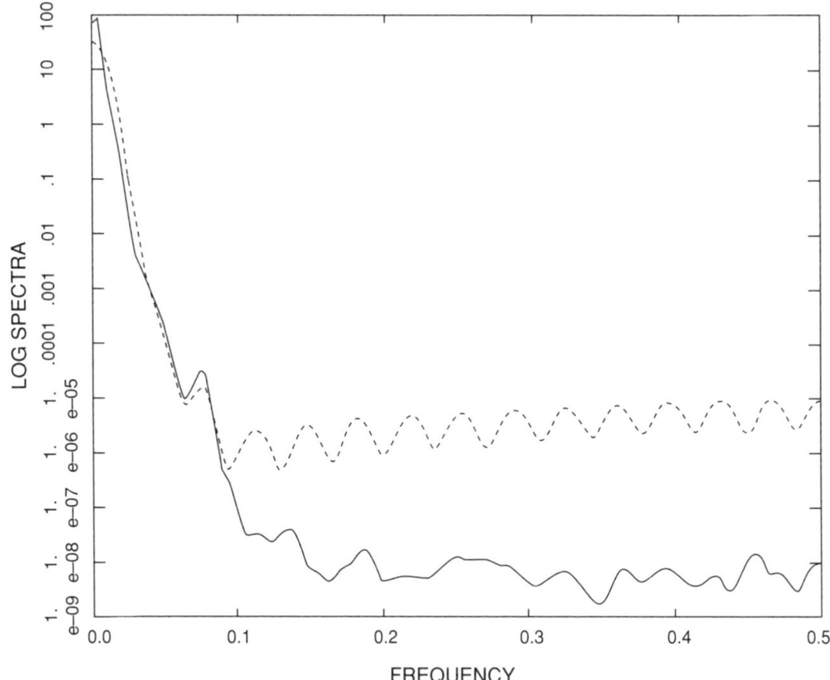

Figure 8.27 Wave-guide data: classical (- - -) and robust (—) spectra. Reproduced from Jour. Royal Statistical Society, B, 41, No. 3, 1979, pp 313–351, Blackwell Publishing, "Robust estimation of power spectra", Kleiner, B., Martin, R. D., and Thomson, D. J., Figures 4A, 4B, 4C. With kind permission of Blackwell Publishing.

are replaced by the robust residual estimates defined as

$$\widetilde{u}_t^* = \widehat{x}_{t|t} - \widehat{\phi}_1\widehat{x}_{t-1|t-1} - \widehat{\phi}_2\widehat{x}_{t-2|t-1} - \ldots - \widehat{\phi}_{\widehat{p}}\widehat{x}_{t-\widehat{p}|t-1},$$

where $\widehat{x}_{t-i|t-1}$, $i = 0, 1, \ldots, \widehat{p}$, are obtained from the robust filter. Note that these robust prediction residuals differ from the robust prediction residuals $\widetilde{u}_t$ (8.76) in Section 8.6.2 in that the latter have $y_t - \mu$ where we have $\widehat{x}_{t|t}$. We make this replacement because we do not want outliers to influence the smoothed periodogram estimate based on the robust residuals. Also, we do not bother with an estimate of μ because as mentioned at the beginning of the section one always works with de-meaned series in spectral analysis.

Note that our approach in this chapter of using robust filtering results in replacing outliers with one-sided predictions based on previous data. It is quite natural to think about improving this approach by using a *robust smoother*, as mentioned at the end of Section 8.6.2. See Martin and Thomson (1982) for the algorithm and its application to spectral density estimation. The authors show, using the wave-guide data, that it can be unsafe to use the robust filter algorithm if the AR order is not sufficiently large or the tuning parameters are changed somewhat, while the robust smoother algorithm results in a more reliable outlier interpolation and associated spectral density estimate (see Figures 24–27 of Martin and Thomson (1982).

Kleiner et al. (1979) also show good results for some examples using a pure robust AR spectral density estimate, i.e., the robust smoothed spectral density estimate $\overline{S}_{\tilde{u}^*,\widehat{p}}(f)$ is replaced with a robust residuals variance estimate $\widehat{\sigma}_u^2$ and a sufficiently high AR order is used. Our feeling is that this approach is only suitable for spectrum analysis contexts where the user is confident that the dynamic range of the spectrum is not very large, e.g., at most two or three decades.

The reader interested in robust spectral density estimation can find more details and several examples in Kleiner et al. (1979) and Martin and Thomson (1982). Martin and Thomson (1982, Section III) point out that small outliers may not only obscure the lower part of the spectrum but also may inflate innovations variance estimates by orders of magnitude.

8.14.7 Robust time-average spectral density estimate

The classic approach to spectral density estimation described in Section 8.14.3 reduces the variability of the periodogram by averaging periodogram values in the frequency domain, as indicated in (8.150). In some applications with large amounts of data it may be advantageous to reduce the variability by averaging the periodogram in the *time* domain, as originally described by Welch (1967). The idea is to break the time series data up into M equal-length contiguous segments of length N, compute the periodogram $\widehat{S}_m(f_k) = \frac{1}{T}|X_m(f_k)|^2$ at each frequency $f_k = k/N$ on the m-th segment, and at each f_k form the smoothed periodogram estimate

$$\overline{S}(f_k) = \frac{1}{M}\sum_{m=1}^{M}\widehat{S}_m(f_k). \tag{8.156}$$

The problem with this estimate is that even a single outlier in the m-th segment can spoil the estimate $\widehat{S}_m(f_k)$ as discussed previously. One way to robustify this estimate is to replace the sample mean in (8.156) with an appropriate robust estimate. One should not use a location M-estimate which assumes a symmetric nominal distribution for the following reason. Under normality the periodogram may be represented by the approximation

$$2\widehat{S}_m(f_k) \approx s_k Y \tag{8.157}$$

where Y is a chi-squared random variable with 2 degrees of freedom and $s_k = \mathrm{E}\widehat{S}_m(f_k) \approx S(f_k)$ for large T. Thus estimation of $S(f_k)$ is equivalent to estimating the scale of an exponential distribution.

Under AO- or RO-type outlier contamination a reasonable approximate model for the distribution of the periodogram $\widehat{S}_m(f_k)$ is the contaminated exponential distribution

$$(1-\varepsilon)\mathrm{Ex}(s_k) + \varepsilon\mathrm{Ex}(s_{c,k}), \tag{8.158}$$

where Ex (α) is the exponential distribution with mean α. Here outliers may result in $s_{c,k} > s_k$, at least at some frequencies f_k. Thus the problem is to find a good robust estimate of s_k in the contaminated exponential model (8.158). It must be kept in mind that the overall data series can have a quite small fraction of contamination and still influence many of the segment estimates $\widehat{S}_m(f_k)$, and hence a high BP estimate of s_k is desirable. Consider a more general model of the form (8.158), in which the contaminating distribution $\mathrm{Ex}(s_{c,k})$ is replaced with the distribution of any positive random variable. As mentioned in Section 5.2.2, the min–max bias estimate of scale for this case is very well approximated by a scaled median (Martin and Zamar, 1989) with scaling constant $(0.693)^{-1}$ for Fisher consistency at the nominal exponential distribution. Thus it is recommended to replace the nonrobust time-average estimate (8.156) with the scaled median estimate

$$\overline{S}(f_k) = \frac{1}{0.693}\mathrm{Med}\left\{\widehat{S}_m(f_k), m = 1, \ldots, M\right\}. \tag{8.159}$$

This estimator can be expected to work well in situations where less than half of the time segments of data contain influential outliers.

The idea of replacing the sample average in (8.156) with a robust estimate of the scale of an exponential distribution was considered by Thomson (1977) and discussed by Martin and Thomson (1982), with a focus on using an asymmetric truncated mean as the robust estimate. See also Chave, Thomson and Ander (1987) for an application.

8.15 Appendix A: heuristic derivation of the asymptotic distribution of M-estimates for ARMA models

To simplify, we replace $\widehat{\sigma}$ in (8.65) by its asymptotic value σ. Generally $\widehat{\sigma}$ is calibrated so that when u_t is normal, then $\sigma^2 = \sigma_u^2 = \mathrm{E}u_t^2$. Differentiating (8.65) we obtain

$$\sum_{t=p+1}^{T} \psi\left(\frac{\widehat{u}_t(\widehat{\boldsymbol{\lambda}})}{\sigma}\right)\frac{\partial \widehat{u}_t(\widehat{\boldsymbol{\lambda}})}{\partial \boldsymbol{\lambda}} = 0. \tag{8.160}$$

We leave it as an exercise (Problem 8.9) to show that

$$\frac{\partial \widehat{u}_t(\boldsymbol{\lambda})}{\partial \mu} = -\frac{1-\phi_1-\ldots-\phi_p}{1-\theta_1-\ldots-\theta_q}, \tag{8.161}$$

$$\frac{\partial \widehat{u}_t(\boldsymbol{\lambda})}{\partial \phi_i} = -\phi^{-1}(B)\widehat{u}_{t-i}(\boldsymbol{\lambda}) \tag{8.162}$$

and

$$\frac{\partial \widehat{u}_t(\boldsymbol{\lambda})}{\partial \boldsymbol{\theta}_j} = \theta^{-1}(B)\widehat{u}_{t-j}(\boldsymbol{\lambda}). \tag{8.163}$$

Let

$$\mathbf{z}_t = \left.\frac{\partial \widehat{u}_t(\boldsymbol{\lambda})}{\partial \boldsymbol{\lambda}}\right|_{\boldsymbol{\lambda}=\boldsymbol{\lambda}_0}, \quad \mathbf{W}_t = \left.\frac{\partial^2 \widehat{u}_t(\boldsymbol{\lambda})}{\partial \boldsymbol{\lambda}^2}\right|_{\boldsymbol{\lambda}=\boldsymbol{\lambda}_0},$$

where $\boldsymbol{\lambda}_0$ is the true value of the parameter. Observe that

$$\mathbf{z}_t = (\mathbf{c}_t, \mathbf{d}_t, \xi)'$$

with ξ defined in (8.63) and

$$\mathbf{c}_t = -\left(\phi^{-1}(B)u_{t-1}, \ldots, \phi^{-1}(B)u_{t-p}\right)',$$

$$\mathbf{d}_t = \left(\theta^{-1}(B)u_{t-1}, \ldots, \theta^{-1}(B)u_{t-q}\right)'.$$

Since $\widehat{u}_t(\boldsymbol{\lambda}_0) = u_t$, a first-order Taylor expansion yields

$$\sum_{t=p+1}^{T} \psi\left(\frac{u_t}{\sigma}\right)\mathbf{z}_t + \left(\frac{1}{\sigma}\sum_{t=p+1}^{T}\psi'\left(\frac{u_t}{\sigma}\right)\mathbf{z}_t\mathbf{z}_t' + \sum_{t=p+1}^{T}\psi\left(\frac{u_t}{\sigma}\right)\mathbf{W}_t\right)(\widehat{\boldsymbol{\lambda}} - \boldsymbol{\lambda}_0) \simeq 0$$

and then

$$T^{1/2}(\widehat{\boldsymbol{\lambda}} - \boldsymbol{\lambda}_0) \simeq \mathbf{B}^{-1}\left(\frac{1}{T^{1/2}}\sum_{t=p+1}^{T}\psi\left(\frac{u_t}{\sigma}\right)\mathbf{z}_t\right) \tag{8.164}$$

with

$$\mathbf{B} = \frac{1}{\sigma T}\sum_{t=p+1}^{T}\psi'\left(\frac{u_t}{\sigma}\right)\mathbf{z}_t\mathbf{z}_t' + \frac{1}{T}\sum_{t=p+1}^{T}\psi\left(\frac{u_t}{\sigma}\right)\mathbf{W}_t.$$

We shall show that

$$\operatorname*{p\,lim}_{T\to\infty} \beta = \frac{1}{\sigma}\mathrm{E}\psi'\left(\frac{u_t}{\sigma}\right)\mathrm{E}\mathbf{z}_t\mathbf{z}_t', \tag{8.165}$$

and that

$$\frac{1}{T^{1/2}}\sum_{t=p+1}^{T}\psi\left(\frac{u_t}{\sigma}\right)\mathbf{z}_t \to_d \mathrm{N}_{p+q+1}\left(\mathbf{0}, \mathrm{E}\psi\left(\frac{u_t}{\sigma}\right)^2\mathrm{E}\mathbf{z}_t\mathbf{z}_t'\right). \tag{8.166}$$

From (8.164), (8.165) and (8.166) we get

$$T^{1/2}(\widehat{\boldsymbol{\lambda}} - \boldsymbol{\lambda}_0) \to_d \mathrm{N}_{p+q+1}(\mathbf{0}, \mathbf{V}_M) \tag{8.167}$$

where

$$\mathbf{V}_M = \frac{\sigma^2\mathrm{E}\psi\,(u_t/\sigma)^2}{(\mathrm{E}\psi'\,(u_t/\sigma))^2}\left(\mathrm{E}\mathbf{z}_t\mathbf{z}_t'\right)^{-1}. \tag{8.168}$$

It is not difficult to show that the terms on the left-hand side of (8.165) are uncorrelated and have the same mean and variance. Hence it follows from the weak law of large numbers that

$$\underset{T\to\infty}{\text{p lim}} \frac{1}{\sigma T} \sum_{t=p+1}^{T} \left(\psi' \left(\frac{u_t}{\sigma} \right) \mathbf{z}_t \mathbf{z}_t' + \psi \left(\frac{u_t}{\sigma} \right) \mathbf{W}_t \right)$$

$$= \left[\frac{1}{\sigma} \mathrm{E} \left(\psi' \left(\frac{u_t}{\sigma} \right) \mathbf{z}_t \mathbf{z}_t' \right) + \mathrm{E} \left(\psi \left(\frac{u_t}{\sigma} \right) \mathbf{W}_t \right) \right].$$

Then, (8.165) follows from the fact that u_t is independent of $\mathbf{z}_t$ and $\mathbf{W}_t$, and (8.66).

Recall that a sequence of random vectors $\mathbf{q}_n \in R^k$ converges in distribution to $\mathrm{N}_k(\mathbf{0}, \mathbf{A})$ if and only if each linear combination $\mathbf{a}'\mathbf{q}_n$ converges in distribution to $\mathrm{N}(0, \mathbf{a}'\mathbf{A}\mathbf{a})$ (Feller, 1971). Then to prove (8.166) it is enough to show that for any $\mathbf{a} \in R^{p+q+1}$

$$\frac{1}{\sqrt{T}} \sum_{t=p+1}^{T} H_t \to_d \mathrm{N}(0, v_0), \tag{8.169}$$

where

$$H_t = \psi(u_t/\sigma)\, \mathbf{a}'\mathbf{z}_t$$

and

$$v_0 = \mathrm{E}(H_t)^2 = \mathbf{a}' \left(\mathrm{E}\psi \left(\frac{u_t}{\sigma} \right)^2 \mathrm{E}(\mathbf{z}_t \mathbf{z}_t') \right) \mathbf{a}.$$

Since the variables are not independent, the standard central limit theorem cannot be applied; but it can be shown that the stationary process H_t satisfies

$$\mathrm{E}(H_t | H_{t-1}, \ldots, H_1) = 0 \text{ a.s.}$$

and is hence a so-called *martingale difference sequence*. Therefore by the central limit theorem for martingales (see Theorem 23.1 of Billingsley, 1968) (8.169) holds, and hence (8.166) is proved.

We shall now find the form of the covariance matrix $\mathbf{V}_M$. Let

$$\phi^{-1}(B)u_t = \sum_{i=0}^{\infty} \pi_i u_{t-i}$$

and

$$\theta^{-1}(B)u_t = \sum_{i=0}^{\infty} \zeta_i u_{t-i},$$

where $\pi_0 = \zeta_0 = 1$. We leave it as an exercise (Problem 8.10) to show that $\mathrm{E}(\mathbf{z}_t \mathbf{z}_t')$ has the following form:

$$\mathrm{E}(\mathbf{z}_t \mathbf{z}_t') = \begin{bmatrix} \sigma_u^2 \mathbf{D} & \mathbf{0} \\ 0 & \xi^2 \end{bmatrix} \tag{8.170}$$

where $\mathbf{D} = \mathbf{D}(\phi, \theta)$ is a symmetric $(p+q)$ matrix with elements

$$D_{i,j} = \sum_{k=0}^{\infty} \pi_k \pi_{k+j-i} \text{ if } i \leq j \leq p$$

$$D_{i,p+j} = \sum_{k=0}^{\infty} \zeta_k \pi_{k+j-i} \text{ if } i \leq p,\ j \leq q,\ i \leq j$$

$$D_{i,p+j} = \sum_{k=0}^{\infty} \pi_k \zeta_{k+i-j} \text{ if } i \leq p,\ j \leq q,\ j \leq i$$

$$D_{p+i,p+j} = \sum_{k=0}^{\infty} \zeta_k \zeta_{k+j-i} \text{ if } i \leq j \leq q.$$

Therefore the asymptotic covariance matrix of $\widehat{\boldsymbol{\lambda}}$ is

$$\mathbf{V}_M = \frac{\sigma^2 \mathrm{E}\psi(u_t/\sigma)^2}{(\mathrm{E}\psi'(u_t/\sigma))^2} \begin{bmatrix} \sigma_u^{-2}\mathbf{D}^{-1} & \mathbf{0} \\ \mathbf{0} & \xi^{-2} \end{bmatrix}.$$

In the case of the LS estimate, since $\psi(u) = 2u$ and $\psi'(u) = 2$ we have

$$\frac{\sigma^2 \mathrm{E}\psi(u_t/\sigma)^2}{(\mathrm{E}\psi'(u_t/\sigma))^2} = \mathrm{E}u_t^2 = \sigma_u^2,$$

and hence the asymptotic covariance matrix is

$$\mathbf{V}_{LS} = \begin{bmatrix} \mathbf{D}^{-1} & \mathbf{0} \\ \mathbf{0} & \sigma_u^2/\xi^2 \end{bmatrix}.$$

In consequence we have

$$\mathbf{V}_M = \frac{\sigma^2 \mathrm{E}\psi(u_t/\sigma)^2}{\sigma_u^2 (\mathrm{E}\psi'(u_t/\sigma))^2} \mathbf{V}_{LS}.$$

In the AR(p) case, the matrix $\sigma_u^2\mathbf{D}$ coincides with the covariance matrix $\mathbf{C}$ of $(y_t, y_{t-1}, \ldots, y_{t-p+1})$ used in (8.49).

8.16 Appendix B: robust filter covariance recursions

The vector $\mathbf{m}_t$ appearing in (8.82) is the first column of the covariance matrix of the state prediction error $\widehat{\mathbf{x}}_{t|t-1} - \mathbf{x}_t$:

$$\mathbf{M}_t = \mathrm{E}(\widehat{\mathbf{x}}_{t|t-1} - \mathbf{x}_t)(\widehat{\mathbf{x}}_{t|t-1} - \mathbf{x}_t)' \tag{8.171}$$

and

$$s_t = \sqrt{M_{t,11}} = \sqrt{m_{t,1}} \tag{8.172}$$

is the standard deviation of the observation prediction error $y_t - \hat{y}_{t|t-1} = y_t - \widehat{x}_{t|t-1}$. The recursion for $\mathbf{M}_t$ is

$$\mathbf{M}_t = \Phi \mathbf{P}_{t-1} \Phi + \sigma_u^2 \mathbf{d}\mathbf{d}', \tag{8.173}$$

where $\mathbf{P}_t$ is the covariance matrix of the state filtering error $\widehat{\mathbf{x}}_{t|t} - \mathbf{x}_t$:

$$\mathbf{P}_t = \mathrm{E}(\widehat{\mathbf{x}}_{t|t} - \mathbf{x}_t)(\widehat{\mathbf{x}}_{t|t} - \mathbf{x}_t)'. \tag{8.174}$$

The recursion equation for $\mathbf{P}_t$ is

$$\mathbf{P}_t = \mathbf{M}_t - \frac{1}{s_t^2} W\left(\frac{\widetilde{u}_t}{s_t}\right) \mathbf{m}_t \mathbf{m}_t'$$

where $W(u) = \psi(u)/u$.

Reasonable initial conditions for the robust filter are $\widehat{\mathbf{x}}_{0|0} = (0, 0, \ldots, 0)'$, and $\mathbf{P}_0 = \hat{\mathbf{P}}_x$ where $\hat{\mathbf{P}}_x$ is a $p \times p$ robust estimate of the covariance matrix for $(y_{t-1}, y_{t-2}, \ldots, y_{t-p})$.

When applying the robust Durbin–Levinson algorithm to estimate an AR(p) model, the above recursions need to be computed for each of a sequence of AR orders $m = 1, \ldots, p$. Accordingly, we shall take $\sigma_u^2 = \sigma_{u,m}^2$, where $\sigma_{u,m}^2$ is the variance of the memory-m prediction error of x_t, i.e.,

$$\sigma_{u,m}^2 = \mathrm{E}(x_t - \widetilde{\phi}_{m,1} x_{t-1} - \ldots - \widetilde{\phi}_{m,m} x_{t-m})^2.$$

Then we need an estimate $\widehat{\sigma}_{u,m}^2$ of $\sigma_{u,m}^2$ for each m. This can be accomplished by using the following relationships:

$$\sigma_{u,1}^2 = (1 - \widetilde{\phi}_{1,1}^2)\sigma_x^2, \tag{8.175}$$

where σ_x^2 is the variance of x_t, and

$$\sigma_{u,m}^2 = (1 - \widetilde{\phi}_{m,m}^2)\sigma_{u,m-1}^2. \tag{8.176}$$

In computing (8.91) for $m = 1$, we use the estimate $\widehat{\sigma}_{u,1}^2$ of $\sigma_{u,1}^2$ parameterized as a function of $\widetilde{\phi} = \widetilde{\phi}_{1,1}$ using (8.175)

$$\widehat{\sigma}_{u,1}^2(\widetilde{\phi}) = (1 - \widetilde{\phi}^2)\widehat{\sigma}_x^2 \tag{8.177}$$

where $\widehat{\sigma}_x^2$ is a robust estimate of σ_x^2 based on the observations y_t. For example, we might use an M- or τ-scale, or the simple estimate $\widehat{\sigma}_x = \mathrm{MADN}(y_t)/0.6745$. Then when computing (8.91) for $m > 1$, we use the estimate $\widehat{\sigma}_{u,m}^2$ of $\sigma_{u,m}^2$ parameterized as a function of $\widetilde{\phi} = \widetilde{\phi}_{m,m}$ using (8.176)

$$\widehat{\sigma}_{u,m}^2(\phi) = (1 - \widetilde{\phi}^2)\widehat{\sigma}_{u,m-1}^2 \tag{8.178}$$

where $\widehat{\sigma}_{u,m-1}$ is the minimized robust scale $\widehat{\sigma}$ in (8.91) for the order-$(m-1)$ fit.

Since the function in (8.91) may have more than one local extrema, the minimization is performed by means of a grid search on (-1,1).

8.17 Appendix C: ARMA model state-space representation

Here we describe the state-space representation (8.102) and (8.104) for an ARMA(p, q) model, and show how to extend it to ARIMA and SARIMA models. We note that a state-space representation of ARMA models is not unique, and the particular representation we chose was that by Ledolter (1979) and Harvey and Phillips (1979). For other representations see Akaike (1974b), Jones (1980) and Chapter 12 of Brockwell and Davis (1991).

Let

$$\phi(B)(x_t - \mu) = \theta(B)u_t.$$

Define $\boldsymbol{\alpha}_t = (\alpha_{1,t}, \ldots, \alpha_{p,t})$, where

$$\alpha_{1,t} = x_t - \mu,$$

$$\alpha_{j,t} = \phi_j(x_{t-1} - \mu) + \ldots + \phi_p(x_{t-p+j-1} - \mu) - \theta_{j-1}u_t - \ldots - \theta_q u_{t-q+j-1},$$
$$j = 2, \ldots, q+1$$

and

$$\alpha_{j,t} = \phi_j(x_{t-1} - \mu) + \ldots + \phi_p(x_{t-p+j-1} - \mu),\ j = q+2, \ldots, p.$$

Then it is left as an exercise to show that the state-space representation (8.102) holds where **d** and Φ are given by (8.103) and (8.104) respectively. In the definition of **d** we take $\theta_i = 0$ for $i > q$.

The case $q \geq p$ is reduced to the above procedure on observing that y_t can be represented as an ARMA($q+1, q$) model where $\phi_i = 0$ for $i \geq p$. Thus, in general the dimension of $\boldsymbol{\alpha}$ is $k = \max(p, q+1)$.

The above state-space representation is easily extended to represent an ARIMA(p, d, q) model (8.109) by writing it as

$$\phi^*(B)(y_t - \mu) = \theta(B)u_t \tag{8.179}$$

where $\phi^*(B) = \phi(B)(1-B)^d$ has order $p^* = p + d$. Now we just proceed as above with the ϕ_i replaced by the ϕ_i^* coefficients in the polynomial $\phi^*(B)$, resulting in the state-transition matrix Φ^*. For example, in the case of an ARIMA($1, 1, q$) model we have $\phi_1^* = 1 + \phi_1$ and $\phi_2^* = -\phi_1$. The order of Φ^* is $k^* = \max(p^*, q+1)$.

The above approach also easily handles the case of a SARIMA model (8.110). One just defines

$$\phi^*(B) = \phi(B)\Phi(B^s)(1-B)^d(1-B^s)^D, \tag{8.180}$$

$$\theta^*(B) = \theta(B)\Theta(B^s) \tag{8.181}$$

and specifies the state-transition matrix Φ^* and vector $\mathbf{d}^*$ based on the coefficients of polynomials $\phi^*(B)$ and $\theta^*(B)$ of order p^* and q^* respectively. The order of Φ^* is now $k = \max(p^*, q^* + 1)$.

8.18 Problems

8.1. Show that $|\widehat{\rho}(1)| \leq 1$ for $\widehat{\rho}(1)$ in (8.3). Also show that if the summation in the denominator in (8.3) ranges only from 1 to $T - 1$, then $|\widehat{\rho}(1)|$ can be larger than one.

8.2. Show that for a "doublet" outlier at t_0 (i.e., $y_{t_0} = A = -y_{t_0+1}$) with $t_0 \in (1, T)$, the limiting value as $A \to \infty$ of $\widehat{\rho}(1)$ in (8.3) is -0.5.

8.3. Show that the limiting value as $A \to \infty$ of $\widehat{\rho}(1)$ defined in (8.4), when there is an isolated outlier of size A, is $-1/T + O\left(1/T^2\right)$.

8.4. Construct a probability model for additive outliers v_t that has non-overlapping patches of length $k > 0$, such that $v_t = A$ within each patch and $v_t = 0$ otherwise, and with $\mathrm{P}(v_t \neq 0) = \varepsilon$.

8.5. Verify the expression for the Yule–Walker equations given by (8.28).

8.6. Verify that for an AR(1) model with parameter ϕ we have $\rho(1) = \phi$.

8.7. Show that the LS estimate of the AR(p) parameters given by (8.26) is equivalent to solving the Yule–Walker equation(s) (8.28) with the true covariances replaced by the sample ones (8.30).

8.8. Prove the orthogonality condition (8.37).

8.9. Verify (8.161)–(8.162)–(8.163).

8.10. Prove (8.170).

8.11. Prove (8.40).

8.12. Verify (8.13) using (8.135).

8.13. Calculate the spectral density for the case that x_{t_0} and x_{t_0+k} are replaced by A and $-A$ respectively.

9

Numerical Algorithms

Computing M-estimates involves function minimization and/or solving nonlinear equations. General methods based on derivatives—like the Newton–Raphson procedure for solving equations—are widely available, but they are inadequate for this type of specific problem, for the reasons given in Section 2.9.5.

In this chapter we treat some details of the iterative algorithms described in the previous chapters to compute M-estimates.

9.1 Regression M-estimates

We shall justify the algorithm in Section 4.5 for solving (4.39); this includes location as a special case. Consider the problem

$$h(\boldsymbol{\beta}) = \min,$$

where

$$h(\boldsymbol{\beta}) = \sum_{i=1}^{n} \rho\left(\frac{r_i(\boldsymbol{\beta})}{\sigma}\right),$$

where $r_i(\boldsymbol{\beta}) = y_i - \mathbf{x}_i'\boldsymbol{\beta}$ and σ is any positive constant.

It is assumed that the $\mathbf{x}_i$'s are not collinear, otherwise there would be multiple solutions. It is assumed that $\rho(r)$ is a ρ-function, that the function $W(x)$ defined in (2.30) is nonincreasing in $|x|$, and that ψ is continuous. These conditions are easily verified for the Huber and the bisquare functions.

Robust Statistics – Theory and Methods Ricardo A. Maronna, R. Douglas Martin and Víctor J. Yohai

It will be proved that h does not increase at each iteration, and that if there is a single stationary point $\boldsymbol{\beta}_0$ of h, i.e., a point satisfying

$$\sum_{i=1}^{n} \psi\left(\frac{r_i(\boldsymbol{\beta}_0)}{\sigma}\right)\mathbf{x}_i = \mathbf{0}, \tag{9.1}$$

then the algorithm converges to it.

For $r \geq 0$ let $g(r) = \rho\left(\sqrt{r}\right)$. It follows from $\rho(r) = g\left(r^2\right)$ that

$$W(r) = 2g'(r^2) \tag{9.2}$$

and hence $W(r)$ is nonincreasing for $r \geq 0$ if and only if g' is nonincreasing.

We claim that

$$g(y) \leq g(x) + g'(x)(y - x), \tag{9.3}$$

i.e., the graph of g lies below the tangent line. To show this, assume first that $y > x$ and note that by the intermediate value theorem,

$$g(y) - g(x) = (y - x)g'(\xi),$$

where $\xi \in [x, y]$. Since g' is nonincreating, $g'(\xi) \leq g'(x)$. The case $y < x$ is dealt with likewise.

A function g with a nonincreasing derivative satisfies for all x, y and all $\alpha \in [0, 1]$

$$g(\alpha x + (1 - \alpha)y) \geq \alpha g(x) + (1 - \alpha)g(y), \tag{9.4}$$

i.e., the graph of g lies above the secant line. Such functions are called *concave*. Conversely, a differentiable function is concave if and only if its derivative is nonincreasing. For twice differentiable functions, concavity is equivalent to having a nonpositive second derivative.

Define the matrix

$$\mathbf{U}(\boldsymbol{\beta}) = \sum_{i=1}^{n} W\left(\frac{r_i(\boldsymbol{\beta})}{\sigma}\right)\mathbf{x}_i\mathbf{x}_i',$$

which is nonnegative definite for all $\boldsymbol{\beta}$, and the function

$$f(\boldsymbol{\beta}) = \arg\min_{\boldsymbol{\gamma}} \sum_{i=1}^{n} W\left(\frac{r_i(\boldsymbol{\beta})}{\sigma}\right)(y_i - \mathbf{x}_i'\boldsymbol{\gamma})^2.$$

The algorithm can then be written as

$$\boldsymbol{\beta}_{k+1} = f(\boldsymbol{\beta}_k). \tag{9.5}$$

A fixed point $\boldsymbol{\beta}_0$, i.e., one satisfying $f(\boldsymbol{\beta}_0) = \boldsymbol{\beta}_0$, is also a stationary point (9.1). Given $\boldsymbol{\beta}_k$, put for simplicity $w_i = W(r_i(\boldsymbol{\beta}_k)/\sigma)$. Note that $\boldsymbol{\beta}_{k+1}$ satisfies

$$\sum_{i=1}^{n} w_i\mathbf{x}_i y_i = \sum_{i=1}^{n} w_i\mathbf{x}_i\mathbf{x}_i'\boldsymbol{\beta}_{k+1} = \mathbf{U}(\boldsymbol{\beta}_k)\boldsymbol{\beta}_{k+1}. \tag{9.6}$$

We shall show that

$$h(\boldsymbol{\beta}_{k+1}) \leq h(\boldsymbol{\beta}_k). \tag{9.7}$$

We have using (9.3) and (9.2)

$$\begin{aligned} h(\boldsymbol{\beta}_{k+1}) - h(\boldsymbol{\beta}_k) &\leq \frac{1}{\sigma^2}\sum_{i=1}^{n} g'\left(\frac{r_i(\boldsymbol{\beta}_k)^2}{\sigma^2}\right)\left(r_i(\boldsymbol{\beta}_{k+1})^2 - r_i(\boldsymbol{\beta}_k)^2\right) \\ &= \frac{1}{2\sigma^2}\sum_{i=1}^{n} w_i\left(r_i(\boldsymbol{\beta}_{k+1}) - r_i(\boldsymbol{\beta}_k)\right)\left(r_i(\boldsymbol{\beta}_{k+1}) + r_i(\boldsymbol{\beta}_k)\right). \end{aligned}$$

But since

$$r_i(\boldsymbol{\beta}_{k+1}) - r_i(\boldsymbol{\beta}_k) = (\boldsymbol{\beta}_k - \boldsymbol{\beta}_{k+1})'\mathbf{x}_i \;\text{ and }\; r_i(\boldsymbol{\beta}_{k+1}) + r_i(\boldsymbol{\beta}_k) = 2y_i - \mathbf{x}_i'\left(\boldsymbol{\beta}_k + \boldsymbol{\beta}_{k+1}\right)$$

we have using (9.6)

$$\begin{aligned} h(\boldsymbol{\beta}_{k+1}) - h(\boldsymbol{\beta}_k) &\leq \frac{1}{2\sigma^2}(\boldsymbol{\beta}_k - \boldsymbol{\beta}_{k+1})'\sum_{i=1}^{n} w_i\mathbf{x}_i\mathbf{x}_i'\left(2\boldsymbol{\beta}_{k+1} - \boldsymbol{\beta}_k - \boldsymbol{\beta}_{k+1}\right) \\ &= \frac{1}{2\sigma^2}(\boldsymbol{\beta}_k - \boldsymbol{\beta}_{k+1})'\mathbf{U}(\boldsymbol{\beta}_k)\left(\boldsymbol{\beta}_{k+1} - \boldsymbol{\beta}_k\right) \leq 0 \end{aligned}$$

since $\mathbf{U}(\boldsymbol{\beta}_k)$ is nonnegative definite. This proves (9.7).

We shall now prove the convergence of $\boldsymbol{\beta}_k$ to $\boldsymbol{\beta}_0$ in (9.1). To simplify the proof we make the stronger assumption that ρ is increasing and hence $W(r) > 0$ for all r. Since the sequence $h(\boldsymbol{\beta}_k)$ is nonincreasing and is bounded from below, it has a limit h_0. Hence the sequence $\boldsymbol{\beta}_k$ is bounded, otherwise there would be a subsequence $\boldsymbol{\beta}_{k_j}$ converging to infinity, and since ρ is increasing, so would $h(\boldsymbol{\beta}_{k_j})$.

Since $\boldsymbol{\beta}_k$ is bounded, it has a subsequence which has a limit $\boldsymbol{\beta}_0$, which by continuity satisfies (9.5) and is hence a stationary point. If it is unique, then $\boldsymbol{\beta}_k \to \boldsymbol{\beta}_0$; otherwise, there would exist a subsequence bounded away from $\boldsymbol{\beta}_0$, which in turn would have a convergent subsequence, which would have a limit different from $\boldsymbol{\beta}_0$, which would also be a stationary point. This concludes the proof of (9.1).

Another algorithm is based on "pseudo-observations". Put

$$\widetilde{y}_i(\boldsymbol{\beta}) = \mathbf{x}_i'\boldsymbol{\beta} + \widehat{\sigma}\psi\left(\frac{r_i(\boldsymbol{\beta})}{\widehat{\sigma}}\right).$$

Then (4.40) is clearly equivalent to

$$\sum_{i=1}^{n}\mathbf{x}_i\left(\widetilde{y}_i(\widehat{\boldsymbol{\beta}}) - \mathbf{x}_i'\widehat{\boldsymbol{\beta}}\right) = 0.$$

Given $\boldsymbol{\beta}_k$, the next step of this algorithm is finding $\boldsymbol{\beta}_{k+1}$ such that

$$\sum_{i=1}^{n}\mathbf{x}_i\left(\widetilde{y}_i(\boldsymbol{\beta}_k) - \mathbf{x}_i'\boldsymbol{\beta}_{k+1}\right) = 0,$$

which is an *ordinary* LS problem. The procedure can be shown to converge (Huber, 1981, Section 7.8) but it is much slower than the reweighting algorithm.

9.2 Regression S-estimates

We deal with the descent algorithm described in Section 5.7.1. As explained there, the algorithm coincides with the one for M-estimates. The most important result is that, if W is nonincreasing, then at each step $\widehat{\sigma}$ does not increase.

To see this, consider at step k the vector $\boldsymbol{\beta}_k$ and the respective residual scale σ_k which satisfies

$$\frac{1}{n}\sum_{i=1}^{n}\rho\left(\frac{r_i\left(\boldsymbol{\beta}_k\right)}{\sigma_k}\right)=\delta.$$

The next vector $\boldsymbol{\beta}_{k+1}$ is obtained from (9.5) (with σ replaced by σ_k), and hence satisfies (9.7). Therefore

$$\frac{1}{n}\sum_{i=1}^{n}\rho\left(\frac{r_i\left(\boldsymbol{\beta}_{k+1}\right)}{\sigma_k}\right)\leq\frac{1}{n}\sum_{i=1}^{n}\rho\left(\frac{r_i\left(\boldsymbol{\beta}_k\right)}{\sigma_k}\right)=\delta. \tag{9.8}$$

Since σ_{k+1} satisfies

$$\frac{1}{n}\sum_{i=1}^{n}\rho\left(\frac{r_i\left(\boldsymbol{\beta}_{k+1}\right)}{\sigma_{k+1}}\right)=\delta, \tag{9.9}$$

and ρ is nondecreasing, it follows from (9.9) and (9.8) that

$$\sigma_{k+1}\leq\sigma_k. \tag{9.10}$$

9.3 The LTS-estimate

We shall justify the procedure in Section 5.7.1. Call $\widehat{\sigma}_1$ and $\widehat{\sigma}_2$ the scales corresponding to $\widehat{\boldsymbol{\beta}}_1$ and $\widehat{\boldsymbol{\beta}}_2$, respectively. For $k=1,2$ let $r_{ik}=y_i-\mathbf{x}_i'\widehat{\boldsymbol{\beta}}_k$ be the respective residuals, and call $r_{(i)k}^2$ the ordered squared residuals. Let $I\subset\{1,\ldots,n\}$ be the set of indices corresponding to the h smallest r_{i1}^2. Then

$$\widehat{\sigma}_2^2=\sum_{i=1}^{h}r_{(i)2}^2\leq\sum_{i\in I}r_{i2}^2\leq\sum_{i\in I}r_{i1}^2=\sum_{i=1}^{h}r_{(i)1}^2=\widehat{\sigma}_1^2.$$

9.4 Scale M-estimates

9.4.1 Convergence of the fixed point algorithm

We shall show that the algorithm (2.78) given for solving (2.54) converges.

Define W as in (2.59). It is assumed again that $\rho(r)$ is a ρ-function of $|r|$. For $r \geq 0$ define

$$g(r) = \rho\left(\sqrt{r}\right). \tag{9.11}$$

It will be assumed that g is *concave* (see below (9.4)). To make things simpler, we assume that g is twice differentiable, and that $g'' < 0$.

The concavity of g implies that W is nonincreasing. In fact, it follows from $W(r) = g(r^2)/r^2$ that

$$W'(r) = \frac{2}{r^3}\left(r^2 g'(r^2) - g(r^2)\right) \leq 0,$$

since (9.3) implies for all t

$$0 = g(0) \leq g(t) + g'(t)(0-t) = g(t) - tg'(t). \tag{9.12}$$

Put for simplicity $\theta = \sigma^2$ and $y_i = x_i^2$. Then (2.54) can be rewritten as

$$\frac{1}{n}\sum_{i=1}^{n} g\left(\frac{y_i}{\theta}\right) = \delta$$

and (2.78) can be rewritten as

$$\theta_{k+1} = h(\theta_k), \tag{9.13}$$

with

$$h(\theta) = \frac{1}{n\delta}\sum_{i=1}^{n} g\left(\frac{y_i}{\theta}\right)\theta. \tag{9.14}$$

It will be shown that h is nondecreasing and concave. It suffices to prove these properties for each term of (9.14). In fact, for all y,

$$\frac{d}{d\theta}\left(\theta g\left(\frac{y}{\theta}\right)\right) = g\left(\frac{y}{\theta}\right) - \frac{y}{\theta} g'\left(\frac{y}{\theta}\right) \geq 0 \tag{9.15}$$

because of (9.12); and

$$\frac{d^2}{d\theta^2}\left(\theta g\left(\frac{y}{\theta}\right)\right) = g''\left(\frac{y}{\theta}\right)\frac{y^2}{\theta^3} \leq 0 \tag{9.16}$$

because $g'' < 0$.

We shall now deal with the resolution of the equation

$$h(\theta) = \theta.$$

Assume it has a unique solution θ_0. We shall show that

$$\theta_k \to \theta_0.$$

Note first that $h'(\theta_0) < 1$. For

$$h(\theta_0) = \int_0^{\theta_0} h'(t)dt,$$

and if $h'(\theta_0) \geq 1$, then $h'(t) > 1$ for $t > 1$, and hence $h(\theta_0) > \theta_0$. Assume first that $\theta_1 > \theta_0$. Since h is nondecreasing, $\theta_2 = h(\theta_1) \geq h(\theta_0) = \theta_0$. We shall prove that $\theta_2 < \theta_1$. In fact,

$$\theta_2 = h(\theta_1) \leq h(\theta_0) + h'(\theta_0)(\theta_1 - \theta_0) < \theta_0 + (\theta_1 - \theta_0) = \theta_1.$$

In the same way, it follows that $\theta_0 < \theta_{k+1} < \theta_k$. Hence the sequence θ_k decreases, and since it is bounded from below, it has a limit. The case $\theta_1 < \theta_0$ is treated likewise.

Actually, the procedure can be accelerated. Given three consecutive values θ_k, θ_{k+1} and θ_{k+2}, the straight line determined by the points (θ_k, θ_{k+1}) and $(\theta_{k+1}, \theta_{k+2})$ intersects the identity diagonal at the point (θ^*,θ^*) with

$$\theta^* = \frac{\theta_{k+1}^2 - \theta_k\theta_{k+2}}{2\theta_{k+1} - \theta_{k+2} - \theta_k}.$$

Then set $\theta_{k+3} = \theta^*$. The accelerated procedure also converges under the given assumptions.

9.4.2 Algorithms for the nonconcave case

If the function g in (9.11) is not concave, the algorithm is not guaranteed to converge to the solution. In this case (2.54) has to be solved by using a general equation-solving procedure. For given $x_1, \ldots, x_n$ let

$$h(\sigma) = \frac{1}{n}\sum_{i=1}^{n} \rho\left(\frac{x_i}{\sigma}\right) - \delta. \tag{9.17}$$

Then we have to solve $h(\sigma) = 0$. Procedures using derivatives, like Newton–Raphson, cannot be used, since the boundedness of ρ implies that h' is not bounded away from zero. Safe procedures without derivatives require locating the solution in an interval $[\sigma_1, \sigma_2]$ such that $\text{sgn}(h(\sigma_1)) \neq \text{sgn}(h(\sigma_2))$. The simplest is the bisection method, but faster ones exist and can be found, for example, in Brent (1973).

To find σ_1 and σ_2, recall that h is nonincreasing. Let $\sigma_0 = \text{Med}(|\mathbf{x}|)$ and set $\sigma_1 = \sigma_0$. If $h(\sigma_1) > 0$, we are done; else set $\sigma_1 = \sigma_1/2$ and continue halving σ_1 until $h(\sigma_1) > 0$. The same method yields σ_2.

9.5 Multivariate M-estimates

Location and covariance will be treated separately for the sake of simplicity. A very detailed treatment of the convergence of the iterative reweighting algorithm for simultaneous estimation was given by Arslan (2004).

Location involves solving

$$h(\boldsymbol{\mu}) = \min,$$

with

$$h(\boldsymbol{\mu}) = \sum_{i=1}^{n} \rho(d_i(\boldsymbol{\mu})),$$

where

$$d_i(\boldsymbol{\mu}) = (\mathbf{x}_i - \boldsymbol{\mu})' \Sigma^{-1} (\mathbf{x}_i - \boldsymbol{\mu}),$$

which implies (6.11). The procedure is as follows. Given $\boldsymbol{\mu}_k$, let

$$\boldsymbol{\mu}_{k+1} = \frac{1}{\sum_{i=1}^{n} w_i} \sum_{i=1}^{n} w_i \mathbf{x}_i,$$

with $w_i = W(d_i(\boldsymbol{\mu}_k))$ and $W = \rho'$. Hence

$$\sum_{i=1}^{n} w_i \mathbf{x}_i = \boldsymbol{\mu}_{k+1} \sum_{i=1}^{n} w_i. \tag{9.18}$$

Assume that W is nonincreasing, which is equivalent to ρ being concave. It will be shown that $h(\boldsymbol{\mu}_{k+1}) \leq h(\boldsymbol{\mu}_k)$. The proof is similar to that of Section 9.1. It is easy to show that the problem can be reduced to $\Sigma = \mathbf{I}$, so that $\mathrm{d}_i(\boldsymbol{\mu}) = \|\mathbf{x}_i - \boldsymbol{\mu}\|^2$. Using the concavity of ρ and then (9.18)

$$\begin{aligned} h(\boldsymbol{\mu}_{k+1}) - h(\boldsymbol{\mu}_k) &\leq \sum_{i=1}^{n} w_i \left[\left\|\mathbf{x}_i - \boldsymbol{\mu}_{k+1}\right\|^2 - \left\|\mathbf{x}_i - \boldsymbol{\mu}_k\right\|^2 \right] \\ &= \left(\boldsymbol{\mu}_k - \boldsymbol{\mu}_{k+1}\right)' \sum_{i=1}^{n} w_i \left(2\mathbf{x}_i - \boldsymbol{\mu}_k - \boldsymbol{\mu}_{k+1}\right) \\ &= \left(\boldsymbol{\mu}_k - \boldsymbol{\mu}_{k+1}\right)' \left(\boldsymbol{\mu}_{k+1} - \boldsymbol{\mu}_k\right) \sum_{i=1}^{n} w_i \leq 0. \end{aligned}$$

The treatment of the covariance matrix is more difficult (Maronna, 1976).

9.6 Multivariate S-estimates

9.6.1 S-estimates with monotone weights

For the justification of the algorithm in Section 6.7.2 we shall show that if the weight function is nonincreasing, and hence ρ is concave, then

$$\widehat{\sigma}_{k+1} \leq \widehat{\sigma}_k. \tag{9.19}$$

Given $\boldsymbol{\mu}_k$ and Σ_k, define $\widehat{\sigma}_{k+1}$, $\boldsymbol{\mu}_{k+1}$ and Σ_{k+1} as in (6.53)–(6.54). It will be shown that

$$\sum_{i=1}^{n} \rho\left(\frac{d(\mathbf{x}_i, \boldsymbol{\mu}_{k+1}, \Sigma_{k+1})}{\widehat{\sigma}_k}\right) \leq \sum_{i=1}^{n} \rho\left(\frac{d(\mathbf{x}_i, \boldsymbol{\mu}_k, \Sigma_k)}{\widehat{\sigma}_k}\right). \tag{9.20}$$

In fact, the concavity of ρ yields (putting w_i for the w_{ki} of (6.53)):

$$\sum_{i=1}^{n} \rho\left(\frac{d(\mathbf{x}_i, \boldsymbol{\mu}_{k+1}, \Sigma_{k+1})}{\widehat{\sigma}_k}\right) - \sum_{i=1}^{n} \rho\left(\frac{d(\mathbf{x}_i, \boldsymbol{\mu}_k, \Sigma_k)}{\widehat{\sigma}_k}\right)$$

$$\leq \frac{1}{\widehat{\sigma}_k} \sum_{i=1}^{n} w_i \left[d(\mathbf{x}_i, \boldsymbol{\mu}_{k+1}, \Sigma_{k+1}) - d(\mathbf{x}_i, \boldsymbol{\mu}_k, \Sigma_k)\right]. \tag{9.21}$$

Note that $\boldsymbol{\mu}_{k+1}$ is the weighted mean of the $\mathbf{x}_i$'s with weights w_i, and hence it minimizes $\sum_{i=1}^{n} w_i\, (\mathbf{x}_i - \boldsymbol{\mu})'\, \mathbf{A}\, (\mathbf{x}_i - \boldsymbol{\mu})$ for any positive definite matrix $\mathbf{A}$. Therefore

$$\sum_{i=1}^{n} w_i d(\mathbf{x}_i, \boldsymbol{\mu}_{k+1}, \Sigma_{k+1}) \leq \sum_{i=1}^{n} w_i d(\mathbf{x}_i, \boldsymbol{\mu}_k, \Sigma_{k+1})$$

and hence the sum on the right-hand side of (9.21) is not larger than

$$\sum_{i=1}^{n} w_i d(\mathbf{x}_i, \boldsymbol{\mu}_k, \Sigma_{k+1}) - \sum_{i=1}^{n} w_i d(\mathbf{x}_i, \boldsymbol{\mu}_k, \Sigma_k)$$

$$= \sum_{i=1}^{n} \mathbf{y}_i \Sigma_{k+1}^{-1} \mathbf{y}_i' - \sum_{i=1}^{n} \mathbf{y}_i \Sigma_k^{-1} \mathbf{y}_i', \tag{9.22}$$

with $\mathbf{y}_i = \sqrt{w_i}\left(\mathbf{x}_i - \boldsymbol{\mu}_k\right)$. Since

$$\Sigma_{k+1} = \frac{\mathbf{C}}{|\mathbf{C}|^{1/p}} \text{ with } \mathbf{C} = \frac{1}{n}\sum_{i=1}^{n} \mathbf{y}_i \mathbf{y}_i',$$

we have that Σ_{k+1} is the sample covariance matrix of the $\mathbf{y}_i$'s normalized to unit determinant, and by (6.35) it minimizes the sum of squared Mahalanobis distances among matrices with unit determinant. Since $|\Sigma_k| = |\Sigma_{k+1}| = 1$, it follows that (9.22) is ≤ 0, which proves (9.20).

Since

$$\frac{1}{n}\sum_{i=1}^{n} \rho\left(\frac{d(\mathbf{x}_i, \boldsymbol{\mu}_{k+1}, \Sigma_{k+1})}{\widehat{\sigma}_{k+1}}\right) = \frac{1}{n}\sum_{i=1}^{n} \rho\left(\frac{d(\mathbf{x}_i, \boldsymbol{\mu}_k, \Sigma_k)}{\widehat{\sigma}_k}\right),$$

the proof of (9.19) follows like that of (9.10).

9.6.2 The MCD

The justification of the "concentration step" in Section 6.7.6 proceeds as in Section 9.3. Put for $k = 1, 2$: $d_{ik} = d(\mathbf{x}_i, \boldsymbol{\mu}_k, \Sigma_k)$ and call $d_{(i)k}$ the respective ordered values and $\widehat{\sigma}_1, \widehat{\sigma}_2$ the respective scales. Let $I \subset \{1, \ldots, n\}$ be the set of indices corresponding to the smallest h values of d_{i1}. Then $\boldsymbol{\mu}_2$ and Σ_2 are the mean and the normalized sample covariance matrix of the set $\{\mathbf{x}_i : i \in I\}$. Hence (6.35) applied to that set implies that

$$\sum_{i \in I} d_{i2} \leq \sum_{i \in I} d_{i1} = \sum_{i=1}^{h} d_{(i)1},$$

and hence

$$\sigma_2^2 = \sum_{i=1}^{h} d_{(i)2} \leq \sum_{i \in I} d_{i2} \leq \sigma_1^2.$$

9.6.3 S-estimates with nonmonotone weights

Note first that if ρ is not concave, the algorithm (2.78) is not guaranteed to yield the scale σ, and hence the approach in Section 9.4.2 must be used to compute σ.

Now we describe the modification of the iterative algorithm for the S-estimate. Call $\left(\widehat{\boldsymbol{\mu}}_N, \widehat{\boldsymbol{\Sigma}}_N\right)$ the estimates at iteration N, and $\sigma\left(\widehat{\boldsymbol{\mu}}_N, \widehat{\boldsymbol{\Sigma}}_N\right)$ the respective scale. Call $\left(\widetilde{\boldsymbol{\mu}}_{N+1}, \widetilde{\boldsymbol{\Sigma}}_{N+1}\right)$ the values given by a step of the reweighting algorithm.

If $\sigma\left(\widetilde{\boldsymbol{\mu}}_{N+1}, \widetilde{\boldsymbol{\Sigma}}_{N+1}\right) < \sigma\left(\widehat{\boldsymbol{\mu}}_N, \widehat{\boldsymbol{\Sigma}}_N\right)$, then we proceed as usual, setting

$$\left(\widehat{\boldsymbol{\mu}}_{N+1}, \widehat{\boldsymbol{\Sigma}}_{N+1}\right) = \left(\widetilde{\boldsymbol{\mu}}_{N+1}, \widetilde{\boldsymbol{\Sigma}}_{N+1}\right).$$

If instead

$$\sigma\left(\widetilde{\boldsymbol{\mu}}_{N+1}, \widetilde{\boldsymbol{\Sigma}}_{N+1}\right) \geq \sigma\left(\widehat{\boldsymbol{\mu}}_N, \widehat{\boldsymbol{\Sigma}}_N\right), \tag{9.23}$$

then for a given $\xi \in R$ put

$$\left(\widehat{\boldsymbol{\mu}}_{N+1}, \widehat{\boldsymbol{\Sigma}}_{N+1}\right) = (1-\xi)\left(\widehat{\boldsymbol{\mu}}_N, \widehat{\boldsymbol{\Sigma}}_N\right) + \xi\left(\widetilde{\boldsymbol{\mu}}_{N+1}, \widetilde{\boldsymbol{\Sigma}}_{N+1}\right). \tag{9.24}$$

Then it can be shown that there exists $\xi \in (0, 1)$ such that

$$\sigma\left(\widehat{\boldsymbol{\mu}}_{N+1}, \widehat{\boldsymbol{\Sigma}}_{N+1}\right) < \sigma\left(\widehat{\boldsymbol{\mu}}_N, \widehat{\boldsymbol{\Sigma}}_N\right). \tag{9.25}$$

The details are given below in Section 9.6.4.

If the situation (9.23) occurs, then the algorithm proceeds as follows. Let $\xi_0 \in (0, 1)$. Set $\xi = \xi_0$ and compute (9.24). If (9.25) occurs, we are done. Else set $\xi = \xi\xi_0$ and repeat the former steps, and so on. At some point we must have (9.25). In our programs we use $\xi = 0.7$.

A more refined method would be a line search; that is, to compute (9.24) for different values of ξ and choose the one yielding minimum σ. Our experiments do not show that this extra effort yields better results.

It must be noted that when the computation is near a local minimum, it may happen that because of rounding errors, no value of $\xi = \xi_0^k$ yields a decrease in σ. Hence it is advisable to stop the search when ξ is less than a small prescribed constant and retain $\left(\widehat{\boldsymbol{\mu}}_N, \widehat{\boldsymbol{\Sigma}}_N\right)$ as the final result.

9.6.4 *Proof of (9.25)

Let $h(\mathbf{z}) : R^m \to R$ be a differentiable function, and call $\mathbf{g}$ its gradient at the point $\mathbf{z}$. Then for any $\mathbf{b} \in R^m$, $h(\mathbf{z}+\xi\mathbf{b}) = h(\mathbf{z}) + \xi\mathbf{g}'\mathbf{b} + o(\xi)$. Hence if $\mathbf{g}'\mathbf{b} < 0$, we have $h(\mathbf{z}+\xi\mathbf{b}) < h(\mathbf{z})$ for sufficiently small ξ.

We must show that we are indeed in this situation. To simplify the exposition, we deal only with $\boldsymbol{\mu}$; we assume $\boldsymbol{\Sigma}$ fixed, and without loss of generality we may take $\boldsymbol{\Sigma} = \mathbf{I}$. Then $d(\mathbf{x}, \boldsymbol{\mu}, \boldsymbol{\Sigma}) = \|\mathbf{x} - \boldsymbol{\mu}\|^2$. Call $\sigma(\boldsymbol{\mu})$ the solution of

$$\frac{1}{n}\sum_{i=1}^{n} \rho\left(\frac{\|\mathbf{x}_i - \boldsymbol{\mu}\|^2}{\sigma}\right) = \delta. \tag{9.26}$$

Call $\mathbf{g}$ the gradient of $\sigma(\boldsymbol{\mu})$ at a given $\boldsymbol{\mu}_1$. Then differentiating (9.26) with respect to $\boldsymbol{\mu}$ yields

$$\sum_{i=1}^{n} w_i \left[2\sigma\left(\mathbf{x} - \boldsymbol{\mu}_1\right) + \left\|\mathbf{x} - \boldsymbol{\mu}_1\right\|^2 \mathbf{g}\right] = 0,$$

with

$$w_i = W\left(\frac{\|\mathbf{x}_i - \boldsymbol{\mu}_1\|^2}{\sigma}\right),$$

and hence

$$\mathbf{g} = -\frac{2\sigma}{\sum_{i=1}^{n} w_i \left\|\mathbf{x} - \boldsymbol{\mu}_1\right\|^2} \sum_{i=1}^{n} w_i \left(\mathbf{x}_i - \boldsymbol{\mu}_1\right). \tag{9.27}$$

Call $\boldsymbol{\mu}_2$ the result of an iteration of the reweighting algorithm, i.e.,

$$\boldsymbol{\mu}_2 = \frac{1}{\sum_{i=1}^{n} w_i} \sum_{i=1}^{n} w_i \mathbf{x}_i.$$

Then

$$\boldsymbol{\mu}_2 - \boldsymbol{\mu}_1 = \frac{1}{\sum_{i=1}^{n} w_i} \sum_{i=1}^{n} w_i (\mathbf{x}_i - \boldsymbol{\mu}_1), \tag{9.28}$$

and it follows from (9.28) and (9.27) that $\left(\boldsymbol{\mu}_2 - \boldsymbol{\mu}_1\right)' \mathbf{g} < 0$.

10

Asymptotic Theory of M-estimates

In order to compare the performances of different estimates, and also to obtain confidence intervals for the parameters, we need their distributions. Explicit expressions exist in some simple cases, such as sample quantiles, which include the median, but even these are in general intractable. It will be necessary to resort to approximating their distributions for large n, the so-called *asymptotic distribution.*

We shall begin with the case of a single real parameter, and we shall consider general M-estimates of a parameter θ defined by equations of the form

$$\sum_{i=1}^{n} \Psi(x_i, \theta) = 0. \tag{10.1}$$

For location, Ψ has the form $\Psi(x, \theta) = \psi(x - \theta)$ with $\theta \in R$; for scale, $\Psi(x, \theta) = \rho(|x| / \theta) - \delta$ with $\theta > 0$. If ψ (or ρ) is nondecreasing then Ψ is nonincreasing in θ.

This family contains maximum likelihood estimates (MLEs). Let $f_\theta(x)$ be a family of densities. The likelihood function for an i.i.d. sample $x_1, \ldots, x_n$ with density f_θ is

$$L = \prod_{i=1}^{n} f_\theta(x_i).$$

If f_θ is everywhere positive, and is differentiable with respect to θ with derivative $\dot{f}_\theta = \partial f_\theta / \partial \theta$, taking logs it is seen that the MLE is the solution of

$$\sum_{i=1}^{n} \Psi_0(x_i, \theta) = 0,$$

Robust Statistics – Theory and Methods Ricardo A. Maronna, R. Douglas Martin and Víctor J. Yohai

with

$$\Psi_0(x,\theta) = -\frac{\partial \log f_\theta(x)}{\partial\theta} = -\frac{\dot{f}_\theta(x)}{f_\theta(x)}. \tag{10.2}$$

10.1 Existence and uniqueness of solutions

We shall first consider the existence and uniqueness of solutions of (10.1). It is assumed that θ ranges in a finite or infinite interval (θ_1, θ_2). For location, $\theta_2 = -\theta_1 = \infty$; for scale, $\theta_2 = \infty$ and $\theta_1 = 0$. Henceforth the symbol ■ means "this is the end of the proof".

Theorem 10.1 *Assume that for each x, $\Psi(x,\theta)$ is nonincreasing in θ and*

$$\lim_{\theta\to\theta_1} \Psi(x,\theta) > 0 > \lim_{\theta\to\theta_2} \Psi(x,\theta) \tag{10.3}$$

(both limits may be infinite). Let

$$g(\theta) = \sum_{i=1}^{n} \Psi(x_i,\theta).$$

Then:
(a) There is at least one point $\widehat{\theta} = \widehat{\theta}(x_1,\ldots,x_n)$ at which g changes sign, i.e.,

$$g(\theta) \geq 0 \; for \; \theta < \widehat{\theta} \text{ and } g(\theta) \leq 0 \; for \; \theta > \widehat{\theta}.$$

(b) The set of such points is an interval.
(c) If Ψ is continuous in θ, then $g(\widehat{\theta}) = 0$.
(d) If Ψ is decreasing, then $\widehat{\theta}$ is unique.

Proof: It follows from (10.3) that

$$\lim_{\theta\to\theta_1} g(\theta) > 0 > \lim_{\theta\to\theta_2} g(\theta); \tag{10.4}$$

and the existence of $\widehat{\theta}$ follows from the monotonicity of g. If two values satisfy $g(\theta) = 0$, then the monotonicity of g implies that any value between them also does, which yields point (b). Statement (c) follows from the intermediate value theorem; and point (d) is immediate. ■

Example 10.1 *If $\Psi(x,\theta) = \text{sgn}(x-\theta)$, which is neither continuous nor increasing, then*

$$g(\theta) = \#(x_i > \theta) - \#(x_i < \theta).$$

The reader can verify that for n odd, $n = 2m-1$, g vanishes only at $\widehat{\theta} = x_{(m)}$, and for n even, $n = 2m$, it vanishes on the interval $(x_{(m)}, x_{(m+1)})$.

Example 10.2 *The equation for scale M-estimation (2.54) does not satisfy (10.3) since* $\rho(0) = 0$ *implies* $\Psi(0, \theta) = -\delta < 0$ *for all* θ. *But the same reasoning shows that (10.4) holds if*

$$\frac{\#(x_i = 0)}{n} < 1 - \frac{\delta}{\rho(\infty)}.$$

Uniqueness may hold without requiring the strict monotonicity of Ψ. For instance, Huber's ψ is not increasing, but the respective location estimate is unique unless there is a large gap in the middle of the data (Problem 10.7). A sufficient condition for the uniqueness of scale estimates is that $\rho(x)$ be increasing for all x such that $\rho(x) < \rho(\infty)$ (Problem 10.6).

Redescending location estimates The above results do not cover the case of location estimates with a redescending ψ. In this case uniqueness requires stronger assumptions than the case of monotone ψ. Uniqueness of the asymptotic value of the estimate requires that the distribution of x, besides being symmetric, is *unimodal*, i.e., it has a density $f(x)$ which for some μ is increasing for $x < \mu$ and decreasing for $x > \mu$.

Theorem 10.2 *Let* x *have a density* $f(x)$ *which is a decreasing function of* $|x|$, *and let* ρ *be any* ρ*-function. Then* $\lambda(\mu) = \mathrm{E}\rho(x - \mu)$ *has a unique minimum at* $\mu = 0$.

Proof: Recall that ρ is even and hence its derivative ψ is odd. Hence the derivative of λ is

$$\lambda'(\mu) = -\int_{-\infty}^{\infty} f(x)\psi(x - \mu)dx$$

$$= \int_{0}^{\infty} \psi(x)\left[f(x - \mu) - f(x + \mu)\right]dx.$$

Since λ is even it is enough to show that $\lambda'(\mu) > 0$ if $\mu > 0$. It follows from the definition of the ρ-function that $\psi(x) \geq 0$ for $x \geq 0$ and $\psi(x) > 0$ if $x \in (0, x_0)$ for some x_0. If x and μ are positive, then $|x - \mu| < |x + \mu|$ and hence $f(x - \mu) > f(x + \mu)$, which implies that the last integral above is positive. ■

If ψ is redescending and f is not unimodal, the minimum need not be unique. Let for instance f be a mixture: $f = 0.5 f_1 + 0.5 f_2$, where f_1 and f_2 are the densities of $\mathrm{N}(k, 1)$ and $\mathrm{N}(-k, 1)$, respectively. Then if k is large enough, $\lambda(\mu)$ has two minima, located near k and $-k$. The reason can be seen intuitively by noting that if k is large, then for $\mu > 0$, $\lambda(\mu)$ is approximately $0.5 \int \rho(x - \mu) f_1(x)\,dx$, which has a minimum at k. Note that instead the asymptotic value of a monotone estimate is uniquely defined for this distribution.

10.2 Consistency

Let $x_1, \ldots, x_n$ now be i.i.d. with distribution F. We shall consider the behavior of the solution $\widehat{\theta}_n$ of (10.1) as a random variable. Recall that a sequence y_n of random variables *tends in probability* to y if $\mathrm{P}(|y_n - y| > \varepsilon) \to 0$ for all $\varepsilon > 0$; this will be denoted by $y_n \to_p y$ or $\text{p}\lim y_n = y$. The sequence y_n tends *almost surely* (a.s.) or *with probability one* to y if $\text{P}(\lim_{n\to\infty} y_n = y) = 1$. The expectation with respect to a distribution F will be denoted by E_F.

We shall need a general result.

Theorem 10.3 (Monotone convergence theorem) *Let y_n be a nondecreasing sequence of random variables such that $\text{E}\,|y_n| < \infty$, and $y_n \to y$ with probability 1. Then*

$$\text{E}y_n \to \text{E}y.$$

The proof can be found in Feller (1971).

Assume that $\text{E}_F|\Psi(x,\theta)| < \infty$ for each θ, and define

$$\lambda_F(\theta) = \text{E}_F \Psi(x,\theta). \tag{10.5}$$

Theorem 10.4 *Assume that $\text{E}_F|\Psi(x,\theta)| < \infty$ for all θ. Under the assumptions of Theorem 10.1, there exists θ_F such that λ_F changes sign at θ_F.*

Proof: Proceeds along the same lines as the proof of Theorem 10.1. The interchange of limits and expectations is justified by the monotone convergence theorem. ■

Note that if λ_F is continuous, then

$$\text{E}_F \Psi(x,\theta_F) = 0. \tag{10.6}$$

Theorem 10.5 *If θ_F is unique, then $\widehat{\theta}_n$ tends in probability to θ_F.*

Proof: To simplify the proof, we shall assume $\widehat{\theta}_n$ is unique. Then it will be shown that for any $\epsilon > 0$,

$$\lim_{n\to\infty} \text{P}(\widehat{\theta}_n < \theta_F - \epsilon) = 0.$$

Let

$$\widehat{\lambda}_n(\theta) = \frac{1}{n}\sum_{i=1}^{n} \Psi(x_i,\theta).$$

Since $\widehat{\lambda}_n$ is nonincreasing in θ and $\widehat{\theta}_n$ is unique, $\widehat{\theta}_n < \theta_F - \epsilon$ implies $\widehat{\lambda}_n(\theta_F - \epsilon) < 0$. Since $\widehat{\lambda}_n(\theta_F - \epsilon)$ is the average of the i.i.d. variables $\Psi(x_i, \theta_F - \epsilon)$, and has expectation $\lambda(\theta_F - \epsilon)$ by (10.5), the law of large numbers implies that

$$\widehat{\lambda}_n(\theta_F - \epsilon) \to_p \lambda(\theta_F - \epsilon) > 0.$$

Hence

$$\lim_{n\to\infty} \text{P}(\widehat{\theta}_n < \theta_F - \epsilon) \le \lim_{n\to\infty} \text{P}(\widehat{\lambda}_n(\theta_F - \epsilon) < 0) = 0.$$

The same method proves that $\text{P}(\widehat{\theta}_n > \theta_F + \epsilon) \to 0$. ■

Example 10.3 *For location $\Psi(x,\theta) = \psi(x-\theta)$. If $\psi(x) = x$, then Ψ is continuous and decreasing, the solution is $\widehat{\theta}_n = \overline{x}$ and $\lambda(\theta) = \text{E}x - \theta$, so that $\theta_F = \text{E}x$;*

convergence occurs only if Ex *exists. If* $\psi(x) = \operatorname{sgn}(x)$, *we have*

$$\lambda(\theta) = \mathrm{P}(x > \theta) - \mathrm{P}(x < \theta);$$

hence θ_F *is a median of* F, *which is unique iff*

$$F(\theta_F + \epsilon) > F(\theta_F - \epsilon)\ \forall\, \epsilon > 0. \tag{10.7}$$

In this case, for $n = 2m$ *the interval* $(x_{(m)}, x_{(m+1)})$ *shrinks to a single point when* $m \to \infty$. *If (10.7) does not hold, the distribution of* $\widehat{\theta}_n$ *does not converge to a point-mass (Problem 10.2).*

Note that for model (2.1), if ψ is odd and $\mathcal{D}(u)$ is symmetric about zero, then $\lambda(\theta) = 0$ so that $\theta_F = \theta$.

For scale, Theorem 10.5 implies that estimates of the form (2.54) tend to the solution of (2.55) if it is unique.

10.3 Asymptotic normality

In Section 2.9.2, the asymptotic normality of M-estimates was proved heuristically, by replacing ψ with its first-order Taylor expansion. This procedure will now be made rigorous.

If the distribution of z_n tends to the distribution H of z, we shall say that z_n *tends in distribution* to z (or to H), and shall denote this by $z_n \to_d z$ (or $z_n \to_d H$). We shall need an auxiliary result.

Theorem 10.6 (Bounded convergence theorem) *Let* y_n *be a sequence of random variables such that* $|y_n| \le z$ *where* E$z < \infty$ *and* $y_n \to y$ *a.s. Then* E$y_n \to$ Ey.

The proof can be found in Feller (1971).

Theorem 10.7 *Assume that* $A = \mathrm{E}\Psi(x, \theta_F)^2 < \infty$ *and that* $B = \lambda'(\theta_F)$ *exists and is nonnull. Let* $\widehat{\theta}_n$ *be a solution of (10.1) such that* $\widehat{\theta}_n \to_p \theta_F$. *Then the distribution of* $\sqrt{n}\,(\widehat{\theta}_n - \theta_F)$ *tends to* N$(0, v)$ *with*

$$v = \frac{A}{B^2}.$$

If $\dot{\Psi}(x, \theta) = \partial\Psi/\partial\theta$ *exists and verifies for all* x, θ

$$\left|\dot{\Psi}\,(x, \theta)\right| \le K(x)\ \ with\ \ \mathrm{E}K(x) < \infty, \tag{10.8}$$

then $B = \mathrm{E}\dot{\Psi}(x, \theta_F)$.

Proof: To make things simpler, we shall make the extra (and unnecessary) assumptions that $\ddot{\Psi}(x, \theta) = \partial^2\Psi/\partial\theta^2$ exists and is bounded, and that $\dot{\Psi}$ verifies (10.8). A completely general proof may be found in Huber (1981, Section 3.2). Note first that the bounded convergence theorem implies $B = \mathrm{E}\dot{\Psi}(x, \theta_F)$. In fact,

$$B = \lim_{\delta \to 0} \mathrm{E}\frac{\Psi(x, \theta_F + \delta) - \Psi(x, \theta_F)}{\delta}.$$

The term in the expectation is $\leq K(x)$ by the mean value theorem, and for each x tends to $\dot{\Psi}(x, \theta_F)$.

A second-order Taylor expansion of Ψ at θ_F yields

$$\Psi(x_i, \widehat{\theta}_n) = \Psi(x_i, \theta_F) + (\widehat{\theta}_n - \theta_F)\dot{\Psi}(x_i, \theta_F) + \frac{1}{2}(\widehat{\theta}_n - \theta_F)^2\ddot{\Psi}(x_i, \theta_i)$$

where θ_i is some value (depending on x_i) between $\widehat{\theta}_n$ and θ_F, and where $\ddot{\Psi} = \partial^2\Psi(x, \theta)/\partial\theta^2$. Averaging over i yields

$$0 = A_n + (\widehat{\theta}_n - \theta_F)B_n + (\widehat{\theta}_n - \theta_F)^2 C_n,$$

where

$$A_n = \frac{1}{n}\sum_{i=1}^{n}\Psi(x_i, \theta_F),\ B_n = \frac{1}{n}\sum_{i=1}^{n}\dot{\Psi}(x_i, \theta_F),\ C_n = \frac{1}{2n}\sum_{i=1}^{n}\ddot{\Psi}(x_i, \theta_i)$$

and hence

$$\sqrt{n}(\widehat{\theta}_n - \theta_F) = -\frac{\sqrt{n}A_n}{B_n + (\widehat{\theta}_n - \theta_F)C_n}.$$

Since the i.i.d. variables $\Psi(x_i, \theta_F)$ have mean 0 (by (10.6)) and variance A, the central limit theorem implies that the numerator tends in distribution to $\mathrm{N}(0, A)$. The law of large numbers implies that $B_n \to_p B$; and since C_n is bounded and $(\widehat{\theta}_n - \theta_F) \to_p 0$, Slutsky's lemma (Section 2.9.3) yields the desired result. ∎

Example 10.4 (location) *For the mean, the existence of A requires that of* $\mathrm{E}x^2$. *In general, if* ψ *is bounded, A always exists. If* ψ' *exists, then* $\lambda'(t) = -\mathrm{E}\,\psi'(x - t)$. *For the median,* ψ *is discontinuous, but if F has a density f, explicit calculation yields*

$$\lambda(\theta) = \mathrm{P}(x > \theta) - \mathrm{P}(x < \theta) = 1 - 2F(\theta),$$

and hence $\lambda'(\theta_F) = -2f(\theta_F)$.

If $\lambda'(\theta_F)$ does not exist, $\widehat{\theta}_n$ tends to θ_F faster than $n^{-1/2}$, and there is no asymptotic normality. Consider for instance the median with F discontinuous. Let $\psi(x) = \mathrm{sgn}\,(x)$, and assume that F is continuous except at zero, where it has its median and a point mass with $\mathrm{P}\,(x = 0) = 2\delta$, i.e.,

$$\lim_{x \uparrow 0} F(x) = 0.5 - \delta,\ \lim_{x \downarrow 0} F(x) = 0.5 + \delta.$$

Then $\lambda(\theta) = 1 - 2F(\theta)$ has a jump at $\theta_F = 0$. We shall see that this entails $\mathrm{P}(\widehat{\theta}_n = 0) \to 1$, and *a fortiori* $\sqrt{n}\,\widehat{\theta}_n \to_p 0$.

Let $N_n = \#(x_i < 0)$, which is binomial $\mathrm{Bi}(n, p)$ with $p = 0.5 - \delta$. Then $\widehat{\theta}_n < 0$ implies $N_n > n/2$, and therefore

$$\mathrm{P}(\widehat{\theta}_n < 0) \leq \mathrm{P}(N_n/n > 0.5) \to 0$$

since the law of large numbers implies $N_n/n \to_p p < 0.5$. The same method yields $\mathrm{P}(\widehat{\theta}_n > 0) \to 0$.

The fact that the distribution of $\widehat{\theta}_n$ tends to a normal $\mathrm{N}(\theta_F, v)$ does *not* imply that the mean and variance of $\widehat{\theta}_n$ tend to θ_F and v (Problem 10.3). In fact, if F is heavy tailed, the distribution of $\widehat{\theta}_n$ will also be heavy tailed, with the consequence that its moments may not exist, or, if they do, they will give misleading information about $\mathcal{D}(\widehat{\theta}_n)$. In extreme cases, they may even not exist for any n. This shows that, as an evaluation criterion, the *asymptotic* variance may be better than the variance. Let $T_n = (\widehat{\theta}_n - \widehat{\theta}_\infty)\sqrt{n/v}$ where $\widehat{\theta}_n$ is the median and v its asymptotic variance under F, so that T_n should be approximately N(0, 1). Figure 10.1 shows for the Cauchy distribution the normal Q–Q plot of T_n, i.e., the comparison between the exact and the approximate quantiles of its distribution, for $n = 5$ and 11. It is seen that although the approximation improves in the middle when n increases, the tails remain heavy.

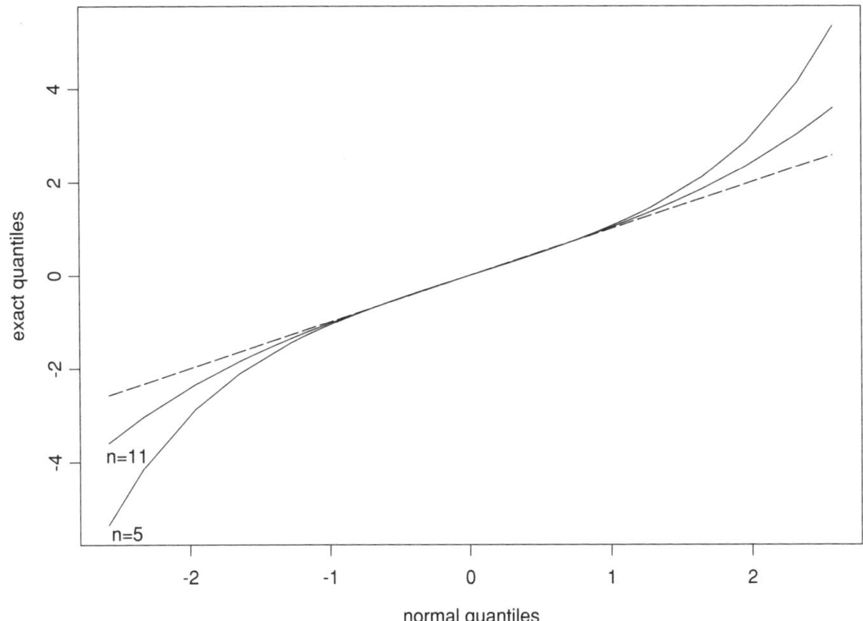

Figure 10.1 Q–Q plot of the sample median for Cauchy data. The dashed line is the identity diagonal

10.4 Convergence of the SC to the IF

In this section we prove (3.5) for general M-estimates (10.1). Call $\widehat{\theta}_n$ the solution of (10.1), and for a given x_0, call $\widehat{\theta}_{n+1}(x_0)$ the solution of

$$\sum_{i=0}^{n} \Psi(x_i, \theta) = 0. \tag{10.9}$$

The sensitivity curve is

$$\mathrm{SC}_n(x_0) = (n+1)\left(\widehat{\theta}_{n+1}(x_0) - \widehat{\theta}_n\right),$$

and

$$\mathrm{IF}_{\widehat{\theta}}(x_0) = -\frac{\Psi(x_0, \theta_F)}{B},$$

with B and θ_F defined in Theorem 10.7 and in (10.6) respectively.

Theorem 10.8 *Assume the same conditions as in Theorem 10.7. Then for each* x_0

$$\mathrm{SC}_n(x_0) \to_p \mathrm{IF}_{\widehat{\theta}}(x_0).$$

Proof: Theorem 10.5 states that $\widehat{\theta}_n \to_p \theta_F$. The same proof shows that also $\widehat{\theta}_{n+1}(x_0) \to_p \theta_F$, since the effect of the term $\Psi(x_0, \theta)$ becomes negligible for large n. Hence

$$\Delta_n =: \widehat{\theta}_{n+1}(x_0) - \widehat{\theta}_n \to_p 0.$$

Using (10.1) and (10.9) and a Taylor expansion yields

$$\begin{aligned} 0 &= \Psi(x_0, \widehat{\theta}_{n+1}(x_0)) + \sum_{i=1}^{n} \left[\Psi(x_i, \widehat{\theta}_{n+1}(x_0)) - \Psi(x_i, \widehat{\theta}_n)\right] \\ &= \Psi(x_0, \widehat{\theta}_{n+1}(x_0)) + \Delta_n \sum_{i=1}^{n} \dot{\Psi}(x_i, \widehat{\theta}_n) + \frac{\Delta_n^2}{2} \sum_{i=1}^{n} \ddot{\Psi}(x_i, \theta_i), \end{aligned} \tag{10.10}$$

where θ_i is some value between $\widehat{\theta}_{n+1}(x_0)$ and $\widehat{\theta}_n$. Put

$$B_n = \frac{1}{n} \sum_{i=1}^{n} \dot{\Psi}(x_i, \widehat{\theta}_n), \quad C_n = \frac{1}{n} \sum_{i=1}^{n} \ddot{\Psi}(x_i, \theta_i).$$

Then C_n is bounded, and the consistency of $\widehat{\theta}_n$, plus a Taylor expansion, show that $B_n \to_p B$. It follows from (10.10) that

$$\mathrm{SC}_n(x_0) = -\frac{\Psi(x_0, \widehat{\theta}_{n+1}(x_0))}{B_n + C_n \Delta_n / 2} \frac{n+1}{n}.$$

And since $\Psi(x_0, \widehat{\theta}_{n+1}(x_0)) \to_p \Psi(x_0, \theta_F)$, the proof follows. ■

10.5 M-estimates of several parameters

We shall need the asymptotic distribution of M-estimates when there are several parameters. This happens in particular with the joint estimation of location and scale in Section 2.6.2, where we have two parameters, which satisfy a system of two equations. This situation also appears in regression (Chapter 4) and multivariate analysis (Chapter 6). Put $\boldsymbol{\theta} = (\mu, \sigma)$, and

$$\Psi_1(x, \boldsymbol{\theta}) = \psi\left(\frac{x-\mu}{\sigma}\right) \text{ and } \Psi_2(x, \boldsymbol{\theta}) = \rho_{\text{scale}}\left(\frac{x-\mu}{\sigma}\right) - \delta.$$

Then the simultaneous location–scale estimates satisfy

$$\sum_{i=1}^{n} \boldsymbol{\Psi}(x_i, \boldsymbol{\theta}) = \mathbf{0}, \qquad (10.11)$$

with $\boldsymbol{\Psi} = (\Psi_1, \Psi_2)$. Here the observations x_i are univariate, but in general they may belong to any set $\mathcal{X} \subset R^q$, and we consider a vector $\boldsymbol{\theta} = (\theta_1, \ldots, \theta_p)'$ of unknown parameters, which ranges in a subset $\Theta \subset R^p$, which satisfies (10.11) where $\Psi = (\Psi_1, \ldots, \Psi_p)$ is function of $\mathcal{X} \times \Theta \rightarrow R^p$. Existence of solutions must be dealt with in each situation. Uniqueness may be proved under conditions which generalize the monotonicity of Ψ in the case of a univariate parameter (as in (d) of Theorem 10.1).

Theorem 10.9 *Assume that for all x and $\boldsymbol{\theta}$, $\Psi(x, \boldsymbol{\theta})$ is differentiable and the matrix $\mathbf{D} = \mathbf{D}(x, \boldsymbol{\theta})$ with elements $\partial\Psi_i/\partial\theta_j$ is negative definite (i.e., $\mathbf{a}'\mathbf{D}\mathbf{a} < 0$ for all $\mathbf{a} \neq \mathbf{0}$). Put for given $x_1, \ldots, x_n$*

$$g(\boldsymbol{\theta}) = \sum_{i=1}^{n} \boldsymbol{\Psi}(x_i, \boldsymbol{\theta}).$$

If there exists a solution of $g(\boldsymbol{\theta}) = \mathbf{0}$, then this solution is unique.

Proof: We shall prove that

$$g(\boldsymbol{\theta}_1) \neq g(\boldsymbol{\theta}_2) \text{ if } \boldsymbol{\theta}_1 \neq \boldsymbol{\theta}_2.$$

Let $\mathbf{a} = \boldsymbol{\theta}_2 - \boldsymbol{\theta}_1$, and define for $t \in R$ the function $h(t) = \mathbf{a}'g(\boldsymbol{\theta}_1 + t\mathbf{a})$, so that $h(0) = \mathbf{a}'g(\boldsymbol{\theta}_1)$ and $h(1) = \mathbf{a}'g(\boldsymbol{\theta}_2)$. Its derivative is

$$h'(t) = \sum_{i=1}^{n} \mathbf{a}'\mathbf{D}(x, \boldsymbol{\theta}_1 + t\mathbf{a})\mathbf{a} < 0 \; \forall t,$$

and hence $h(0) > h(1)$, which implies $g(\boldsymbol{\theta}_1) \neq g(\boldsymbol{\theta}_2)$. ∎

To treat consistency, assume the x_i's are i.i.d. with distribution F, and put

$$\widehat{\lambda}_n(\boldsymbol{\theta}) = \frac{1}{n} \sum_{i=1}^{n} \boldsymbol{\Psi}(x_i, \boldsymbol{\theta}) \qquad (10.12)$$

and

$$\boldsymbol{\lambda}(\boldsymbol{\theta}) = \mathrm{E}_F \boldsymbol{\Psi}(x, \boldsymbol{\theta}). \tag{10.13}$$

Let $\widehat{\boldsymbol{\theta}}_n$ be any solution of (10.11); it is natural to conjecture that, if there is a unique solution θ_F of $\boldsymbol{\lambda}(\boldsymbol{\theta}) = \mathbf{0}$, then as $n \to \infty$, $\widehat{\boldsymbol{\theta}}_n$ tends in probability to θ_F. General criteria are given in Huber (1981, Section 6.2). However, their application to each situation must be dealt with separately.

For asymptotic normality we can generalize Theorem 10.7. Assume that $\widehat{\boldsymbol{\theta}}_n \to_p \theta_F$ and that $\boldsymbol{\lambda}$ is differentiable at θ_F, and call $\mathbf{B}$ the matrix of derivatives with elements

$$B_{jk} = \left.\frac{\partial \lambda_j}{\partial \theta_k}\right|_{\theta=\theta_F}. \tag{10.14}$$

Assume $\mathbf{B}$ is nonsingular. Then under general assumptions (Huber, 1981, Section 6.3)

$$\sqrt{n}\left(\widehat{\boldsymbol{\theta}}_n - \boldsymbol{\theta}_F\right) \to_d \mathrm{N}_p\left(\mathbf{0}, \mathbf{B}^{-1}\mathbf{A}\mathbf{B}^{-1\prime}\right) \tag{10.15}$$

where

$$\mathbf{A} = \mathrm{E}\boldsymbol{\Psi}(x, \theta_F)\boldsymbol{\Psi}(x, \theta_F)', \tag{10.16}$$

and $\mathrm{N}_p(\mathbf{t}, \mathbf{V})$ denotes the p-variate normal distribution with mean $\mathbf{t}$ and covariance matrix $\mathbf{V}$.

If $\dot{\Psi}_{jk} = \partial \Psi_j / \partial \theta_k$ exists and verifies for all $x, \boldsymbol{\theta}$

$$\left|\dot{\Psi}_{jk}(x, \boldsymbol{\theta})\right| \leq K(x) \text{ with } \mathrm{E}K(x) < \infty, \tag{10.17}$$

then $\mathbf{B} = \mathrm{E}\dot{\Psi}(x, \theta_F)$, where $\dot{\Psi}$ is the matrix with elements $\dot{\Psi}_{jk}$.

The intuitive idea behind the result is like that of (2.86)–(2.87): we take a first-order Taylor expansion of Ψ around $\boldsymbol{\theta}_F$ and drop the higher-order terms. Before dealing with the proof of (10.15), let us see how it applies to simultaneous M-estimates of location–scale. Conditions for existence and uniqueness of solutions are given in Huber (1981, Section 6.4) and Maronna and Yohai (1981). They may hold without requiring monotonicity of ψ. This holds in particular for the Student MLE. As can be expected, under suitable conditions they tend in probability to the solution (μ_0, σ_0) of the system of equations (2.72)–(2.73). The joint distribution of $\sqrt{n}(\widehat{\mu} - \mu_0, \hat{\sigma} - \sigma_0)$ tends to the bivariate normal with mean 0 and covariance matrix

$$\mathbf{V} = \mathbf{B}^{-1}\mathbf{A}(\mathbf{B}^{-1})', \tag{10.18}$$

where

$$\mathbf{A} = \begin{bmatrix} a_{11} & a_{12} \\ a_{21} & a_{22} \end{bmatrix}, \ \mathbf{B} = \frac{1}{\sigma}\begin{bmatrix} b_{11} & b_{12} \\ b_{21} & b_{22} \end{bmatrix},$$

with

$$a_{11} = \mathrm{E}\psi(r)^2, \ a_{12} = a_{21} = \mathrm{E}(\rho_{\text{scale}}(r) - \delta)\psi(r), \ a_{22} = \mathrm{E}\left(\rho_{\text{scale}}(r) - \delta\right)^2,$$

where

$$r = \frac{x - \mu_0}{\sigma_0},$$

and

$$b_{11} = \mathrm{E}\psi'(r), \ b_{12} = \mathrm{E}r\psi'(r),$$

$$b_{21} = \mathrm{E}\rho'_{\text{scale}}(r), \ b_{22} = \mathrm{E}r\rho'_{\text{scale}}(r).$$

If ψ is odd, ρ_{scale} is even, and F is symmetric, the reader can verify (Problem 10.5) that $\mathbf{V}$ is diagonal,

$$\mathbf{V} = \begin{bmatrix} v_{11} & 0 \\ 0 & v_{22} \end{bmatrix},$$

so that $\widehat{\mu}$ and $\hat{\sigma}$ are asymptotically independent, and their variances take on a simple form:

$$v_{11} = \sigma_0^2 \frac{a_{11}}{a_{22}^2}, \ v_{22} = \sigma_0^2 \frac{b_{11}}{b_{22}^2};$$

that is, the asymptotic variance of each estimate is calculated as if the other parameter were constant.

We shall now prove (10.15) under much more restricted assumptions. We shall need an auxiliary result.

Theorem 10.10 ("Multivariate Slutsky's lemma") *Let $\mathbf{u}_n$ and $\mathbf{v}_n$ be two sequences of random vectors and $\mathbf{W}_n$ a sequence of random matrices such that for some constant vector $\mathbf{u}$, random vector $\mathbf{v}$ and random matrix $\mathbf{W}$*

$$\mathbf{u}_n \rightarrow_p \mathbf{u}, \ \mathbf{v}_n \rightarrow_d \mathbf{V}, \ \mathbf{W}_n \rightarrow_p \mathbf{W}.$$

Then

$$\mathbf{u}_n + \mathbf{v}_n \rightarrow_d \mathbf{u} + \mathbf{v} \text{ and } \mathbf{W}_n\mathbf{v}_n \rightarrow_d \mathbf{W}\mathbf{v}.$$

Now we proceed with the proof of asymptotic normality under more restricted assumptions. Let $\widehat{\boldsymbol{\theta}}_n$ be any solution of (10.11).

Theorem 10.11 *Assume that $\widehat{\boldsymbol{\theta}}_n \rightarrow_p \boldsymbol{\theta}_F$ where $\boldsymbol{\theta}_F$ is the unique solution of $\boldsymbol{\lambda}_F(\boldsymbol{\theta}) = \mathbf{0}$. Let $\boldsymbol{\Psi}$ be twice differentiable with respect to $\boldsymbol{\theta}$ with bounded derivatives, and satisfying also (10.17). Then (10.15) holds.*

Proof: The proof follows that of Theorem 10.7. For each j, call $\ddot{\boldsymbol{\Psi}}_j$ the matrix with elements $\partial\Psi_j/\partial\theta_k\partial\theta_l$, and $\mathbf{C}_n(x,\boldsymbol{\theta})$ the matrix with its j-th row equal to $\left(\widehat{\boldsymbol{\theta}}_n - \boldsymbol{\theta}_F\right)' \ddot{\Psi}_j(x,\boldsymbol{\theta})$. By a Taylor expansion

$$\mathbf{0} = \widehat{\boldsymbol{\lambda}}_n(\widehat{\boldsymbol{\theta}}_n) = \sum_{i=1}^{n} \left\{ \boldsymbol{\Psi}(x_i, \boldsymbol{\theta}_F) + \dot{\Psi}(x_i, \boldsymbol{\theta}_F)\left(\widehat{\boldsymbol{\theta}}_n - \boldsymbol{\theta}_F\right) + \frac{1}{2}\mathbf{C}_n(x_i, \boldsymbol{\theta}_i)\left(\widehat{\boldsymbol{\theta}}_n - \boldsymbol{\theta}_F\right) \right\}.$$

That is,

$$\mathbf{0} = \mathbf{A}_n + \left(\mathbf{B}_n + \overline{\mathbf{C}}_n\right)\left(\widehat{\boldsymbol{\theta}}_n - \boldsymbol{\theta}_F\right),$$

with

$$\mathbf{A}_n = \frac{1}{n}\sum_{i=1}^{n} \boldsymbol{\Psi}(x_i, \boldsymbol{\theta}_F),\ \mathbf{B}_n = \frac{1}{n}\sum_{i=1}^{n} \dot{\Psi}(x_i, \boldsymbol{\theta}_F),\ \overline{\mathbf{C}}_n = \frac{1}{2n}\sum_{i=1}^{n} \mathbf{C}_n(x_i, \boldsymbol{\theta}_i),$$

i.e., $\overline{\mathbf{C}}_n$ is the matrix with its j-th row equal to $\left(\widehat{\boldsymbol{\theta}}_n - \boldsymbol{\theta}_F\right)' \ddot{\Psi}_j^-$, where

$$\ddot{\Psi}_j^- = \frac{1}{n}\sum_{i=1}^{n} \ddot{\Psi}_j(x_i, \boldsymbol{\theta}_i),$$

which is bounded; since $\widehat{\boldsymbol{\theta}}_n - \boldsymbol{\theta}_F \to_p \mathbf{0}$, this implies that also $\overline{\mathbf{C}}_n \to_p \mathbf{0}$. We have

$$\sqrt{n}\left(\widehat{\boldsymbol{\theta}}_n - \boldsymbol{\theta}_F\right) = -(\mathbf{B}_n + \overline{\mathbf{C}}_n)^{-1}\sqrt{n}\mathbf{A}_n.$$

Note that for $i = 1, 2, \ldots,$ the vectors $\boldsymbol{\Psi}(x_i, \boldsymbol{\theta}_F)$ are i.i.d. with mean $\mathbf{0}$ (since $\boldsymbol{\lambda}(\boldsymbol{\theta}_F) = \mathbf{0}$) and covariance matrix $\mathbf{A}$, and the matrices $\dot{\Psi}(x_i, \boldsymbol{\theta}_F)$ are i.i.d. with mean $\mathbf{B}$. Hence when $n \to \infty$, the law of large numbers implies $\mathbf{B}_n \to_p \mathbf{B}$, which implies $\mathbf{B}_n + \overline{\mathbf{C}}_n \to_p \mathbf{B}$, which is nonsingular; and the multivariate central limit theorem implies $\sqrt{n}\mathbf{A}_n \to_d \mathrm{N}_p(\mathbf{0}, \mathbf{A})$; hence (10.15) follows by the multivariate version of Slutsky's lemma. ■

10.6 Location M-estimates with preliminary scale

We shall consider the asymptotic behavior of solutions of (2.65). For each n let $\widehat{\sigma}_n$ be a dispersion estimate, and call $\widehat{\mu}_n$ the solution (assumed unique) of

$$\sum_{i=1}^{n} \psi\left(\frac{x_i - \mu}{\hat{\sigma}_n}\right) = 0. \tag{10.19}$$

For consistency, it will be assumed that

A1 ψ is monotone and bounded with a bounded derivative
A2 $\sigma = \operatorname{p\,lim} \widehat{\sigma}_n$ exists
A3 the equation $\mathrm{E}\psi\left((x - \mu)/\sigma\right) = 0$ has a unique solution μ_0.

Theorem 10.12 *If A1–A2–A3 hold, then* $\widehat{\mu}_n \to_p \mu_0$.

The proof follows along the lines of Theorem 10.5, but the details require much more care, and are hence omitted.

Now define $u_i = x_i - \mu_0$ and

$$a = \mathrm{E}\psi\left(\frac{u}{\sigma}\right)^2,\ b = \mathrm{E}\psi'\left(\frac{u}{\sigma}\right),\ c = \mathrm{E}\left(\frac{u}{\sigma}\right)\psi'\left(\frac{u}{\sigma}\right). \tag{10.20}$$

For asymptotic normality, assume

A4 the quantities defined in (10.20) exist and $b \neq 0$
A5 $\sqrt{n}\,(\widehat{\sigma}_n - \sigma)$ converges to some distribution
A6 $c = 0$.

Theorem 10.13 *Under A4–A5–A6, we have*

$$\sqrt{n}(\widehat{\mu}_n - \mu_0) \to_d \mathrm{N}(0, v) \text{ with } v = \sigma^2 \frac{a}{b^2}. \tag{10.21}$$

Note that if x has a symmetric distribution, then μ_0 coincides with its center of symmetry, and hence the distribution of u is symmetric about zero, which implies (since ψ is odd) that $c = 0$.

Adding the assumption that ψ has a bounded second derivative, the theorem may be proved along the lines of Theorem 10.5, but the details are somewhat more involved. We shall content ourselves with a heuristic proof of (10.21) to exhibit the main ideas.

Put for brevity

$$\widehat{\Delta}_{1n} = \widehat{\mu}_n - \mu_0, \; \widehat{\Delta}_{2n} = \widehat{\sigma}_n - \sigma.$$

Then expanding ψ as in (2.86) yields

$$\psi\left(\frac{x_i - \widehat{\mu}_n}{\widehat{\sigma}_n}\right) = \psi\left(\frac{u_i - \widehat{\Delta}_{1n}}{\sigma + \widehat{\Delta}_{2n}}\right)$$
$$\approx \psi\left(\frac{u_i}{\sigma}\right) - \psi'\left(\frac{u_i}{\sigma}\right)\frac{\widehat{\Delta}_{1n} + \widehat{\Delta}_{2n} u_i/\sigma}{\sigma}.$$

Inserting the right-hand side of this expression in (2.65) and dividing by n yields

$$0 = A_n - \frac{1}{\sigma}\left(\widehat{\Delta}_{1n} B_n + \widehat{\Delta}_{2n} C_n\right),$$

where

$$A_n = \frac{1}{n}\sum_{i=1}^n \psi\left(\frac{u_i}{\sigma}\right), \; B_n = \frac{1}{n}\sum_{i=1}^n \psi'\left(\frac{u_i}{\sigma}\right), \; C_n = \frac{1}{n}\sum_{i=1}^n \left(\frac{u_i}{\sigma}\right)\psi'\left(\frac{u_i}{\sigma}\right),$$

and hence

$$\sqrt{n}\widehat{\Delta}_{1n} = \sigma \frac{\sqrt{n}A_n - \widehat{\Delta}_{2n}\sqrt{n}C_n}{B_n}. \tag{10.22}$$

Now A_n is the average of i.i.d. variables with mean 0 (by (10.19)) and variance a, and hence the central limit theorem implies that $\sqrt{n}A_n \to_d \mathrm{N}(0, a)$; the law of large numbers implies that $B_n \to_p b$. If $c = 0$, then $\sqrt{n}C_n$ tends to a normal by the central limit theorem, and since $\widehat{\Delta}_{2n} \to_p 0$ by hypothesis, Slutsky's lemma yields (10.21).

If $c \neq 0$, the term $\widehat{\Delta}_{2n}\sqrt{n}C_n$ does not tend to zero, and the asymptotic variance of $\widehat{\Delta}_{1n}$ will depend on that of $\widehat{\sigma}_n$ and also on the correlation between $\widehat{\sigma}_n$ and $\widehat{\mu}_n$.

10.7 Trimmed means

Although the numerical computing of trimmed means—and in general of L-estimates—is very simple, their asymptotic theory is much more complicated than that of M-estimates; even heuristic derivations are involved.

It is shown (see Huber, 1981, Section 3.3) that under suitable regularity conditions, $\widehat{\theta}_n$ converges in probability to

$$\theta_F = \frac{1}{1-2\alpha} \mathrm{E}_F x \mathrm{I}(k_1 \leq x \leq k_2), \tag{10.23}$$

where

$$k_1 = F^{-1}(\alpha), \ k_2 = F^{-1}(1-\alpha). \tag{10.24}$$

Let $F(x) = F_0(x-\mu)$ with F_0 symmetric about zero; then $\theta_F = \mu$ (Problem 10.4).

If F is as above, then $\sqrt{n}(\widehat{\theta} - \mu) \rightarrow_d \mathrm{N}(0, v)$ with

$$v = \frac{1}{(1-2\alpha)^2} \mathrm{E}_F \, \psi_k(x-\mu)^2, \tag{10.25}$$

where ψ_k is Huber's function with $k = F_0^{-1}(1-\alpha)$, so that the asymptotic variance coincides with that of an M-estimate.

10.8 Optimality of the MLE

It can be shown that the MLE is "optimal" in the sense of minimizing the asymptotic variance, in a general class of asymptotically normal estimates (Shao, 2003). Here its optimality will be shown within the class of M-estimates of the form (10.1).

The MLE is an M-estimate, which under the conditions of Theorem 10.5 is Fisher-consistent, i.e., verifies (3.31). In fact, assume $\dot{f}_\theta = \partial f / \partial \theta$ is bounded. Then differentiating

$$\int_{-\infty}^{\infty} f_\theta(x) dx = 1$$

with respect to θ yields

$$0 = \int_{-\infty}^{\infty} \dot{f}_\theta(x) dx = -\int_{-\infty}^{\infty} \Psi_0(x, \theta) f_\theta(x) dx \ \forall \theta, \tag{10.26}$$

so that (10.6) holds (the interchange of integral and derivative is justified by the bounded convergence theorem).

Under the conditions of Theorem 10.7, the MLE has asymptotic variance

$$v_0 = \frac{A_0}{B_0^2},$$

with

$$A_0 = \int_{-\infty}^{\infty} \Psi_0^2(x,\theta) f_\theta(x)dx, \ \ B_0 = \int_{-\infty}^{\infty} \dot{\Psi}_0(x,\theta) f_\theta(x)dx,$$

with $\dot{\Psi}_0 = \partial \Psi_0 / \partial \theta$. The quantity A_0 is called *Fisher information*.

Now consider another M-estimate of the form (10.1) which is Fisher-consistent for θ, i.e., such that

$$\int_{-\infty}^{\infty} \Psi(x,\theta) f_\theta(x)dx = 0 \ \forall \theta, \tag{10.27}$$

and has asymptotic variance

$$v = \frac{A}{B^2},$$

with

$$A = \int_{-\infty}^{\infty} \Psi^2(x,\theta) f_\theta(x)dx, \ \ B = \int_{-\infty}^{\infty} \dot{\Psi}(x,\theta) f_\theta(x)dx.$$

It will be shown that

$$v_0 \leq v. \tag{10.28}$$

We shall show first that $B_0 = A_0$, which implies

$$v_0 = \frac{1}{A_0}. \tag{10.29}$$

In fact, differentiating the last member of (10.26) with respect to θ yields

$$0 = B_0 + \int_{-\infty}^{\infty} \Psi_0(x,\theta) \frac{\dot{f}_\theta(x)}{f_\theta(x)} f_\theta(x)dx = B_0 - A_0.$$

By (10.29), (10.28) is equivalent to

$$B^2 \leq A_0 A. \tag{10.30}$$

Differentiating (10.27) with respect to θ yields

$$B - \int_{-\infty}^{\infty} \Psi(x,\theta)\Psi_0(x,\theta) f_\theta(x)dx = 0.$$

The Cauchy–Schwarz inequality yields

$$\left(\int_{-\infty}^{\infty} \Psi(x,\theta)\Psi_0(x,\theta) f_\theta(x)dx \right)^2 \leq \left(\int_{-\infty}^{\infty} \Psi(x,\theta)^2 f_\theta(x)dx \right) \times \left(\int_{-\infty}^{\infty} \Psi_0^2(x,\theta) f_\theta(x)dx \right),$$

which proves (10.30).

10.9 Regression M-estimates

10.9.1 Existence and uniqueness

From now on it will be assumed that $\mathbf{X}$ has full rank, and that $\widehat{\sigma}$ is fixed or estimated previously (i.e., it does not depend on $\widehat{\beta}$). We first establish the existence of solutions of (4.39) for monotone estimates.

Theorem 10.14 *Let $\rho(r)$ be a continuous nondecreasing unbounded function of $|r|$. Then there exists a solution of (4.39).*

Proof: Here $\widehat{\sigma}$ plays no role, so that we may put $\widehat{\sigma} = 1$. Since ρ is bounded from below, so is the function

$$R(\beta) = \sum_{i=1}^{n} \rho\left(y_i - \mathbf{x}_i'\beta\right). \tag{10.31}$$

Call L its infimum, i.e., the larger of its lower bounds. We must show the existence of β_0 such that $R(\beta_0) = L$. It will be shown first that $R(\beta)$ is bounded away from L if $\|\beta\|$ is large enough. Let

$$a = \min_{\|\beta\|=1} \max_{i=1,\ldots,n} \left|\mathbf{x}_i'\beta\right|.$$

Then $a > 0$, since otherwise there would exist $\beta \neq \mathbf{0}$ such that $\mathbf{x}_i'\beta = 0$ for all i, which contradicts the full rank property. Let $b_0 > 0$ be such that $\rho(b_0) > 2L$, and b such that $ba - \max_i |y_i| \geq b_0$. Then $\|\beta\| > b$ implies $\max_i \left|y_i - \mathbf{x}_i'\beta\right| \geq b_0$, and hence $R(\beta) > 2L$. Thus minimizing R for $\beta \in R^p$ is equivalent to minimizing it on the closed ball $\{\|\beta\| \leq b\}$. A well-known result of analysis states that a function which is continuous on a closed bounded set attains its minimum in it. Since R is continuous, the proof is complete. ∎

Now we deal with the uniqueness of monotone M-estimates. Again we may take $\widehat{\sigma} = 1$.

Theorem 10.15 *Assume ψ is nondecreasing. Put for given $(\mathbf{x}_i, y_i)$*

$$\mathbf{L}(\beta) = \sum_{i=1}^{n} \psi\left(\frac{y_i - \mathbf{x}_i'\beta}{\widehat{\sigma}}\right)\mathbf{x}_i.$$

Then (a) all solutions of $\mathbf{L}(\beta) = \mathbf{0}$ minimize $R(\beta)$ defined in (10.31) and (b) if furthermore ψ has a positive derivative, then $\mathbf{L}(\beta) = \mathbf{0}$ has a unique solution.

Proof: (a) The equivariance of the estimate implies that without loss of generality we may assume that $\mathbf{L}(\mathbf{0}) = \mathbf{0}$. For a given β let $H(t) = R(t\beta)$ with R defined in (10.31). We must show that $H(1) \geq H(0)$. Since $dH(t)/dt = \beta'\mathbf{L}(t\beta)$, we have

$$H(1) - H(0) = R(\beta) - R(\mathbf{0}) = \sum_{i=1}^{n} \int_0^1 \psi(t\mathbf{x}_i'\beta - y_i)(\mathbf{x}_i'\beta)dt.$$

If $\mathbf{x}_i'\beta > 0$ (resp. < 0), then for $t > 0$ $\psi(t\mathbf{x}_i'\beta - y_i)$ is greater (smaller) than $\psi(-y_i)$. Hence

$$\psi(t\mathbf{x}_i'\beta - y_i)(\mathbf{x}_i'\beta) \geq \psi(-y_i)(\mathbf{x}_i'\beta)$$

which implies that

$$R(\beta) - R(\mathbf{0}) \geq \sum_{i=1}^{n} \psi(-y_i)(\mathbf{x}_i'\beta) = \beta'\mathbf{L}(\mathbf{0}).$$

(b) The matrix of derivatives of $\mathbf{L}$ with respect to β is

$$\mathbf{D} = -\frac{1}{\widehat{\sigma}}\sum_{i=1}^{n} \psi'\left(\frac{y_i - \mathbf{x}_i'\beta}{\widehat{\sigma}}\right)\mathbf{x}_i\mathbf{x}_i',$$

which is negative definite; and the proof proceeds as that of Theorem 10.9. ■

The above results do not cover MM- or S-estimates, since they are not monotonic. As was shown for location in Theorem 10.2, uniqueness holds for the asymptotic value of the estimate under the model $y_i = \mathbf{x}_i'\beta + u_i$ if the u_i's have a symmetric unimodal distribution.

10.9.2 Asymptotic normality: fixed X

Now, to treat the asymptotic behavior of the estimate, we consider an infinite sequence $(\mathbf{x}_i, y_i)$ described by model (4.4). Call $\widehat{\beta}_n$ the estimate (4.40) and $\widehat{\sigma}_n$ the scale estimate. Call $\mathbf{X}_n$ the matrix with rows $\mathbf{x}_i'$ $(i = 1, \ldots, n)$, which is assumed to have full rank. Then

$$\mathbf{X}_n'\mathbf{X}_n = \sum_{i=1}^{n} \mathbf{x}_i\mathbf{x}_i'$$

is positive definite, and hence it has a "square root", i.e., a (nonunique) $p \times p$ matrix $\mathbf{R}_n$ such that

$$\mathbf{R}_n'\mathbf{R}_n = \mathbf{X}_n'\mathbf{X}_n. \tag{10.32}$$

Call λ_n the smallest eigenvalue of $\mathbf{X}_n'\mathbf{X}_n$, and define

$$h_{in} = \mathbf{x}_i'\left(\mathbf{X}_n\mathbf{X}_n'\right)^{-1}\mathbf{x}_i \tag{10.33}$$

and

$$M_n = \max\{h_{in} : i = 1, \ldots, n\}.$$

Define v as in (4.44) and $\mathbf{R}_n$ as in (10.32). Assume

B1 $\lim_{n\to\infty} \lambda_n = \infty$

B2 $\lim_{n\to\infty} M_n = 0.$

Then we have

Theorem 10.16 *Assume conditions A1–A2–A3 of Section 10.6. If B1 holds, then* $\widehat{\beta}_n \to_p \beta$. *If also B2 and A4–A5–A6 hold, then*

$$\mathbf{R}_n\left(\widehat{\beta}_n - \beta\right) \to_d \mathrm{N}_p(\mathbf{0}, v\mathbf{I}), \tag{10.34}$$

with v *given by (10.21).*

The proof in a more general setting can be found in Yohai and Maronna (1979). For large n, the left-hand side of (10.34) has an approximate $\mathrm{N}_p(\mathbf{0}, v\mathbf{I})$ distribution, and from this, (4.43) follows since $\mathbf{R}_n^{-1}\mathbf{R}_n^{-1'} = \left(\mathbf{X}_n'\mathbf{X}_n\right)^{-1}$.

When $p = 1$ (fitting a straight line through the origin), condition B1 means that $\sum_{i=1}^n x_i^2 \to \infty$, which prevents the x_i's from clustering around the origin; and condition B2 becomes

$$\lim_{n\to\infty} \frac{\max\left\{x_i^2 : i = 1, \ldots, n\right\}}{\sum_{i=1}^n x_i^2} = 0,$$

which means that none of the x_i^2's dominates the sum in the denominator, i.e., there are no leverage points.

Now we consider a model with an intercept, namely (4.4). Let

$$\underline{\overline{\mathbf{x}}}_n = \mathrm{ave}_i(\underline{\mathbf{x}}_i) \text{ and } \mathbf{C}_n = \sum_{i=1}^n \left(\underline{\mathbf{x}}_i - \underline{\overline{\mathbf{x}}}_n\right)\left(\underline{\mathbf{x}}_i - \underline{\overline{\mathbf{x}}}_n\right)'.$$

Let $\mathbf{T}_n$ be any square root of $\mathbf{C}_n$, i.e.,

$$\mathbf{T}_n'\mathbf{T}_n = \mathbf{C}_n.$$

Theorem 10.17 *Assume conditions A1, A2, A4, A5, B1 and B2, and* $E\psi(u_i/\sigma) = 0$. *Then*

$$\mathbf{T}_n\left(\widehat{\beta}_{n1} - \beta_1\right) \to_d \mathrm{N}_{p-1}(\mathbf{0}, v\mathbf{I}). \tag{10.35}$$

We shall give a heuristic proof for the case of a straight line, i.e.,

$$y_i = \beta_0 + \beta_1 x_i + u_i, \tag{10.36}$$

so that $\mathbf{x}_i = (1, x_i)$. Put

$$\overline{x}_n = \mathrm{ave}_i(x_i),\ \ \overset{*}{x}_{in} = x_i - \overline{x}_n,\ \ C_n = \sum_{i=1}^n \overset{*2}{x}_{in}.$$

Then condition B1 is equivalent to

$$C_n \to \infty \text{ and } \frac{\overline{x}_n^2}{C_n} \to 0. \tag{10.37}$$

The first condition prevents the x_i's from clustering around a point.

An auxiliary result will be needed. Let v_i, $i = 1, 2, \ldots$, be i.i.d. variables with a finite variance, and for each n let $a_{1n}, \ldots, a_{nn}$ be a set of constants. Let

$$V_n = \sum_{i=1}^{n} a_{in} v_i, \ \gamma_n = \mathrm{E} V_n, \ \tau_n^2 = \mathrm{Var}(V_n).$$

Then $W_n = (V_n - \gamma_n)/\tau_n$ has zero mean and unit variance, and the central limit theorem asserts that if for each n we have $a_{11} = \ldots = a_{nn}$, then $W_n \to_d \mathrm{N}(0, 1)$. It can be shown that this is still valid if the a_{in}'s are such that no term in V_n "dominates the sum", in the following sense:

Lemma 10.18 *If the a_{in}'s are such that*

$$\lim_{n\to\infty} \frac{\max\left\{a_{in}^2 : i = 1, \ldots, n\right\}}{\sum_{i=1}^{n} a_{in}^2} = 0 \tag{10.38}$$

then $W_n \to_d \mathrm{N}(0, 1)$.

This result is a consequence of the so-called Lindeberg theorem (Feller, 1971). To see the need for condition (10.38), consider the v_i's having a non-normal distribution G with unit variance. Take for each n: $a_{1n} = 1$ and $a_{in} = 0$ for $i > 1$. Then $V_n/\tau_n = v_1$ which has distribution G for all n, and hence does not tend to the normal.

To demonstrate Theorem 10.17 for the model (10.36), let

$$T_n = \sqrt{C_n}, \ \ z_{in} = \frac{x_{in}^*}{T_n},$$

so that

$$\sum_{i=1}^{n} z_{in} = 0, \ \sum_{i=1}^{n} z_{in}^2 = 1. \tag{10.39}$$

Then (10.33) becomes

$$h_{in} = \frac{1}{n} + z_{in}^2,$$

so that condition B2 is equivalent to

$$\max\left\{z_{in}^2 : i = 1, \ldots, n\right\} \to 0. \tag{10.40}$$

We have to show that for large n, $\widehat{\beta}_{1n}$ is approximately normal with variance v/C_n, i.e., that $T_n(\widehat{\beta}_{1n} - \beta_1) \to_p \mathrm{N}(0, v)$.

The estimating equations are

$$\sum_{i=1}^{n} \psi\left(\frac{r_i}{\widehat{\sigma}_n}\right) = 0, \tag{10.41}$$

$$\sum_{i=1}^{n} \psi\left(\frac{r_i}{\widehat{\sigma}_n}\right) x_i = 0, \tag{10.42}$$

with $r_i = y_i - (\widehat{\beta}_{0n} + \widehat{\beta}_{1n} x_i)$. Combining both equations yields

$$\sum_{i=1}^{n} \psi\left(\frac{r_i}{\widehat{\sigma}_n}\right) x_{in}^* = 0. \tag{10.43}$$

Put

$$\widehat{\Delta}_{1n} = \widehat{\beta}_{1n} - \beta_1, \ \widehat{\Delta}_{0n} = \widehat{\beta}_{0n} - \beta_0, \ \widehat{\Delta}_{2n} = \widehat{\sigma}_n - \sigma.$$

The Taylor expansion of ψ at t is

$$\psi(t + \varepsilon) = \psi(t) + \varepsilon\psi'(t) + o(\varepsilon), \tag{10.44}$$

where the last term is a higher-order infinitesimal. Writing

$$r_i = u_i - \left(\widehat{\Delta}_{0n} + \widehat{\Delta}_{1n}\left(x_{in}^* + \overline{x}_n\right)\right),$$

expanding ψ at u_i/σ and dropping the last term in (10.44) yields

$$\psi\left(\frac{r_i}{\widehat{\sigma}_n}\right) = \psi\left(\frac{u_i - \left(\widehat{\Delta}_{0n} + \widehat{\Delta}_{1n}\left(x_{in}^* + \overline{x}_n\right)\right)}{\sigma + \widehat{\Delta}_{2n}}\right)$$

$$\approx \psi\left(\frac{u_i}{\sigma}\right) - \psi'\left(\frac{u_i}{\sigma}\right)\frac{\widehat{\Delta}_{0n} + \widehat{\Delta}_{1n}\left(x_{in}^* + \overline{x}_n\right) + \widehat{\Delta}_{2n}u_i/\sigma}{\sigma}. \tag{10.45}$$

Inserting (10.45) in (10.43), multiplying by σ and dividing by T_n, and recalling (10.39), yields

$$\sigma A_n = \left(T_n\widehat{\Delta}_{1n}\right)\left(B_n + C_n\frac{\overline{x}_n}{T_n}\right) + \widehat{\Delta}_{0n}C_n + \widehat{\Delta}_{2n}D_n, \tag{10.46}$$

where

$$A_n = \sum_{i=1}^{n} \psi\left(\frac{u_i}{\sigma}\right) z_{in}, \ B_n = \sum_{i=1}^{n} \psi'\left(\frac{u_i}{\sigma}\right) z_{in}^2,$$

$$C_n = \sum_{i=1}^{n} \psi'\left(\frac{u_i}{\sigma}\right) z_{in}, \ D_n = \sum_{i=1}^{n} \psi'\left(\frac{u_i}{\sigma}\right)\frac{u_i}{\sigma} z_{in}.$$

Put

$$a = \mathrm{E}\psi\left(\frac{u}{\sigma}\right)^2, \ b = \mathrm{E}\psi'\left(\frac{u}{\sigma}\right), \ e = \mathrm{Var}\left(\psi'\left(\frac{u}{\sigma}\right)\right).$$

Applying Lemma 10.18 to A_n (with $v_i = \psi(u_i/\sigma)$) and recalling (10.39) and (10.40) yields $A_n \to_d \mathrm{N}(0, a)$. The same procedure shows that C_n and D_n have normal limit distributions. Applying (10.39) to B_n yields

$$\mathrm{E}B_n = b, \ \mathrm{Var}(B_n) = e\sum_{i=1}^{n} z_{in}^4.$$

Now by (10.39) and (10.40)

$$\sum_{i=1}^{n} z_{in}^4 \leq \max_{1\leq i\leq n} z_{in}^2 \to 0,$$

and then Tchebychev's inequality implies that $B_n \to_p b$. Recall that $\widehat{\Delta}_{2n} \to_p 0$ by hypothesis, and $\widehat{\Delta}_{0n} \to_p 0$ by Theorem 10.16. Since $\overline{x}_n/T_n \to 0$ by (10.37), application of Slutsky's lemma to (10.46) yields the desired result.

The asymptotic variance of $\widehat{\beta}_0$ may be derived by inserting (10.45) in (10.41). The situation is similar to that of Section 10.6: if $c = 0$ (with c defined in (10.20)) the proof can proceed, otherwise the asymptotic variance of $\widehat{\beta}_{0n}$ depends on that of $\widehat{\sigma}_n$.

10.9.3 Asymptotic normality: random X

Since the observations $\mathbf{z}_i = (\mathbf{x}_i, y_i)$ are i.i.d., this situation can be treated with the methods of Section 10.5. A regression M-estimate is a solution of

$$\sum_{i=1}^{n} \Psi(\mathbf{z}_i, \boldsymbol{\beta}) = \mathbf{0}$$

with

$$\Psi(\mathbf{z}, \boldsymbol{\beta}) = \mathbf{x}\psi\left(y - \mathbf{x}'\boldsymbol{\beta}\right).$$

We shall prove (5.14) for the case of σ known and equal to one. It follows from (10.15) that the asymptotic covariance matrix of $\widehat{\boldsymbol{\beta}}$ is $\nu \mathbf{V}_{\mathbf{x}}^{-1}$ with ν given by (5.15) with $\sigma = 1$, and $\mathbf{V}_{\mathbf{x}} = \mathrm{E}\mathbf{x}\mathbf{x}'$. In fact, the matrices $\mathbf{A}$ and $\mathbf{B}$ in (10.16)-(10.14) are

$$\mathbf{A} = \mathrm{E}\psi(y - \mathbf{x}'\boldsymbol{\beta})^2\mathbf{x}\mathbf{x}', \ \mathbf{B} = -\mathrm{E}\psi'(y - \mathbf{x}'\boldsymbol{\beta})\mathbf{x}\mathbf{x}',$$

and their existence is ensured by assuming that ψ and ψ' are bounded, and that $\mathrm{E}\,||\mathbf{x}||^2 < \infty$. Under the model (5.1)-(5.2) we have

$$\mathbf{A} = \mathrm{E}\psi(u)^2\mathbf{V}_{\mathbf{x}}, \mathbf{B} = -\mathrm{E}\psi'(u)\mathbf{V}_{\mathbf{x}},$$

and the result follows immediately

10.10 Nonexistence of moments of the sample median

We shall show that there are extremely heavy-tailed distributions for which the sample median has no finite moments of any order.

Let the sample $\{x_1, \ldots, x_n\}$ have a continuous distribution function F and an odd sample size $n = 2m + 1$. Then its median $\widehat{\theta}_n$ has distribution function G such that

$$\mathrm{P}\left(\widehat{\theta}_n > t\right) = 1 - G(t) = \sum_{j=0}^{m} \binom{n}{j} F(t)^j \left(1 - F(t)\right)^{n-j}. \tag{10.47}$$

In fact, let $N = \#(x_i < t)$ which is binomial $\mathrm{Bi}(n, F(t))$. Then $\widehat{\theta}_n > t$ iff $N \leq m$, which yields (10.47).

It is easy to show using integration by parts that if T is a nonnegative variable with distribution function G, then

$$\mathrm{E}\left(T^k\right) = k\int_0^\infty t^{k-1}\left(1 - G(t)\right)dt. \tag{10.48}$$

Now let

$$F(x) = \left(1 - \frac{1}{\log x}\right)\mathrm{I}(x \geq e).$$

Since for all positive r and s

$$\lim_{t\to\infty}\frac{t^r}{\log t^s} = \infty,$$

it follows from (10.48) that $\mathrm{E}\widehat{\theta}_n^k = \infty$ for all positive k.

10.11 Problems

10.1. Let $x_1, \ldots, x_n$ be i.i.d. with continuous distribution function F. Show that the distribution function of the order statistic $x_{(m)}$ is

$$G(t) = \sum_{k=m}^{n}\binom{n}{k}F(t)^k\left(1 - F(t)\right)^{n-k}$$

[hint: for each t, the variable $N_t = \#\{x_i \leq t\}$ is binomial and verifies $x_{(m)} \leq t \Leftrightarrow N_t \geq m$].

10.2. Let F be such that $F(a) = F(b) = 0.5$ for some $a < b$. If $x_1, \ldots, x_{2m-1}$ are i.i.d. with distribution F, show that the distribution of $x_{(m)}$ tends to the average of the point masses at a and b.

10.3. Let $F_n = (1 - n^{-1})\mathrm{N}(0, 1) + n^{-1}\delta_{n^2}$ where δ_x is the point mass at x. Verify that $F_n \to \mathrm{N}(0, 1)$, but its mean and variance tend to infinity.

10.4. Verify that if x is symmetric about μ, then (10.23) is equal to μ.

10.5. Verify that if ψ is odd, ρ is even, and F is symmetric, then $\mathbf{V}$ in (10.18) is diagonal; and compute the asymptotic variances of $\widehat{\mu}$ and $\widehat{\sigma}$.

10.6. Show that scale M-estimates are uniquely defined if $\rho(x)$ is increasing for all x such that $\rho(x) < \rho(\infty)$ [to make things easier, assume ρ is differentiable].

10.7. Show that the location estimate with Huber's ψ_k and previous dispersion $\widehat{\sigma}$ is uniquely defined unless there exists a solution $\widehat{\mu}$ of $\sum_{i=1}^n \psi_k\left((x_i - \widehat{\mu})/\widehat{\sigma}\right) = 0$ such that $|x_i - \widehat{\mu}| > k\widehat{\sigma}$ for all i.

11

Robust Methods in S-PLUS

In this chapter we describe the implementation in S-PLUS of the robust methods considered in this book, and the steps needed for the reader to reproduce the examples therein. Further material on this subject can be found in Marazzi (1993) and in the user's guide of the S-PLUS robust library.

11.1 Location M-estimates: function *Mestimate*

Location M-estimates can be computed with the function *Mlocation* supplied by the authors.

The location M-estimate uses the MAD as scale. It is computed using the iterative weighted means algorithm described in Section 2.7.

The call to this function is

location = Mestimate(x,fun = 2, cons = "NULL", err0 = .0001)

where

x is the name of the variable.

fun selects the ψ-function: *fun = 1* or *fun = 2* selects the Huber or bisquare functions, respectively.

cons Tuning constant for the ψ-function. The default values are *cons = 1.345* if *fun = 1* and *cons = 4.685* if *fun = 2*. These values correspond to an asymptotic efficiency of 0.95 for normal samples.

err0 The algorithm stops when the relative difference between two consecutive weighted means is smaller than *err0*.

The components of the output are

location value of the location estimate

Robust Statistics – Theory and Methods Ricardo A. Maronna, R. Douglas Martin and Víctor J. Yohai

scale value of the scale estimate
sdevloc standard error of the location estimate.

11.2 Robust regression

11.2.1 A general function for robust regression: *lmRob*

The main S-PLUS function to compute a robust regression is *lmRob,* in library *robust.* We shall describe only the main features of this function but shall not explore all of its possibilities and options. For a more complete description the reader can use the HELP feature of the robust library or the manual that can be found in the directory of the robust library.

The function *lmRob* computes an initial and a final estimate. There are two options for the final estimate: an MM-estimate described in Section 5.5 or the adaptive Gervini–Yohai estimate mentioned in Section 5.6.3. In the first case the final estimate is a redescending M-estimate that uses an M-scale of the residuals of the initial estimate. This M-estimate is computed using the reweighted least-squares algorithm starting with the initial estimate. In case that the final estimate is the adaptive one, the initial estimate is used to compute estimates of the error scale and of the error distribution.

The initial estimate depends on the number and type of explanatory variables used in the regression. There are four cases:

1. If all explanatory variables are quantitative and their number does not exceed 15, then the initial estimate is an S-estimate.
2. If all explanatory variables are qualitative, then the initial estimate is an M-estimate with Huber's ψ.
3. If there are both quantitative and qualitative variables and their total number does not exceed 15, then the initial estimate is the alternating SM procedure described in Section 5.15.
4. If the total number of explanatory variables exceeds 15 and at least one is quantitative, then the initial estimate is obtained using Peña and Yohai's fast procedure mentioned in Section 4.3.

We now describe the use of the function *lmRob*. Before using this function it is necessary to load the *robust* library using the command

library(robust, first = T)

Then *lmRob* can be called as

robust.reg = lmRob(formula, data, weights, subset, na.action, robust.control)

where

robust.reg is the name of the object with the regression output (we can use any name for this object).

formula is an object that indicates the variables used in the regression. A *formula* object has the form *V1* $\sim$ *V2* + *V3* + ... + *VP*, where *V1* is the name of the dependent variable, and *V2*, ..., *VP* are the names of the explanatory variables.

data This argument is used only if the variables belong to a dataframe, in which case *data* is the name of the dataframe.

weights The vector of observation weights; if supplied, the algorithm minimizes the sum of a function of the square root of the weights multiplied into the residuals. For example, in the case of heteroskedastic errors, the weights should be inversely proportional to the variances of the residuals. The default value of weights is a vector of ones.

subset is an expression indicating the observations that we want to use in the regression. For example, *subset* = *(V1 > 0)* means that only those observations with V1 > 0 are used in the regression. The default value is the set of all observations.

na.action indicates which action should be taken when the data have missing observations. The default value is *na.action* = *na.fail*, which gives an error message if any missing values are found. The other possible value is *na.action* = *na.exclude*, which deletes all observations with at least one missing variable.

robust.control specifies some optional parameters of the regression. Only four of these options will be considered here. The options not specified in this parameter take the default values. If we omit *robust.control* in the call, all options are taken equal to the default variables.

The options in *robust.control* considered here are: *estim*, *weight*, *efficiency* and *final.alg*, where

- *estim* determines the type of estimate to be computed. If *estim* = *"Initial"*, only the initial estimate is computed; if *estim* = *"Final"*, then the final estimate is returned. The default is *"Final"*.
- *weight* indicates the family of ρ-functions used for the initial S-estimate and the final M-estimate when computing an MM-estimate. There are two possibilities: *"Bisquare"* and *"Optimal"*. The default option is that both ρ-functions belong to the optimal family described in Section 5.9.1. To indicate that we want both families to be the bisquare we have to make *weight* = *c("Bisquare", "Bisquare")*. It is recommended to use the same family for both estimates. Observe that this option *weight* is not related to the input *weights* described above.
- *efficiency* indicates the relative efficiency of the final estimate under normal errors and no outliers. The default value is *efficiency* = *0.90*. As we explained in Section 5.9, there is a trade-off between efficiency and robustness: when we increase the efficiency, the robustness of the estimate decreases. We recommend *efficiency* = *0.85*.
- *final.alg* defines the final estimate. It can take two values: *"MM"* and *"Adaptive"*, and the default is *"MM"*. When *final.alg* = *"Adaptive"* the final estimate is the Gervini–Yohai one mentioned in Section 5.6.3.

The optional parameters are defined using the function *lmRob.robust.control*. Suppose that we want to save the options in a variable called *control*. Then, to compute as final estimate an MM-estimate with efficiency 0.85 and initial and final estimates computed using the bisquare function, the command is

control = lmRob.robust.control(weight = c("Bisquare", "Bisquare"), efficiency = 0.85)

Then when calling *lmRob* we make *robust.control = control.*

Observe that when defining *control* the parameters *estimate* and *final.alg* are omitted because we are using the default values.

The main components of the output are

coefficients is the vector of coefficients for the robust regression, which are final or initial estimates according to *estim = "final"* or *estim = "initial"*.
scale is the scale estimate computed using the initial estimates.
residuals The residual vector corresponding to the estimates returned in *coefficients.*
fitted.values The vector of fitted values corresponding to the estimates returned in *coefficients.*
cov Estimated covariance matrix of the estimates as in Section 5.8.
r.squared Fraction of variation in y explained by the robust regression on $\mathbf{x}$ corresponding to the final MM-estimates in coefficients, if applicable (see Section 5.16.8).
test for bias Test for the null hypothesis that the sample does not contain outliers. There are two tests: the first, proposed by Yohai et al. (1991), rejects the null hypothesis when the differences between the initial S-estimate and the final M-estimate are significant. When the result of this test is significant it is advisable to recompute the MM-estimate using a lower efficiency. The second test compares the LS estimate with the MM-estimate. Since its power is very low we recommend ignoring this test.

The results saved on *robust.reg* are displayed with the command

summary (robust.reg)

Example

Consider the data in Example 4.1. The model is fitted with the commands

Cont = lmRob.robust.control(weight = c("Bisquare", "Bisquare"), efficiency = 0.85)
ratsrob = lmRob(formula = V2~V1, data = rats, robust.control = Cont)

The first command defines the object *Cont*, where the optional parameters are saved. Only the parameters *weight* and *efficiency* take values different from default.

The results are displayed with the command

summary (ratsrob)

and the resulting output is

```
Call: lmRob(formula = V2 ~V1, data = rats, robust.control = Cont)

Residuals:
 Min        1Q        Median   3Q      Max
 -1.874     -0.1678   0.2469   1.622   7.593

Coefficients:
               Value    Std Error  t value    Pr(>|t|)
(Intercept)    7.6312   0.4683     16.7225    0.0000
V1            -0.4082   0.0498     -8.1911    0.0000
Residual standard error: 0.9408 on 14 degrees of freedom

Multiple R-Squared: 0.4338

Correlation of Coefficients:
               (Intercept)     V1
(Intercept)    1.0000
V1             -0.9123         1.0000

Test for Bias:
Statistics P-value
M-estimate 0.121 0.941
LS-estimate 0.912 0.634
```

11.2.2 Categorical variables: functions *as.factor* and *contrasts*

lmRob recognizes that a variable is categorical if it is defined as a factor. To define a variable as a factor, the function *as.factor* is used. For example, to define the variable X as a factor, we use the command

X = as.factor(X)

If initially X is numeric, after the command *as.factor* is applied, each of the numeric values is transformed to a label. To use as an explanatory variable a variable X which is a factor with k different levels $a_1, \ldots, a_k$, is equivalent to using the k numeric explanatory variables $Z_1, \ldots, Z_k$, where Z_i is one or zero according to $Z_i = a_i$ or $Z_i \neq a_i$ respectively. Since $\sum_{i=1}^k Z_i = 1$, the corresponding coefficients $\beta_1, \ldots, \beta_k$ are not identifiable. To make them identifiable it is necessary to add a restriction of the form $\sum_{i=1}^k c_i \beta_i = 0$. S-PLUS has several ways of choosing these coefficients. The simplest is the one that takes $c_1 = 1$, and $c_i = 0$ for $i > 1$, i.e., the one that makes $\beta_1 = 0$. This is achieved with the command

contrasts(X) = contr.treatment(k)

Example

Consider the dataframe *scheffem* used in Example 4.2, which contains three variables: *yield*, *variety* and *block*. The first one is qualitative and the two others are quantitative.

Since there are eight varieties and five blocks, the variable *variety* takes on the values 1 to 8 and the variable *block* takes 1 to 5. To interpret these values as levels we need the commands

scheffem$variety = as.factor(scheffem$variety)
scheffem$block = as.factor(scheffem$block)

The required contrasts are declared with

contrasts(scheffem$variety) = contr.treatment(8)
contrasts(scheffem$block) = contr.treatment(5)

To fit an analysis of variance model using an MM-estimate with bisquare ρ-function and efficiency 0.85, we use

scheffemrob = lmRob(yield˜variety + block, data = scheffem, robust.control = Cont)

(The object *Cont* was defined above.)
The results are displayed with the command

summary(scheffemrob)

and the resulting output is

```
Coefficients:
                Value      Std Error  t value    Pr(>|t|)
(Intercept)   394.5172     27.9493    14.1155    0.0000
variety2       12.9688     30.3800     0.4269    0.6727
variety3       -2.5049     27.6948    -0.0904    0.9286
variety4      -73.6916     28.7255     2.5654    0.0160
variety5       50.3809     32.4551     1.5523    0.1318
variety6      -32.8943     29.1655    -1.1278    0.2690
variety7      -24.1806     29.8412    -0.8103    0.4246
variety8       -1.4947     31.4475    -0.0475    0.9624
block2        -27.9634     24.7026    -1.1320    0.2672
block3         -1.3075     24.3514    -0.0537    0.9576
block4        -61.3369     29.6143    -2.0712    0.0477
block5        -64.1934     26.1610    -2.4538    0.0206
Residual standard error: 38.89 on 28 degrees of freedom

Multiple R-Squared: 0.4031
```

Remark 1: The function *lmRob* has some bugs. To overcome this problem it is necessary to change the function *lmrob.fit.compute* called by *lmRob*. To make this change, the first time that S-PLUS is run in a working directory, use the command *source("\\path1\\.path2\\...\\lmrob.fit.compute)*, where \path1\path2\... is the path where *lmrob.fit.compute* is saved. The fixed function *lmrob.fit.compute* is provided on the book's web site.

11.2.3 Testing linear assumptions: function *rob.linear.test*

To test robustly a linear assumption $H_0 : \gamma = \mathbf{A}\beta = \gamma_0$ as explained in Section 4.7, we have to fit two models with *lmRob* using the option *final.alg* = *"MM"*. The first fit should be done *without* the constraint $\gamma = \gamma_0$; let the result be saved in object *lmrob1*. The second fit should be computed *with* the constraint $\gamma = \gamma_0$; let the result be saved in *lmrob2*. The two fits should have the same options in *robust.control*. The robust likelihood ratio-type test of H_0 described in Section 4.7 is performed by the function *rob.linear.test* supplied by the authors. The S-PLUS robust library contain two functions to test linear hypothesis *anova.lmRob y aovrob*, but the results are not reliable. The call to *rob.linear.test* is

robtest = rob.linear.test (lmrob1,lmrob2)

The components of the output are

test the chi-squared statistic
chisq.pvalue the p-value of the chi-squared approximation
f.pvalue p-value of the F-approximation
df degrees of freedom of numerator and denominator.

Example

Consider Example 4.2 again. Recall that we have already fit the complete model and the output is saved in the object *scheffemrob*. Suppose that we want to test the null hypothesis of no variety effect. Then, we fit a second model with the function *lmRob*, deleting the variable *variety* by means of

scheffemrob.variety = lmRob(yield~block, data = scheffem,
robust.control = Cont)

Then we use the function *rob.linear.test* by entering the command

test.variety = rob.linear.test(scheffemrob,scheffemrob.variety)

The output is displayed with the command

test.variety

and the displayed output is

```
$test:
[1] 38.88791
$chisq.pvalue:
[1] 2.053092e-006
$f.pvalue:
[1] 0.0004361073
$df:
[1] 7 28
```

A similar test can be performed for the hypothesis of no block effects.

11.2.4 Stepwise variable selection: function *step*

To perform a backward stepwise variable selection procedure using the RFPE criterion as described in Section 5.12 we proceed as follows. First, using *lmRob*, a linear model is fit with all the variables and with the option *final.alg* = *"MM"*.

Suppose that the output is saved in the object *regrob*. Then we use the command

stepresults = *step(regrob)*

The output is the same as the one of *lmRob* applied to the selected regression.

Example

For the example in Section 5.12 we first fit the full model with the command

simrob = lmRob(formula = simdata.1 ˜simdata.2 + simdata.3 + simdata.4 + simdata.5 + simdata.6 + simdata.7, robust.control = Cont, data = simdata)

Then the stepwise selection procedure is performed with

simrob.step = step(simrob)

The output is displayed with

summary(simrob.step)

with the following results:

```
Call: lmRob(formula = simdata.1 ~simdata.2 + simdata.3 +
simdata.4, data = simdata, robust.control = Cont)

Residuals:
Min      1Q        Median     3Q        Max
-28.68   -0.6878   0.001164   0.5552    19.91

Coefficients:
             Value     Std. Error   t value   Pr(>|t|)
(Intercept)  0.0938    0.1473       0.6370    0.5269
simdata.2    1.2340    0.1350       9.1416    0.0000
simdata.3    0.7906    0.1468       5.3874    0.0000
simdata.4    0.8339    0.1650       5.0529    0.0000

Residual standard error: 1.016 on 52 degrees of freedom

Multiple R-Squared: 0.4094
```

Warning: Before running the function *step* do not forget to change the function *lmrob.fit.compute* as explained at the end of Remark 1, otherwise the results will be wrong.

11.3 Robust multivariate location and dispersion

To compute a robust estimate of multivariate location and dispersion we can use the function *covRob* in the robust library of S-PLUS and two functions supplied by the authors to compute S-estimates: *cov.Sbic* for the S-estimate with bisquare ρ and *cov.SRocke* for the SR estimate defined in Section 6.4.4. These two functions start from the improved MVE estimate described in Section 6.7.3.

11.3.1 A general function for computing robust location–dispersion estimates: *covRob*

The call to this function is

cov = covRob(data, corr = F, center = T, distance = T, na.action = na.fail, estim = "auto", control = covRob.control(estim, ...), ...)

where

data is the data set for computing the dispersion matrix, which may be a matrix or a dataframe. Columns represent variables, and rows represent observations.
corr A logical flag: if *corr* = *T* the estimated correlation matrix is returned.
center A logical flag or numeric vector containing the location about which the dispersion is to be taken. If *center* = *T* then a robust estimate of the center is computed; if *center* = *F* then no centering takes place and the center is set equal to the zero vector. This argument is used only by the Stahel–Donoho estimate (which is not the default estimate)
distance A logical flag: if *distance* = *T*, the Mahalanobis distances are computed.
na.action A function to filter missing data. The default (*na.fail*) is to create an error if any missing values are found. A possible alternative is *na.omit*, which deletes observations that contain one or more missing values.
estim The robust estimator used by *covRob*. The choices are: *"mcd"* for the fast MCD algorithm in Section 6.7.6; *"donostah"* for the Stahel–Donoho estimate in Section 6.7.7; *"M"* for the S-estimate with Rocke's "translated bisquare" ρ-function; *"pairwiseQC"* and *"pairwiseGK"* correspond to Maronna and Zamar's nonequivariant estimate in Section 6.9.1, where the pairwise covariances are respectively the quadrant and the Gnanadesikan–Kettenring covariances defined in that section. The default *"auto"* selects from *"donostah"*, *"mcd"* and *"pairwiseQC"* according to the numbers of observations and of variables. For the reasons given in the remark in Section 6.8, for high-dimensional data we recommend use of the function *cov.Rocke* described below, and not Rocke's translated bisquare.
control is a list of control parameters to be used in the numerical algorithms. See the help on *covRob.control* for the possible parameters and their default settings.

The main components of the output are

cov robust estimate of the covariance/correlation matrix
center robust estimate (or the specification, depending on *center*) of the location vector
dist Mahalanobis distances computed with the chosen estimate; returned only if $distance = T$
evals eigenvalues of the covariance or the correlation matrix, according to $corr = F$ or $corr = T$ respectively.

11.3.2 The SR-α estimate: function *cov.SRocke*

The call to this function is

$$cov = cov.SRocke\ (x,nsub = 500,maxit = 5,alpha = 0.05)$$

where

x is the data set for computing the dispersion matrix, which may be a matrix or a dataframe. Columns represent variables, and rows represent observations.
nsub is the number of subsamples.
maxit is the maximum number of iterations.
alpha the parameter α in (6.40).

The components of the output are

center is the MVE location estimate.
cov is the MVE dispersion matrix.
dist is the vector of Mahalanobis distances using the MVE estimates.

11.3.3 The bisquare S-estimate: function *cov.Sbic*

The call to this function is

$$cov = cov.Sbic\ (x,nsub = 500)$$

where

x is the data set for computing the dispersion matrix, which may be a matrix or a dataframe. Columns represent variables, and rows represent observations.
nsub is the number of subsamples.
maxit is the maximum number of iterations.

The components of the output are the same as in *cov.SRocke*

11.4 Principal components

To estimate the principal components the authors supply two programs: namely, *prin.comp.rob*, which uses the method of Locantore et al. (1999) described in

Section 6.10.2, and *princomp.cov*, which uses the SR-α or the bisquare S-dispersion matrix computed with the programs *cov.SRocke* and *cov.Sbic*, respectively.

11.4.1 Spherical principal components: function *prin.comp.rob*

The call to this function is

princ = princ.comp.rob (x, corr = F, delta = 0.001)

where

x is the data set for computing the principal components, which may be a matrix or a dataframe. Columns represent variables, and rows represent observations.
cor A logical flag: if *corr = T* the estimated robust correlation matrix is returned and if *corr = F* the robust covariance matrix is computed.
delta An accuracy parameter.

The main components of the output are

loadings Orthogonal matrix containing the loadings. The first column is the linear combination of columns of **x** defining the first principal component etc.
eigenvalues The vector with the squares of the component scales
scores The matrix with scores of the principal components
plot Object of class *princomp* to use as input with the S-PLUS function *biplot*. The use of this function produce a biplot graph. It has the same structure as the output of the S-PLUS function *princomp*.

11.4.2 Principal components based on a robust dispersion matrix: function *princomp.cov*

The call to this function is

princ = princomp.cov (x, estim = "SRocke", alpha = 0.05, nsub = 500, corr = F)

where

x is the data set for computing the principal components, which may be a matrix or a dataframe. Columns represent variables, and rows represent observations.
estim The dispersion matrix used. The S-estimate with SR-α or bisquare ρ is used according to *estim = "SRocke"* or *"Sbic"*.
alpha Same as in function *cov.SRocke*. Required only if *estim = "SRocke"*.
nsub Same as in function *cov.SRocke*.
corr A logical flag: if *corr = T* the estimated robust correlation matrix is returned and if *corr = F* the robust covariance matrix is computed.

The components of the output are the same as in *prin.comp.rob*.

11.5 Generalized linear models

We describe three S-PLUS functions that can be used for generalized linear models: *BYlogreg*, *WBYlogreg* and *glmRob*. The first two, developed by Christophe Croux, give robust estimates of the logistic model; and the third, which belongs to the S-PLUS robust library, can be used to fit logistic and Poisson regression models.

11.5.1 M-estimate for logistic models: function *BYlogreg*

BYlogreg computes an M-estimate for the logistic model using the procedure described in Section 7.2.2 with the ψ-function given in (7.14). It is called with the command

logireg = BYlogreg(x0,y,initwml = T,const = 0.5,kmax = 1000, maxhalf = 10)

The input arguments are

x0 The matrix of explanatory variables, where each column is a variable. A column of ones is automatically added for the intercept.
y The vector of binomial responses (0 or 1).
initwml Logical value for selecting one of the two possible methods for computing the initial value of the optimization process. If *initwml* = *T* (default), a weighted MLE defined in Section 7.2.1 is computed; the dispersion matrix used for the weights is the fast MCD estimate. If *initwml* = *F*, a classical ML fit is performed.
const Tuning constant used in the computation of the estimate (default = 0.5).
kmax Maximum number of iterations before convergence (default = 1000).
maxhalf Maximum number of "step-halving", a parameter related with the stopping rule of the iterative optimization algorithm (default = 10).

The components of the output are

convergence T or F according to the convergence of the algorithm being achieved or not
objective the value of the objective function at the minimum
coef the vector of parameter estimates
sterror standard errors of the parameters (if *convergence* = *T*).

Example

Consider Example 7.1. To fit a logistic model with *BYlogreg* we use the command

leukBY = BYlogreg(Xleuk, yleuk)

Here *Xleuk* is a 33×2 matrix whose two columns are the regressors *wbc* and *ag*, and *yleuk* is a vector with the response binary variable.

The output is displayed with

leukBY

and the results are

```
leukBY
$convergence:
[1] T
$objective:
[1] 0.5436513
$coef:
(Intercept)  x0wbc            x0ag
0.1594679    -0.0001773792    1.927563
$sterror:
1.6571165099  0.0002334576    1.1632861341
```

11.5.2 Weighted M-estimate: function *WBYlogreg*

This function computes a weighted M-estimate defined by (7.17). The ψ-function used is the one given in (7.14) and the dispersion matrix used to compute the weights is the MCD. The input and output of *WBYlogreg* are similar to those of *BYlogreg*, the only difference being that this function does not require the parameter *initwml*, since it always uses the MCD to compute the weights.

The call to this function is

logireg = WBYlogreg(x0,y,const = 0.5,kmax = 1000,maxhalf = 10)

Consider Example 7.1 again. A logistic model is fitted with *WBYlogreg* by the command

leukWBY = WBYlogreg(Xleuk, yleuk)

The output is displayed with

leukWBY

and the results are

```
$convergence:
[1] T
$objective:
[1] 0.5159552
$coef:
(Intercept)  x0wbc            x0ag
0.19837      -0.0002206507    2.397789
$sterror:
[1] 1.19403021700 0.00009822409 1.29831401203
```

11.5.3 A general function for generalized linear models: *glmRob*

This function computes conditionally unbiased bounded influence (CUBIF) estimates for the logistic and Poisson regression models, described in Section 7.3. For the logistic model, it also computes the weighted MLE as described in Section 7.2.1. We do not give a complete description of this function. For a complete description of all optional parameters consult the S-PLUS help.

The call to this function is

glmRob(formula, family = binomial, data, subset, na.action, fit.method = "cubif", estim = "mcd")

The parameters *formula*, *data*, *subset* and *na.action* are similar to those in *lmrob*. The other parameters are

family can take two values: *"binomial"* and *"poisson"*. In the case of the binomial family it fits a logistic model.

fit.method The two options that are documented here are *"cubif"* and *"mallows"*. The option *"mallows"* computes a weighted MLE in Section 7.2.1; it is available only when *family = "binomial"*.

estim indicates the robust dispersion matrix used to compute the weights when using *fit.method = "mallows"*. The options for *estim* are the same as for the function *covRob* described below and the default is the MCD estimate.

The main components of the output are

coefficients Parameter estimates

linear.predictors Linear fit, given by the product of the model matrix and the coefficients.

fitted.values Fitted mean values, obtained by transforming *linear.predictors* using the inverse link function.

residuals Residuals from the final fit, also known as "working residuals". They are typically not interpretable.

deviance Up to a constant, minus twice the log likelihood evaluated at the final coefficients. Similar to the residual sum of squares.

null.deviance Deviance corresponding to the model with no predictors.

weights Weights from the final fit.

Example

Consider the epilepsy data in Example 7.3. To fit a Poisson regression model to this data we enter

breslowrob = glmRob(formula = sumY ~Age10 + Base4 Trt, family = poisson, data = breslow.dat)*

The output is displayed with

summary(breslowrob)

and the results are

```
Call: glmRob(formula = sumY ~Age10 + Base4 * Trt,
family = poisson, data = breslow.dat)

Deviance Residuals:
Min 1Q Median 3Q Max
Min      1Q       Median     3Q      Max
-55.94   -1.458   -0.03073   1.063   9.476

Coefficients:
              Value       Std. Error   t value
(Intercept)   1.62935     0.27091      6.0145
Age10         0.12862     0.07888      1.6305
Base4         0.14723     0.02238      6.5774
Trt           -0.22113    0.11691      -1.8914
Base4:Trt     0.01529     0.02212      0.6913
(Dispersion Parameter for Poisson family taken to be 1)

Null Deviance: 2872.921 on 58 degrees of freedom

Residual Deviance: 3962.335 on 54 degrees of freedom

Number of Iterations: 5
```

11.6 Time series

We shall describe the function *ar.gm* in S-PLUS for computing GM-estimates for AR models and the function *arima.rob* of the FinMetric module of S-PLUS, which compute filtered τ-estimates for ARIMA and REGARIMA models and detect outliers and level shifts.

11.6.1 GM-estimates for AR models: function *ar.gm*

This function computes a Mallows-type GM-estimate for an AR model as described in Section 5.11. The function w_1 may be based on a Huber or bisquare ψ-function. The first iterations use w_2 based on a Huber function and the last ones on a bisquare function. We shall describe only the main features of this function, but shall not explore all of its possibilities and options. For a more complete description the reader can use the HELP feature of S-PLUS or the S-PLUS manual for Windows, *Guide to Statistics*, Volume 2.

The call to this function is

armodel = ar.gm(x,order,effgm,effloc,b,iterh,iterb)

where

x Univariate time series or a real vector. Missing values are not allowed.
order Integer giving the order of the autoregression.
effgm Desired asymptotic efficiency of GM-estimates of AR coefficients for Gaussian data (based on first-order AR theory). The default is 0.87.
effloc Desired asymptotic efficiency of the M-estimate of location used for centering. This efficiency is also a component in *effgm*. The default is 0.96.
b Logical parameter determining whether a bisquare or Huber ψ-function is to be used to form w_2 weights. If $b = T$, a bisquare is used; otherwise $b = F$ and a Huber function is used.
iterh The number of iterations with w_2 based on a Huber ψ-function. Use *iterh = 0* for least squares.
iterb The number of iterations with w_2 based on a bisquare ψ-function. Use *iterb = 0* for least squares.

VALUE:
The main components of the output are

ar A vector of length order containing the GM-estimates of the AR coefficients.
sinnov A vector of innovations scale estimates for the AR models of orders *1* through *order.*
rmu Robust location estimate for **x**, the sample mean if *iterh = iterb = 0.*
sd Robust scale estimate for **x**; gives the standard deviation if *iterh = iterb = 0.*
effgm The value of *effgm* used for the estimate.
effloc The value of *effloc*.

11.6.2 Fτ-estimates and outlier detection for ARIMA and REGARIMA models: function *arima.rob*

We only consider here the main options for this program. For a complete description see the help and manual of the FinMetrics module.

The call to this function is

arirob = arima.rob(formula, data, start = NULL, end = NULL, p = 0, q = 0, d = 0, sd = 0, freq = 1, sfreq = NULL, sma = F, max.p = NULL, auto.ar = F, n.predict = 20, tol = 10^(−6), max.fcal = 2000, innov.outlier = F, critv = NULL, iter = F)

where

formula For a REGARIMA model, the same as in *lmRob.* For an ARIMA model, *formula* should be *x ~1*, where *x* is the observed series.

start A character string which can be passed to the *timeDate* function to specify the starting date for the estimation. This can only be used if the data argument is a "*timeSeries*" dataframe. The default is NULL.

end A character string which can be passed to the *timeDate* function to specify the ending date for the estimation. This can only be used if the data argument is a "*timeSeries*" dataframe. The default is NULL.

p Autoregressive order of the errors model. The default is 0.

q Moving-average order of the errors model. The default is 0.

d The number of regular differences in the ARIMA model. It must be 0, 1 or 2. The default is 0.

sd The number of seasonal differences. It must be 0, 1 or 2. The default is 0.

freq Frequency of data. The default is 1.

sfreq Seasonality frequency of data. If NULL, it is set to be equal to *freq*. The default is NULL.

sma A logical flag: if TRUE, the errors model includes a seasonal moving-average parameter. The default is FALSE.

auto.ar A logical flag: if TRUE, an AR(p) model is selected automatically using a robust AIC. The default is FALSE.

max.p Maximum order of the autoregressive stationary model that approximates the ARMA stationary model. If NULL, then $max.p = \max(p+q, 5)$. If $q = 0$, then *max.p* is not necessary. The default is NULL.

n.predict Maximum number of future periods for which we wish to compute the predictions. The default is 20.

innov.outlier A logical flag: if TRUE, the function *arima.rob* looks for innovation outliers in addition to additive outliers and level shifts; otherwise, *arima.rob* only looks for additive outliers and level shifts. The default is FALSE.

critv Critical value for detecting outliers. If NULL, it assumes the following default values: $critv = 3$ if the length of the time series is less than 200; $critv = 3.5$ if it is between 200 and 500; and $critv = 4$ if it is greater than 500.

iter A logical flag or the number of iterations to reestimate the model after the outliers and level shifts are detected and their effects are eliminated. If $iter = F$ the procedure is not iterated, if $iter = T$ one iteration is performed and if $iter = n$, where n is a positive integer, n iterations are performed. It is recommended to perform at most one iteration.

The main components of the output are

regcoef Estimates of regression coefficients. When we fit a pure ARIMA model, this variable contains only the intercept which is the mean of the differenced series.

regcoef.cov Estimated covariance matrix of the regression coefficients.

innov The vector of the estimated innovations.

innov.acf A series whose autocorrelations or partial autocorrelations are the robust estimates of the innovation or the partial autocorrelations.

regresid Estimated regression residuals cleaned of additive outliers by the robust filter.

regresid.acf A series whose autocorrelations or partial autocorrelations are the robust estimates of the autocorrelations or partial autocorrelations of the differenced regression residuals.

sigma.innov A robust estimate of the innovation scale.

sigma.regresid An estimate of the scale of the differenced regression residuals.

sigma.first An initial estimate of the innovation scale based only on the scale of the differenced model and the ARMA parameters.

tuning.c Bandwidth of the robust filter.

y.robust Response series cleaned of outliers by the robust filter.

y.cleaned Response series cleaned of additive outliers and level shifts after the outlier detection procedure.

predict.error Fitted and predicted regression errors.

predict.scales Standard deviations of the fitted and predicted regression errors.

n.predict The number of predicted observations, which is equal to the *n.predict* argument passed to the *arima.rob* function that produced the "arima.rob" object.

tauef The inverse of the estimated efficiency factor of the τ-estimate with respect to the LS estimate.

inf Information about the outcome of the last optimization procedure: $inf = 1$ and $inf = 0$ indicate that the procedure did or did not converge, respectively.

model Includes the number of regular and seasonal differences, the seasonal frequency, the AR and MA coefficients and the MA seasonal parameter.

innov.outlier A logical flag, the same as the *innov.outlier* argument passed to the *arima.rob* function that produced the "*arima.rob*" object.

outliers An object of class "*outliers*", which contains all the detected outliers (and level shifts).

outliers.iter Optionally, a list of objects of class "*outliers*", if the *iter* argument passed to the *arima.rob* function that produced the "*arima.rob*" object is nonzero.

n0 The number of missing innovations at the beginning.

call An image of the call that produced the object, but with all arguments named, and with the actual formula included as the formula argument.

To show the main components of the outcome including the standard deviations of the regression and ARMA coefficients use the command

summary(arirob)

where arirob is the name of the output object.

Warning: When either d or sd is greater than zero, the interpretation of the intercept in the formula is different from its usual one: it represents the coefficient of the lowest-order power of the time trend that can be identified. For example, if $d = 2$ and $sd = 0$, the intercept represents the coefficient of the term t^2, where t is the period.

11.7 Public-domain software for robust methods

In this section we give some references for freely available implementations of robust procedures.

Some robust methods are available in the standard versions on R. Claudio Agostinelli's site

http://www.dst.unive.it/~claudio/robusta/links.html

which contains links to several sites containing software for robustness, such as the Matlab implementation of robust procedures by Verboven and Hubert (2005).

The site

http://hajek.stat.ubc.ca/~matias/soft.html

contains Matias Salibian-Barrera's implementation in R of regression M-estimates and of the fast S-estimate described in Section 5.7.3.

The site www.iumsp.ch (Plan du site–Software for robust statistics) contains a large number of robust methods in S-PLUS developed by Alfio Marazzi, in particular the library ROBETH.

12

Description of Data Sets

We describe below the data sets used in the book.

Alcohol

The solubility of alcohols in water is important in understanding alcohol transport in living organisms. This data set from Romanelli et al. (2001) contains physicochemical characteristics of 44 aliphatic alcohols. The aim of the experiment was the prediction of the solubility on the basis of molecular descriptors. The columns are:

1. SAG = solvent accessible surface-bounded molecular volume
2. V = volume
3. log PC (PC = octanol–water partitions coefficient)
4. P = polarizability
5. RM = molar refractivity
6. Mass
7. log(Solubility) (response)

Algae

This data set is part of a larger one (http://kdd.ics.uci.edu/databases/coil/coil.html) which comes from a water quality study where samples were taken from sites on different European rivers over a period of approximately one year. These samples were analyzed for various chemical substances. In parallel, algae samples were collected to determine the algae population distributions. The columns are:

1. season (1,2,3,4 for winter, spring, summer and autumn)
2. river size (1,2,3 for small, medium and large)

Robust Statistics – Theory and Methods Ricardo A. Maronna, R. Douglas Martin and Víctor J. Yohai

3. fluid velocity (1,2,3 for low, medium and high)
4.–11. content of nitrogen in the form of nitrates, nitrites and ammonia, and other chemical compounds

The response is the abundance of a type of algae (type 6 in the complete file). For simplicity we deleted the rows with missing values, or with null response values, and took the logarithm of the response.

Aptitude

There are three variables observed on 27 subjects:

Score: numeric, represents scores on an aptitude test for a course
Exp: numeric, represents months of relevant previous experience
Pass: binary response, 1 if the subject passed the exam at the end of the course and 0 otherwise.

The data may be downloaded as data set 6.2 from the site: http://www.jeremymiles.co.uk/regressionbook/data/

Bus

This data set from the Turing Institute, Glasgow, Scotland, contains measures of shape features extracted from vehicle silhouettes. The images were acquired by a camera looking downward at the model vehicle from a fixed angle of elevation

The following features were extracted from the silhouettes:

1. compactness
2. circularity
3. distance circularity
4. radius ratio
5. principal axis aspect ratio
6. maximum length aspect ratio
7. scatter ratio
8. elongatedness
9. principal axis rectangularity
10. maximum length rectangularity
11. scaled variance along major axis
12. scaled variance along minor axis
13. scaled radius of gyration
14. skewness about major axis
15. skewness about minor axis
16. kurtosis about minor axis
17. kurtosis about major axis
18. hollows ratio

Glass

This is part of a file donated by Vina Speihler, describing the composition of glass pieces from cars.

The columns are:

1. RI refractive index
2. Na_2O sodium oxide (unit measurement: weight percent in corresponding oxide, as are the rest of attributes)
3. MgO magnesium oxide
4. Al_2O_3 aluminum oxide
5. SiO_2 silcon oxide
6. K_2O potassium oxide
7. CaO calcium oxide

Hearing

Prevalence rates in percent for men aged 55–64 with hearing levels 16 decibels or more above the audiometric zero,

The rows correspond to different frequencies and to normal speech.

1. 500 hertz
2. 1000 hertz
3. 2000 hertz
4. 3000 hertz
5. 4000 hertz
6. 6000 hertz
7. Normal speech

The columns classify the data in seven occupational groups:

1. professional–managerial
2. farm
3. clerical sales
4. craftsmen
5. operatives
6. service
7. laborers

Image

The data were supplied by A. Frery. They are part of a synthetic aperture satellite radar image corresponding to a suburb of Munich.

Krafft

The Krafft point is an important physical characteristic of the compounds called *surfactants*, establishing the minimum temperature at which a surfactant can be used. The purpose of the experiment was to estimate the Krafft point of compounds as a function of their molecular structure.

The columns are:

1. Randiç index
2. Volume of tail of molecule
3. Dipole moment of molecule
4. Heat of formation
5. Krafft point (response)

Neuralgia

The data come from a study on the effect of iontophoretic treatment of elderly patients complaining of post-herpetic neuralgia. There were 18 patients in the study, who were interviewed six weeks after the initial treatment and were asked if the pain had been reduced.

There are 18 observations on five variables:

Pain: binary response: 1 if the pain eased, 0 otherwise.
Treatment: binary variable: 1 if the patient underwent treatment, 0 otherwise.
Age: the age of the patient in completed years.
Gender: M (male) or F (female).
Duration: pretreatment duration of symptoms (in months).

The data may be downloaded from the site: http://www.sci.usq.edu.au/staff/dunn/Datasets/Books/Hand/Hand-R/neural-R.html

Oats

Yield of grain in grams per 16-foot row for each of eight varieties of oats in five replications in a randomized-block experiment.

Solid waste

The original data are the result of a study on production waste and land use by Golueke and McGauhey (1970), and contain nine variables. Here we consider the following six:

1. industrial land (acres)
2. fabricated metals (acres)
3. trucking and wholesale trade (acres)

4. retail trade (acres)
5. restaurants and hotels (acres)
6. solid waste (millions of tons), response

Stack loss

The columns are:

1. air flow
2. cooling water inlet temperature (°C)
3. acid concentration (%)
4. Stack loss, defined as the percentage of ingoing ammonia that escapes unabsorbed (response)

Toxicity

The aim of the experiment was to predict the toxicity of carboxylic acids on the basis of several molecular descriptors. The attributes for each acid are:

1. $\log(IGC_{50}^{-1})$: Aquatic toxicity (response)
2. log Kow: Partition coefficient
3. pKa: Dissociation constant
4. ELUMO: Energy of the lowest unoccupied molecular orbital
5. Ecarb: Electrotopological state of the carboxylic group
6. Emet: Electrotopological state of the methyl group
7. RM: Molar refractivity
8. IR: Refraction index
9. Ts: Surface tension
10. P: Polarizability

Wine

This data set, which is part of a larger one donated by Riccardo Leardi, gives the composition of several wines. The attributes are:

1. Alcohol
2. Malic acid
3. Ash
4. Alkalinity of ash
5. Magnesium
6. Total phenols
7. Flavanoids

8. Nonflavanoid phenols
9. Proanthocyanins
10. Color intensity
11. Hue
12. OD280/OD315 of diluted wines
13. Proline

Bibliography

Abraham, B. and Chuang, A. (1989), Expectation–maximization algorithms and the estimation of time series models in the presence of outliers, *Journal of Time Series Analysis*, **14**, 221–234.

Adrover, J.G. (1998), Minimax bias-robust estimation of the dispersion matrix of a multivariate distribution, *The Annals of Statistics*, **26**, 2301–2320.

Adrover, J.G., Maronna, R.A. and Yohai, V.J. (2002), Relationships between maximum depth and projection regression estimates, *Journal of Statistical Planning and Inference*, **105**, 363–375.

Adrover, J.G., and Yohai, V.J. (2002), Projection estimates of multivariate location, *The Annals of Statistics*, **30**, 1760–1781.

Agulló, J. (1996), Exact iterative computation of the multivariate minimum volume ellipsoid estimator with a branch and bound algorithm, *Proceedings of the 12th Symposium in Computational Statistics* (COMPSTAT 12), 175–180.

Agulló, J. (1997), Exact algorithms to compute the least median of squares estimate in multiple linear regression, *L_1-Statistical Procedures and Related Topics*, Y. Dodge (ed.), Institute of Mathematical Statistics Lecture Notes-Monograph Series, vol. 31, 133–146.

Agulló, J. (2001), New algorithms for computing the least trimmed squares regression estimator, *Computational Statistics and Data Analysis*, **36**, 425–439.

Akaike, H. (1970), Statistical predictor identification, *Annals of the Institute of Statistical Mathematics*, **22**, 203–217.

Akaike, H. (1973), Information theory and an extension of the maximum likelihood principle, in *International Symposium on Information Theory*, B.N. Petran and F. Csaki (eds.) 2nd Edition, 267–281, Budapest: Akademiai Kiadi.

Akaike, H. (1974a), A new look at the statistical model identification, *IEEE Transactions on Automatic Control*, **19**, 716–723.

Akaike, H. (1974b), Markovian representation of stochastic processes and its application to the analysis of autoregressive moving average processes, *Annals of the Institute of Statistical Mathematics*, **26**, 363–387.

Robust Statistics – Theory and Methods Ricardo A. Maronna, R. Douglas Martin and Víctor J. Yohai

Albert, A. and Anderson, J.A. (1984), On the existence of maximum likelihohod estimates in logistic regression models, *Biometrika*, **71**, 1–10.

Alqallaf, F.A., Konis, K.P., Martin, R.D. and Zamar, R.H. (2002), Scalable robust covariance and correlation estimates for data mining, *Proceedings of SIGKDD 2002, Edmonton, Alberta, Canada, Association of Computing Machinery (ACM).*

Analytical Methods Committee (1989), Robust statistics – How not to reject outliers, *Analyst*, **114**, 1693–1702.

Anderson, T.W. (1994), *The Statistical Analysis of Time Series*, New York: John Wiley & Sons, Inc.

Arslan, O. (2004), Convergence behavior of an iterative reweighting algorithm to compute multivariate M-estimates for location and scatter, *Journal of Statistical Planning and Inference*, **118**, 115–128.

Bai, Z.D. and He, X. (1999), Asymptotic distributions of the maximal depth estimators for regression and multivariate location, *The Annals of Statistics*, **27**, 1616–1637.

Barnett, V. and Lewis, T. (1998), *Outliers in Statistical Data*, 3rd Edition, New York: John Wiley & Sons, Inc.

Barrodale, I. and Roberts, F.D.K. (1973), An improved algorithm for discrete l_1 linear approximation, *SIAM Journal of Numerical Analysis*, **10**, 839–848.

Belsley, D.A., Kuh, E. and Welsch, R.E. (1980), *Regression Diagnostics*, New York: John Wiley & Sons, Inc.

Berrendero, J.R., Mendes, B.V.M. and Tyler, D.E. (2005), On the maximum bias functions of MM-estimates and constrained M-estimates of regression, Unpublished manuscript.

Berrendero, J.R. and Zamar, R.H. (2001), The maxbias curve of robust regression estimates, *The Annals of Statistics*, **29**, 224–251.

Bianco, A.M. and Boente, G.L. (2002), On the asymptotic behavior of one-step estimates in heteroscedastic regression models, *Statistics and Probability Letters*, **60**, 33–47.

Bianco, A. and Yohai, V.J. (1996), Robust estimation in the logistic regression model, *Robust Statistics, Data Analysis and Computer Intensive Methods, Proceedings of the workshop in honor of Peter J. Huber*, H. Rieder (ed.), Lecture Notes in Statistics **109**, 17–34, New York: Springer.

Bianco, A.M., Boente, G.L. and Di Rienzo, J. (2000), Some results of GM-based estimators in heteroscedastic regression models, *Journal of Statistical Planning and Inference*, **89**, 215–242.

Bianco, A.M., García Ben, M., Martínez, E.J. and Yohai, V.J. (1996), Robust procedures for regression models with ARIMA errors, *COMPSTAT 96, Proceedings in Computational Statistics*, Albert Prat (ed.), 27–38, Heidelberg: Physica-Verlag.

Bianco, A.M., García Ben, M., Martínez, E.J. and Yohai, V.J. (2001), Outlier detection in regression models with ARIMA errors using robust estimates, *Journal of Forecasting*, **20**, 565–579.

Bianco, A.M., García Ben, M. and Yohai, V.J. (2005), Robust estimation for linear regression with asymmetric errors, *Canadian Journal of Statistics*, **33**, 511–528.

Bickel, P.J. and Doksum, K.A. (2001), *Mathematical Statistics: Basic Ideas and Selected Topics*, vol. I, 2nd Edition, London; Prentice Hall.

Billingsley, P. (1968), *Convergence of Probability Measures*, New York: John Wiley & Sons, Inc.

Blackman, R.B. and Tukey, J.W. (1958), *The Measurement of Power Spectra*, New York: Dover.

Bloomfield, P. (1976), *Fourier Analysis of Time Series: An Introduction*, New York: John Wiley & Sons, Inc.

Bloomfield, P. and Staiger, W.L. (1983), *Least Absolute Deviations: Theory, Applications and Algorithms*, Basle: Birkhäuser.

Boente, G.L. (1983), Robust methods for principal components (in Spanish), Ph.D. thesis, University of Buenos Aires.

Boente, G.L. (1987), Asymptotic theory for robust principal components, *Journal of Multivariate Analysis*, **21**, 67–78.

Boente, G.L. and Fraiman, R. (1999), Discussion of Locantore et al., 1999, *Test*, **8**, 28–35.

Boente, G.L., Fraiman, R. and Yohai, V.J. (1987), Qualitative robustness for stochastic processes, *The Annals of Statistics*, **15**, 1293–1312.

Bond, N.W. (1979), Impairment of shuttlebox avoidance-learning following repeated alcohol withdrawal episodes in rats, *Pharmacology, Biochemistry and Behavior*, **11**, 589–591.

Box, G.E.P., Hunter, W.G. and Hunter, J.S. (1978), *Statistics for Experimenters*, New York: John Wiley & Sons, Inc.

Brandt, A. and Künsch, H.R. (1988), On the stability of robust filter-cleaners, *Stochastic Processes and their Applications*, **30**, 253–262.

Breiman, L., Friedman, J.H., Olshen, R.A. and Stone, C.J. (1984), *Classification and Regression Trees*, Belmont, CA: Wadsworth.

Breslow, N.E. (1996), Generalized linear models: Checking assumptions and strengthening conclusions, *Statistica Applicata*, **8**, 23–41.

Brent, R. (1973), *Algorithms for Minimisation Without Derivatives*, Englewood Cliffs, NJ: Prentice Hall.

Brockwell, P.J. and Davis, R.A. (1991), *Introduction to Time Series and Forecasting*, New York: Springer.

Brockwell, P.J. and Davis, R.A. (1996), *Time Series: Theory and Methods*, 2nd Edition, New York: Springer.

Brownlee, K.A. (1965), *Statistical Theory and Methodology in Science and Engineering*, 2nd Edition, New York: John Wiley & Sons, Inc.

Brubacher, S.R. (1974), Time series outlier detection and modeling with interpolation, Bell Laboratories Technical Memo, Murray Hill, NJ.

Bruce, A.G. and Martin, R.D. (1989), Leave-k-out diagnostics for time series (with discussion), *Journal of the Royal Statistical Society (B)*, **51**, 363–424.

Bustos, O.H. (1982), General M-estimates for contaminated pth-order autoregressive processes: Consistency and asymptotic normality. Robustness in autoregressive processes. *Zeitschrift für Wahrscheinlichkeitstheorie und Verwandte Gebiete*, **59**, 491–504.

Bustos, O.H. and Yohai, V.J. (1986), Robust estimates for ARMA models, *Journal of the American Statistical Association*, **81**, 491–504.

Butler, R.W., Davies, P.L. and Jhun, M. (1993), Asymptotics for the Minimum Covariance Determinant Estimator, *The Annals of Statistics*, **21**, 1385–1400.

Campbell, N.A. (1980), Robust procedures in multivariate analysis I: Robust covariance estimation, *Applied Statistics*, **29**, 231–237.

Campbell, N.A. (1989), Bushfire mapping using NOAA AVHRR data, Technical Report, CSIRO, Australia.

Cantoni, E. and Ronchetti, E. (2001), Robust inference for generalized linear models, *Journal of the American Statistical Association*, **96**, 1022–1030.

Carroll, R.J. and Pederson, S. (1993), On robustness in the logistic regression model, *Journal of the Royal Statistical Society (B)*, **55**, 693–706.

Carroll, R.J. and Ruppert, D. (1982), Robust estimation in heteroscedastic linear models, *The Annals of Statistics*, **10**, 429–441.

Chambers, J. (1977), *Computational Methods for Data Analysis*. New York: John Wiley & Sons, Inc.

Chang, I., Tiao, G.C. and Chen, C. (1988), Estimation of time series parameters in the presence of outliers, *Technometrics*, **30**, 193–204.

Chatterjee, S. and Hadi, A.S. (1988), *Sensitivity Analysis in Linear Regression*, New York: John Wiley & Sons, Inc.

Chave, A.D., Thomson, D.J. and Ander, M.E. (1987), On the robust estimation of power spectra, coherence, and transfer functions, *Journal of Geophysical Research*, **92**, 633–648.

Chen, C. and Liu, L.M. (1993), Joint estimation of the model parameters and outlier effects in time series, *Journal of the American Statistical Association*, **88**, 284–297.

Chen, Z. and Tyler, D.E. (2002), The influence function and maximum bias of Tukey's median, *The Annals of Statistics*, **30**, 1737–1759.

Clarke, B.R. (1983), Uniqueness and Fréchet differentiability of functional solutions to maximum likelihood type equations, *The Annals of Statistics*, **11**, 1196–1205.

Cook, R.D. and Hawkins, D.M. (1990), Comment to "Unmasking multivariate outliers and leverage points" by P. Rousseeuw and B. van Zomeren, *Journal of the American Statistical Association*, **85**, 640–644.

Cook, R.D. and Weisberg, S (1982), *Residuals and Influence in Regression*, London: Chapman and Hall.

Croux, C. (1998), Limit behavior of the empirical influence function of the median, *Statistics and Probability Letters*, **37**, 331–340.

Croux, C. and Dehon, C. (2003), Estimators of the multiple correlation coefficient: Local robustness and confidence intervals, *Statistical Papers*, **44**, 315–334.

Croux, C, Dhaene, G. and Hoorelbeke, D. (2003), Robust standard errors for robust estimators, Discussion Papers Series 03.16, K.U. Leuven, CES.

Croux, C., Flandre, C. and Haesbroeck, G. (2002), The breakdown behavior of the Maximum Likelihood Estimator in the logistic regression model, *Statistics and Probability Letters*, **60**, 377–386.

Croux, C. and Haesbroeck, G. (2000), Principal component analysis based on robust estimators of the covariance or correlation matrix: Influence functions and efficiencies, *Biometrika*, **87**, 603–618.

Croux, C. and Haesbroeck, G. (2003), Implementing the Bianco and Yohai estimator for logistic regression, *Computational Statistics and Data Analysis*, **44**, 273–295.

Croux, C. and Rousseeuw, P.J. (1992), Time-efficient algorithms for two highly robust estimators of scale, *Computational Statistics*, **2**, 411–428.

Croux, C. and Ruiz-Gazen, A. (1996), A fast algorithm for robust principal components based on projection pursuit, *COMPSTAT: Proceedings in Computational Statistics*, A. Prat (ed.), 211–216, Heidelberg: Physica-Verlag.

Daniel, C. (1978), Patterns in residuals in the two-way layout, *Technometrics*, **20**, 385–395.

Davidson, J. (1994), *Stochastic Limit Theory: An Introduction for Econometricians*, New York: Oxford University Press.

Davies, P.L. (1987), Asymptotic behavior of S-estimators of multivariate location parameters and dispersion matrices, *The Annals of Statistics*, **15**, 1269–1292.

Davies, P.L. (1990), The asymptotics of S-estimators in the linear regression model, *The Annals of Statistics*, **18**, 1651–1675.

Davies, P.L. (1992), The asymptotics of Rousseeuw's minimum volume ellipsoid estimator, *The Annals of Statistics*, **20**, 1828–1843.

Davies, P.L. (1993), Aspects of robust linear regression, *The Annals of Statistics*, **21**, 1843–1899.

Davis, R.A. (1996), Gauss-Newton and M-estimation for ARMA processes with infinite variance, *Stochastic Processes and their Applications*, **63**, 75–95.

Davis, R.A., Knight, K. and Liu, J. (1992), M-estimation for autoregression with infinite variance, *Stochastic Processes and their Applications*, **40**, 145–180.

Davison, A.C. and Hinkley, D.V. (1997), *Bootstrap Methods and their Application*, Cambridge: Cambridge University Press.

de Luna, X. and Genton, M.G. (2001), Robust simulation-based estimation of ARMA models, *Journal of Computational and Graphical Statistics*, **10**, 370–387.

Denby, L. and Martin, R.D. (1979), Robust estimation of the first-order autoregressive parameter, *Journal of the American Statistical Association*, **74**, 140–146.

de Vel, O., Aeberhard, S. and Coomans, D. (1993), Improvements to the classification performance of regularized discriminant analysis, *Journal of Chemometrics*, **7**, 99–115.

Devlin, S.J., Gnanadesikan, R. and Kettenring, J.R. (1981), Robust estimation of dispersion matrices and principal components, *Journal of the American Statistical Association*, **76**, 354–362.

Donoho, D.L. and Gasko, M. (1992), Breakdown properties of location estimators based on halfspace depth and projected outlyingness, *The Annals of Statistics*, **20**, 1803–1827.

Donoho, D.L. (1982), Breakdown properties of multivariate location estimators, Ph.D. Qualifying paper, Harvard University.

Donoho, D.L. and Huber, P.J. (1983), The notion of breakdown point, *A Festschrift for Erich L. Lehmann*, P.J. Bickel, K.A. Doksum and J.L. Hodges, (eds.), 157–184, Belmont, CA: Wadsworth.

Draper, N.R. and Smith, H. (2001), *Applied Regression Analysis*, 3rd Edition, New York: John Wiley & Sons, Inc.

Durrett, R. (1996), *Probability Theory and Examples*, 2nd Edition, Belmont, CA: Duxbury Press.

Dutter, R. (1975), Robust regression: Different approaches to numerical solutions and algorithms. Research Report No. 6, Fachgruppe für Statistik, Eidgenössische Technische Hochschule, Zurich.

Dutter, R. (1977), Algorithms for the Huber estimator in multiple regression, *Computing*, **18**, 167–176.

Efron, B. and Tibshirani, R.J. (1993), *An Introduction to the Bootstrap*, New York: Chapman and Hall.

Ellis, S.P. and Morgenthaler, S. (1992), Leverage and breakdown in L_1 regression, *Journal of the American Statistical Association*, **87**, 143–148.

Feller, W. (1971), *An Introduction to Probability Theory and its Applications, Vol. II*, 2nd Edition, New York: John Wiley & Sons, Inc.

Fernholz, L.T. (1983), *Von Mises Calculus for Statistical Functionals*, Lecture Notes in Statistics No. 19, New York: Springer.

Finney, D.J. (1947), The estimation from individual records of the relationship between dose and quantal response, *Biometrika*, **34**, 320–334.

Fox, A.J. (1972), Outliers in time series, *Journal of the Royal Statistical Society (B)*, **34**, 350–363.

Fraiman, R., Yohai, V.J. and Zamar, R.H. (2001), Optimal M-estimates of location, *The Annals of Statistics*, **29**, 194–223.

Frery, A. (2005), Personal communication.

García Ben, M. and Yohai, V.J. (2004), Quantile-quantile plot for deviance residuals in the Generalized Linear Model, *Journal of Computational and Graphical Statistics*, **13**, 36–47.

Gather, U. and Hilker, T. (1997), A note on Tyler's modification of the MAD for the Stahel-Donoho estimator, *The Annals of Statistics*, **25**, 2024–2026.

Genton, M.G. and Lucas, A. (2003), Comprehensive definitions of breakdown-points for independent and dependent observations, *Journal of the Royal Statistical Society (B)*, **65**, 81–94.

Genton, M.G. and Ma, Y. (1999), Robustness properties of dispersion estimators, *Statistics and Probability Letters*, **44**, 343–350.

Gervini, D. and Yohai, V.J. (2002), A class of robust and fully efficient regression estimators, *The Annals of Statistics*, **30**, 583–616.

Giltinan, D.M., Carroll, R.J. and Ruppert, D. (1986), Some new estimation methods for weighted regression when there are possible outliers, *Technometrics*, **28**, 219–230.

Gnanadesikan, R. and Kettenring, J.R. (1972), Robust estimates, residuals, and outlier detection with multiresponse data, *Biometrics*, **28**, 81–124.

Golueke, C.G. and McGauhey, P.H. (1970), *Comprehensive Studies of Solid Waste Management*, US Department of Health, Education and Welfare, Public Health Services Publication No. 2039.

Grenander, U. (1981), *Abstract Inference*, New York: John Wiley & Sons, Inc.

Grenander, U. and Rosenblatt, M. (1957), *Statistical Analysis of Stationary Time Series*, New York: John Wiley & Sons, Inc.

Hampel, F.R. (1971), A general definition of qualitative robustness, *The Annals of Mathematical Statistics*, **42**, 1887–1896.

Hampel, F.R. (1974), The influence curve and its role in robust estimation, *The Annals of Statistics*, **69**, 383–393.

Hampel, F.R. (1975), Beyond location parameters: Robust concepts and methods, *Bulletin of the International Statistical Institute*, **46**, 375–382.

Hampel, F.R., Ronchetti, E.M., Rousseeuw, P.J. and Stahel, W.A. (1986), *Robust Statistics: The Approach Based on Influence Functions*. New York: John Wiley & Sons, Inc.

Hannan, E.J. and Kanter, M. (1977), Autoregressive processes with infinite variance, *Journal of Applied Probability*, **14**, 411–415.

Harvey, A.C. and Phillips, G.D.A. (1979), Maximum likelihood estimation of regression models with autoregressive moving average disturbances, *Biometrika*, **66**, 49–58.

Hastie, T., Tibshirani, R. and Friedman, J. (2001), *The Elements of Statistical Learning*, New York: Springer.

Hawkins, D.M. (1993), A feasible solution algorithm for the minimum volume ellipsoid estimator, *Computational Statistics*, **9**, 95–107.

Hawkins, D.M. (1994), The feasible solution algorithm for least trimmed squares regression, *Computational Statistics and Data Analysis*, **17**, 185–196.

He, X. (1997), Quantile curves without crossing, *The American Statistician*, **51**, 186–192.

He, X. and Portnoy, S. (1992), Reweighted LS estimators converge at the same rate as the initial estimator, *The Annals of Statistics*, **20**, 2161–2167.

Hennig, C. (1995), Efficient high-breakdown-point estimators in robust regression: Which function to choose?, *Statistics and Decisions*, **13**, 221–241.

Hettich, S. and Bay, S.D. (1999), The UCI KDD Archive [http://kdd.ics.uci.edu], Irvine, CA: University of California, Department of Information and Computer Science.

Hössjer, O. (1992), On the optimality of S-estimators, *Statistics and Probability Letters*, **14**, 413–419.

Huber, P.J. (1964), Robust estimation of a location parameter, *The Annals of Mathematical Statistics*, **35**, 73–101.

Huber, P.J. (1965), A robust version of the probability ratio test, *The Annals of Mathematical Statistics*, **36**, 1753–1758.

Huber, P.J. (1967), The behavior of maximum likelihood estimates under nonstandard conditions, *Proceedings of the Fifth Berkeley Symposium on Mathematics and Statistics Probability*, **1**, 221–233, University of California Press.

Huber, P.J. (1968), Robust confidence limits, *Zeitschrift für Wahrscheinlichkeitstheorie und Verwandte Gebiete*, **10**, 269–278.

Huber, P.J. (1973), Robust regression: Asymptotics, conjectures and Monte Carlo, *The Annals of Statistics*, **1**, 799–821.

Huber, P.J. (1981), *Robust Statistics*, New York: John Wiley & Sons, Inc.

Huber, P.J. (1984), Finite sample breakdown of M- and P-estimators, *The Annals of Statistics*, **12**, 119–126.

Huber-Carol, C. (1970), Étude asymptotique des tests robustes, Ph.D. thesis, Eidgenössische Technische Hochschule, Zurich.

Hubert, M. and Rousseeuw, P.J. (1996), Robust regression with a categorical covariable, *Robust Statistics, Data Analysis, and Computer Intensive Methods*, H. Rieder (ed.), Lecture Notes in Statistics No. 109, 215–224, New York: Springer.

Hubert, M. and Rousseeuw, P.J. (1997), Robust regression with both continuous and binary regressors, *Journal of Statistical Planning and Inference*, **57**, 153–163.

Jalali-Heravi, M. and Knouz, E. (2002), Use of quantitative structure-property relationships in predicting the Krafft point of anionic surfactants, *Electronic Journal of Molecular Design*, **1**, 410–417.

Johnson, R.A. and Wichern, D.W. (1998), *Applied Multivariate Statistical Analysis*, Upper Saddle River, NJ: Prentice Hall.

Johnson, W. (1985), Influence measures for logistic regression: Another point of view, *Biometrika*, **72**, 59–65.

Jones, R.H. (1980), Maximum likelihood fitting of ARMA models to time series with missing observations, *Technometrics*, **22**, 389–396.

Kandel, R. (1991), *Our Changing Climate*, New York: McGraw-Hill.

Kanter, M. and Steiger, W.L. (1974), Regression and autoregression with infinite variance, *Advances in Applied Probability*, **6**, 768–783.

Kent, J.T. and Tyler, D.E. (1991), Redescending M-estimates of multivariate location and scatter, *The Annals of Statistics*, **19**, 2102–2119.

Kent, J.T. and Tyler, D.E. (1996), Constrained M-estimation for multivariate location and scatter, *The Annals of Statistics*, **24**, 1346–1370.

Kim, J. and Pollard, D. (1990), Cube root asymptotics, *The Annals of Statistics*, **18**, 191–219.

Klein, R. and Yohai, V.J. (1981), Iterated M-estimators for the linear model, *Communications in Statistics, Theory and Methods*, **10**, 2373–2388.

Kleiner, B., Martin, R.D. and Thomson, D.J. (1979), Robust estimation of power spectra (with discussion), *Journal of the Royal Statistical Society (B)*, **41**, 313–351.

Knight, K. (1987), Rate of convergence of centered estimates of autoregressive parameters for infinite variance autoregressions, *Journal of Time Series Analysis*, **8**, 51–60.

Knight, K. (1989), Limit theory for autoregressive-parameter estimates in an infinite variance, random walk, *Canadian Journal of Statistics*, **17**, 261–278.

Koenker, R. and Bassett, G.J. (1978), Regression quantiles, *Econometrica*, **46**, 33–50.

Koenker, R., Hammond, P. and Holly, A. (eds.) (2005), *Quantile Regression*, Cambridge: Cambridge University Press.

Krasker, W.S. and Welsch, R.E. (1982), Efficient bounded influence regression estimation, *Journal of the American Statistical Association*, **77**, 595–604.

Künsch, H. (1984), Infinitesimal robustness for autoregressive processes, *The Annals of Statistics*, **12**, 843–863.

Künsch, H.R., Stefanski, L.A. and Carroll, R.J. (1989), Conditionally unbiased bounded-influence estimation in general regression models, with applications to generalized linear models, *Journal of the American Statistical Association*, **84**, 460–466.

Ledolter, J. (1979), A recursive approach to parameter estimation in regression and time series models, *Communications in Statistics*, **A8**, 1227–1245.

Ledolter, J. (1991), Outliers in time series analysis: Some comments on their impact and their detection, *Directions in Robust Statistics and Diagnostics, Part I*, W. Stahel and S. Weisberg (eds.), 159–165, New York: Springer.

Lee, C.H. and Martin, R.D. (1986), Ordinary and proper location M-estimates for autoregressive-moving average models, *Biometrika*, **73**, 679–686.

Lehmann, E.L. and Casella, G. (1998), *Theory of Point Estimation*, 2nd Edition, Springer Texts in Statistics, New York: Springer.

Li, B. and Zamar, R.H. (1991), Min-max asymptotic variance when scale is unknown, *Statistics and Probability Letters*, **11**, 139–145.

Li, G. and Chen, Z. (1985), Projection-pursuit approach to robust dispersion matrices and principal components: Primary theory and Monte Carlo, *Journal of the American Statistical Association*, **80**, 759–766.

Liu, R.Y. (1990), On a notion of data depth based on random simplices, *The Annals of Statistics*, 18, 405–414.

Locantore, N., Marron, J.S., Simpson, D.G., Tripoli, N., Zhang, J.T. and Cohen, K.L. (1999), Robust principal components for functional data, *Test*, **8**, 1–28.

Lopuhaä, H.P. (1989), On the relation between S-estimators and M-estimators of multivariate location and covariance, *The Annals of Statistics*, **17**, 1662–1683.

Lopuhaä, H.P. (1991), Multivariate τ-estimators for location and scatter, *Canadian Journal of Statistics*, **19**, 307–321.

Lopuhaä, H.P. (1992), Highly efficient estimators of multivariate location with high breakdown point, *The Annals of Statistics*, **20**, 398–413.

Lopuhaä, H.P. and Rousseeuw, P.J. (1991), Breakdown properties of affine-equivariant estimators of multivariate location and covariance matrices, *The Annals of Statistics*, **19**, 229–248.

Ma, Y. and Genton, M.G. (2000), Highly robust estimation of the autocovariance function, *Journal of Time Series Analysis*, **21**, 663–684.

Maguna, F.P., Núñez, M.B., Okulik, N.B. and Castro, E.A. (2003), Improved QSAR analysis of the toxicity of aliphatic carboxylic acids, *Russian Journal of General Chemistry*, **73**, 1792–1798.

Mallows, C.L. (1975), On some topics in robustness, Unpublished memorandum, Bell Telephone Laboratories, Murray Hill, NJ.

Mancini, L., Ronchetti, E. and Trojani, F. (2005), Optimal conditionally unbiased bounded-influence inference in dynamic location and scale models, *Journal of the American Statistical Association*, **100**, 628–641.

Manku, G.S., Rajagopalan, S. and Lindsay, B. (1999), Random sampling techniques for space efficient online computation of order statistics of large data sets, ACM SGIMOD Record 28.

Marazzi, A. (1993), *Algorithms, Routines, and S Functions for Robust Statistics*, Pacific Grove, CA: Wadsworth & Brooks/Cole.

Marazzi, A., Paccaud, F., Ruffieux, C. and Beguin, C. (1998), Fitting the distributions of length of stay by parametric models, *Medical Care*, **36**, 915–927.

Maronna, R.A. (1976), Robust M-estimators of multivariate location and scatter, *The Annals of Statistics*, **4**, 51–67.

Maronna, R.A. (2005), Principal components and orthogonal regression based on robust scales, *Technometrics*, **47**, 264–273.

Maronna, R.A., Bustos, O.H. and Yohai, V.J. (1979), Bias- and efficiency-robustness of general M-estimators for regression with random carriers, *Smoothing techniques for curve estimation*, T. Gasser and J.M. Rossenblat (eds.), Lecture Notes in Mathematics **757**, 91–116, New York: Springer.

Maronna, R.A., Stahel, W.A. and Yohai, V.J. (1992), Bias-robust estimators of multivariate scatter based on projections, *Journal of Multivariate Analysis*, **42**, 141–161.

Maronna, R.A. and Yohai, V.J. (1981), Asymptotic behavior of general M-estimates for regression and scale with random carriers, *Zeitschrift für Wahrscheinlichkeitstheorie und Verwandte Gebiete*, **58**, 7–20.

Maronna, R.A. and Yohai, V.J. (1991), The breakdown point of simultaneous general M-estimates of regression and scale, *Journal of the American Statistical Association*, **86**, 699–703.

Maronna, R.A. and Yohai, V.J. (1993), Bias-robust estimates of regression based on projections, *The Annals of Statistics*, **21**, 965–990.

Maronna, R.A. and Yohai, V.J. (1995), The behavior of the Stahel-Donoho robust multivariate estimator, *Journal of the American Statistical Association*, **90**, 330–341.

Maronna, R.A. and Yohai, V.J. (1999), Robust regression with both continuous and categorical predictors, Technical Report, Faculty of Exact Sciences, University of Buenos Aires. (Available by anonymous ftp at: ulises.ic.fcen.uba.ar).

Maronna, R.A. and Yohai, V.J. (2000), Robust regression with both continuous and categorical predictors, *Journal of Statistical Planning and Inference*, **89**, 197–214.

Maronna, R.A. and Zamar, R.H. (2002), Robust estimation of location and dispersion for high-dimensional data sets, *Technometrics*, **44**, 307–317.

Martin, R.D. (1979), Approximate conditional mean type smoothers and interpolators, *Smoothing Techniques for Curve Estimation*, T. Gasser and M. Rosenblatt (eds.), 117–143, Berlin: Springer.

Martin, R.D. (1980), Robust estimation of autoregressive models, *Directions in Time Series*, D.R. Billinger and G.C. Tiao (eds.), 228–254, Haywood, CA: Institute of Mathematical Statistics.

Martin, R.D. (1981), Robust methods for time series, *Applied Time Series Analysis II*, D.F. Findley (ed.), 683–759, New York: Academic Press.

Martin, R.D. and Jong, J.M. (1977), Asymptotic properties of robust generalized M-estimates for the first-order autoregressive parameter, Bell Labs Technical Memo, Murray Hill, NJ.

Martin, R.D. and Lee, C.H. (1986), Ordinary and proper location M-estimates for ARMA models, *Biometrika*, **73**, 679–686.

Martin, R.D., Samarov, A. and Vandaele, W. (1983), Robust methods for ARIMA models, *Applied Time Series Analysis of Economic Data*, E. Zellner (ed.), 153–177, Washington, DC: Bureau of the Census.

Martin, R.D. and Su, K.Y. (1985), Robust filters and smoothers: Definitions and design. Technical Report No. 58, Department of Statistics, University of Washington.

Martin, R.D. and Thomson, D.J. (1982), Robust-resistant spectrum estimation, *IEEE Proceedings*, **70**, 1097–1115.

Martin, R.D. and Yohai, V.J. (1985), Robustness in time series and estimating ARMA models, *Handbook of Statistics, Volume 5: Time Series in the Time Domain*, E.J. Hannan, P.R. Krishnaiah and M.M. Rao (eds.), Amsterdam: Elsevier.

Martin, R.D. and Yohai, V.J. (1986), Influence functionals for time series (with discussion), *The Annals of Statistics*, **14**, 781–818.

Martin, R.D. and Yohai, V.J. (1991), Bias robust estimation of autoregression parameters, *Directions in Robust Statistics and Diagnostics Part I*, W. Stahel and S. Weisberg (eds.), IMA Volumes in Mathematics and its Applications, vol. 30, Berlin: Springer.

Martin, R.D., Yohai, V.J. and Zamar, R.H. (1989), Min–max bias robust regression, *The Annals of Statistics*, **17**, 1608–1630.

Martin, R.D. and Zamar, R.H. (1989), Asymptotically min-max bias-robust M-estimates of scale for positive random variables, *Journal of the American Statistical Association*, **84**, 494–501.

Martin, R.D. and Zamar, R.H. (1993a), Efficiency-constrained bias-robust estimation of location, *The Annals of Statistics*, **21**, 338–354.

Martin, R.D. and Zamar, R.H. (1993b), Bias robust estimates of scale, *The Annals of Statistics*, **21**, 991–1017.

Masarotto, G. (1987), Robust and consistent estimates of autoregressive-moving average parameters, *Biometrika*, **74**, 791–797.

Masreliez, C.J. (1975), Approximate non-gaussian filtering with linear state and observation relations, *IEEE-Transactions on Automatic Control*, **AC-20**, 107–110.

Masreliez, C.J. and Martin, R.D. (1977), Robust Bayesian estimation for the linear model and robustifying the Kalman filter, *IEEE-Transactions on Automatic Control*, **AC-22**, 361–371.

Meinhold, R.J. and Singpurwalla, N.D. (1989), Robustification of Kalman filter models, *Journal of the American Statistical Association*, **84**, 479–88.

Mendes, B.V.M. and Tyler, D.E. (1996), Constrained M-estimation for regression, *Robust Statistics, Data Analysis and Computer Intensive Methods (Schloss Thurnau, 1994)*, 299–320, Lecture Notes in Statistics 109, New York: Springer.

Menn, C. and Rachev, S.T. (2005), A GARCH option pricing model with alpha-stable innovations, *European Journal of Operational Research*, **163**, 201–209.

Mikosch, T., Gadrich, T., Kluppelberg, C. and Adler, R.J. (1995), Parameter estimation for ARMA models with infinite variance innovations, *The Annals of Statistics*, **23**, 305–326.

Miles, J. and Shevlin, M. (2000), *Applying Regression and Correlation: A guide for students and researchers*, London: Sage.

Mili, L., Cheniae, M.G., Vichare, N.S. and Rousseeuw, P.J. (1996), Robust state estimation based on projection statistics, *IEEE Transactions on Power Systems*, **11**, 1118–1127.

Mili, L. and Coakley, C.W. (1996), Robust estimation in structured linear regression, *The Annals of Statistics*, **24**, 2593–2607.

Miller, A.J. (1990), *Subset Selection in Regression*, London: Chapman and Hall.

Montgomery, D.C., Peck, E.A. and Vining, G.G. (2001), *Introduction to Linear Regression Analysis*, 3rd Edition, New York: John Wiley & Sons, Inc.

Muirhead, R.J. (1982), *Aspects of Multivariate Statistical Theory*, New York: John Wiley & Sons, Inc.

Peña, D. (1987), Measuring the importance of outliers in ARIMA models, M.L. Puri, J.P. Vilaplana and W. Wertz (eds.), *New Perspectives in Theoretical and Applied Statistics*, 109–118, New York: John Wiley & Sons, Inc.

Peña, D. (1990), Influential observations in time series, *Journal of Business and Economic Statistics*, **8**, 235–241.

Peña, D. and Prieto, F.J. (2001), Robust covariance matrix estimation and multivariate outlier rejection, *Technometrics*, **43**, 286–310.

Peña, D. and Prieto, F.J. (2004), Combining random and specific directions for robust estimation of high-dimensional multivariate data, Working Paper, Universidad Carlos III, Madrid.

Peña, D. and Yohai, V.J. (1999), A fast procedure for outlier diagnostics in large regression problems, *Journal of the American Statistical Association*, **94**, 434–445.

Percival, D.B. and Walden, A.T. (1993), *Spectral Analysis for Physical Applications: Multitaper and Conventional Univariate Techniques*, Cambridge: Cambridge University Press.

Piegorsch, W.W. (1992), Complementary log regression for generalized linear models, *The American Statistician*, **46**, 94–99.

Pires, R.C., Simões Costa, A. and Mili, L. (1999), Iteratively reweighted least squares state estimation through Givens rotations, *IEEE Transactions on Power Systems*, **14**, 1499–1505.

Portnoy, S. and Koenker, R. (1997), The Gaussian hare and the Laplacian tortoise: Computability of squared-error versus absolute-error estimators, *Statistical Science*, **12**, 299–300.

Pregibon, D. (1981), Logistic regression diagnostics, *The Annals of Statistics*, **9**, 705–724.

Pregibon, D. (1982), Resistant fits for some commonly used logistic models with medical applications, *Biometrics*, **38**, 485–498.

Qian, G. and Künsch, H.R. (1998), On model selection via stochastic complexity in robust linear regression, *Journal of Statistical Planning and Inference*, **75**, 91–116.

Rachev, S. and Mittnik, S. (2000), *Stable Paretian Models in Finance*, New York: John Wiley & Sons, Inc.

Rieder, H. (1978), A robust asymptotic testing model, *The Annals of Statistics*, **6**, 1080–1094.

Rieder, H. (1981), Robustness of one- and two-sample rank tests against gross errors, *The Annals of Statistics*, **9**, 245–265.

Roberts, J. and Cohrssen, J. (1968), Hearing levels of adults, US National Center for Health Statistics Publications, Series 11, No. 31.

Rocke, D.M. (1996), Robustness properties of S-estimators of multivariate location and shape in high dimension, *The Annals of Statistics*, **24**, 1327–1345.

Romanelli, G.P., Martino, C.M. and Castro, E.A. (2001), Modeling the solubility of aliphatic alcohols via molecular descriptors, *Journal of the Chemical Society of Pakistan*, **23**, 195–199.

Ronchetti, E., Field, C. and Blanchard, W. (1997), Robust linear model selection by cross-validation, *Journal of the American Statistical Association*, **92**, 1017–1023.

Ronchetti, E. and Staudte, R.G. (1994), A robust version of Mallow's C_p, *Journal of the American Statistical Association*, **89**, 550–559.

Rousseeuw, P.J. (1984), Least median of squares regression, *Journal of the American Statistical Association*, **79**, 871–880.

Rousseeuw, P.J. (1985), Multivariate estimation with high breakdown point, *Mathematical Statistics and its Applications* (*vol. B*), W. Grossmann, G. Pflug, I. Vincze and W. Wertz (eds.), 283–297, Dordrecht: Reidel.

Rousseeuw, P.J. and Croux, C. (1993), Alternatives to the median absolute deviation, *Journal of the American Statistical Association*, **88**, 1273–1283.

Rousseeuw, P.J and Hubert, M. (1999), Regression depth, *Journal of the American Statistical Association*, **94**, 388–402.

Rousseeuw, P.J. and Leroy, A.M. (1987), *Robust Regression and Outlier Detection*, New York: John Wiley & Sons, Inc.

Rousseeuw, P.J. and van Driessen, K. (1999), A fast algorithm for the minimum covariance determinant estimator, *Technometrics*, **41**, 212–223.

Rousseeuw, P.J. and van Driessen, K. (2000), An algorithm for positive-breakdown regression based on concentration steps, *Data Analysis: Modeling and Practical Applications*, W. Gaul, O. Opitz and M. Schader (eds.), 335–346, New York: Springer.

Rousseeuw, P.J. and van Zomeren, B.C. (1990), Unmasking multivariate outliers and leverage points, *Journal of the American Statistical Association*, **85**, 633–639.

Rousseeuw, P.J. and Wagner, J. (1994), Robust regression with a distributed intercept using least median of squares, *Computational Statistics and Data Analysis*, **17**, 65–76.

Rousseeuw, P.J and Yohai, V.J. (1984), Robust regression by means of S-estimators, *Robust and Nonlinear Time Series*, J. Franke, W. Härdle and R.D. Martin (eds.), Lectures Notes in Statistics **26**, 256–272, New York: Springer.

Ruppert, D. (1992), Computing S-estimators for regression and multivariate location/dispersion, *Journal of Computational and Graphical Statistics*, **1**, 253–270.

Salibian-Barrera, M. and Yohai, V.J. (2005), A fast algorithm for S-regression estimates, *Journal of Computational and Graphical Statistics (to appear).*

Salibian-Barrera, M. and Zamar, R.H. (2002), Bootstrapping robust estimates of regression, *The Annals of Statistics*, **30**, 556–582.

Samarakoon, D.M. and Knight, K. (2005), A note on unit root tests with infinite variance noise, Unpublished manuscript.

Scheffé, H. (1959), *The Analysis of Variance*, New York: John Wiley & Sons, Inc.

Schweppe, F.C., Wildes, J. and Rom, D.B. (1970), Power system static-state estimation, Parts I, II and III, *IEEE Transactions on Power Apparatus and Systems*, **PAS-89**, 120–135.

Seber, G.A.F. (1984), *Multivariate Observations*, New York: John Wiley & Sons, Inc.

Seber, G.A.F. and Lee, A.J. (2003), *Linear Regression Analysis*, 2nd Edition, New York: John Wiley & Sons, Inc.

Shao, J. (2003), *Mathematical Statistics*, 2nd Edition, New York: Springer.

Siebert, J.P. (1987), Vehicle recognition using rule based methods, Turing Institute Research Memorandum TIRM-87-018.

Simpson, D.G., Ruppert, D. and Carroll, R.J. (1992), On one-step GM-estimates and stability of inferences in linear regression, *Journal of the American Statistical Association*, **87**, 439–450.

Smith, R.E., Campbell, N.A. and Lichfield, A. (1984), Multivariate statistical techniques applied to pisolitic laterite geochemistry at Golden Grove, Western Australia, *Journal of Geochemical Exploration*, **22**, 193–216.

Sposito, V.A. (1987), On median polish and L_1 estimators, *Computational Statistics and Data Analysis*, **5**, 155–162.

Stahel, W.A. (1981), Breakdown of covariance estimators, Research Report 31, Fachgruppe für Statistik, ETH. Zurich.

Stapleton, J.H. (1995), *Linear Statistical Models*, New York: John Wiley & Sons, Inc.

Staudte, R.G. and Sheather, S.J. (1990), *Robust Estimation and Testing*, New York: John Wiley & Sons, Inc.

Stigler, S. (1973), Simon Newcomb, Percy Daniell, and the history of robust estimation 1885–1920, *Journal of the American Statistics Association*, **68**, 872–879.

Stigler, S.M. (1977), Do robust estimators deal with *real* data?, *The Annals of Statistics*, **5**, 1055–1098.

Stigler, S.M. (1986), *The History of Statistics: The Measurement of Uncertainty before 1900*, Cambridge, MA, and London: Belknap Press of Harvard University Press.

Stromberg, A.J. (1993a), Computation of high breakdown nonlinear regression parameters, *Journal of the American Statistical Association*, **88**, 237–244.

Stromberg, A.J. (1993b), Computing the exact least median of squares estimate and stability diagnostics in multiple linear regression, *SIAM Journal of Scientific Computing*, **14**, 1289–1299.

Svarc, M., Yohai, V.J. and Zamar, R.H. (2002), Optimal bias-robust M-estimates of regression, *Statistical Data Analysis Based on the L_1 Norm and Related Methods*, Yadolah Dodge (ed.), 191–200, Basle: Birkhäuser.

Tatsuoka, K.S. and Tyler, D.E. (2000), On the uniqueness of S-functionals and M-functionals under nonelliptical distributions, *The Annals of Statistics*, **28**, 1219–1243.

Thomson, D.J. (1977), Spectrum estimation techniques for characterization and development of WT4 waveguide, *Bell System Technical Journal*, **56**, 1769–1815 and 1983–2005.

Tsay, R.S. (1988), Outliers, level shifts and variance changes in time series, *Journal of Forecasting*, **7**, 1–20.

Tukey, J.W. (1960), A survey of sampling from contaminated distributions, *Contributions to Probability and Statistics*, I. Olkin (ed.), Stanford, CA: Stanford University Press.

Tukey, J.W. (1962), The future of data analysis, *The Annals of Mathematical Statistics* **33**, 1–67.

Tukey, J.W. (1967), An introduction to the calculations of numerical spectrum analysis, *Proceedings of the Advanced Seminar on Spectral Analysis of Time Series*, B. Harris (ed.), 25–46, New York: John Wiley & Sons, Inc.

Tukey, J.W. (1975a), Useable resistant/robust techniques of analysis, *Proceedings of the First ERDA Symposium*, Los Alamos, New Mexico, 11–31.

Tukey, J.W. (1975b), Comments on "Projection pursuit", *The Annals of Statistics*, **13**, 517–518.

Tukey, J.W. (1977), *Exploratory Data Analysis*, Reading, MA: Addison-Wesley.

Tyler, D.E. (1983), Robustness and efficiency properties of scatter matrices, *Biometrika*, **70**, 411–420.

Tyler, D.E. (1987), A distribution-free M-estimator of multivariate scatter, *The Annals of Statistics*, **15**, 234–251.

Tyler, D.E. (1990), Breakdown properties of the M-estimators of multivariate scatter, Technical Report, Department of Statistics, Rutgers University.

Tyler, D.E. (1991), Personal communication.

Tyler, D.E. (1994), Finite-sample breakdown points of projection-based multivariate location and scatter statistics, *The Annals of Statistics*, **22**, 1024–1044.

Verboven, S. and Hubert, M. (2005), LIBRA: A MATLAB library for robust analysis, *Chemometrics and Intelligent Laboratory Systems*, **75**, 127–136.

Wedderburn, R.W.M. (1974), Quasi-likelihood functions, generalized linear models, and the Gauss–Newton method, *Biometrika*, **61**, 439–447.

Weisberg, S. (1985), *Applied Linear Regression*, 2nd Edition, New York: John Wiley & Sons, Inc.

Welch, P.D. (1967), The use of the fast Fourier transform for estimation of spectra: A method based on time averaging over short, modified periodograms, *IEEE Transactions on Audio and Electroacoustics*, **15**, 70–74.

West, M. and Harrison, P.J. (1997), *Bayesian Forecasting and Dynamic Models*, 2nd Edition, Berlin: Springer.

Whittle, P. (1962), Gaussian estimation in stationary time series, *Bulletin of the International Statistical Institute*, **39**, 105–129.

Wold, H. (1954), *A Study in the Analysis of Stationary Time Series*, 2nd Edition, Stockholm: Almqvist and Wiksell.

Woodruff, D.L. and Rocke, D.M. (1994), Computable robust estimation of multivariate location and shape in high dimension using compound estimators, *Journal of the American Statistical Association*, **89**, 888–896.

Yohai, V.J. (1987), High breakdown-point and high efficiency estimates for regression, *The Annals of Statistics*, **15**, 642–656.

Yohai, V.J. and Maronna, R.A. (1976), Location estimators based on linear combinations of modified order statistics, *Communications in Statistics (Theory and Methods)*, **A5**, 481–486.

Yohai, V.J. and Maronna, R.A. (1977), Asymptotic behavior of least-squares estimates for autoregressive processes with infinite variance, *The Annals of Statistics*, **5**, 554–560.

Yohai, V.J. and Maronna, R.A. (1979), Asymptotic behavior of M-estimates for the linear model, *The Annals of Statistics*, **7**, 258–268.

Yohai, V.J. and Maronna, R.A. (1990), The maximum bias of robust covariances, *Communications in Statistics (Theory and Methods)*, **A19**, 3925–3933.

Yohai, V.J., Stahel, W.A. and Zamar, R.H. (1991), A procedure for robust estimation and inference in linear regression, *Directions in Robust Statistics and Diagnostics (Part II)*, W. Stahel and S. Weisberg (eds.), The IMA Volumes in Mathematics and its Applications, 365–374, New York: Springer.

Yohai, V.J. and Zamar, R.H. (1988), High breakdown estimates of regression by means of the minimization of an efficient scale, *Journal of the American Statistical Association*, **83**, 406–413.

Yohai, V.J. and Zamar, R.H. (1997), Optimal locally robust M-estimates of regression, *Journal of Statistical Planning and Inference*, **57**, 73–92.

Yohai, V.J. and Zamar, R.H. (2004), Robust nonparametric inference for the median, *The Annals of Statistics*, **5**, 1841–1857.

Zhao, Q. (2000), Restricted regression quantiles, *Journal of Multivariate Analysis*, **72**, 78–99.

Zivot, E. and Wang, J. (2005), *Modeling Financial Time Series with S-PLUS*, 2nd Edition, New York: Springer.

Zuo, Y., Cui, H. and He, X. (2004a), On the Stahel-Donoho estimator and depth-weighted means of multivariate data, *The Annals of Statistics*, **32**, 167–188.

Zuo, Y., Cui, H. and Young, D. (2004b), Influence function and maximum bias of projection depth based estimators, *The Annals of Statistics*, **32**, 189–218.

Zuo, Y. and Serfling, R. (2000), General notions of statistical depth function, *The Annals of Statistics*, **28**, 461–482.

Index

WILEY SERIES IN PROBABILITY AND STATISTICS

ESTABLISHED BY WALTER A. SHEWHART AND SAMUEL S. WILKS

The ***Wiley Series in Probability and Statistics*** is well established and authoritative. It covers many topics of current research interest in both pure and applied statistics and probability theory. Written by leading statisticians and institutions, the titles span both state-of-the-art developments in the field and classical methods.

Reflecting the wide range of current research in statistics, the series encompasses applied, methodological and theoretical statistics, ranging from applications and new techniques made possible by advances in computerized practice to rigorous treatment of theoretical approaches.

This series provides essential and invaluable reading for all statisticians, whether in academia, industry, government, or research.

ABRAHAM and LEDOLTER · Statistical Methods for Forecasting
AGRESTI · Analysis of Ordinal Categorical Data
AGRESTI · An Introduction to Categorical Data Analysis
AGRESTI · Categorical Data Analysis, *Second Edition*
ALTMAN, GILL, and McDONALD · Numerical Issues in Statistical Computing for the Social Scientist
AMARATUNGA and CABRERA · Exploration and Analysis of DNA Microarray and Protein Array Data
ANDĚL · Mathematics of Chance
ANDERSON · An Introduction to Multivariate Statistical Analysis, *Third Edition*
*ANDERSON · The Statistical Analysis of Time Series
ANDERSON, AUQUIER, HAUCK, OAKES, VANDAELE, and WEISBERG · Statistical Methods for Comparative Studies
ANDERSON and LOYNES · The Teaching of Practical Statistics
ARMITAGE and DAVID (editors) · Advances in Biometry
ARNOLD, BALAKRISHNAN, and NAGARAJA · Records
*ARTHANARI and DODGE · Mathematical Programming in Statistics
*BAILEY · The Elements of Stochastic Processes with Applications to the Natural Sciences
BALAKRISHNAN and KOUTRAS · Runs and Scans with Applications
BARNETT · Comparative Statistical Inference, *Third Edition*
BARNETT · Environmental Statistics: Methods & Applications

*Now available in a lower priced paperback edition in the Wiley Classics Library.

BARNETT and LEWIS · Outliers in Statistical Data, *Third Edition*
BARTOSZYNSKI and NIEWIADOMSKA-BUGAJ · Probability and Statistical Inference
BASILEVSKY · Statistical Factor Analysis and Related Methods: Theory and Applications
BASU and RIGDON · Statistical Methods for the Reliability of Repairable Systems
BATES and WATTS · Nonlinear Regression Analysis and Its Applications
BECHHOFER, SANTNER, and GOLDSMAN · Design and Analysis of Experiments for Statistical Selection, Screening, and Multiple Comparisons
BELSLEY · Conditioning Diagnostics: Collinearity and Weak Data in Regression
BELSLEY, KUH, and WELSCH · Regression Diagnostics: Identifying Influential Data and Sources of Collinearity
BENDAT and PIERSOL · Random Data: Analysis and Measurement Procedures, *Third Edition*
BERNARDO and SMITH · Bayesian Theory
BERRY, CHALONER, and GEWEKE · Bayesian Analysis in Statistics and Econometrics: Essays in Honor of Arnold Zellner
BHAT and MILLER · Elements of Applied Stochastic Processes, *Third Edition*
BHATTACHARYA and JOHNSON · Statistical Concepts and Methods
BHATTACHARYA and WAYMIRE · Stochastic Processes with Applications
BIEMER, GROVES, LYBERG, MATHIOWETZ and SUDMAN · Measurement Errors in Surveys
BILLINGSLEY · Convergence of Probability Measures, *Second Edition*
BILLINGSLEY · Probability and Measure, *Third Edition*
BIRKES and DODGE · Alternative Methods of Regression
BLISCHKE and MURTHY (editors) · Case Studies in Reliability and Maintenance
BLISCHKE and MURTHY · Reliability: Modeling, Prediction, and Optimization
BLOOMFIELD · Fourier Analysis of Time Series: An Introduction, *Second Edition*
BOLLEN · Structural Equations with Latent Variables
BOLLEN and CURRAN · Latent Curve Models: A Structural Equation Perspective
BOROVKOV · Ergodicity and Stability of Stochastic Processes
BOULEAU · Numerical Methods for Stochastic Processes
BOX · Bayesian Inference in Statistical Analysis
BOX · R. A. Fisher, the Life of a Scientist
BOX and DRAPER · Empirical Model-Building and Response Surfaces
*BOX and DRAPER · Evolutionary Operation: A Statistical Method for Process Improvement
BOX, HUNTER, and HUNTER · Statistics for Experimenters: An Introduction to Design, Data Analysis, and Model Building
BOX, HUNTER, and HUNTER · Statistics for Experimenters: Design, Innovation and Discovery, *Second Edition*
BOX and LUCEÑO · Statistical Control by Monitoring and Feedback Adjustment

*Now available in a lower priced paperback edition in the Wiley Classics Library.

BRANDIMARTE · Numerical Methods in Finance: A MATLAB-Based Introduction
BROWN and HOLLANDER · Statistics: A Biomedical Introduction
BRUNNER, DOMHOF, and LANGER · Nonparametric Analysis of Longitudinal Data in Factorial Experiments
BUCKLEW · Large Deviation Techniques in Decision, Simulation, and Estimation
CAIROLI and DALANG · Sequential Stochastic Optimization
CASTILLO, HADI, BALAKRISHNAN and SARABIA · Extreme Value and Related Models with Applications in Engineering and Science
CHAN · Time Series: Applications to Finance
CHATTERJEE and HADI · Sensitivity Analysis in Linear Regression
CHATTERJEE and PRICE · Regression Analysis by Example, *Third Edition*
CHERNICK · Bootstrap Methods: A Practitioner's Guide
CHERNICK and FRIIS · Introductory Biostatistics for the Health Sciences
CHILÈS and DELFINER · Geostatistics: Modeling Spatial Uncertainty
CHOW and LIU · Design and Analysis of Clinical Trials: Concepts and Methodologies, *Second Edition*
CLARKE and DISNEY · Probability and Random Processes: A First Course with Applications, *Second Edition*
*COCHRAN and COX · Experimental Designs, *Second Edition*
CONGDON · Applied Bayesian Modelling
CONGDON · Bayesian Statistical Modelling
CONGDON · Bayesian Models for Categorical Data
CONOVER · Practical Nonparametric Statistics, *Second Edition*
COOK · Regression Graphics
COOK and WEISBERG · Applied Regression Including Computing and Graphics
COOK and WEISBERG · An Introduction to Regression Graphics
CORNELL · Experiments with Mixtures, Designs, Models, and the Analysis of Mixture Data, *Third Edition*
COVER and THOMAS · Elements of Information Theory
COX · A Handbook of Introductory Statistical Methods
*COX · Planning of Experiments
CRESSIE · Statistics for Spatial Data, *Revised Edition*
CSÖRGÖ and HORVÁTH · Limit Theorems in Change Point Analysis
DANIEL · Applications of Statistics to Industrial Experimentation
DANIEL · Biostatistics: A Foundation for Analysis in the Health Sciences, *Sixth Edition*
*DANIEL · Fitting Equations to Data: Computer Analysis of Multifactor Data, *Second Edition*
DASU and JOHNSON · Exploratory Data Mining and Data Cleaning
DAVID and NAGARAJA · Order Statistics, *Third Edition*
*DEGROOT, FIENBERG, and KADANE · Statistics and the Law
DEL CASTILLO · Statistical Process Adjustment for Quality Control

*Now available in a lower priced paperback edition in the Wiley Classics Library.

DEMARIS · Regression with Social Data: Modeling Continuous and Limited Response Variables
DEMIDENKO · Mixed Models: Theory and Applications
DENISON, HOLMES, MALLICK, and SMITH · Bayesian Methods for Nonlinear Classification and Regression
DETTE and STUDDEN · The Theory of Canonical Moments with Applications in Statistics, Probability, and Analysis
DEY and MUKERJEE · Fractional Factorial Plans
DILLON and GOLDSTEIN · Multivariate Analysis: Methods and Applications
DODGE · Alternative Methods of Regression
*DODGE and ROMIG · Sampling Inspection Tables, *Second Edition*
*DOOB · Stochastic Processes
DOWDY and WEARDEN, and CHILKO · Statistics for Research, *Third Edition*
DRAPER and SMITH · Applied Regression Analysis, *Third Edition*
DRYDEN and MARDIA · Statistical Shape Analysis
DUDEWICZ and MISHRA · Modern Mathematical Statistics
DUNN and CLARK · Applied Statistics: Analysis of Variance and Regression, *Second Edition*
DUNN and CLARK · Basic Statistics: A Primer for the Biomedical Sciences, *Third Edition*
DUPUIS and ELLIS · A Weak Convergence Approach to the Theory of Large Deviations
EDLER and KITSOS (editors) · Recent Advances in Quantitative Methods in Cancer and Human Health Risk Assessment
*ELANDT-JOHNSON and JOHNSON · Survival Models and Data Analysis
ENDERS · Applied Econometric Time Series
ETHIER and KURTZ · Markov Processes: Characterization and Convergence
EVANS, HASTINGS, and PEACOCK · Statistical Distributions, *Third Edition*
FELLER · An Introduction to Probability Theory and Its Applications, Volume I, *Third Edition*, Revised; Volume II, *Second Edition*
FISHER and VAN BELLE · Biostatistics: A Methodology for the Health Sciences
FITZMAURICE, LAIRD, and WARE · Applied Longitudinal Analysis
*FLEISS · The Design and Analysis of Clinical Experiments
FLEISS · Statistical Methods for Rates and Proportions, *Second Edition*
FLEMING and HARRINGTON · Counting Processes and Survival Analysis
FULLER · Introduction to Statistical Time Series, *Second Edition*
FULLER · Measurement Error Models
GALLANT · Nonlinear Statistical Models
GELMAN and MENG (editors): Applied Bayesian Modeling and Casual Inference from Incomplete-data Perspectives
GEWEKE · Contemporary Bayesian Econometrics and Statistics
GHOSH, MUKHOPADHYAY, and SEN · Sequential Estimation

*Now available in a lower priced paperback edition in the Wiley Classics Library.

GIESBRECHT and GUMPERTZ · Planning, Construction, and Statistical Analysis of Comparative Experiments
GIFI · Nonlinear Multivariate Analysis
GIVENS and HOETING · Computational Statistics
GLASSERMAN and YAO · Monotone Structure in Discrete-Event Systems
GNANADESIKAN · Methods for Statistical Data Analysis of Multivariate Observations, *Second Edition*
GOLDSTEIN and LEWIS · Assessment: Problems, Development, and Statistical Issues
GREENWOOD and NIKULIN · A Guide to Chi-Squared Testing
GROSS and HARRIS · Fundamentals of Queueing Theory, *Third Edition*
*HAHN and SHAPIRO · Statistical Models in Engineering
HAHN and MEEKER · Statistical Intervals: A Guide for Practitioners
HALD · A History of Probability and Statistics and their Applications Before 1750
HALD · A History of Mathematical Statistics from 1750 to 1930
HAMPEL · Robust Statistics: The Approach Based on Influence Functions
HANNAN and DEISTLER · The Statistical Theory of Linear Systems
HEIBERGER · Computation for the Analysis of Designed Experiments
HEDAYAT and SINHA · Design and Inference in Finite Population Sampling
HELLER · MACSYMA for Statisticians
HINKELMAN and KEMPTHORNE: · Design and Analysis of Experiments, Volume 1: Introduction to Experimental Design
HINKELMANN and KEMPTHORNE · Design and analysis of experiments, Volume 2: Advanced Experimental Design
HOAGLIN, MOSTELLER, and TUKEY · Exploratory Approach to Analysis of Variance
HOAGLIN, MOSTELLER, and TUKEY · Exploring Data Tables, Trends and Shapes
*HOAGLIN, MOSTELLER, and TUKEY · Understanding Robust and Exploratory Data Analysis
HOCHBERG and TAMHANE · Multiple Comparison Procedures
HOCKING · Methods and Applications of Linear Models: Regression and the Analysis of Variance, *Second Edition*
HOEL · Introduction to Mathematical Statistics, *Fifth Edition*
HOGG and KLUGMAN · Loss Distributions
HOLLANDER and WOLFE · Nonparametric Statistical Methods, *Second Edition*
HOSMER and LEMESHOW · Applied Logistic Regression, *Second Edition*
HOSMER and LEMESHOW · Applied Survival Analysis: Regression Modeling of Time to Event Data
HUBER · Robust Statistics
HUBERTY · Applied Discriminant Analysis
HUNT and KENNEDY · Financial Derivatives in Theory and Practice, *Revised Edition*

*Now available in a lower priced paperback edition in the Wiley Classics Library.

HUSKOVA, BERAN, and DUPAC · Collected Works of Jaroslav Hajek—with Commentary

HUZURBAZAR · Flowgraph Models for Multistate Time-to-Event Data

IMAN and CONOVER · A Modern Approach to Statistics

JACKSON · A User's Guide to Principle Components

JOHN · Statistical Methods in Engineering and Quality Assurance

JOHNSON · Multivariate Statistical Simulation

JOHNSON and BALAKRISHNAN · Advances in the Theory and Practice of Statistics: A Volume in Honor of Samuel Kotz

JOHNSON and BHATTACHARYYA · Statistics: Principles and Methods, *Fifth Edition*

JUDGE, GRIFFITHS, HILL, LU TKEPOHL, and LEE · The Theory and Practice of Econometrics, *Second Edition*

JOHNSON and KOTZ · Distributions in Statistics

JOHNSON and KOTZ (editors) · Leading Personalities in Statistical Sciences: From the Seventeenth Century to the Present

JOHNSON, KOTZ, and BALAKRISHNAN · Continuous Univariate Distributions, Volume 1, *Second Edition*

JOHNSON, KOTZ, and BALAKRISHNAN · Continuous Univariate Distributions, Volume 2, *Second Edition*

JOHNSON, KOTZ, and BALAKRISHNAN · Discrete Multivariate Distributions

JOHNSON, KOTZ, and KEMP · Univariate Discrete Distributions, *Second Edition*

JUREČKOVÁ and SEN · Robust Statistical Procedures: Asymptotics and Interrelations

JUREK and MASON · Operator-Limit Distributions in Probability Theory

KADANE · Bayesian Methods and Ethics in a Clinical Trial Design

KADANE and SCHUM · A Probabilistic Analysis of the Sacco and Vanzetti Evidence

KALBFLEISCH and PRENTICE · The Statistical Analysis of Failure Time Data, *Second Edition*

KARIYA and KURATA · Generalized Least Squares

KASS and VOS · Geometrical Foundations of Asymptotic Inference

KAUFMAN and ROUSSEEUW · Finding Groups in Data: An Introduction to Cluster Analysis

KEDEM and FOKIANOS · Regression Models for Time Series Analysis

KENDALL, BARDEN, CARNE, and LE · Shape and Shape Theory

KHURI · Advanced Calculus with Applications in Statistics, *Second Edition*

KHURI, MATHEW, and SINHA · Statistical Tests for Mixed Linear Models

*KISH · Statistical Design for Research

KLEIBER and KOTZ · Statistical Size Distributions in Economics and Actuarial Sciences

KLUGMAN, PANJER, and WILLMOT · Loss Models: From Data to Decisions

*Now available in a lower priced paperback edition in the Wiley Classics Library.

KLUGMAN, PANJER, and WILLMOT · Solutions Manual to Accompany Loss Models: From Data to Decisions
KOTZ, BALAKRISHNAN, and JOHNSON · Continuous Multivariate Distributions, Volume 1, *Second Edition*
KOTZ and JOHNSON (editors) · Encyclopedia of Statistical Sciences: Volumes 1 to 9 with Index
KOTZ and JOHNSON (editors) · Encyclopedia of Statistical Sciences: Supplement Volume
KOTZ, READ, and BANKS (editors) · Encyclopedia of Statistical Sciences: Update Volume 1
KOTZ, READ, and BANKS (editors) · Encyclopedia of Statistical Sciences: Update Volume 2
KOVALENKO, KUZNETZOV, and PEGG · Mathematical Theory of Reliability of Time-Dependent Systems with Practical Applications
KUROWICKA and COOKE · Uncertainty Analysis with High Dimensional Dependence Modelling
LACHIN · Biostatistical Methods: The Assessment of Relative Risks
LAD · Operational Subjective Statistical Methods: A Mathematical, Philosophical, and Historical Introduction
LAMPERTI · Probability: A Survey of the Mathematical Theory, *Second Edition*
LANGE, RYAN, BILLARD, BRILLINGER, CONQUEST, and GREENHOUSE · Case Studies in Biometry
LARSON · Introduction to Probability Theory and Statistical Inference, *Third Edition*
LAWLESS · Statistical Models and Methods for Lifetime Data, *Second Edition*
LAWSON · Statistical Methods in Spatial Epidemiology, *Second Edition*
LE · Applied Categorical Data Analysis
LE · Applied Survival Analysis
LEE and WANG · Statistical Methods for Survival Data Analysis, *Third Edition*
LEPAGE and BILLARD · Exploring the Limits of Bootstrap
LEYLAND and GOLDSTEIN (editors) · Multilevel Modelling of Health Statistics
LIAO · Statistical Group Comparison
LINDVALL · Lectures on the Coupling Method
LINHART and ZUCCHINI · Model Selection
LITTLE and RUBIN · Statistical Analysis with Missing Data, *Second Edition*
LLOYD · The Statistical Analysis of Categorical Data
LOWEN and TEICH · Fractal-Based Point Processes
MAGNUS and NEUDECKER · Matrix Differential Calculus with Applications in Statistics and Econometrics, *Revised Edition*
MALLER and ZHOU · Survival Analysis with Long Term Survivors
MALLOWS · Design, Data, and Analysis by Some Friends of Cuthbert Daniel
MANN, SCHAFER, and SINGPURWALLA · Methods for Statistical Analysis of Reliability and Life Data

*Now available in a lower priced paperback edition in the Wiley Classics Library.

MANTON, WOODBURY, and TOLLEY · Statistical Applications Using Fuzzy Sets
MARCHETTE · Random Graphs for Statistical Pattern Recognition
MARDIA and JUPP · Directional Statistics
MARONNA, MARTIN, and YOHAI · Robust Statistics: Theory and Methods
MASON, GUNST, and HESS · Statistical Design and Analysis of Experiments with Applications to Engineering and Science, *Second Edition*
MCCULLOCH and SEARLE · Generalized, Linear, and Mixed Models
MCFADDEN · Management of Data in Clinical Trials
MCLACHLAN · Discriminant Analysis and Statistical Pattern Recognition
MCLACHLAN, DO, and AMBROISE · Analyzing Microarray Gene Expression Data
MCLACHLAN and KRISHNAN · The EM Algorithm and Extensions
MCLACHLAN and PEEL · Finite Mixture Models
MCNEIL · Epidemiological Research Methods
MEEKER and ESCOBAR · Statistical Methods for Reliability Data
MEERSCHAERT and SCHEFFLER · Limit Distributions for Sums of Independent Random Vectors: Heavy Tails in Theory and Practice
MICKEY, DUNN and CLARK · Applied Statistics: Analysis of Variance and Regression, *Third Edition*
*MILLER · Survival Analysis, *Second Edition*
MONTGOMERY, PECK, and VINING · Introduction to Linear Regression Analysis, *Third Edition*
MORGENTHALER and TUKEY · Configural Polysampling: A Route to Practical Robustness
MUIRHEAD · Aspects of Multivariate Statistical Theory
MURRAY · X-STAT 2.0 Statistical Experimentation, Design Data Analysis, and Nonlinear Optimization
MURTHY, XIE, and JIANG · Weibull Models
MYERS and MONTGOMERY · Response Surface Methodology: Process and Product Optimization Using Designed Experiments, *Second Edition*
MYERS, MONTGOMERY, and VINING · Generalized Linear Models. With Applications in Engineering and the Sciences
†NELSON · Accelerated Testing, Statistical Models, Test Plans, and Data Analyses
†NELSON · Applied Life Data Analysis
NEWMAN · Biostatistical Methods in Epidemiology
OCHI · Applied Probability and Stochastic Processes in Engineering and Physical Sciences
OKABE, BOOTS, SUGIHARA, and CHIU · Spatial Tesselations: Concepts and Applications of Voronoi Diagrams, *Second Edition*
OLIVER and SMITH · Influence Diagrams, Belief Nets and Decision Analysis
PALTA · Quantitative Methods in Population Health: Extensions of Ordinary Regressions

*Now available in a lower priced paperback edition in the Wiley Classics Library.
†Now available in a lower priced paperback edition in the Wiley – Interscience Paperback Series.

PANKRATZ · Forecasting with Dynamic Regression Models
PANKRATZ · Forecasting with Univariate Box-Jenkins Models: Concepts and Cases
*PARZEN · Modern Probability Theory and It's Applications
PEÑA, TIAO, and TSAY · A Course in Time Series Analysis
PIANTADOSI · Clinical Trials: A Methodologic Perspective
PORT · Theoretical Probability for Applications
POURAHMADI · Foundations of Time Series Analysis and Prediction Theory
PRESS · Bayesian Statistics: Principles, Models, and Applications
PRESS · Subjective and Objective Bayesian Statistics, *Second Edition*
PRESS and TANUR · The Subjectivity of Scientists and the Bayesian Approach
PUKELSHEIM · Optimal Experimental Design
PURI, VILAPLANA, and WERTZ · New Perspectives in Theoretical and Applied Statistics
PUTERMAN · Markov Decision Processes: Discrete Stochastic Dynamic Programming
QIU · Image Processing and Jump Regression Analysis
*RAO · Linear Statistical Inference and Its Applications, *Second Edition*
RAUSAND and HØYLAND · System Reliability Theory: Models, Statistical Methods and Applications, *Second Edition*
RENCHER · Linear Models in Statistics
RENCHER · Methods of Multivariate Analysis, *Second Edition*
RENCHER · Multivariate Statistical Inference with Applications
RIPLEY · Spatial Statistics
RIPLEY · Stochastic Simulation
ROBINSON · Practical Strategies for Experimenting
ROHATGI and SALEH · An Introduction to Probability and Statistics, *Second Edition*
ROLSKI, SCHMIDLI, SCHMIDT, and TEUGELS · Stochastic Processes for Insurance and Finance
ROSENBERGER and LACHIN · Randomization in Clinical Trials: Theory and Practice
ROSS · Introduction to Probability and Statistics for Engineers and Scientists
ROSSI, ALLENBY and MCCULLOCH · Bayesian Statistics and Marketing
ROUSSEEUW and LEROY · Robust Regression and Outlier Detection
RUBIN · Multiple Imputation for Nonresponse in Surveys
RUBINSTEIN · Simulation and the Monte Carlo Method
RUBINSTEIN and MELAMED · Modern Simulation and Modeling
RYAN · Modern Regression Methods
RYAN · Statistical Methods for Quality Improvement, *Second Edition*
SALTELLI, CHAN, and SCOTT (editors) · Sensitivity Analysis
*SCHEFFE · The Analysis of Variance
SCHIMEK · Smoothing and Regression: Approaches, Computation, and Application
SCHOTT · Matrix Analysis for Statistics

*Now available in a lower priced paperback edition in the Wiley Classics Library.

SCHOUTENS · Levy Processes in Finance: Pricing Financial Derivatives
SCHUSS · Theory and Applications of Stochastic Differential Equations
SCOTT · Multivariate Density Estimation: Theory, Practice, and Visualization
*SEARLE · Linear Models
SEARLE · Linear Models for Unbalanced Data
SEARLE · Matrix Algebra Useful for Statistics
SEARLE, CASELLA, and McCULLOCH · Variance Components
SEARLE and WILLETT · Matrix Algebra for Applied Economics
SEBER · Multivariate Observations
SEBER and LEE · Linear Regression Analysis, *Second Edition*
SEBER and WILD · Nonlinear Regression
SENNOTT · Stochastic Dynamic Programming and the Control of Queueing Systems
*SERFLING · Approximation Theorems of Mathematical Statistics
SHAFER and VOVK · Probability and Finance: Its Only a Game!
SILVAPULLE and SEN · Constrained Statistical Inference: Inequality, Order, and Shape Restrictions
SMALL and MCLEISH · Hilbert Space Methods in Probability and Statistical Inference
SRIVASTAVA · Methods of Multivariate Statistics
STAPLETON · Linear Statistical Models
STAUDTE and SHEATHER · Robust Estimation and Testing
STOYAN, KENDALL, and MECKE · Stochastic Geometry and Its Applications, *Second Edition*
STOYAN and STOYAN · Fractals, Random Shapes and Point Fields: Methods of Geometrical Statistics
STYAN · The Collected Papers of T. W. Anderson: 1943–1985
SUTTON, ABRAMS, JONES, SHELDON, and SONG · Methods for Meta-Analysis in Medical Research
TANAKA · Time Series Analysis: Nonstationary and Noninvertible Distribution Theory
THOMPSON · Empirical Model Building
THOMPSON · Sampling, *Second Edition*
THOMPSON · Simulation: A Modeler's Approach
THOMPSON and SEBER · Adaptive Sampling
THOMPSON, WILLIAMS, and FINDLAY · Models for Investors in Real World Markets
TIAO, BISGAARD, HILL, PEÑA, and STIGLER (editors) · Box on Quality and Discovery: with Design, Control, and Robustness
TIERNEY · LISP-STAT: An Object-Oriented Environment for Statistical Computing and Dynamic Graphics
TSAY · Analysis of Financial Time Series

*Now available in a lower priced paperback edition in the Wiley Classics Library.

UPTON and FINGLETON · Spatial Data Analysis by Example, Volume II: Categorical and Directional Data
VAN BELLE · Statistical Rules of Thumb
VAN BELLE, FISHER, HEAGERTY, and LUMLEY · Biostatistics: A Methodology for the Health Sciences, *Second Edition*
VESTRUP · The Theory of Measures and Integration
VIDAKOVIC · Statistical Modeling by Wavelets
VINOD and REAGLE · Preparing for the Worst: Incorporating Downside Risk in Stock Market Investments
WALLER and GOTWAY · Applied Spatial Statistics for Public Health Data
WEERAHANDI · Generalized Inference in Repeated Measures: Exact Methods in MANOVA and Mixed Models
WEISBERG · Applied Linear Regression, *Second Edition*
WELSH · Aspects of Statistical Inference
WESTFALL and YOUNG · Resampling-Based Multiple Testing: Examples and Methods for p-Value Adjustment
WHITTAKER · Graphical Models in Applied Multivariate Statistics
WINKER · Optimization Heuristics in Economics: Applications of Threshold Accepting
WONNACOTT and WONNACOTT · Econometrics, *Second Edition*
WOODING · Planning Pharmaceutical Clinical Trials: Basic Statistical Principles
WOOLSON and CLARKE · Statistical Methods for the Analysis of Biomedical Data, *Second Edition*
WU and HAMADA · Experiments: Planning, Analysis, and Parameter Design Optimization
YANG · The Construction Theory of Denumerable Markov Processes
*ZELLNER · An Introduction to Bayesian Inference in Econometrics
ZELTERMAN · Discrete Distributions: Applications in the Health Sciences
ZHOU, OBUCHOWSKI, and McCLISH · Statistical Methods in Diagnostic Medicine

*Now available in a lower priced paperback edition in the Wiley Classics Library.